A MANUAL of CALIFORNIA VEGETATION

A MANUAL of CALIFORNIA VEGETATION

— *by* —

John O. Sawyer

Professor of Botany
Humboldt State University

— *and* —

Todd Keeler-Wolf

Vegetation Ecologist
Natural Heritage Division
California Department of Fish and Game

California Native Plant Society

Printed in the United States of America

Library of Congress Catalog Card Number: 95-071912
ISBN: 0-943460-26-2   (softcover)
ISBN: 0-943460-25-5   (hardcover)

Cover and photo inset design: Beth Hansen

Front cover photograph: Fall colors, Rock Creek Road, Inyo National
Forest. Tommy Dodson/East of Eden.

Back cover photograph: Wildflowers in the Temblor Range, San Luis
Obispo County. Frank Balthis.

Table of Contents

The Plant Communities Committee

Chairman
Michael Barbour, University of California, Davis

Members
Sarah Allen, US National Park Service
Barbara Allen-Diaz, University of California, Berkeley
Bruce Bingham, US Forest Service
Mark Borchert, US Forest Service
Dan Cheatham, CNPS
David Chipping, CNPS
Charles Convis, Environmental Systems Research Institute
Bruce Delgado, US Bureau of Land Management
Wayne Ferren, University of California, Santa Barbara
Chris Gillespie, US Department of Defense
David Graber, US National Biological Service
Steve Hartman, CNPS
Sylvia Haultain, University of California, Davis
Deborah Hillyard, California Department of Fish and Game
Mark Hoshovsky, California Department of Fish and Game
John Hunter, University of California, Davis
Jim Jokerst, Jones and Stokes
Todd Keeler-Wolf, California Department of Fish and Game
Corky Matthews, CNPS
Bruce Orr, EA Engineering, Science, and Technology
Oren Pollak, The Nature Conservancy
J. Robert Haller, University of California, Santa Barbara
Cynthia Roye, California Department of Parks and Recreation
Connie Rutherford, US Fish and Wildlife Service
John Sawyer, Humboldt State University
Jim Shevock, CNPS
Janine Stenback, California Department of Forestry
Joan Stewart, CNPS
Scott White, Tierra Madre Consultants
Roy Woodward, Bechtel Corporation

Acknowledgments

This manual has benefited from the assistance and advice of many people. Their knowledge of vegetation, philosophical views of classification, and hard mental and physical labor on this project has made this manual better than it would have been without them. No one person knows enough about California vegetation to write a book of this nature by themselves. We have relied upon many people for contributions of data, review of drafts, and distribution of the drafts. We would like to thank them individually:

Annette Albert, Sarah Allen, Barbara Allen-Diaz, Michael Barbour, Ed Biery, Bruce Bingham, Roxanne Bittman, Mignonne Bivin, Mark Borchert, David Bramlet, Dan Cheatham, David Chipping, Jane Cole, Charles Convis, Max Creasy, Ellen Cypher, Bruce Delgado, Wayne Ferren, JoAnn Fites, Chris Gillespie, Claire Tipple Golec, Hazel Gordon, Jim Griffin, Robert Haller, Steve Hartman, Duane Haselfeld, Sylvia Haultain, Deborah Hillyard, Ken Himes, Julie Horenstein, Marc Hoshovsky, John Hunter, Tom Jimerson, Jim Jokerst, Michelle Kanoart, Jon Keeley, Melody Lardner, Lori Lawrentz, Kari Lewis, Sally Manning, Corky Matthews, Richard Minnich, Marlyce Myers, Tom Oberbauer, Dennis Odion, Bruce Orr, Oren Pollak, Don Potter, Robert Preston, Cynthia Roye, Connie Rutherford, Andy Sanders, Jim Shevock, Jim Smith, Sydney Smith, Janine Stenback, Joan Stewart, Neil Sugihara, Kathy Thomas, Kenneth Turner, Felicity Wasser, Scott White, Roy Woodward, Paul Zedler.

Special appreciation goes to Michael Barbour, Bruce Bingham, Mark Borchert, Hazel Gordon, Joan Stewart, and Scott White for detailed reviews of various drafts and to Jim Shevock for reviews, production, and layout assistance.

We dedicate this book to Jane, Ginger, Evan, and Amos.

Introduction

The California Native Plant Society (CNPS) is an organization of amateurs and professionals united by an interest in the plants of California. Its chief aims are to preserve the native flora and to add to our knowledge of it. Its members participate in monitoring rare and endangered plants, fostering public education, supporting legislation that protects native plants, and providing expert testimony to government bodies. In 1974 CNPS published *Inventory of Rare and Endangered Vascular Plants of California*, the first of five successive editions, which have documented the occurrences of rare species. In 1985 CNPS entered into a formal partnership with the California Department of Fish and Game to continue a process of rare plant monitoring. Over time, more information became available about habitats and plant communities with rare species.

In 1990, CNPS decided that the Society's activities in the protection of rare species could be enhanced by a complementary focus on the protection of rare plant communities. The Board of Directors, responding to a suggestion from a small *ad hoc* subcommittee of the Rare Plant Program, proposed a parallel program to develop comparable information about California's plant communities.

Michael Barbour was invited to chair the committee, and he established a committee of approximately 25 individuals from academia, conservation organizations, environmental consulting companies, and state and federal agencies. The group met for the first time on February 9, 1991.

An early objective of the Committee was to foster adoption of a uniform vegetation classification among private, state, and federal resource agencies with jurisdiction over land management. At that time, several conflicting systems were being used, making it difficult for biologists to communicate. The adoption of a common classification allows conventions, descriptions, and names to be consistent. A uniform vocabulary permits a longer-term objective to be met, the legislative recognition and protection of rare, threatened, or endangered plant communities across administrative boundaries.

By developing quantitative, defensible definitions of rare and threatened communities, we can invoke the California Environmental Quality Act (CEQA) to help conserve them. CEQA specifically calls for the preservation of examples of plant and animal communities within the state. Before working with rare and threatened communities, we need to create a systematic classification of all communities, including the common and extensive, as well as the rare ones.

By providing a common language and clear definitions, the Committee hoped also to facilitate a number of processes linked with the assessment of development projects that must be reviewed under CEQA. These processes include the identification and ranking of plant communities by conservation priority within a project's boundaries. It is anticipated that conservation efforts can now be more accurately addressed through planning documents via CEQA, the National Environmental Policy Act (NEPA), and other laws, regulations, and policies.

The Committee, now numbering 32 members and renamed the Plant Communities Committee, celebrates the start of its fourth year of existence with the publication of this book, *A Manual of California Vegetation*.

The classification scheme adopted in the *Manual*, the kind of information included, text format, and the review process were collectively agreed upon by committee members. John Sawyer served initially as editor, but it become evident that a more efficient approach was for one person to write community descriptions. Todd Keeler-Wolf became increasingly emersed in the project at all levels.

The Start of an Interactive Relationship among Users and Future Editions

This edition is the culmination of a tremendous amount of conservation insight, work, and commitment. The Board of CNPS decided early on that to make a meaningful contribution toward the conservation and appropriate uses of California vegetation, this manual must be based on the best scientific information available.

The *Manual* also serves as a foundation to shift conservation emphasis from a single species approach to a landscape approach that encompasses groups of species, plant communities, and ecosystems while continuing to incorporate the need for rare and endangered species conservation and management. Viewing and describing landscapes within the framework of a unified vegetational classification provides the common language necessary for managers to make informed decisions. It is hoped that the *Manual* will be widely used by landuse planners at the county, state, and federal government levels, environmental consultants, conservationists, and natural resource specialists, including botanists, ecologists, foresters, range managers, and wildlife biologists.

California is a wonderfully diverse state. Its flora has been studied for over 100 years and yet new species, even genera, are still being described. It should not surprise anyone who has worked with vegetation in this state that the need to provide a framework and systematic approach to describe and quantify California vegetation is long overdue.

CNPS realizes that many conceivable vegetation series, our formal name of a plant community, are not included in this first edition, because we lack documentation to effectively describe them. Similarly, the distribution of many series throughout the state is inadequately understood. We rely on the cumulative knowledge and submission of new data sets by the users to prepare future editions. As this manual goes to press a computerized database version is being developed that will facilitate easy updating and customized usage of information.

If you have data, literature, or proposed types or vegetation information that could assist the CNPS Plant Communities Committee in describing and validating a new series, or add relevant information to existing series and/or their distribution, send the information directly to: CNPS
1722 J Street, Suite 17
Sacramento, CA 95814
attn: Plant Communities Database.
Call: (916) 447-2677

A copy of the CNPS plant committees field data form and instructions for using the CNPS sampling protocol are included in the Appendix. This form can be photocopied or additional forms may be requested from CNPS. If you have reconnaissance information on any of the rare vegetation types described in this manual please use the NDDB California Natural Community Survey Form included in the Appendix. For more forms contact:

Natural Diversity Data Base
California Department of Fish and Game
1416 Ninth Street
Sacramento, CA 95814
Call: (916) 324-6857

The Plant Communities Committee requests comments from the users on ways to improve the utility of this classification. Hearing from users is the best interactive way to improve future editions. In addition, the Plant Communities Committee is seeking quality photographs of series, stands, habitats, and vernal pools for future editions of the *Manual* and for CD-ROM applications.

What is Vegetation?

Vegetation is a large part of our state's unique signature. The dark, fog-shrouded cathedrals of its coastal redwood forests, the sunny open foothill woodlands of blue oak, and the low gray-green aromatic sage scrubs of the central and southern coasts are each as representative of California as its scenic coastline, Mediterranean climate, and Yosemite Valley.

Vegetation – all the plant species in a region and the way they are arranged – is a part of the natural world that humans learn to recognize at an early age. It is our nature to differentiate among a grassland and a shrubland or a meadow and an adjacent forest. The differences among types of vegetation we see initially are strengthened by their diverse resource values that we exploit. We are all vegetation classifiers to some degree.

Unlike an artist's contrasting brush strokes, differences among vegetation types are not strongly defined. Although we can agree that a dense lodgepole pine forest is qualitatively different from an adjacent wet mountain meadow, the closer we look, the harder it is to discern where the meadow begins and the forest ends. We see individual sedges, asters, and buttercups trailing off in decreasing density into the forest as it closes overhead, and young pines and scattered mature trees stationed well out into the meadow. The problem of fuzzy boundaries is characteristic of vegetation science.

Also unlike the permanence of a master's brush stroke, vegetation patterns change. Because plants are alive, they create a transitory scene. At any location trees, shrubs, and herbs form a unique pattern that shifts subtly from season to season and year to year through the growth, death, and reproduction of individual plants. Patterns can also change profoundly over time

following disturbance. However, there is a tendency for a given landscape to have a limited set of vegetation types. Vegetation following fire, seasonal or periodic drought, avalanche, or disease often approaches conditions similar to those before the disturbance.

Some changes, however, are not within the adaptive repertoire of the plants, forcing permanent shifts in the scene. New vegetation types may arise within a geologic blink of an eye. For example, a wet meadow may change to a dry grassland after a single flood. Rapid downcutting of a stream can drop the water table below the reach of the roots of meadow plants.

Pattern in vegetation can be thought of as being driven by environmental variation. Differences in such important environmental factors as moisture, soil chemistry, temperature, and time since disturbance exert powerful influences. For example, the moisture content of the soil ranging from a streamside terrace to adjacent upland slopes has a direct influence on the types of vegetation occupying these areas. These environmental variables may be viewed as gradients, and vegetation changes subtly along these gradients. Conversely, vegetation may change dramatically where there is a "break" in gradients such as between a chemically harsh serpentine soil and a more benign, sandstone soil. The study of how this environmental variation influences vegetation is an important part of vegetation science.

Community-unit Versus Continuum View of Vegetation

Early in this century the lines became drawn in an ideological debate over the fundamental way vegetation is interpreted. Frederick Clements (1916, 1920) metaphorically equated units of vegetation with organisms. He saw that clusters of species repeatedly associated together. He believed that at least some of these species had obligatory relationships with the community much like that of organs within the living creature. In Clements' vision of vegetation, its units had little overlap as many species were confined to a single type of vegetation. The boundaries among adjacent clusters of plants were narrow, with very little overlap of species ranges, except for a few widespread plants.

Henry Gleason (1926, 1939) published his ideas on the continuum view of vegetation soon after those of Clements. His contention was that vegetation was the result of two major factors: the fluctuating arrival and departure of plants, and the equally fluctuating environment. In contrast to Clements' view of plant associations as discrete and interdependent combinations of species, Gleason viewed species ordered along environmental gradients in ways prescribed by the *individual* requirements of species. Overlap in species distribution was considered largely coincidental and independent. Gleason regarded the clusters of species predicted by the discrete community view of Clements and his followers as merely artifacts of the investigator's perception or of sampling and analysis methodologies.

Recent studies have continued to support the Gleason's individualistic view of vegetation. For example, even understory species in several forest types have been shown to be influenced not so much by the species of trees that shade them in the canopy, but by the presence of shade itself (McCormick & Platt 1980, Parker & Leopold 1983, Rodgers 1980, 1981). Smith & Huston (1989) suggest that the vast majority of spatial and temporal structure in vegetation can be attributed to competition among species for light and water as plants grow larger.

As a direct result of these investigations and theories, vegetation ecologists today are cautious about acknowledging the existence of plant communities. Wilson (1991) suggests that there is very little evidence to support their existence. Others (Dale 1994, Keddy 1993, Palmer & White 1994) suggest that definitions of ecological communities need to be changed before their existence can be refuted or substantiated. Keddy (1993) points out that there are a number of things we define as existing that are difficult to prove empirically (*e.g.*, electrons, individual human beings, individual species). Yet these terms are useful. Palmer & White (1994) suggest that community ecologists should define "community" operationally and thus remove themselves from the ontological dilemma of whether or not communities exist.

Sophisticated techniques have been devised to define associations among plant species and their environmental correlates (Bonham 1989, Gauch 1982, Kent & Coker 1992, Jongman *et al.* 1987, Mueller-Dombois & Ellenberg 1974). Although these techniques distinguish groupings of plants following quantitative field sampling and analysis, we recognize that plant communities erected using these techniques are arbitrarily drawn. Community boundaries depend on the investigator, the sampling and analysis methods as well as the sharpness of the environmental gradients in the landscape. For this reason, we use the word **vegetation type** when referring to plant assemblages.

Vegetation Change

Clements (1916) popularized the idea that a community grew, matured, died, and reproduced itself though a progression of serial stages, each depending on the previous one. This concept of succession, like his community theory, has been challenged. Egler (1954), McCormick (1968), and others conducted studies in the eastern United States that suggested that succession depends on initial floristic composition, not serial stages. Progression from dominance by shortlived plants to one of longer-lived plants has elements of chance that can create several possible avenues as plants grow and mature. Clements' predictable model of succession to a single persistent state is not realistic.

Recent ideas applied to non-forested environments (*e.g.*, Westoby *et al.* 1989, Huntsinger & Bartolome 1992) suggest that for major kinds of vegetation, for example, grassland or oak woodland, there is a catalog of possible states and transitions among them. The states represent visually distinct vegetation types, grassland, grassland with shrub seedlings, and dense cover of shrubs with little grass, all in the same area. The transitions represent change from one state to another brought about by changes in environmental conditions, disturbance events, or by the inevitable growth of new species overtopping previously taller plants.

Forest ecologists (*e.g.*, Oliver & Larsen 1990) go as far as to replace the term succession with a new one, stand dynamics, when talking about the way forests change with age. Their domain of interest is not the plant community, but a stand of plants that have the same history. Stands of seedlings become established after a disturbance. As the seedlings grow to be trees, the stand goes through structurally, and possibly floristically, distinct stages. A landscape is seen as a matrix of a few to many stands with different histories and ages.

As the need for conservation and management of vegetation continues, so does the practical need to understand how vegetation develops and how it is defined. Operational views of change and of vegetation classes are destined to replace the metaphorical concepts that initially described them. However, the accuracy of science will never be a substitute for our aesthetic appreciation of the beauty of vegetation patterns across the landscape. It behooves us to pause and reflect.

Vegetation Classification

A classification is a language created to bring order out of apparent chaos. Because it is a language, the basic goal of a classification is to solve a communication problem. A vegetation classification is, therefore, a language developed to meet the need for a single commonly accepted terminology to discuss various kinds of vegetation.

A single classification cannot be all things for all people. The numerous vegetation systems that have been developed reflect a variety of descriptive scales, philosophies, and purposes. We recognize that all classifications are artificial in that the units are subjectively defined and described to meet a particular need.

Reasons for developing a vegetation classification are varied and can include assisting in resource inventory, land use planning, conservation, illustration of ecological relationships, or building a framework for understanding vegetation dynamics. Likewise, the vegetation units created vary and can include: functional resource management criteria, such as timber or range types; descriptions of vegetation associated with landscape units, ecological units, or animal habitats; emphasis on vegetation structure; emphasis on floristic assemblages; or emphasis on units recognizable with aerial photographs.

Scale and Hierarchy in Classifications

Any vegetation classification recognizes a particular level of detail. For example, land use cover classifications and accompanying maps published by the U.S. Department of Agriculture or various California county governments list general units, "grassland, conifer forest, scrub, or marsh." These same classifications may put more emphasis on distinguishing among types of agricultural crops than among natural vegetation units. Wall maps of the vegetation of California, (Küchler 1977, Parker & Matayas 1979, Wieslander & Jensen 1946), show gross patterns of vegetation, but are not accurate at a fine, local scale. In contrast, a vegetation classification and map produced for a small nature preserve or a state park exhibit detail at very fine scale.

The concept of hierarchy has proved valuable in the theoretical organization of vegetation throughout the world, in much the same way that it has been used to express relationships among plant or animal taxa. Several well-defined vegetation classification hierarchics exist. For example, the UNESCO classification (Muller-Dombois & Ellenberg 1974) or the classification developed by Barry (1989) for the California State Park System incorporate many levels of scale, from the individual plant to the biosphere.

In traditional approaches to vegetation classification, the broadest units are defined by vegetational differences correlated with basic environmental differences (*e.g.*, aquatic and terrestrial). Lower levels are arranged around physiognomic features of plants including height and dominant life form (*e.g.*, tree, shrub, herb), canopy cover (open to closed) and its foliage characteristics (*e.g.*, evergreen, deciduous, broadleaf, needleleaf). It is not until the classifier moves well down in the hierarchy that classes are defined by dominant species: the formation-type level in Barry's (1989) scheme, the alliance level in Braun-Blanquet's (1932) system, and the series level in Daubenmire's (1939) classification. Beneath that dominant species level – whether it is called an alliance, formation-type, or series – is another level, the association, that is defined by secondary species.

Hierarchical approaches to classification of natural environments exist that do not emphasize vegetation. One of the most widely used of these is the wetland and deep water habitat classification of Cowardin *et al.* (1979). In the Cowardin approach, physical features of the environment are arranged hierarchically. These upper levels relate solely to physical characteristics, such as substrate type, water chemistry, and water regime. Only the lowest level, the dominance-type, is biologically defined. In the case of wetlands, biological dominance may be either botanical or zoological, mussel beds in the marine intertidal zone for case. Ferren *et al.* (1994) in a recent classification of central and southern California wetlands use a modified version of the Cowardin system to describe hundreds of dominance types, many of which are analogous to the series, alliances, or associations of traditional vegetation systems.

Hierarchial arrangements of ecosystems, for example, the USDA Forest Service ECOMAP project (ECOMAP 1993), utilize synthetic ecological units derived from abiotic (primarily soils, climatic zones, and landforms) and biotic classifications (primarily vegetation) at regional, subregional, landscape, and land-unit levels. The basic landscape and land-unit levels of the ECOMAP project are called "landtypes."

The CNPS Approach to Classification

Our Focus is on Floristics and Rarity

Floristic components of a classification are those that refer to the plant taxa making up the vegetation of a given area. We believe that the most important units of conservation in any hierarchy are the floristically-based lower units of the **series** or **association**. Broad physiognomic units are largely synthetic, including units without common species. For example, open stands of singleleaf pinyon pine or Engelmann oak are both described as woodlands, yet they share few if any plant taxa. Recognizing both units as woodlands does not express the level of biological diversity that we are interested in classifying.

Although we recognize the value of higher levels of vegetation, they are not treated in this manual. The current hierarchies developed by UNESCO (Muller-Dombois and Ellenberg 1974), USDA Forest Service (ECOMAP 1993), The Nature Conservancy (1994), and others can be easily blended with this system.

Some examples of rare California series are:

Alkali sacaton series	Ashy ryegrass series
California oatgrass series	California walnut series
Darlingtonia series	Engelmann oak series
Giant sequoia series	Grand fir series
Idaho fescue series	Ione manzanita series
McNab cypress series	Native dunegrass series
Parry pinyon series	Santa Lucia fir series
Sargent cypress series	Sitka spruce series
Teddy-bear cholla series	Valley oak series
Washoe pine series	Water birch series

Some of these rare series are representative of extensive series that occur beyond California's borders, while others are endemic. Some were once extensive, now reduced to a small part of their original range; others were never exten-

sive. Some are diverse and include several associations, while others may be represented by a single association.

The process of quantitative vegetation classification is often long and detailed. For example, recent work by the California Natural Diversity Data Base on the California sycamore series required two seasons of field sampling and several months of data analysis. It was only then possible to quantitatively define and map three associations, two of which are very rare (Keeler-Wolf *et al.* 1994). We hope that such sampling and analysis efforts will initially focus on describing the rarest communities to help protect them.

The focus on rarity does not imply that we ignore vegetation types that are common and extensive. The only way we can understand the rare ones is in the context of all of the state's vegetation. Yet, how do we build a quantitative, data-driven classification without collecting and analyzing data from all vegetation? We have, by necessity, chosen a compromise. We established basic rules of dominance and nomenclature for the larger units of floristic composition that we call series. Based on our knowledge of the distribution of a series, distinct subdivisions can be targeted for study.

For example, we know that the Valley oak series is widespread. Although clearing of oak woodland for agriculture and urban development has reduced acreage over the past 150 years, there is still a considerable area dominated by valley oaks. However, acreage of closed-canopy valley oak forest in the riparian zone is very small today. It has been reduced substantially by agricultural clearing, gravel mining, and other stream bed alterations. Because this subtype of Valley oak series is visually distinct, it has been treated as a distinct vegetation type called Great Valley valley oak riparian forest (Holland 1986). By targeting stands dominated by valley oak and others not dominated by it in the riparian zone for sampling and analysis, we are able to ascertain if valley oak riparian forest is a unique vegetation type. If it is, we are able to offer a quantitative description of it, map its extent, and definitively distinguish it from other types of vegetation.

Our Views of Dominance, Existing Vegetation, and Refinement of this Classification

Although the general rules of dominance and cover apply to all series in this manual, there are many cases where a series is not yet quantitatively derived through sampling and analysis. Dominance can be visually estimated and in many cases it is an obvious enough trait to make detailed sampling and analysis to prove the point unnecessary. In some cases you will see series named by a dominant species, while in other cases you will see them named by a dominant genus. This has been done in the case where species in the genus have similar ecological requirements. This practice is in keeping with the way series have been defined in earlier efforts (Parker & Matayas 1979, Paysen *et al.* 1982).

Series are defined using the dominance rule. Vegetation ecologists have found that certain species other than the dominant upper layer species may be informative in defining ecological groupings of vegetation types. These species are described as "representative," "diagnostic," and "characteristic" (Mueller-Dombois & Ellenberg 1974, Kent & Coker 1992). Types that have the same characteristic species are described as belonging to an "alliance" rather than to a series (The Nature Conservancy 1994). A few of the series in this manual, the Woollyleaf manzanita series for example, are defined in terms of characteristic species rather than the dominant ones.

Our classification views vegetation in terms of what exists today. There is no presupposition on our part that stands of one series will naturally always stay in a series or be replaced by another one. The vegetation dynamics are currently being worked out for a number of forest types in the state, but our principal focus in this manual is to identify any vegetation type that currently exists as a visually distinct entity. At the same time, we recognize that vegetation change and development are real, and that there is value in understanding vegetational states of a particular area. Understanding allows for wise resource management, including that of endangered species requiring special habitat conditions.

In comparison, most of the quantitative classifications developed for California by the Forest Service (*e.g.*, Fites 1993, Jimerson 1994) are based on the concept of a "potential natural community," not on one of the existing community or vegetation. A potential natural community is one "that would be established if all successional sequences of its ecosystem were completed without human-caused disturbance, under present environment conditions" (Forest Service Manual). We cannot provide a uniform view of such a potential for the state's series. We also believe that other concepts of stand dynamics and state-transition concepts may afford a better way of understanding much of California's vegetation.

Conservation and Management in the Present and Future

How Description of Vegetation Can Help with Conservation and Ecosystem Management

Those who have spent time in California have heard it described by superlatives in comparison to other states in the country – highest and lowest points, greatest number of plant species, greatest number of endemic plant taxa, greatest number of climatic zones, most complex geology, greatest number of endangered species (an unfortunate superlative), the list continues. California also contains the most diverse and complicated patterning of vegetation of any area of comparable size in North America. The combination of complex geology, climatic conditions, natural disturbance regimes, and topography, coupled with its high plant species diversity, has created an extremely diverse hodgepodge of vegetation types.

Classifiers of California vegetation are faced with an urgent task. Its seriousness is underscored because CNPS believes that many of the rare and endangered communities in California are being destroyed. Yet, without quantitative descriptions of California's vegetation types, we cannot distinguish the rare or endangered series from other more common ones. Further, we cannot justify their protection in terms of CEQA if we cannot clearly define them.

As we encounter economic pressures that alter and destroy our vegetation, the amount of time to distinguish its many forms appears frighteningly short. The California Natural Diversity Data Base (NDDB) has determined that 135 out of the 280 vegetation types listed as end points in the Holland (1986) classification are rare enough to warrant concern and some level of protection. At least 50 of these are so rare that it is believed that there are

fewer than 2000 acres of high quality habitat for each of them.

We are at a crossroad. The federal and state endangered species acts are undergoing more scrutiny and challenge than ever before. The legal tools for conservation of biodiversity were not explicitly developed for conservation of vegetation types and ecosystems. There are only vague inferences and brief intent language that address conservation of habitats and ecosystems in CEQA legislation. Yet, conservationists recognize that to sustain the natural values of California there must be something more useful and comprehensive than single species-based conservation. CNPS hopes eventually to secure explicit legislative protection for rare and endangered vegetation types, but we recognize that we must now focus on seeking protection through existing mechanisms. These include CEQA, NEPA, and a number of innovative programs. Recently, some programs based on vegetation types have begun. These efforts pertain to southern California's coastal sage scrub, a pilot project initiated by the 1991 Natural Communities Conservation Planning Act (NCCPA), and multispecies conservation plans for the Sacramento Valley, San Joaquin Valley, western Mojave Desert, and other areas.

Conceptually, these programs have the advantage of establishing habitat reserves for key portions of an ecological region, while allowing areas of lesser ecological importance to be modified or developed, satisfying both the conservationist and developer. However, these programs are scientifically compromised by the lack of information. These programs have, by default, focused on individual threatened or endangered indicator species within broadly conceived ecosystems. Chief among the reasons for this compromise is insufficient information on the vegetational building blocks of these areas. CNPS believes that by developing quantitative vegetation descriptions, critical habitat for numerous targeted rare species can be better defined.

As an example, not all areas within the zone described for the NCCPA activities are prime habitat for the California gnatcatcher, the bird species motivating much of the conservation work in southern California. Classification of vegetation can play an important role in resolving conflicts. Broad conceptual views of coastal sage scrub and chaparral in the vein of the Munz & Keck (1949, 1950) or Holland (1986) classification may under or overestimate the value of areas to gnatcatcher habitat. In all likelihood, the bird is responding to distinct mixes of plant species composition with special structure, as recently pointed out by Read (1994) and White (1994). Only a detailed, quantitatively based vegetation classification can specify such important resource differences.

Using carefully crafted quantitative vegetation descriptions can aid to multispecies conservation. There are, however, vegetation types that are in and of themselves rare. These may occur within areas not slated for broad scale habitat planning, and may exist without containing a single threatened or endangered species. Yet, we recognize the uniqueness of the vegetation type and the threats to its existence. If we want to develop a framework for the protection of these types, we need to develop defensible definitions for them.

The conservation-based Holland classification currently used by NDDB is insufficient to answer repeated questions about the relevance of many of its vegetation types because of the lack of quantitative data. Does southern maritime chaparral differ significantly from ad-jacent chamise chaparral or southern mixed chaparral? How can we differentiate sycamore alluvial woodland from southern sycamore-alder riparian woodland? Such questions arise because NDDB has identified some communities as rare and endangered. Therefore, landowners and resource managers want to know how to identify and protect, or more skeptically, how to demonstrate or discredit the uniqueness of these vegetation types. Rigorous definitions and mapping remain a problem because until very recently we have not chosen to quantify salient components of these types.

We treat Holland (1986) high priority communities in various ways as they are translated into series. The following list identifies some of Holland "rare" communities in need of precise sampling, and their relationships to the new system.

Holland communities represented by several series:
  Coastal sage scrub, Coastal dune scrub, Freshwater swamps, Great Basin grassland, Great Valley riparian forest, Maritime chaparral, Native coastal grassland, Sitka spruce-grand fir forest, Southern riparian forest.

Holland rare communities considered as associations:
  Gabbroic northern mixed chaparral, Great Valley chenopod scrub, Maritime coast range ponderosa pine forest, North Coast black cottonwood riparian forest, Sycamore alluvial woodland.

Holland communities considered as series:
  California walnut forest and woodland, Ione manzanita chaparral, Washoe pine-fir forest, Western hemlock forest.

Holland communities treated as stands in which a species, not the community, is of interest:
  Cherry forest, Island ironwood forest, Torrey pine forest.

How the CNPS Classification Fits into a National Vegetation Classification

The enormity of the task of vegetation classification in California is multiplied several times when we consider classifying the vegetation of the entire country. Yet, an integrated and complete classification of vegetation for our country would afford the most comprehensive approach to conservation and management of our nation's natural ecosystems. The need has been long recognized. Classification efforts have proceeded in two ways: from the top down in the form of physiognomically based classifications, and from the bottom up in the form of small regional systems that may be pieced together to form an as yet incomplete, fine-scale national classification quilt.

The top-down physiognomic hierarchies of Driscoll *et al.* (1984) and Bailey (1994) are frameworks that have been built upon by projects, for example, the USDA Forest Service current (ECOMAP 1993) effort, which nationally describes a hierarchy of regional to landscape levels of physiognomically defined vegetation and ecosystems. However, the lower floristically defined building blocks are far from completely described.

The most extensive integration of the floristic vegetation units at the national level is being conducted by The Nature Conservancy (TNC) Heritage Task Force ecologists. By integrating existing floristic-level classifications from the USDA Forest Service and thousands of other agency and individual studies much progress has been made on assembling a national system. However, due to the somewhat inconsistent methods of classification, certain regions of the country are less well integrated than others. The most successfully organized region is the western United States, where quantitative definition of vegetation associations has been most complete. TNC's heritage ecologists have recently produced a preliminary classification of the western United States (Bourgeron & Engelking 1994). This work relates the quantitatively defined associations to alliances. It does not include California at this time, a situation that this CNPS effort will change due to the comparability of approaches.

That conservation and resource management classifications are now speaking the same language is an important step forward. Specificity is important in both a practical manager's and an idealistic conservationist's lexicon. The work that the Forest Service and other land management agencies have done to attain this level of understanding is of direct value to conservation.

In 1994, TNC western heritage ecologists and members of CNPS Plant Community Committee met and agreed that the time was right to meld the California and western United States classifications. The first collaborative effort was recently published (Grossman *et al.* 1994). This initial survey of rare plant communities of the conterminous United States was possible because the classifications are at compatible scales. In addition, the efforts were able to use comparable ranking criteria for rarity.

Recently Michael Barbour, on behalf of the Ecological Society of America (ESA), has agreed to chair a special subcommittee on vegetation classification. Its principal charge is to develop a united system for use in the nation (Barbour 1994).

Conservation of Vegetation in the Future

Beyond the level of quantitative definition of series and associations, serious questions may arise about the basic premise of the CNPS plant communities program: to identify and to protect rare plant communities in California. As more and more associations are defined, we may be faced with an enormous number of rare plant associations. How can we justify protecting them in a political, social, and economic climate where it is becoming nearly impossible even to legally "list" strongly substantiated rare species? Can we effectively justify supporting legal protection for potentially hundreds of rare assemblages of plants? In addition to the refinement of classification, there must come a refinement of concepts of threat, endangerment, and different levels of biodiversity value for vegetation. Two points may serve as examples of the complexity of issues involved.

We already know that several quantitatively defined associations of Blue oak series in the southern Coast Ranges are restricted to just a few sites of limited acreage (Borchert *et al.* 1993). However, they have no rare species associated with them. Instead they are uncommon assemblages of widespread species limited by topographic position and soil type. Are these limited assemblages as important as plant associations that are dominated by or contain a mixture of rare species that occur nowhere else? To evaluate such situations, might we not incorporate wildlife values, watershed protection values, and other non-vegetation traits into our assessment and deliberations?

Secondly, as a result of direct human disturbances, including livestock grazing and invasion by exotic plants, excellent examples of common vegetation types are rare in themselves. Do we need to become more interested in identifying and quantifying the best remaining examples of some of our more common plant assemblages? Ranking and prioritizing sites based directly on human disturbances underscores another need for quantitative assessment that goes beyond standard vegetation sampling techniques.

Conservation strategies will surely arise that we do not fully comprehend at this time. They may enable us to view more clearly the value of individual vegetation associations in terms of larger watersheds and landscapes. We may be able to envision and sustain a set of vegetational states and a patchwork of ecosystems that enables us to foster processes necessary to maintain biodiversity without the legal protection of any specific plant associations. In the interim, the least we can do is commit ourselves to analysis and study of vegetation so we can provide some of the scientific building blocks for these larger goals.

The History of Vegetation Classification in California

The early explorers left records describing California's vegetation in broadest of terms. In the late 1800s, descriptions by Brewer (*in* Farquahr 1966), Sudworth (1908), Kellogg and Greene (1889) and others mentioned oak woodlands, grasslands, chaparral, and conifer forests, but these accounts lacked precision.

In the early 1900s many vegetation classification systems in the United States were associated with Clements' concept of stratifying the world's vegetation into large-scaled climatic zones (Clements 1916, 1920, Weaver & Clements 1938). Cooper (1922) was interested in characterizing California's communities dominated by plants with relatively stiff thick leaves with a waxy cuticle (the so-called sclerophylls) occurring in its Mediterranean-type climate. Within the two main sclerophyll groups, the California chaparral formation and the Broad sclerophyll forest formation, he made further divisions into units representing distinctive categories. These units he called associations or consociations, if the unit was dominated by a single species. Because it is a useful way of accounting for vegetation without the need to know the identity of the smallest units, Cooper's approach has been followed and extended over the years. This approach can be naturally extended to mapping and is seen in the system developed by Wieslander in his Vegetation Type Map (VTM) Survey of California (Wieslander 1935, Critchfield 1971).

The Wieslander mapping effort was a remarkable achievement. Between 1928 and 1940 almost half of the state (about 40 million acres) was mapped by USDA Forest Service field crews who combed California using a specific protocol. Because this project was initiated before the widespread availability of aerial photos, the crews worked from the prominent peaks and ridges to gain the perspective needed to draw and label the individual patches of vegetation. In addition to sketching these types, the VTM survey collected data from thousands of individual plots which further characterized and documented

---

The Difference Between a Vegetation Classification and a Vegetation Map

Vegetation mapping units (the individual patches depicted on a map) and vegetation classification units (the vegetation type described at a given hierarchical level of a classification) are not necessarily one and the same. A vegetation map is a symbolic representation of visually distinct groupings of plants. A classification can afford to be much more detailed and descriptive. It can involve more floristic and structural details than that perceptible on aerial photographs or easily depicted in maps.

Another important difference between a map and a classification is that a map is limited by scale, while a classification need not have this restriction. The development of a quantitative classification system allows the user to benefit from the knowledge that all the pieces of vegetation can be arranged from the smallest identified units to the largest. In a system that starts with stands of vegetation, the classes at a given level in the hierarchy are grouped according to prescribed similarities to form classes at the next level. This approach affords a consistent interpretation that is driven by the known framework. For example, if the classification is based on species cover, rules to that effect can be established and consistently followed. Conversely, in a vegetation map, the matter of scale will always compromise a dominance-based system. Naturally occurring types smaller than the minimum mapping unit will not be represented and must be included within other often unrelated, but more extensive types. In addition, species that are visually similar to one another may need to be lumped as a single map unit, obliterating important ecological differences.

The best vegetation maps are presented in the context of a classification. The mapping units are often elegantly presented in categories of increasing generalized classes. Each class in the hierarchy can be consistently described in detail in a text that accompanies the map.

---

this dominance-based system. These data, inmany cases over 60 years old, have provided a valuable resource for further studies of the relationships of vegetation in cismontane California. Recent work by Allen *et al.* (1990) has taken advantage of the VTM vegetation plots to develop classification systems of California's oak and other hardwood rangelands. Griffin & Critchfield (1976) used the species identifications of the VTM survey and the succeeding Soil-Vegetation Survey maps (another USDA Forest Service project with less detail on species composition) to develop an atlas for forest trees in California. Minnich *et al.* (1994) showed the effects of 60 years of fire suppression on mixed coniferous forests in southern California by resampling plots established in the 1930s by the VTM crews.

The multifaceted values of the mapping project data were clearly envisioned by Wieslander (1935). He states, in his description of the project, that the maps of broadly defined plant associations (his high hierarchical units) are useful to engineers, foresters, and others charged with the management of wildlands. They represent an attempt to group subtypes having similar fire hazard characteristics and qualities of economic importance. Wieslander considered his more detailed subtype units as providing basic information on vegetation cover desired by the research worker in various fields of botany, ecology, or forestry.

Despite the great effort involved, no systematic attempt was made to synthesize the Wieslander vegetation units into a classification that also described the mapping units. In 1980, Glen Holstein, the first ecologist with the California Natural Diversity Data Base, catalogued the vegetation polygons on the VTM maps and came up with about 1800 individual cover types. Because of the large number of types, the incomplete total coverage, the somewhat inconsistent listing of subdominant species, and the lumping of some vegetation mosaics into larger polygons, it has proved difficult to synthesize the mapping data in a bottom-up

hierarchy meaningful for the state. The ancillary plot data have proven much more useful in this regard (*see* Allen *et al.* 1990, Allen-Diaz & Holzman 1991), although limitations of geographic coverage and lack of detailed understory species composition pose constraints.

Jensen (1947) developed a methodology for classifying California's vegetation in several ways based on the VTM survey. Jensen presents three classifications useful for different purposes. The first section is a vegetation cover and land status classification including 12 categories, such as chaparral, commercial conifers, bare ground, and cultivated areas. The second section is a system of the "species units," including single species, mixed species, or "mosaic mixtures" where two or more elements are involved. These species units are analogous to the mapping units in the original VTM survey. The third section is a synthesis of the first two, grouping the 32 main types, including commercial conifer, non-commercial conifer, hardwood, and non-tree types. This latter synthesis was developed into a statewide vegetation map (Wieslander & Jensen 1946). Both Wieslander and the Jensen efforts were based on maps of a certain scale (15' USGS 1:62,500), thus limiting the size of the vegetation units depicted to no less than about 40 acres (Jensen 1947).

In contrast to the map-based efforts of Jensen and Wieslander, the next California vegetation classification developed was based on broad climatically inferred plant communities (Munz & Keck 1949, 1950). It was meant to aid in the discussion of habitat and distribution of species in the book, *A California Flora* (Munz 1959). This was another Clementsian-based classification that created 29 plant communities within 11 vegetation types, distributed among five biotic provinces. The names, familiar to many contemporary botanists, use names of dominant species (red fir, redwood, and bristlecone pine forests) while others invoke habitat characteristics or physiognomy (foothill woodland, valley grassland, coastal strand, and alkali sink). Although the Munz and Keck plant communities form a useful tool for understanding the statewide distribution of plants, the system omits numerous ecologically distinctive habitat types.

In the 1970s, heightened interest in vegetation within California resulted in two books (Barbour & Major 1977, Ornduff 1974), a number of descriptions of local vegetation (*e.g.*, Hanes 1976, Minnich 1976, Sawyer & Thornburgh 1977, Vogl 1976), and two statewide classification systems that refined the Munz and Keck approach to finer levels and developed many new types (Cheatham & Haller 1975, Thorne 1976). The two books describe California's vegetation on various scales of knowledge and were not intended as a formal classification.

Thorne's system was published by CNPS for purposes of plant conservation. Cheatham and Haller's system was never published and was specifically designed for the identification of University of California natural reserves. The Thorne system is more detailed than Munz and Keck's, but is more general than Cheatham and Haller's. Both divide broad habitats, including dunes, scrub and chaparral, bogs and marshes, riparian, woodlands, and forests into finer divisions based on species composition and geographic location.

Arising directly from the Cheatham and Haller system was the classification used by the California Natural Diversity Data Base (NDDB) to identify rare natural communities. The unpublished Holland (1986) classification, a refined version of the NDDB system, is structurally identical with Cheatham and Haller, but it defines more types and, for the first time, assigns rarity ranks and conservation status to units of the hierarchy.

An advantage of the Cheatham and Haller and Holland classifications is that they are specific and differentiate among similar, but geographically isolated types. A disadvantage lies in the lack of uniform criteria in distinguishing units. Some are defined by location, some by structure, others by specific taxa. Although descriptions of most types indicate dominant and characteristic species and distribution, the accounts are general and tend to overlap. Although the Holland system is hierarchically arranged, it is of uneven resolution with coarse as well as fine scales of vegetation defined at the lowest units of the system (Keeler-Wolf 1993).

Another concern with classifications, as those of Cheatham and Haller and Holland, arises with the increasing importance of vegetation units as indicators of ecosystems. We need to unequivocally define these units, and to relate them to similar units across state boundaries so that a broad network of ecosystem conservation may become established. We also need to ratify and validate these units so that they gain acceptance in scientific and land management circles. This situation requires quantitatively defensible definitions for vegetation types as the most important units of conservation.

A system to classify California wildlife habitats was developed in the early 1980s (Mayer & Laudenslayer 1988). This is the California Wildlife Habitat Relationships System (WHR), and it consists of about 50 types intermediate in scale and definition between Munz and Keck's and Thorne's community types. The WHR system also consists of some non-native vegetation types, including eucalyptus groves and several types of agricultural and developed habitats not treated in other classifications.

The WHR system was developed primarily to classify and predict habitat value for the vertebrate animals. It is one of the few systems that identifies structural stages in various tree, shrub, and herb-dominated types. Because we establish cover class and size rules, the system has use in predicting habitat value based on management practices.

WHR has been less successful in differentiating between vegetation types. Because the habitat types are inconsistently defined, a broad familiarity of its detailed descriptions is needed to differentiate among types of similar structure. Although mappers have constructed rules for discriminating among types, difficulties still remain because species dominance varies substantially within some types and broad overlaps in dominant plants occur among types. Other problems arise due to the small number of classes and the inconsistencies in scale among them.

At approximately the same time that Thorne's and Cheatham and Haller's classifications were introduced, the USDA Forest Service was coming to grips with the need to manage their California lands in a way that would sustain natural resources and underscore the potential of different environments to support timber removal, grazing, recreational use, and other human activities. To accomplish this goal they recognized that classification of ecosystems was a high priority. Much of their effort in the 1970s and 1980s was directed towards classification (Allen 1987, Hunter & Paysen 1986, Paysen 1982, Parker & Matayas 1979, Paysen *et al.* 1980, 1982).

The purpose of these efforts was to develop a land classification that could be applied to both research and management activities. Their approach differs fundamentally from earlier classifications of California's vegetation because it relies on field samples (plots) of the vegetation to build the bottom layers of the hierarchy. Consequently the basic taxonomic units of the system were not identified by arbitrarily assigning a name to what is general-

ly perceived as a unique type. Instead, the basic units are defined *after* a number of plots are analyzed over a large area. This classification is quantitative and driven by the availability of data, rather than qualitative and anecdotal.

California has lagged behind Oregon, Washington, Idaho, and Montana in developing data-driven vegetation classifications. Efforts of Daubenmire (1952), Daubenmire & Daubenmire (1968), Franklin *et al.* (1971), Pfister & Arno (1980), and others set the standards for the USDA Forest Service, Pacific Southwest Region classification of forests and chaparral in California. The California data, however, can be added to a standardized system that can be related to a national land use system (Driscoll *et al.* 1984, ECOMAP 1993).

The larger floristic units of these efforts became known as series and the smaller basic units as associations. Series were identified by the dominant plants in the overstory, while the associations were identified by characteristic species in the understory layers. There is an ecological basis for grouping associations into a series. For example, although there are many associations of white fir forest in the Sierra Nevada, all occupy sites that are warmer than the red fir-dominated forests. Red fir typically occurs at higher elevations or on cooler exposures, so the dominant overstory species reflect broadscale environmental differences. The presence of certain understory species reflect more localized differences related to microclimate and soil.

The framework of the classification produced by the USDA Forest Service in California is described by Allen (1987). This project entails several Forest Service zone ecologists developing classifications of targeted areas. The work includes extensive sampling of the vegetation, soils, and other environmental characteristics.

Classifications are completed for Port Orford-cedar forests (Jimerson 1994) in the North Coast and Klamath regions, for blue oak woodlands (Borchert *et al.* 1993) and redwood forests (Borchert *et al.* 1988) in the Central Coast region, for mixed conifer forests (Fites 1994) and red fir forests (Potter (1994) of the Sierra Nevada, for eastside pine forests of the Cascade Range, Modoc Plateau, and Sierra Nevada (Smith 1994), and for chaparral types of the Transverse and Peninsular Ranges (Gordon & White 1994).

In the past few years there have been many local and regional vegetation mapping efforts. A number of land management and resource agencies have been funding these projects, including the California Department of Forestry and Fire Protection, the USDA Forest Service, the USDI Fish and Wildlife Service, the USDI Bureau of Land Management, and various individual county governments. Most of these recent efforts have relied on sophisticated state-of-the-art mapping techniques based on computer-assisted interpretation of satellite images and aerial photos. Computerized Geographic Information Systems (GIS) are used to associate detailed information with each mapping unit and allow easy updating of products. These projects have used various vegetation classifications.

The most prominent and extensive of these recent mapping projects is the California Gap Analysis Project (GAP) (Scott *et al.* 1993). This is part of a national effort by the Fish and Wildlife Service to identify gaps in the preservation of species and habitats. The distribution of species and known significant natural areas are overlain on a map of land ownership. Significant elements of biodiversity that are not protected within nature reserves, parks, or other preserves are then identified and targeted for protection. The underlying rationale for GAP is to gain a fuller understanding of the current mosaic of vegetation types. Since an up-to-date vegetation map for California does not exist, the project is developing one. The team members are currently piecing together a map on a region-by-region basis using the 10 main floristic provinces in *The Jepson Manual* (Hickman 1993) as a guide. Currently the South Coast, Colorado Desert, and Mojave Desert are completed and other regions are in varying stages of completion. The classification is based on the CALVEG series-level classification (Parker and Matayas 1979). In many cases the mapping effort is flexible enough to translate to Holland (1986), WHR, and other systems. This flexibility is afforded by decision rules incorporating species dominance as a translator among classifications. The principal problem with the GAP map is one of scale. The mapping is based on a scale of 1:100,000 [1 cm^2 on the map, 100 ha on the ground] with a minimum polygon size of 100 ha (240 acres). Consequently it is insufficient to depict small vegetation units.

How to Use this Manual

California's vegetation is classified into a set of **series**. Keys allow quick access to the individual series descriptions, which are alphabetically arranged after each key.

Some vegetation cannot be classified using series rules. Some vegetation is better thought of as **unique stands** because they are found in only one occurrence, are defined by the presence of a rare plant, or are structurally distinctive. Some vegetation types are better defined by **habitats**. **Vernal pools** offer a special challenge in that they are defined as much by physical conditions as by plants. These categories are treated separately in sections following series descriptions.

The body of the *Manual* is arranged into six sections.

> Herb-dominated series
> Shrub-dominated series
> Tree-dominated series
> Unique stands
> Habitats
> Vernal pools

How to Use the Keys

The keys are written as if the user were located in a stand and wanted to identify the series to which it belongs. To do that, read *both* opposing couplet statements with the same number. The statements pick one character (a given species, for example) of the stand. Each character has various states (whether a species is dominant, important, common, rare). Pick the state that best matches the condition in the stand. Continue this process until you reach a series name.

Read the series description before you decide whether you have correctly keyed out the stand. If you have, the stand's structure, spe-cies composition, habitat, and location should agree with the description. If they do not match, try the key again, remembering that series-level vegetation types vary, and that some stands are transitional to other series. Some series may be missing from this classification, as well.

The key primarily asks the user whether a species is dominant or important in the layer with the greatest amount of cover: the tree layer in forests and woodlands, the shrub layer in shrublands, and the ground layer in herbaceous vegetation. The words dominate and dominant refer to the extent to which plant crowns of a species or all species in a genus spread over (cover) the area, and how often the species or genus is encountered in the stand. It would be ideal to furnish cover values for each species listed in a series description, but few of California's plant vegetation types have sufficient data to assign average values. This is why these terms are not quantitatively defined here.

If the crown cover of a species spreads over much of the stand and individual plants are often encountered, the species dominates or is a dominant. Species that are not dominant have much lower cover and occurrence. In some series two or three species have comparably high cover and occurrence; they are described as important. Species that are not important have much lower cover. In some cases the user is asked whether a species is present or absent. The existence of the species in the stand is sufficient to answer "present." Such species are often called a "character" or "characteristic" species. In these instances whether a species is dominant, important, or rare is not necessary in identifying the series.

We Present Three Keys.

Key 1: Series dominated by herbaceous plants
         (page 28)
Key 2: Series dominated by shrubs (page 91)
Key 3: Series dominated by trees (page 214)

Key Design

The shrub and tree keys are long and involved, so they are separated into subkeys to make use easier. After a few tries, the user can quickly jump directly to subkeys. The same series can occur several times in these keys.

To shorten the key, we avoided obvious strings of "yes or no" steps by establishing groups with informal names such as NEEDLEGRASS GRASSLANDS. These terms do not imply hierarchical relationships, but are only meant as convenient groupings. Many series are not easily placed in common categories: desert scrub, chaparral, coastal scrub, or many kinds of forests and woodlands. For example, is the Chamise-white sage series better placed in chaparral or coastal scrub? For this reason, these terms are not given formal status.

Series descriptions follow the keys and are arranged alphabetically within the three key groups: Series dominated by herbaceous plants, Series dominated by shrubs, and Series dominated by trees. In a few cases genera and subspecies are diagnostic of a series.

Terms Used in the Keys:

Abundant: a species that is very likely to be encountered; it need not be dominant.
Chaparral: a shrubland dominated by species having evergreen, leathery leaves such as chamise, manzanita, or scrub oaks.

Chenopod scrub: a shrubland dominated by species which are members of the Chenopodiaceae, such as greasewood, iodine bush, or saltbush.
Coastal scrub [coastal sage scrub]: a shrubland dominated by species having evergreen or deciduous, non-leathery leaves, such as California buckwheat, California sagebrush, coyote brush, or sages.
Desert scrub: a shrubland of Colorado or Mojave Desert taxa other than chenopod scrub.
Dominant [dominance]: an abundant species with high crown cover, especially in relation to other species in the stand.
Exotic [alien, introduced]: a species that is judged to be a non-native member of the California flora.
Herb [herbaceous]: Plant lacking woody stems above ground; may be annual or perennial. Includes aquatics, forbs, and grasses.
Important [importance]: two or more species with similar abundance and crown cover in relation to other species in the stand.
Shrub: a woody plant with a short ulitmate height, commonly with 2+ stems from the base.
Shrublands: areas where shrubs dominate, including chaparrals, chenopod scrubs, coastal scrubs, and desert scrubs.
Similarity: used with term "important" to indicate species with equal abundance and crown cover in a stand.
Stand: an actual piece of California's vegetation in which plant composition and structure are uniform.
Subshrub: a plant with woody lower stems and herbaceous upper stems that die back seasonally.
Tree: a woody plant with a tall final height; commonly with one stem [trunk] from the base.

Format of Vegetation Type Descriptions

Format [refer to a series description such as the Purple needlegrass series]

The first paragraph describes the compositional and structural features of a series, unique stand, habitat, or vernal pool type. The first sentence specifies the composition of the most continuous layer by listing the sole, dominant, and important species before a semi-colon. This phrase defines the type. The alphabetically listed species following the semi-colon are commonly present as well, but they may be absent with other species present due to habitat and regional variation. They are listed to describe the type, but are not part of the definition. The remaining sentences summarize the structure of lower layers; presence, degree of canopy closure, and height. Species are not included for the secondary layers, except for typical emergent plants, those that are taller than plants of the most continuous layer.

In the definition three adjectives (**sole, dominant, important**) modify the species listed in the layer with the greatest cover: the *tree layer* in forests and woodlands, the *shrub layer* in shrublands, and the *ground layer* in herbaceous vegetation. Herbs are non-woody plants with either broad leaves [forbs] or narrow, grasslike leaves.

The words **sole, dominant,** and **important** refer to the *extent* to which the plant *crowns* of a species or all species in a genus spread over (*cover*) the area, and how often the species or life form is encountered (*frequency*) in the stand. If crowns of a species spread over all or much of the stand and individuals are often encountered, then the species is a **sole** or a **dominant**. Species that do not dominate have low cover and are less frequent.

In some types two or three species have comparably high cover and frequency; they are described as **important**. Species that are *not* important have low cover and frequency relative to dominant or important ones. It would be ideal to furnish cover and frequency values for each species listed in a type description, but few of California's plant communities have sufficient data to assign average values. This is why these terms are not quantitatively defined here.

The second paragraph relates habitat factors of the type: topography, soil, and other characteristics of the environment. Wetland and upland settings for the type are characterized as well.

The third paragraph summarizes the geographic range and elevation. Provinces and regions are detailed in the California geography section. If the type occurs outside the state, its range is indicated by listing state names by two-letter codes. The range of a type in these states is not described. In some cases, we use more extensive terms such as "inter-West" or "Asia."

The sections below the line in each type description include supporting information.

The **NDDB**/Holland type and status section catalog the California Department of Fish and Game Natural Diversity Data Base (**NDDB**) element code(s) and Heritage Program status (see next page).

The **Other types** section lists synonymy categories in other vegetation classifications. The **General references** section lists commonly available treatments in the literature.

The **Comments** section is distinctive to a type, so format varies. We do not attempt to exhaustively review the ecology of the type at this time. Relationships among types are suggested in some cases. Botanical names generally conform to nomenclature of *The Jepson Manual*. We use the following alternate names:

Alaska yellow-cedar, *Chamaecyparis nootkatensis.*
Bluebunch wheatgrass, *Elymus spicatus.*
Cuyamaca cypress, *Cupressus stephensonii.*
Desert needlegrass, *Stipa speciosa.*
Foothill needlegrass, *Stipa lepida.*
Indian ricegrass, *Oryzopsis hymemoides.*
Needle-and-thread, *Stipa comata.*
Nodding needlegrass, *Stipa cernua.*
Port Orford-cedar, *Chamaecyparis lawsoniana.*
Purple needlegrass, *Stipa pulchra.*
Western wheatgrass, *Elymus smithii.*

The paragraph of **Plot-based descriptions** catalog associations included in the type. The author of the association is cited after the association name. For example, the Cordgrass series contains a Dense-flowered cordgrass association described in Eicher (1987).

Synonymy is presented if two or more authors describe the same vegetation type. The earlier publication is given priority. For example, in the Native dunegrass series, Paker described a Native dunegrass-sea rocket association (1974) and LaBanca also characterized a *Elymus mollis* association (1993). The data presented by the authors suggest that the investigators sampled the same association.

Kinds of plot-derived data other than association definitions exist in the literature. Many times authors **describe** stand composition and structure without intent to classify; instead they present stand attributes as species present and their density, cover, or frequency. Some authors **report** specific detailed measurements for the most continuous layer as tree density, basal area, and frequency for forests. These papers are reported in this section along with association names.

The Nature Conservancy Heritage Program Status Ranks

Global ranks

G1:	Fewer than 6 viable occurrences worldwide and/or 2000 acres
G2:	6–20 viable occurrences worldwide and/or 2000–10,000 acres
G3:	21-100 viable occurrences worldwide and/or 10,000–50,000 acres
G4:	Greater than 100 viable occurrences worldwide and/or greater than 50,000 acres
G5:	Community demonstrably secure due to worldwide abundance

State ranks

S1:	Fewer than 6 viable occurrences statewide and/or less than 2000 acres
S2:	6–20 viable occurrences statewide and/or 2000–10,000 acres
S3:	21–100 viable occurrences statewide and/or 10,000–50,000 acres
S4:	Greater than 100 viable occurrences statewide and/or greater than 50,000 acres
S5:	Community demonstrably secure statewide

Threat ranks

0.1:	Very threatened
0.2:	Threatened
0.3:	No current threats known

We have reviewed the Forest Service Research Natural Area references with the coordinator for accuracy.

The shaded box titled **Species mentioned in text** lists the species mentioned in the text. Botanical equivalents for the common names are given there. The Comments section reports both names if the one used in the box differs from the name in *The Jepson Manual.*

Conventions

References: We use the following labels for commonly cited references in series descriptions. Most are other vegetation classifications mentioned in the Other types section.

Barry type: *A hierarchical vegetation classification system with emphasis on California plant communities* by W.J. Barry, 1989.

Brown Lowe Pase type: *A digitized systematic classification of ecosystems with an illustrated summary of the natural vegetation of North America* by D.E. Brown et al., 1980.

Cheatham & Haller type: *An annotated list of California habitat types*, by N.D. Cheatham and J. R. Haller, 1975.

Cowardin class: *Classification of wetlands and deepwater habitats of the United States*, by L.M. Cowardin et al., 1979.

Holland type: *Preliminary descriptions of the terrestrial natural communities of California*, by R. F. Holland, 1986.

Jones & Stokes type: *Methods used to survey the vegetation of Orange County parks and open space areas and the Irvine Company property*, by Jones and Stokes, 1993.

PSW-45 type: *A vegetation classification system applied to Southern California*, by T.F. Paysen et al., 1980.

Rangeland type: *Rangeland cover types of the United States*, by T.N. Shiflet, 1994.

Thorne type: The vascular plant communities of California by R.F. Thorne in *Plant communities of southern California*, 1976.

The Jepson Manual: *The Jepson Manual* edited by J.C. Hickman, 1993.

WHR type: *A guide to wildlife habitats of California*, by K.E. Mayer and W.F. Laudenslayer, Jr., 1988.

Plant names: We chose to use common names for type and their description at the recommendation of the Plant Communities Committee. Those who disagreed suggested that common names were inconsistent in their application to scientific names; others argued that common names were more stable. Some sentiment may be in response to the different treatments in *A California Flora* and *The Jepson Manual.*

We realize that a single species may have many common names and that none is correct. In most cases, we used the most commonly used names in the ecological literature. In some cases, personal bias is no doubt involved.

Once a name is chosen, the form that it takes varies greatly in the literature as well. Capitalization, endings, separating words, and use of a hyphen are common differences. Birchleaf mountain-mahogany offers an example:

Why not Birchleaf Mountain Mahogany? In the botanical literature the more common rule is not to capitalize common names unless they refer to a person. Douglas-fir is preferred over douglas-fir.

The birchleaf part of the name can take several forms as birch-leaf, birchleaved or birch-leaved. We use the shorter "leaf" form over "leaved" and combine terms.

The name mountain-mahogany follows a customary rule that certain common names are associated with certain genera. Douglas-fir is not a member of the genus *Abies,* the firs, as indicated by the hyphen. *Cercocarpus* is not a member of the genus of commercial mahogany.

Should the possessive form be used for those involving people's names? The common name is most often Douglas-fir not Douglas'-fir. For some reason, tree names generally lack the possessive ending: Jeffrey pine, McNab cypress, etc. Yet the ending is used for shrubs, and especially herbs: Brewer's sedge, Davidson's penstemon, and King's ricegrass. For consistency, the possessive is **not used** for common names throughout the *Manual*.

The combining of two words is another place where inconsistences abound. Is black bush better than blackbush? In this case, literature also offers black brush and blackbrush, as well. In this manual, the most commonly used form is generally chosen.

The Association

Variation in species composition among stands can occur in a series. Tree canopies may have a consistent makeup among stands, but the shrub layer and ground layer composition may vary. This variation may be related to habitat as well. For example, a stand with a dense shrub layer and few ferns and herbs on the ground layer may occur at the top of a hill. At the bottom of a hill, a stand may have few shrubs and a luxuriant ground layer of herbs and ferns. Both stands have tree canopies of Douglas-fir.

This variation can be handled by considering the series to be composed of several vegetation types called **associations.** This approach is equivalent to thinking of a plant family as being composed of a set of genera, or a genus being a collection of species. In the earlier example, the Douglas-fir tree canopy is consistent. Variation in understory species defines the associations of the series. The convention of listing members of two layers easily separates series and association-level names. The Douglas-fir forest series includes Douglas-fir/huckleberry and Douglas-fir/sword fern associations. This "/" convention will be used for associations with more than one layer. If only one layer exists, then a hyphen will be used to separate important species. In some chaparral series, parentheses are used to distinguish associations where series names involve more than one species.

Associations found in the literature are based on the dominance rules used to define series. The association's name, author, and citation are listed in the plot-data descriptions section. In some cases, an association is assigned to more than one series, and the association's name is changed to conform with our conventions.

PLATE 1

1. Bulrush series, Monterey Co. *T. Keeler-Wolf*

2. Cordgrass series, Marin Co. *P. Faber*

3. Saltgrass series, Marin Co. *T. Keeler-Wolf*

4. Pickleweed series, Marin Co. *P. Faber*

5. Spikerush series, Siskiyou Co. *T. Keeler-Wolf*

PLATE 1

PLATE 2

1.

2.

3.

1. Bur-reed series, Mendocino Co. *T. Keeler-Wolf*

2. Mosquito fern series, Kern Co. *J. Shevock*

3. Pondweeds with floating leaves series, Mendocino Co. *T. Keeler-Wolf*

4. Darlingtonia series, Trinity Co. *R. Bittman*

5. Yellow pond-lily series, Mendocino Co. *C. Tipple-Golec*

5.

4.

PLATE 2

PLATE 3

1.

2.

3.

4.

5.

1. Nebraska sedge series, Tulare Co. *J. Shevock*

2. Tufted hairgrass series, Inyo Co. *T. Keeler-Wolf*

3. Cattail series, Alameda Co. *T. Keeler-Wolf*

4. Beaked sedge series, Siskiyou Co. *T. Keeler-Wolf*

5. Sedge series, Modoc Co. *T. Keeler-Wolf*

PLATE 3

PLATE 4

1. Bluebunch wheatgrass series, Lassen Co. *T. Keeler-Wolf*

2. Native dunegrass series, Humboldt Co. *T. Keeler-Wolf*

3. Alkali sacaton series, Merced Co. *R. Bittman*

4. Big galleta series, San Bernardino Co. *T. Keeler-Wolf*

5. Pacific reedgrass series, Sonoma Co. *T. Keeler-Wolf*

PLATE 4

PLATE 5

1.

2.

3.

1. Nodding needlegrass series, Monterey Co. *T. Keeler-Wolf*

2. California annual grassland series, San Luis Obispo Co. *J. Shevock*

3. Idaho fescue series, Modoc Co. *T. Keeler-Wolf*

4. Introduced perennial grassland series, Sonoma Co. *T. Keeler-Wolf*

5. California oatgrass series, Humboldt Co. *T. Keeler-Wolf*

4.

5.

PLATE 5

PLATE 6

1. Desert sand-verbena series, Imperial Co. *T. Keeler-Wolf*

2. Purple needlegrass series, Solano Co. *R. Bittman*

3. Ashy ryegrass series, Modoc Co. *T. Keeler-Wolf*

4. Sand-verbena—beach bursage series, Humboldt Co. *T. Keeler-Wolf*

5. Desert needlegrass series, Kern Co. *T. Keeler-Wolf*

6. Common reed series, Inyo Co. *L. Norris*

PLATE 6

PLATE 7

1.

1. Sitka alder series, Siskiyou Co. *T. Keeler-Wolf*

2. Salal—black huckleberry series, Mendocino Co. *T. Keeler-Wolf*

3. Tamarisk series, Riverside Co. *J. Shevock*

4. Buttonbush series, Sacramento Co. *T. Keeler-Wolf*

5. Mountain alder series, Siskiyou Co. *T. Keeler-Wolf*

2.

3.

5.

4.

PLATE 7

PLATE 8

1.

2.

3.

4.

5.

6.

1. Leather oak series, Tehama Co. *T. Keeler-Wolf*

2. Chamise series, Tulare Co. *J. Shevock*

3. Chamise–Eastwood manzanita series, San Diego Co. *T. Keeler-Wolf*

4. Narrowleaf willow series, San Bernardino Co. *T. Keeler-Wolf*

5. Greenleaf manzanita series, Tulare Co. *J. Shevock*

6. Eastwood manzanita series, San Diego Co. *T. Keeler-Wolf*

PLATE 8

PLATE 9

1.

1. Mountain whitethorn series, Tulare Co. *J. Shevock*

2. Red shank series, Los Angles Co. *K. Himes*

3. Chamise—bigberry manzanita series, Los Angles Co.
   T. Keeler-Wolf

4. Bush chinquapin series, Tulare Co. *J. Shevock*

5. Red shank—chamise series, Riverside Co. *M. Lardner*

2.

3.

4.

5.

PLATE 9

PLATE 10

1.

1. Whiteleaf manzanita series, Kern Co. *J. Shevock*

2. Hoaryleaf ceanothus series, San Bernardino Co. *T. Keeler-Wolf*

3. Holodiscus series, San Mateo Co. *K. Himes*

4. Ione manzanita series, Amador Co. *K. Himes*

5. Deer brush series, Riverside Co. *T. Keeler-Wolf*

2.

3.

5.

4.

PLATE 10

PLATE 11

1.

2.

3.

5.

4.

1. Wedgeleaf ceanothus series, Tulare Co. *J. Shevock*

2. Purple sage series, Los Angles Co. *T. Keeler-Wolf*

3. Chaparral whitethorn series, Kern Co. *J. Shevock*

4. Coast prickly-pear series, Ventura Co. *T. Keeler-Wolf*

5. Scrub oak—chaparral whitethorn series, San Diego Co.
   J. Shevock

PLATE 11

PLATE 12

1.

2.

3.

4.

5.

6.

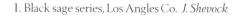

1. Black sage series, Los Angles Co. *J. Shevock*

2. California sagebrush series, Marin Co. *T. Keeler-Wolf*

3. White sage series, San Bernardino Co. *T. Keeler-Wolf*

4. California sagebrush—California buckwheat series, Los Angles Co.
   T. Keeler-Wolf

5. Dune lupine—goldenbush series, San Francisco Co. *T. Keeler-Wolf*

6. Coyote brush series, Alameda Co. *T. Keeler-Wolf*

PLATE 12

PLATE 13

1. Winter fat series, Kern Co. *J. Shevock*

2. California encelia series, Ventura Co. *T. Keeler-Wolf*

3. Black bush series, Kern Co. *J. Shevock*

4. Bitterbrush series, Lassen Co. *T. Keeler-Wolf*

5. Big sagebrush series, Tulare Co. *J. Shevock*

1.

2.

3.

4.

5.

PLATE 13

PLATE 14

1.

1. Rubber rabbitbrush series, Lassen Co. *T. Keeler-Wolf*

2. Parry rabbitbrush series, Mono Co. *T. Keeler-Wolf*

3. Low sagebrush series, Modoc Co. *T. Keeler-Wolf*

4. Rothrock sagebrush series, Tulare Co. *J. Shevock*

5. Shadscale series, Modoc Co. *T. Keeler-Wolf*

2.

4.

5.

PLATE 14

PLATE 15

1. Spinescale series, Inyo Co.
   T. Keeler-Wolf

2. Teddy-bear cholla series, San
   Diego Co. *R. Bittman*

3. Catclaw acacia series, San Diego
   Co. *T. Keeler-Wolf*

4. Creosote bush—white bursage
   series, San Bernardino Co.
   T. Keeler-Wolf

5. Creosote bush series, San
   Bernardino Co. *T. Keeler-Wolf*

1.

2.

3.

4.

5.

PLATE 15

PLATE 16

1.

2.

4.

3.

5.

1. Joshua tree series, Kern Co. *J. Shevock*

2. Nolina series, San Bernardino Co. *T. Keeler-Wolf*

3. Ocotillo series, San Diego Co. *J. Shevock*

4. Mojave yucca series, San Bernardino Co. *T. Keeler-Wolf*

5. Hop-sage series, Kern Co. *J. Shevock*

PLATE 16

PLATE 17

1. Brittlebush series, San Bernardino Co. *T. Keeler-Wolf*

2. Allscale series, Kern Co. *T. Keeler-Wolf*

3. Scalebroom series, San Bernardino Co. *T. Keeler-Wolf*

1.

2.

3.

4.

5.

4. Greasewood series, Modoc Co. *T. Keeler-Wolf*

5. Foothill-palo verde–saguaro series, San Bernardino Co. *T. Keeler-Wolf*

PLATE 17

PLATE 18

1.

2.

3.   5.

4.

1. Arroyo willow series, Contra Costa Co. *T. Keeler-Wolf*

2. Red alder series, San Mateo Co. *K. Himes*

3. Hooker willow series, Mendocino Co. *T. Keeler-Wolf*

4. Fan palm series, Riverside Co. *R. Bittman*

5. White alder series, Tulare Co. *T. Keeler-Wolf*

PLATE 18

PLATE 19

2.

1.

1. Water birch series, Inyo Co. *T. Keeler-Wolf*

2. Fremont cottonwood series, Los Angles Co. *T. Keeler-Wolf*

3. Aspen series, Mono Co. *L. Norris*

4. Mixed willow series, Kern Co. *J. Shevock*

5. California sycamore series, Monterey Co. *T. Keeler-Wolf*

3.

5.

4.

PLATE 20

1.

2.

3.

1. Coast live oak series, Contra Costa Co. *T. Keeler-Wolf*

2. Blue oak series, Tehama Co. *T. Keeler-Wolf*

3. California walnut series, Los Angles Co. *T. Keeler-Wolf*

4. Mixed oak series, Contra Costa Co. *T. Keeler-Wolf*

5. Valley oak series, Monterey Co. *T. Keeler-Wolf*

4.   5.

PLATE 20

PLATE 21

1.

2.

3.

4.

5.

1. California bay series, Contra Costa Co. *T. Keeler-Wolf*

2. Santa Lucia fir series, Monterey Co. *J. Griffin*

3. Oregon white oak series, Mendocino Co. *C. Tipple-Golec*

4. Black oak series, Shasta Co. *T. Keeler-Wolf*

5. California buckeye series, San Mateo Co. *K. Himes*

PLATE 21

PLATE 22

2.

3.

4.

1.

5.

1. Blue palo verde–ironwood–smoke tree series, San Diego Co.
   T. Keeler-Wolf

2. Singleleaf pinyon–Utah juniper series, Inyo Co. *T. Keeler-Wolf*

3. California juniper series, Kern Co. *T. Keeler-Wolf*

4. Singleleaf pinyon series, Kern Co. *L. Norris*

5. Mesquite series, San Bernardino Co. *T. Keeler-Wolf*

PLATE 22

PLATE 23

2.

1.

3.

4.

5.

1. Grand fir series, Mendocino Co. *T. Keeler-Wolf*

2. Knobcone pine series, Santa Cruz Co. *J. Shevock*

3. Monterey pine series, Monterey Co. *T. Keeler-Wolf*

4. Douglas-fir series, Del Norte Co. *T. Keeler-Wolf*

5. Beach pine series, Humboldt Co. *T. Keeler-Wolf*

PLATE 23

PLATE 24

1.

1. Port Orford-cedar series, Siskiyou Co.
   T. Keeler-Wolf

2. Western hemlock series, Del Norte Co.
   T. Keeler-Wolf

3. Sitka spruce series, Humboldt Co.
   T. Keeler-Wolf

4. Redwood series, Humboldt Co.
   T. Jimerson

5. Douglas-fir—tanoak series, Del Norte Co.
   T. Jimerson

2.

3.

4.

5.

PLATE 24

PLATE 25

1.

1. Curlleaf mountain-mahogany series, Lassen Co. *T. Keeler-Wolf*

2. Pygmy cypress series, Mendocino Co. *T. Keeler-Wolf*

3. Western juniper series, Lassen Co. *J. Shevock*

4. Sargent cypress series, Colusa Co. *D. Cheatham*

5. Whitebark pine series, Fresno Co. *L. Norris*

PLATE 25

PLATE 26

1. Ponderosa pine series, Siskiyou Co. *T. Keeler-Wolf*

2. Coulter pine series, Riverside Co. *T. Keeler-Wolf*

3. Lodgepole pine series, Mono Co. *T. Keeler-Wolf*

4. Washoe pine series, Sierra Co. *T. Keeler-Wolf*

5. Jeffrey pine series, Tulare Co. *J. Shevock*

2.

1.

3.

4.

5.

PLATE 26

PLATE 27

1.

2.

3.

4.

5.

1. White fir series, Siskiyou Co. *T. Keeler-Wolf*

2. Western white pine series, El Dorado Co. *T. Keeler-Wolf*

3. Giant sequoia series, Tulare Co. *T. Keeler-Wolf*

4. Red fir series, Tulare Co. *J. Shevock*

5. Mixed conifer series, Tuolumne Co. *T. Keeler-Wolf*

PLATE 28

1.

2.

1. Engelmann spruce series, Siskiyou Co. *T. Keeler-Wolf*

2. Mountain hemlock series, Siskiyou Co. *T. Keeler-Wolf*

3. Mixed subalpine forest series, Tulare Co. *J. Shevock*

4. Bristlecone pine series, Inyo Co. *D. Cheatham*

5. Foxtail pine series, Tulare Co. *L. Norris*

3.

4.

5.

PLATE 28

PLATE 29

1.

1. Cuyamaca cypress stands, San Diego Co. *T. Keeler-Wolf*

2. Stands on San Benito Mountain, San Benito Co. *K. Himes*

3. Piute cypress stands, Kern Co. *J. Shevock*

4. Santa Cruz cypress stands, Santa Cruz Co. *K. Himes*

5. Baker cypress stands, Shasta Co. *T. Keeler-Wolf*

2.

3.

5.

4.

PLATE 29

PLATE 30

1. Crucifixion-thorn stands, San Bernardino Co. *T. Keeler-Wolf*

2. Elephant tree stands, San Diego Co. *L. Norris*

1.

2.

3.

4.

3. Torrey pine stands, San Diego Co. *T. Keeler-Wolf*

4. Catalina ironwood stands, Santa Barbara Co. (Santa Cruz Island) *T. Keeler-Wolf*

5. Enriched stands in the Klamath Mountains, Siskiyou Co. *T. Keeler-Wolf*

5.

PLATE 30

PLATE 31

1. Subalpine meadow habitat, Tuolumne Co. *T. Keeler-Wolf*

2. Alpine habitat, Inyo Co. *T. Keeler-Wolf*

3. Subalpine wetland shrub habitat, Tuolumne Co. *T. Keeler-Wolf*

4. Subalpine upland shrub habitat, Inyo Co. *T. Keeler-Wolf*

5. Montane wetland shrub habitat, Tuolumne Co. *T. Keeler-Wolf*

1.

2.

3.

4.

5.

PLATE 31

PLATE 32

1.

2.

3.

1. Northern claypan vernal pools, Tulare Co.
   J. Shevock

2. San Diego mesa vernal pools, San Diego Co.
   J. Shevock

3. Fen habitat, Plumas Co. *J. Shevock*

4. Montane meadow habitat, Alpine Co.
   T. Keeler-Wolf

5. Northern basalt flow vernal pools, Siskiyou
   Co. *T. Keeler-Wolf*

4.

5.

PLATE 32

Geography of California

The geographical regions used to describe the locations of series in this manual are similar to those used in *The Jepson Manual*. However, because this is a manual of vegetation and not of flora, we choose to emphasize and recognize certain geographic areas as regions rather than subregions. For example, the Channel Islands, Peninsular Ranges, South Coast, and Transverse Ranges are treated as separate regions rather than being lumped as part of a Southwest California Region (Hickman 1993).

Lands west of the Cascade-Sierra Nevada-Peninsular range crest are designated cismontane California [**Cis-CA**], and are comparable to the California Floristic Province in *The Jepson Manual*. The area to the east is called transmontane California [**Trans-CA**], and it is equal to the Great Basin and Desert provinces in *The Jepson Manual*. The word "eastside" is also used to make this distinction in many papers on California vegetation.

We divide the **Cis-CA** into northern California [**Nor-CA**] and southern California [**So-CA**] subprovinces; **Trans-CA** into the Great Basin [**GB**] and Desert [**DES**] subprovinces.

The four subprovinces are divided into the physiographic regions or subregions used in *The Jepson Manual*. Six regions occur in **Nor-CA**, and four in **So-CA**.

Nor-CA includes five mountain regions surrounding the Central Valley [**CenV**]. The Klamath Ranges [**KlaR**] separates the coastal North Coast [**NorCo**] and Central Coast [**CenCo**] from the inland Cascade Range [**CasR**] and Sierra Nevada [**SN**].

So-CA includes two mountain regions, the Transverse Ranges [**TraR**] and Peninsular Ranges [**PenR**], and two low elevation regions, the South Coast [**SoCo**] and Channel Islands [**ChaI**].

Subdivisions of **NorCo**, **CenCo**, **TraR**, **SoCo** recognize climatic and vegetational differences between coastal or outer [**o-**] and inner [**i-**] locations.

Where mountains are high enough to show vegetation zonation, the following zones are possible: foothills [**f-**] surrounding **CenV**, low elevation [**l-**] canyons and valleys, montane [**m-**], subalpine [**su-**], and alpine [**a-**] zones.

We recognize in **CenV** the delta [**d-CenV**], Sacramento Valley [**sac-CenV**], and San Joaquin Valley [**sj-CenV**].

In **GB**, the area east of the Cascade Range [**trans-CASR**] distinguishes from that east of the Sierra Nevada [**trans-SN**], and in **DES**, the Mojave Desert [**Moj-D**] is differentiated from the Colorado Desert [**Col-D**]. **Trans-CAS** and **trans-SN**, and **Moj-D** each have two regions, while **Col-D** has one. In **trans-CASR**, we recognize the Modoc Plateau [**ModP**] with its Warner Range [**WarR**]. The **trans-SN**, involves the White, Inyo, and Sweetwater Ranges [**WIS**] and the valleys to their west plus the eastside of the Sierra Nevada [**TraSN**].

DES includes the Mojave Desert as a region [**Moj-D**] with the desert floor [**MojD**] and its Desert Ranges [**DesR**]. The Colorado Desert as a region [**Col-D**] has one area [**ColD**].

Modifiers indicating the northern [**n.**] central [**c.**], southern [**s.**], western [**w.**], and eastern [**e.**] parts of a region may be used.

California Geography

I. **Cis-CA** Cismontane California
A. **Nor-CA** Northern California
1. **NorCo** North Coast
  a.   **o-NorCo** outer North Coast
  b.   **i-NorCo** inner North Coast
  c.   **m-NorCo** montane North Coast Ranges
2. **CenCo** Central Coast
  a.   **o-CenCo** outer Central Coast
  b.   **i-CenCo** inner Central Coast
  c.   **m-CenCo** montane Central Coast Ranges
3. **CenV** Central Valley
  a.   **d-CenV** Delta
  b.   **sac-CenV** Sacramento Valley
  c.   **sj-CenV** San Joaquin Valley
4. **KlaR** Klamath Ranges
  a.   **l-KlaR** low elevations of the Klamath Ranges
  b.   **f-KlaR** Klamath foothills
  c.   **m-KlaR** montane Klamath Ranges
  d.   **su-KlaR** subalpine Klamath Ranges
5. **CasR** Cascade Range
  a.   **f-CasR** Cascade Range foothills
  b.   **m-CasR** montane Cascade Range
  c.   **su-CasR** subalpine Cascade Range
  d.   **a-CasR** alpine Cascade Range
6. **SN** Sierra Nevada
  a.   **f-SN** Sierra Nevada foothills
  b.   **m-SN** montane Sierra Nevada
  c.   **su-SN** subalpine Sierra Nevada
  d.   **a-SN** alpine Sierra Nevada
B. **So-CA** Southern California
1. **SoCo** South Coast
  a.   **o-SoCo** outer South Coast
  b.   **i-SoSo** inner South Coast
2. **TraR** Transverse Ranges
  a.   **o. TraR** outer Transverse Ranges
  (1) **o. l-TraR** Low elevation Transverse Ranges
  (2) **o. m-TraR** outer montane Transverse Ranges
  (3) **o. su-TraR** outer subalpine Transverse Ranges

  b.   **i. TraR** inner Transverse Ranges
  (1) **i. m-TraR** inner montane Transverse Ranges
  (2) **i. su-TraR** inner subalpine Transverse Ranges
  (3) **i. a-TraR** inner alpine Transverse Ranges
3. **PenR** Peninsular Ranges
  a.   **m-PenR** montane Peninsular Ranges
  b.   **su-PenR** subalpine Peninsular Ranges
  c.   **a-PenR** alpine Peninsular Ranges
4. **ChaI** Channel Islands

II. **Trans-CA** Transmontane California
A. **GB** Great Basin
1. **trans-CASR** Transmontane Cascade Range
  a.   **ModP** Modoc Plateau
  b.   **WarR** Warner Range
  (1) **m-WarR** montane Warner Range
  (2) **su-WarR** subalpine Warner Range
2. **trans-SN** Transmontane Sierra Nevada
  a.   **TraSN** eastside Sierra and valleys
  b.   **WIS** White, Inyo, Sweetwater Ranges
  (1) **m-WIS** montane White, Inyo, Sweetwater Ranges
  (2) **su-WIS** subalpine White, Inyo, Sweetwater Ranges
  (3) **a-WIS** alpine White, Inyo, Sweetwater Ranges
B. **DES** Deserts
1. **Moj-D** Mojave Desert area
  a.   **MojD** Mojave Desert
  b.   **DesR** Desert Ranges
  (1) **m-DesR** montane Desert Ranges
  (2) **su-DesR** subalpine Desert Ranges
2. **Col-D** Colorado Desert area
  a.   **ColD** Colorado Desert

Modifiers for any region: **n.** northern part, **c.** central part, **s.** southern part, **w.** western part, **e.** eastern part. Other geographical areas: States are referred to by the postal codes, inter-West = intermountain West as defined in the *Intermountain Flora* (Cronquist *et al.* 1989).

Key I: Series Dominated by Herbaceous Plants

1   Grasses dominant 2
1' Grasses not dominant 6

    2   A needlegrass dominant   NEEDLEGRASS GRASSLANDS
                                                Desert needlegrass series
                                              Foothill needlegrass series
                                              Needle-and-thread series
                                              Nodding needlegrass series
                                              Purple needlegrass series

    2' A needlegrass not dominant 3
3   A ryegrass dominant   *Ashy ryegrass series*
                                            Creeping ryegrass series

3' A ryegrass not dominant 4
    4   A reedgrass dominant   *Pacific reedgrass series*
                                        Shorthair reedgrass series

    4' A reedgrass not dominant 5
5   Introduced annual or perennial grass species dominant
               GRASSLANDS DOMINATED BY NON-NATIVE SPECIES
                       California annual grassland series
                                        Cheatgrass series
                                   Crested wheatgrass series
                               European beachgrass series
                                         Giant reed series
                  Introduced perennial grassland series
                                 Kentucky bluegrass series
                                     Pampas grass series

    5' A native grass species dominant   NATIVE GRASSLANDS
                                     Alkali sacaton series
                                         Big galleta series
                             Bluebunch wheatgrass series
                               California oatgrass series
                                     Common reed series
                                       Cordgrass series
                                     Green fescue series
                                     Idaho fescue series
                                   Indian ricegrass series
                                   Native dunegrass series
                                 One-sided bluegrass series
                                       Saltgrass series
                                   Tufted hairgrass series

        6   Sedges dominant   SEDGE MEADOWS
                                       Beaked sedge series

Nebraska sedge series
Rocky Mountain sedge series
Sedge series
Shorthair sedge series

6'   Sedges not dominant 7

7   Spikerushes dominant   EMERGENT AQUATICS

Spikerush series

7'   Spikerushes not dominant 8

   8   Bulrushes and cattails similarly important   *Bulrush-cattail series*

   8'   Bulrushes and cattails not similarly important 9

9   Bulrushes dominant   *Bulrush series*

9'   Bulrushes not dominant 10

   10   Cattails dominant   *Cattail series*

   10'   Cattails not dominant   11

11   Darlingtonia dominant   *Darlingtonia series*

11'   Darlingtonia not dominant 12

   12   Pickleweed dominant   *Pickleweed series*

   12'   Pickleweed not dominant 13

13   Dominant plants small < 1 cm in diameter   FLOATING AND SUBMERGED AQUATICS

Duckweed series
Mosquito fern series

13'   Dominant plants not small < 1 cm in diameter 14

   14   Dominant plants having floating or submerged leaves   *Bur-reed series*

Ditch-grass series
Pondweeds with submerged leaves series
Pondweeds with floating leaves series
Quillwort series
Yellow pond-lily series

   14'   Dominant plants not having floating or submerged leaves 15

15   Ice plant dominant   *Iceplant series*

15'   Sand-verbena important   DUNELANDS

Desert sand-verbena series
Sand-verbena - beach bursage series

Alkali sacaton series

Alkali sacaton sole or dominant grass in the ground layer; one-side bluegrass and/or saltgrass may be present. Emergent iodine bush and/or bush seepweed may be present. Grass < 1 m; cover open (Plate 4).

Wetlands: habitat intermittently flooded, saturated. Water chemistry: saline. Valley bottoms, lower portions of alluvial slopes. Cowardin class: Palustrine emergent saline wetland. The national list of wetland plants (Reed 1988) lists alkali sacaton as a FAC+.

Distribution: o-CenCo, d-CenV, sj-CenV, So-CA, TraSN, MojD, ColD.
Elevation: sea level-2100 m.

Species mentioned in text	
Alkali sacaton	*Sporobolus airoides*
Ashy ryegrass	*Leymus cinereus*
Baltic rush	*Juncus balticus*
Bush seepweed	*Suaeda moquinii*
Creeping ryegrass	*Leymus triticoides*
Iodine bush	*Allenrolfea occidentalis*
One-sided bluegrass	*Poa secunda*
Rubber rabbitbrush	*Chrysothamnus nauseosus*
Saltgrass	*Distichlis spicata*

Plot-based descriptions: Ferren & Davis define a Rubber rabbitbrush-alkali sacaton association, Odion *et al.* (1992) an Alkali sacaton-iodine bush association [as *Sporobolus-Allenrolfea* association] At Fish Slough in TraSN.

NDDB/Holland type and status: **Valley and foothill grasslands** (42000), **Meadows and seeps** (45000).
    Valley sacaton grassland (42120) G1 S1.1.
    Alkali meadow (45310 *in part*) G3 S2.1.

Other types:
Barry type: G7412311 BSPAI00.
Cheatham & Haller type: Great Basin native grassland.
PSW-45 type: Sacaton series.
Thorne type: Alkali meadow and aquatic.
WHR type: Wet meadow.

General references: Bittman (1985), Brown (1982e), Griggs (1980), Paysen *et al.* (1980), Thorne (1982), Werschkull *et al.* (1984).

Comments: Many regional descriptions include this series in an alkali meadows category. Meadows of this habitat can be dominated by this or other species, so several series handle this variety [*see* Ashy ryegrass series, Creeping ryegrass series, Saltgrass series, Spikerush series]. Stands of this series form a fine scale mosaic with Iodine bush, Bush seepweed, and California annual grassland series, and with Saltbrush series at a coarser scale.

Ashy ryegrass series

Ashy ryegrass sole or dominant grass in ground layer; bluebunch wheatgrass, cheatgrass, Indian ricegrass, Junegrass, needle-and-thread, one-sided bluegrass, squirreltail, Thurber needlegrass, and/or western wheatgrass may be present. Emergent shrubs may be present. Grass < 1.5 m; cover open (Plate 6).

Wetlands: habitat permanently saturated with shallow water table. Water chemistry: saline. Cowardin class: Palustrine persistent emergent saline wetland. **Uplands:** north-facing slopes, lower portions of alluvial slopes, valley bottoms. The national list of wetland plants (Reed 1988) lists ashy ryegrass as a NI.

Distribution: m-CasR, m-SN, ModP, TraSN, m-DesR.
Elevation: 1400-3000 m.

NDDB/Holland type and status: Great Basin grasslands (43000 *in part*) G1 S1.1.

Other types:
Barry type: G7411331 BELCI20.
Cheatham & Haller type: Cismontane introduced grasses.
PSW-45 type: Wild rye series.
Thorne type: Great Basin sagebrush scrub.
WHR type: Perennial grass.

General references: Heady (1977), Paysen *et al.* (1980).

Comments: Many regional descriptions include this series in an alkali meadows category. Meadows of this habitat can be dominated by this or other species, so several series exist to handle this variety [*see* Alkali sacaton series, Creeping ryegrass series, Saltgrass series, Spikerush series].

In the ecological literature ashy ryegrass is referred to as *Elymus cinereus. The Jepson Manual* places ashy ryegrass in the genus *Leymus*.

Species mentioned in text	
Alkali sacaton	*Sporobolus airoides*
Ashy ryegrass	*Leymus cinereus*
Bluebunch wheatgrass	*Elymus spicatus*
Cheatgrass	*Bromus tectorum*
Creeping ryegrass	*Leymus triticoides*
Indian ricegrass	*Oryzopsis hymenoides*
Junegrass	*Koeleria macrantha*
Needle-and-thread	*Stipa comata*
One-sided bluegrass	*Poa secunda*
Saltgrass	*Distichlis spicata*
Squirreltail	*Elymus elymoides*
Thurber needlegrass	*Stipa thurberiana*
Western wheatgrass	*Elymus smithii*

Plot-based descriptions: NDDB has plot data on file for Modoc Co. and for Sheldon National Antelope Refuge in NV.

Beaked sedge series

Beaked sedge sole or dominant herb in ground canopy; bulrushes, rushes, sedges, and/or spikerushes may be present. Herbs < 1 m; canopy continuous (Plate 3).

Wetlands: habitat seasonally flooded, semipermanently flooded, permanently saturated, seasonally saturated. Water chemistry: fresh. Wet meadows, ponds, depressions; seeps; swales. Cowardin class: Palustrine persistent emergent freshwater wetland. The national list of wetland plants (Reed 1988) lists beaked sedge as a OBL.

Distribution: NorCo, CenCo, KlaR, m-CasR, su-CasR, m-SN, su-SN, TraR, TraSN.
Elevation: sea level-3400 m.

NDDB/Holland type and status: Meadows and seeps (45000).
　　Wet montane meadow (45110 *in part*) G3 S3.
　　Freshwater seep (45400 *in part*) G4 S4.

Other types:
Barry type: G7411331.
Cheatham & Haller type: Meadows and swamps.
PSW-45 type: Sedge series.
Rangeland type: SRM 213.
Thorne type: Freshwater aquatic.
WHR type: Wet meadow.

General references: Allen-Diaz (1994), Hermann (1970), Mason (1957), Paysen *et al.* (1980).

Comments: This series occurs in inter-West mountains. If Beaked sedge is present in a stand, but not dominant, the stand is a member of the Sedge series [*see* also Nebraska sedge series, Rocky Mountain sedge series]. In many manuals beaked sedge is referred to as *Carex rostrata*.

Plot-based descriptions: Halpern (1986) defines one association at Sequoia National Park in Fresno Co.

m-SN. Ratliff (1985) takes up montane and subalpine meadows for the central SN, especially Yosemite National Park and surrounding areas.

Species mentioned in text	
Beaked sedge	*Carex utriculata*
Bulrushes	*Scirpus* species
Joshua tree	*Yucca brevifolia*
Nebraska sedge	*Carex nebrascensis*
Rocky Mountain sedge	*Carex scopulorum*
Rushes	*Juncus* species
Sedges	*Carex* species
Spikerushes	*Eleocharis* species

Associations:
Halpern (1986):
　　Beaked sedge association [included in Beaked sedge vegetative series (Ratliff 1985)].

Big galleta series

Big galleta sole or dominant grass in ground layer; black gramma, fluff grass, foxtail chess, galleta, Indian ricegrass, and/or white bursage may be present. Emergent shrubs or Joshua tree may be present. Grass < 1 m; cover open (Plate 4).

Uplands: flat ridges, lower slopes, stabilized sand dunes.

Distribution: MojD, ColD.
Elevation: 75 below sea level-1400 m.

NDDB/Holland type and status: **Sonoran desert scrub** (33000), **Mojavean desert scrub** (34000).
Sonoran creosote bush scrub (33100 *in part*) G4 S4.
Mojave creosote bush scrub (34100 *in part*) G4 S4.
Mojave mixed steppe (34220 *in part*) G3 S2.2.
Mojave yucca scrub and steppe (34230 *in part*) G3 S3.2.

Other types:
Barry type: G7411331 BHIRI00.
Cheatham & Haller type: Creosote bush scrub.
PSW-45 type: Galleta grass series.
Thorne type: Creosote bush scrub.
WHR type: Desert scrub.

General references: Brown (1982e), Burk (1977), Paysen *et al.* (1980), Vasek & Barbour (1977b).

Comments: Many regional descriptions include this series in Creosote bush scrub. Stands of this series often form fine grain mosaics with stands of Creosote bush or Joshua tree series as big galleta may be a common ground layer species in shrub or tree dominated stands. In addition, scattered shrubs or trees may be emergent on stands dominated by big galleta. Determination of which layer has the highest cover will determine the proper assignment of the stand [see Black bush series, Creosote bush series, Joshua tree series, Mojave yucca series].

In the ecological literature big galleta is referred to as *Hilaria rigida. The Jepson Manual* places big galleta in the genus *Pleuraphis.*

Species mentioned in text	
Big galleta	*Pleuraphis rigida*
Black gramma	*Bouteloua eriopoda*
Creosote bush	*Larrea tridentata*
Fluff grass	*Erioneuron pulchellum*
Foxtail chess	*Bromus madritensis*
Galleta	*Pleuraphis jamesii*
Indian ricegrass	*Oryzopsis hymenoides*
Joshua tree	*Yucca brevifolia*
White bursage	*Ambrosia dumosa*

Plot-based descriptions: Minnich *et al.* (1993) define a Creosote bush/big galleta association [as Creosote bush scrub with galleta grass] at Marine Corps Airground Combat Center in San Bernardino Co. MojD; NDDB has plot data on file for the East Mojave Scenic Area in San Bernardino Co. MojD; Anza Borrego Desert State Park in San Diego Co. ColD.

Bluebunch wheatgrass series

Bluebunch wheatgrass sole or dominant grass in ground layer; ashy ryegrass, cheatgrass, Indian ricegrass, Junegrass, needle-and-thread, one-sided bluegrass, squirreltail, and/or western wheatgrass may be present. Emergent shrubs, especially sagebrush species, may be present. Grass < 1.5 m; cover open (Plate 4).

Uplands: all topographic locations. Soils commonly rocky. The national list of wetland plants (Reed 1988) lists bluebunch wheatgrass as a FACU*.

Distribution: ModP, m-WarR, TraSN, inter-West.
Elevation: 800-1650 m.

NDDB/Holland type and status: **Great Basin grasslands** (43000 *in part*) G1 S1.1.

Other types:
Barry type: G7411331 BAGSP00.
Brown Lowe Pase type: 142.211.
Cheatham & Haller type: Great Basin native grassland.
Rangeland type: SRM 101.
Thorne type: Great Basin sagebrush scrub.
WHR type: Perennial grass.

General references: Daubenmire (1970), Young *et al.* (1977), Stoddart *et al.* (1975), Tisdale (1994), Turner & Brown (1982).

Comments: This series is more common outside California. Stands often include non-native, annual species mixed with the perennial grasses and herbs. Cheatgrass and medusa-head are common. In the ecological literature bluebunch wheatgrass is referred to as *Agropyron spicatum*. *The Jepson Manual* places bluebunch wheatgrass in the genus *Pseudoroegneria*. Conservative treatments place it in the genus *Elymus*.

Species mentioned in text	
Ashy ryegrass	*Leymus cinereus*
Bluebunch wheatgrass	*Elymus spicatus*
Cheatgrass	*Bromus tectorum*
Indian ricegrass	*Oryzopsis hymenoides*
Junegrass	*Koeleria macrantha*
Medusa-head	*Taeniatherum caput-medusae*
Needle-and-thread	*Stipa comata*
One-sided bluegrass	*Poa secunda*
Squirreltail	*Elymus elymoides*
Western wheatgrass	*Elymus smithii*

Plot-based descriptions: NDDB has plot data on file for stands in Lassen and Modoc Cos.

Bulrush series

Bulrushes sole or dominant in herb canopy; broadleaf cattail, California bulrush, common three-square, common tule, narrowleaf cattail, Nevada bulrush, river bulrush, saltgrass, saltmarsh bulrush, slenderbeaked sedge, southern cattail, umbrella flatsedge, water-plantain, and/or yerba mansa may be present. Herbs < 4 m; cover continuous, intermittent, or open (Plate 1).

Wetlands: habitat permanently flooded, regularly flooded, semipermanently flooded, seasonally flooded, irregularly flooded, irregularly exposed. Water chemistry: fresh, mixohaline, hyperhaline, mixosaline, hypersaline. Bay, estuary, dune swale, slough terrace edges; berm, backwater, bank, bottomland margins of rivers; channel, creek, ditch margins; lake beds; lagoon, pond, reservoir margins; along geologic faults. Soils peaty. Cowardin classes: Estuarine intertidal persistent emergent wetland, Lacustrine littoral emergent wetland, Palustrine nonpersistent emergent saline wetland, Palustrine persistent emergent freshwater wetland, Palustrine persistent emergent saline wetland. The national list of wetland plants (Reed 1988) lists dominant bulrushes as a OBL.

Distribution: Cis-CA, Trans-CA, North America.
Elevation: sea level-2100 m.

NDDB/Holland type and status: **Marsh and swamp** (52000).
  Coastal brackish marsh (52200 *in part*) G2 S2.1.
  Cismontane alkali marsh (52310 *in part*) G1 S1.1.
  Transmontane alkali marsh (52320 *in part*) G3 S2.1.
  Coast and valley freshwater marsh (52410 *in part*) G3 S2.1.

Transmontane freshwater marsh (52420 *in part*) G3 S2.2.
Montane freshwater marsh (52430 *in part*) G3 S3.

Other types:
Barry type: G7412311 BSCAM20.
Cheatham & Haller type: Coastal brackish marsh, Alkali marshes, Freshwater marshes.
PSW-45 type: Bulrush series.
Rangeland type: SRM 217.
Thorne type: Freshwater aquatic.
WHR type: Fresh emergent wetland, Saline emergent wetland.

General references: Barnhart *et al.* (1992), MacDonald (1977), Paysen *et al.* (1980), Thorne (1982), Marshall (1948), Zedler (1982).

Species mentioned in text	
Broadleaf cattail	*Typha latifolia*
Bulrushes	*Scirpus* species
California bulrush	*Scirpus californicus*
Common three-square	*Scirpus americanus*
Common tule	*Scirpus acutus*
Littlebeak spikerush	*Eleocharis rostellata*
Narrowleaf cattail	*Typha angustifolia*
Nevada bulrush	*Scirpus nevadensis*
River bulrush	*Scirpus fluviatilis*
Saltgrass	*Distichlis spicata*
Saltmarsh bulrush	*Scirpus maritimus*
Slender-beaked sedge	*Carex athrostachya*
Southern cattail	*Typha domingensis*
Tules	*Scirpus* species
Umbrella flatsedge	*Cyperus eragrostis*
Water-plantain	*Alisma plantago-aquatica*
Yerba mansa	*Anemopsis californica*

Comments: This series occurs in alkali, brackish, or freshwater marshes dominated by tall bulrushes or tules. Barnhart *et al.* (1992) describe Humboldt Bay marshes in Humboldt Co. o-NorCo; MacDonald (1977) the San Francisco Bay complex, Ferren (1989) the Santa Barbara area marshes o-CenCo; MacDonald (1977) South Coast marshes o-SoCo. The Bulrush series, Cattail series, and Bulrush-cattail series grow in similar habitats [*see* Cattail series, Bulrush-cattail series].

Plot-based descriptions: Atwater *et al.* (1979) define one association for Contra Costa, Solano Cos. d-CenV; Bradley (1970) one association at Death Valley National Monument in Inyo Co. MojD; Odion *et al.* (1992) one association at Fish Slough in Inyo Co. TraSN.

Associations:
Atwater *et al.* (1979):
    California bulrush association.
Bradley (1970):
    Common three-square association
    [as Bulrush vegetation with *Scirpus olneyi*, see *The Jepson Manual*].
Odion *et al.* (1992):
    Common three-square - littlebeak spikerush association [as *Scirpus-Eleocharis* association].

Bulrush-cattail series

Bulrushes and cattails important herbs emerging from water; broadleaf cattail, California bulrush, common three-square, common tule, narrowleaf cattail, Nevada bulrush, river bulrush, saltgrass, saltmarsh bulrush, slender-beaked sedge, southern cattail, umbrella flatsedge, water-plantain, and/or yerba mansa may be present. Herbs < 4 m; cover continuous, intermittent, or open.

Wetlands: habitat permanently flooded, regularly flooded, semipermanently flooded, seasonally flooded, irregularly flooded, irregularly exposed. Water chemistry: fresh, mixohaline, hyperhaline, mixosaline, hypersaline. Bay, estuary, dune swale, slough terrace edges; berm, backwater, bank, bottomland, mouth margins of rivers; channel, creek, ditch margins; lake beds; lagoon, pond, reservoir margins; along geologic faults. Soils peaty. Cowardin classes: Estuarine intertidal persistent emergent wetland, Lacustrine littoral emergent wetland, Palustrine nonpersistent emergent saline wetland, Palustrine persistent emergent freshwater wetland, Palustrine persistent emergent saline wetland. The national list of wetland plants (Reed 1988) lists dominant species as a OBL.

Distribution: Cis-CA, Trans-CA, North America.
Elevation: sea level-2100 m.

NDDB/Holland type and status: Marsh and swamp (52000).
    Coastal brackish marsh (52200 *in part*) G2 S2.1.
    Cismontane alkali marsh (52310 *in part*) G1 S1.1.
    Transmontane alkali marsh (52320 *in part*) G3 S2.1.

Coast and valley freshwater marsh (52410 *in part*) G3 S2.1.
Transmontane freshwater marsh (52420 *in part*) G3 S2.1.
Montane freshwater marsh (52430 *in part*) G3 S3.

Other types:
Barry type: G7412311.
Cheatham & Haller type: Coastal brackish marsh, Alkali marshes, Freshwater marshes.
Rangeland type: SRM 217.
Thorne type: Freshwater aquatic.
WHR type: Fresh emergent wetland, Saline emergent wetland.

Species mentioned in text

Broadleaf cattail	*Typha latifolia*
Bulrushes	*Scirpus* species
California bulrush	*Scirpus californicus*
Cattails	*Typha* species
Common three-square	*Scirpus americanus*
Common tule	*Scirpus acutus*
Narrowleaf cattail	*Typha angustifolia*
Nevadan bulrush	*Scirpus nevadensis*
River bulrush	*Scirpus fluviatilis*
Saltgrass	*Distichlis spicata*
Saltmarsh bulrush	*Scirpus maritimus*
Slender-beaked sedge	*Carex athrostachya*
Southern cattail	*Typha domingensis*
Tules	*Scirpus* species
Umbrella flatsedge	*Cyperus eragrostis*
Water-plantain	*Alisma plantago-aquatica*
Yerba mansa	*Anemopsis californica*

General references: Ferren (1989), Marshall (1948), MacDonald (1977), Thorne (1982), Zedler (1982).

Comments: This series occurs in alkali, brackish, or freshwater marshes. Barnhart *et al.* (1992) describe Humboldt Bay marshes in Humboldt Co. o-NorCo; MacDonald (1977) the San Francisco Bay complex, Ferren (1989) the Santa Barbara area marshes o-CenCo; MacDonald (1977) South Coast marshes o-SoCo. The Bulrush series and Cattail series occur in similar habitats, but stands of Bulrush-cattail series lack a dominant species [*see* Bulrush series, Cattail series].

Plot-based descriptions: Cuneo (1987) describes San Francisco and San Pablo bays in NorCo; Atwater *et al.* (1979) define one association in Contra Costa, Solano Cos. d-CenV; Odion *et al.* (1992) one association at Fish Slough in TraSN.

Associations:
Atwater *et al.* (1979):
    Bulrush-cattail association.
Odion *et al.* (1992):
    Common tule-southern cattail association
    [as *Scirpus-Typha* association].

Bur-reed series

Bur-reeds, a floating-leaved, anchored hydrophyte, sole or dominant emergent or submergent plant; beaked sedge, bulrushes, cattails, pondweeds, narrowleaf bur-reed, simpleleaf bur-reed, small bur-reed, water-shield, and/or yellow pond-lily may be present. Emergent plants may be present. Plants < 1 m in size; cover continuous, intermittent, or open (Plate 2).

Wetlands: habitat permanently flooded, intermittently exposed. Water chemistry: fresh. Ditches; lakes; ponds; slow streams. Cowardin classes: Lacustrine littoral aquatic bed wetland, Palustrine aquatic bed wetland, Riverine aquatic bed wetland. The national list of wetland plants (Reed 1988) lists bur-reeds as a OBL.

Distribution: o-NorCo, m-KlaR, su-KlaR, m-CasR, su-CasR, m-SN, su-SN, TraR, ModP. **Elevation:** sea elevation-3700 m.

NDDB/Holland type and status: **Marsh and swamp** (52000).
Montane freshwater marsh (52430 *in part*) G3 S3.

Other types: Barry type: G7421321.
Cheatham & Haller type: Freshwater marsh.
Thorne type: Freshwater aquatic.
WHR type: Freshwater emergent wetland.
General references: Mason (1957).

Comments: Five species of bur-reed grow in California. Narrowleaf bur-reed is the most conspicuous one, often present as monospecific stands in mountain lakes. The ecological relationships between narrowleaf bur-reed and other bur-reed species needs clarification. Currently, all bur-reeds are placed in this series. Some species are emergent, others have floating leaves. Small bur-reed (a CNPS List 4 plant) is in m-SN.

Species mentioned in text	
Beaked sedge	*Carex utriculata*
Bulrushes	*Scirpus* species
Bur-reeds	*Sparganium* species
Cattails	*Typha* species
Narrowleaf bur-reed	*Sparganium angustifolium*
Pondweeds	*Potamogeton* species
Simpleleaf bur-reed	*Sparganium emersum*
Small bur-reed	*Sparganium natans*
Water-shield	*Brasenia schreberi*
Yellow pond-lily	*Nuphar luteum*

Plot-based descriptions: none available.

California annual grassland series

Annual grasses and herbs dominant in ground layer; bromes, California poppy, filarees, goldfields, lupines, mustards, oats, owl's-clovers, ryegrasses, and/or star-thistles may be present. Emergent shrubs and trees may be present. Grass < 1 m; continuous or open (Plate 5).

Uplands: all topographic locations.

Distribution: NorCo, i-CenCo, CenV, l-KlaR, f-CasR, f-SN, SoCo, m-TraR, m-PenR, ChaI, w. MojD, Baja CA.
Elevation: sea elevation-1200 m.

NDDB/Holland type and status: **Valley and foothill grasslands** (42000)
      Non-native grassland (42200) G4 S4.
      Wildflower field (42300) G2 S2.2.

Other types:
Barry type: G7411332.
Brown Lowe Pase type: 143.12.
Cheatham & Haller type: Cismontane introduced grasses
PSW-45 type: Bromegrass series, Wild oats series.
Rangeland type: SRM 215.
Thorne type: Great Valley and Coast Range grassland.
WHR type: Annual grass.

General references: Barry (1972), Bartolome (1989), Bartolome *et al.* (1980), Bartolome & Brown (1994), Brown (1982d), Burcham (1957), George *et al.* (1992), Heady (1977), Huenneke & Mooney (1989), Keeley (1989), Parsons (1981), Paysen *et al.* (1980), Sims (1988), Stoddart *et al.* (1975), Turner & Brown (1982), Webster (1981), Witham (1976), Young & Evans (1989).

Comments: This extensive series is composed of many alien and native annual species; composition varies among stands. Many species beyond those listed above may be present. Ripgut, soft chess, and foxtail chess are common bromes - *Erodium botrys* and *E. cicutarium* common filarees -*Lasthenia cali-*

forica a common goldfields - *Lupinus bicolor* a common lupine - *Avena barbata* and *A. fatua* common oats - *Lolium multiflorum* a common ryegrass. *The Jepson Manual* places owl's-clover and butter-and-eggs in the genera *Castilleja* and *Triphysaria*.

Fall temperatures and precipitation are major factors determining grassland composition, along with light intensity affected by shading from plants and litter, and differences in microtopography (Evans & Young 1989). The fine scale variation in temporal and spatial structure found in the California annual grassland suggests that recognition of many series is not useful (Bartolome 1989).

Species mentioned in text	
Bromes	*Bromus* species
Butter-and-eggs	*Triphysaria eriantha*
California poppy	*Eschscholzia californica*
Dogtail	*Cynosurus echinatus*
European hairgrass	*Aira caryophyllea*
Filaree	*Erodium* species
Foxtail chess	*Bromus madritensis*
Goldfields	*Lasthenia* species
Lupines	*Lupinus* species
Mustards	*Brassica* species
Oats	*Avena* species
Owl's-clovers	*Castilleja* species
Purple owl's-clover	*Castilleja exserta*
Rattail fescue	*Vulpia hirsuta*
Ripgut	*Bromus diandrus*
Ryegrass	*Lolium* species
Slender oats	*Avena barbata*
Soft chess	*Bromus hordeaceus*
Star-thistles	*Centaurea* species
Storkbill	*Erodium botrys*
Wild oats	*Avena fatua*

Plot-based descriptions: Saenz & Sawyer (1986) report on sites grazed for the full 1982 season. They were dominated by dogtail, in what now is Redwood National Park in Humboldt Co., Foin & Hektner (1986) define a Stable meadow in Sonoma Co. o-NorCo.; Schlising & Sanders (1982) two associations associated with vernal pools in Butte Co. sac-CenV; Parsons & Stohlgren (1989) one Slender oat association at Sequoia National Park in Tulare Co. f-SN; Kopecko & Lathrop (1975) two types associated with vernal pools at Santa Rosa Plateau in Riverside Co. o-SoCo.

Associations:
Kopecko & Lathrop (1975):
    Slender oat-soft brome association
    [as Dry grassland zone],
    Soft brome-rattail fescue
    [as Vernally moist zone].
Schlising & Sanders (1982):
    European hairgrass association,
    Soft brome-storkbill association.

California oatgrass series

California oatgrass sole or dominant grass in ground layer; bracken, California melic, foothill sedge, Idaho fescue, one-sided bluegrass, purple needlegrass, red fescue, tall-oatgrass, velvet grass, and/or vernal grass may be present. Emergent shrubs and trees may be present. Grass < 1 m; cover open (Plate 5).

Wetlands: habitat seasonally or permanently saturated with shallow water table. Water chemistry: fresh, saline. Valley bottoms, lower portions of alluvial slopes. Cowardin class: Palustrine emergent wetland. **Uplands:** coastal bluffs, terraces, slopes and ridges. The national list of wetland plants (Reed 1988) lists California oatgrass as a FACW.

Distribution: NorCo, CenCo, m-CasR, m-SN, m-TraR, m-PenR, ModP, OR.
Elevation: sea level-2200 m.

NDDB/Holland type and status: **Coastal prairies** (41000).
    Coastal terrace prairie (41100 *in part*) G2 S2.2.
    Bald Hills prairie (41200 *in part*) G2 S2.1.
    Great Basin grassland (43000) G1 S1.1

Other types:
Barry type: G7411331 BDECA00.
Cheatham & Haller type: Coastal prairie.
Rangeland type: SRM 214.
Thorne type: Northern coastal prairie.
WHR type: Perennial grass.

General references: Bartolome *et al.* (1994), Heady *et al.* (1977b).

Comments: Often considered part of coastal prairie, this series extends from coastal terraces to inland bald hills. Species dominance varies at a fine scale [*see* Idaho fescue series, Pacific reedgrass series, Tufted hairgrass series]. It often mixes with tree

series at a coarser scale [*see* Douglas-fir - tanoak series, Oregon white oak series, Redwood series].

Sawyer *et al.* (1978b) in Keeler-Wolf (1990e) qualitatively describes *Danthonia unispicata* meadows at Preacher Meadows candidate RNA in Trinity Co. m-KlaR.

Species mentioned in text	
Bracken	*Pteridium aquilinum*
California melic	*Melica californica*
California oatgrass	*Danthonia californica*
Douglas-fir	*Pseudotsuga menziesii*
Foothill sedge	*Carex tumulicola*
Idaho fescue	*Festuca idahoensis*
One-sided bluegrass	*Poa secunda*
Oregon white oak	*Quercus garryana*
Pacific reedgrass	*Calamagrostis nutkaensis*
Pull-up muhly	*Muhlenbergia filiformis*
Purple needlegrass	*Stipa pulchra*
Red fescue	*Festuca rubra*
Redwood	*Sequoia sempervirens*
Squirreltail	*Elymus elymoides*
Tall-oatgrass	*Arrhenatherum elatius*
Tanoak	*Lithocarpus densiflora*
Tufted hairgrass	*Deschampsia cespitosa*
Velvet grass	*Holcus lanatus*
Vernal grass	*Anthoxanthum odoratum*

Plot-based descriptions: Heady *et al.* (1977b) report cover for California oatgrass type in Sonoma Co.; NDDB has plot data on file for stands in Sonoma and Humboldt Cos., Grenier (1989) defines one association at Redwood National Park in Humboldt Co. o-NorCo; Stuart *et al.* (1992) an association at Castle Crags State Park in Siskiyou Co. l-KlaR; Helms & Ratliff (1987) an association in Yosemite National Park m-SN.

Associations:
Grenier (1989):
    California oatgrass - tall-oatgrass association.
Stuart *et al.* (1992):
    Squirreltail-California oatgrass association.
Helms & Ratliff (1987):
    California oatgrass - pull-up muhly association.

Cattail series

Cattails sole or dominant herb emerging from water; broadleaf cattail, California bulrush, common three-square, common tule, narrowleaf cattail, Nevada bulrush, river bulrush, saltgrass, saltmarsh bulrush, slender-beaked sedge, southern cattail, umbrella flatsedge, water-plantain, and/or yerba mansa may be present. Herbs < 4 m; cover continuous, intermittent, or open (Plate 3).

Wetlands: habitat permanently flooded, regularly flooded, semipermanently flooded, seasonally flooded, irregularly flooded, irregularly exposed. Water chemistry: fresh, mixohaline, hyperhaline, mixosaline, hypersaline. Bay, estuary, dune swale, slough terrace edges; berm, backwater, bank, bottomland, mouth margins of rivers; channel, creek, ditch margins; lake beds; lagoon, pond, reservoir margins; along geologic faults. Soils peaty. Cowardin classes: Estuarine intertidal persistent emergent wetland, Lacustrine littoral emergent wetland, Palustrine nonpersistent emergent saline wetland, Palustrine persistent emergent freshwater wetland, Palustrine persistent emergent saline wetland. The national list of wetland plants (Reed 1988) lists cattails as a OBL.

Distribution: Cis-CA, Trans-CA, North America.
Elevation: sea level-2000 m

NDDB/Holland type and status: Marsh and swamp (52000).

> Coastal brackish marsh (52200 *in part*) G2 S2.1.
> Cismontane alkali marsh (52310 *in part*) G1 S1.1.
> Transmontane alkali marsh (52320 *in part*) G3 S2.1.

> Coast and valley freshwater marsh (52410 *in part*) G3 S2.1.
> Transmontane freshwater marsh (52420 *in part*) G3 S2.2.
> Montane freshwater marsh (52430 *in part*) G3 S3.

Other types:
Barry type: G7412311 CTULA00.
Cheatham & Haller type: Coastal brackish marsh, Alkali marshes, Freshwater marshes.
PSW-45 type: Cattail series.
Thorne type: Freshwater aquatic.
WHR type: Fresh emergent wetland, Saline emergent wetland.

General references: Ferren (1989), Marshall (1948), MacDonald (1977), Paysen *et al.* (1980), Thorne (1982), Zedler (1982).

Species mentioned in text	
Broadleaf cattail	*Typha latifolia*
California bulrush	*Scirpus californicus*
Cattails	*Typha* species
Common three-square	*Scirpus americanus*
Common tule	*Scirpus acutus*
Narrowleaf cattail	*Typha angustifolia*
Nevada bulrush	*Scirpus nevadensis*
River bulrush	*Scirpus fluviatilis*
Saltgrass	*Distichlis spicata*
Saltmarsh bulrush	*Scirpus maritimus*
Slender-beaked sedge	*Carex athrostachya*
Southern cattail	*Typha domingensis*
Umbrella flatsedge	*Cyperus eragrostis*
Water-plantain	*Alisma plantago-aquatica*
Yerba mansa	*Anemopsis californica*

Comments: This series occurs in alkali, brackish, or freshwater marshes. Barnhart *et al.* (1992) describe Humboldt Bay marshes in Humboldt Co. o-NorCo; MacDonald (1977) the San Francisco Bay complex, Ferren (1989) the Santa Barbara area marshes o-CenCo; MacDonald (1977) South Coast marshes o-SoCo. The Bulrush series, Cattail series, and Bulrush-cattail series grow in similar habitats [*see* Bulrush series, Bulrush-cattail series]. Baker (1972) qualitatively describes stands as Fringing emergent at Ingelnook Fen in Mendocino Co. o-NorCo.
Plot-based descriptions: none available.

Cheatgrass series

Cheatgrass sole or dominant grass in ground layer; crested wheatgrass, Indian ricegrass, needle-and-thread, squirreltail, and/or western wheatgrass may be present. Emergent shrubs may be present. Grass < 1 m; cover open.

Uplands: all topographic locations.

Distribution: Cis-CA, Trans-CA, inter-West.
Elevation: sea level-2200 m.

NDDB/Holland type and status: **Great Basin grassland** (43000).

Other types: none.

General references: Beatley (1966), Mack (1981).

Comments: Cheatgrass is a Eurasian introduction that occurs throughout CA, especially on disturbed sites. It dominates over large expanses, especially in Trans-CA.

```
┌─────────────────────────────────────────────┐
│         Species mentioned in text           │
│  Cheatgrass              Bromus tectorum     │
│  Crested wheatgrass      Agropyron desertorum│
│  Indian ricegrass        Oryzopsis hymenoides│
│  Needle-and-thread          Stipa comata     │
│  Ripgut                  Bromus diandrus     │
│  Squirreltail            Elymus elymoides    │
│  Western wheatgrass        Elymus smithii    │
└─────────────────────────────────────────────┘
```

Plot-based descriptions: Stuart *et al.* (1993) defines an association at Castle Crags State Park in Siskiyou Co. l-KlaR; White (1994a) one association at Cleghorn Canyon candidate RNA in San Bernardino Co. i. m-TraR.

Associations:
Stuart *et al.* (1993):
    Cheatgrass association.
White (1994a):
    Cheatgrass-ripgut association.

Common reed series

Common reed sole dominant in ground layer; few other species present. Emergent shrubs and trees may be present. Grass < 4 m; cover continuous (Plate 6).

Wetlands: habitat permanently saturated with shallow water table. Water chemistry: fresh, mixohaline, mixosaline. Cowardin class: Palustrine persistent emergent wetland. The national list of wetland plants (Reed 1988) lists giant reed as a FACW.

Distribution: Cis-CA, Trans-CA, North America.
Elevation: sea level-1600 m.

NDDB/Holland type and status: none.

Other types: none.

General references: Marks *et al.* (1994).

Comments: Common reed may the most extensive seed-bearing plant species in the world. Marks *et al.* (1994) report that in many areas it is considered part of the historic vegetation and is unchanging in population size and extent. In other areas it is currently expanding and reducing native species diversity in the eastern U.S, possibly from invasive genotypes introduced from the Old World. In the western U.S. some populations appear to be decreasing.

Species mentioned in text	
Common reed	*Phragmites australis*

Plot-based descriptions: none available.

Cordgrass series

Cordgrasses sole or dominant grasses with herbs in ground canopy; alkali heath, arrow-grasses, common pickleweed, dodder, and/or saltgrass may be present. Plants < 1.5 m; canopy continuous or intermittent (Plate 1).

Wetlands: habitat regularly flooded, irregularly flooded. Water chemistry: mixohaline, euhaline. Estuaries: mudflats, banks, berms, and margins of bays, deltas, sandbars. Cowardin class: Estuarine intertidal persistent emergent wetland. The national list of wetland plants (Reed 1988) lists cordgrasses as a OBL or FACW.

Distribution: NorCo, CenCo, SoCo, TraSN, MojD.
Elevation: sea level.

NDDB/Holland type and status: Marsh and swamp (52000).
Northern coastal salt marsh (52110 *in part*) G3 S3.3
Southern coastal salt marsh (52120 *in part*) G2 S2.1.

Other types:
Barry type: G7412311 FSPFO00.
Cheatham & Haller type: Coastal salt marshes.
PSW-45 type: Cordgrass series.
Rangeland type: SRM 217.
Thorne type: Coastal salt marsh.
WHR type: Saline emergent wetland.

General references: Chapman (1977), Henrickson (1976), MacDonald (1977), MacDonald & Barbour (1974), Paysen *et al.* (1980), Vogl (1966), Zedler *et al.* (1992).

Comments: Five species of *Spartina* are recognized in *The Jepson Manual*; two are native. California cordgrass grows in CenCo and SoCo salt marshes; Alkali cordgrass in alkali meadows and marshes in TraSN, MojD. Three cordgrasses are introduced.

Dense-flowered cordgrass is locally important in CenCo (San Francisco Bay) and NorCo (Humboldt Bay) salt marshes; saltmeadow cordgrass in CenCo salt marshes. The introduction of exotic cordgrasses has altered historic vegetation patterns (Spicher & Josselyn 1985).

CA salt marshes are represented by this, Pickleweed series, and Saltgrass series. MacDonald (1977) describes regional differences in salt marshes. Barnhart *et al.* (1992) details salt marshes at Humboldt Bay; Josselyn (1983) at San Francisco Bay o-CenCo; Zedler (1982) So-CA ones.

Species mentioned in text	
Alkali cordgrass	*Spartina gracilis*
Alkali heath	*Frankenia salina*
Alkali sacaton	*Sporobolus airoides*
Arrow-grasses	*Triglochin* species
California cordgrass	*Spartina foliosa*
Common pickleweed	*Salicornia virginica*
Cordgrasses	*Spartina* species
Dense-flowered cordgrass	*Spartina densiflora*
Dodder	*Cuscuta salina*
Salt-meadow cordgrass	*Spartina patens*
Salt-water cordgrass	*Spartina alterniflora*
Saltgrass	*Distichlis spicata*

Plot-based descriptions: Eicher (1987) defines an association at Humboldt Bay in Humboldt Co., Cuneo (1987) describes marshes in San Francisco and San Pablo bays NorCo; Atwater *et al.* (1979) defines one association in Contra Costa Co. d-CenV; Peinado *et al.* (1994) one association for CA coastal salt marshes; Odion *et al.* (1992) one association at Fish Sough in TraSN.

Associations:
Atwater *et al.* (1979):
California cordgrass association
[= Spartinetum foliosae (Peinado *et al.* 1994)].
Eicher (1987):
Dense-flowered cordgrass association.
Odion *et al.* (1992):
Alkali cordgrass-alkali sacaton association
[as *Spartina-Sporobolus* association].

Creeping ryegrass series

Creeping ryegrass sole or dominant grass in ground layer; bromes, needle-and-thread, oats, one-spike oatgrass, red three-awn, and/or squirreltail may be present. Emergent shrubs may be present. Grass < 1 m; cover open.

Wetlands: habitat permanently saturated with shallow water table. Water chemistry: fresh or saline. Valley bottoms, lower portions of alluvial slopes. Cowardin class: Palustrine persistent emergent saline wetland. The national list of wetland plants (Reed 1988) lists creeping ryegrass as a NI.

Distribution: Cis-CA, Trans-CA, inter-West. **Elevation:** sea level-2300 m.

NDDB/Holland type and status: **Valley and foothill grasslands** (42000).
    Valley wildrye grassland (42140) G2 S2.1.

Other types:
Barry type: G7411331 CELT30.
Cheatham & Haller type: Cismontane native grassland, Great Basin native grassland.
PSW-45 type: Wild rye series.
Thorne type: Great Valley and Coast Range grassland.
WHR type: Perennial grass.

General references: Brown (1982e), Heady (1977), Paysen *et al.* (1980).

Comments: Many regional descriptions include this series in an alkali meadows category. Meadows of this habitat are dominated by this or other species [*see* Alkali sacaton series, Ashy ryegrass series, Saltgrass series, Spikerush series]. This widespread species is a component of many series, and is a common understory plant in riparian forests [*see* Fremont cottonwood series].

In the ecological literature creeping ryegrass is referred to as *Elymus triticoides. The Jepson Manual* places creeping ryegrass in the genus *Leymus.*

Species mentioned in text	
Alkali sacaton	*Sporobolus airoides*
Ashy ryegrass	*Leymus cinereus*
Bromes	*Bromus* species
Creeping ryegrass	*Leymus triticoides*
Needle-and-thread	*Stipa comata*
Oats	*Avena* species
One-sided bluegrass	*Poa secunda*
One-spike oatgrass	*Danthonia unispicata*
Red three-awn	*Aristida purpurea*
Saltgrass	*Distichlis spicata*
Spikerushes	*Eleocharis* species
Squirreltail	*Elymus elymoides*

Plot-based descriptions: Odion *et al.* (1992) define a Creeping wild rye - one-sided bluegrass association [as *Elymus-Poa* association] in Inyo Co. TraSN.

Crested wheatgrass series

Crested wheatgrass sole or dominant grass in ground layer; cheatgrass, Indian ricegrass, squirreltail, and/or western wheatgrass may be present. Emergent shrubs may be present. Grass < 1 m; cover open.

Uplands: all topographic locations.

Distribution: m-CasR, n. m-SN, ModP, m-WarR, TraSN, inter-West.
Elevation: 600-1500 m.

NDDB/Holland type and status: **Great Basin grassland** (43000).

Other types:
Barry type: G7411331.

General references: not as a vegetation type.

Comments: Crested wheatgrass is a European introduction that occurs in many series in Trans-CA, especially on disturbed sites. It is commonly planted, and has also become naturalized.

Species mentioned in text	
Cheatgrass	*Bromus tectorum*
Crested wheatgrass	*Agropyron desertorum*
Indian ricegrass	*Oryzopsis hymenoides*
Squirreltail	*Elymus elymoides*
Western wheatgrass	*Elymus smithii*

Plot-based descriptions: not available.

Darlingtonia series

Darlingtonia dominant herb in ground canopy; butterwort, California lady-slipper, cone flower, grass-of-Parnassus, Labrador-tea, lilies, sedges, sundew, and/or western azalea may be present. Emergent shrubs and trees such as Port Orford-cedar may be present. Plants to < 20 m; canopy continuous, intermittent, or open (Plate 2).

Wetlands: habitat saturated with running water, may or may not have peat. Typically on peridotite, but also on other parent materials. The national list of wetland plants (Reed 1988) lists darlingtonia as a OBL.

Distribution: KlaR, n. m-SN, OR.
Elevation: sea level-2700 m.

NDDB/Holland type and status: **Bog and fen** (51000). Darlingtonia bog (51120) G4 S3.

Other types:
Barry type: G7412411.
Cheatham & Haller type: Bogs.
Thorne type: Darlingtonia bog.
WHR type: Wet meadow.

General references: Bittman (1985), Cheatham (1976), Gore (1983), Kruckeberg (1984), Sawyer (1986a), Keeler-Wolf (1990e).

Comments: Darlingtonia occurs in areas with running or standing water. It is rarely associated with sphagnum in KlaR where the series is most extensive on serpentine rocks. For these reasons, Sawyer (1986a) recommends the term Darlingtonia seep as more descriptive than Darlingtonia bog. Bogs are ombrogenous and have sphagnum (Gore 1983).

Keeler-Wolf (1986) in Keeler-Wolf (1990e) qualitatively describes Darlingtonia seeps with an overstory of Port Orford-cedar, at the L. E. Horton candidate RNA, Del Norte Co. l-KlaR; Keeler-Wolf

(1982, 1987) in Keeler-Wolf (1990e) at Cedar Basin candidate RNA in Siskiyou Co. su-KlaR, Sawyer *et al.* (1978a) in Keeler-Wolf (1990e) at Preacher Meadows candidate RNA in Trinity Co. m-KlaR; Cheatham (1976), Knight *et al.* (1970) describe Butterfly Valley Botanical Area in Plumas Co. m-SN.

Species mentioned in text	
Butterwort	*Pinguicula macroceras*
California lady-slipper	*Cypripedium californicum*
Cone flower	*Rudbeckia californica*
Darlingtonia	*Darlingtonia californica*
Grass-of-Parnassus	*Parnassia palustris*
Labrador-tea	*Ledum glandulosum*
Lilies	*Lilium* species
Port Orford-cedar	*Chamaecyparis lawsoniana*
Sedges	*Carex* species
Sundew	*Drosera rotundifolia*
Western azalea	*Rhododendron occidentale*

Plot-based descriptions: not available.

Desert needlegrass series

Desert needlegrass sole, dominant, or important grass in ground layer; foothill needlegrass, Indian ricegrass, nodding needlegrass, one-sided bluegrass, and/or squirreltail may be present. Emergent shrubs, as cheesebush, may be present. Grass < 1 m; cover open (Plate 6).

Uplands: flat ridges, lower slopes. Soils sandy to rocky.

Distribution: MojD.
Elevation: 600-1000 m.

NDDB/Holland type and status: **Valley and foothill grassland** (42000).
    Valley needlegrass grassland (42110 *in part*) G1 S1.1.

Other types:
Cheatham & Haller type: Creosote bush scrub.
Thorne type: Creosote bush scrub.
WHR type: Desert scrub.

General references: Holland (1986).

Comments: Many regional descriptions include this series in Creosote bush scrub. This series is present in a few areas in western Antelope Valley of Kern and Los Angeles Cos. Desert needlegrass is the unifying species in these stands as associates vary.

In the ecological literature desert needlegrass is referred to as *Stipa speciosum. The Jepson Manual* places desert needlegrass in the genus *Achnatherum*.

Species mentioned in text	
Cheesebush	*Hymenoclea salsola*
Desert needlegrass	*Stipa speciosum*
Foothill needlegrass	*Stipa lepida*
Indian ricegrass	*Oryzopsis hymenoides*
Nodding needlegrass	*Stipa cernua*
One-sided bluegrass	*Poa secunda*
Squirreltail	*Elymus elymoides*

Plot-based descriptions: NDDB has plot data on file from Kern and Los Angeles Cos.

Desert sand-verbena series

Scattered forbs and grasses in the ground canopy; birdcage, California croton, desert sand-verbena, desert gold, desert dicoria, Indian ricegrass, big galleta, naked cleome, evening-primrose, and/or wild rhubarb may be present. Individual emergent shrubs may be present as Algodones buckwheat, creosote bush, fourwing saltbush, and/or white bursage. Shrubs < 3 m. Ground layer open; annuals seasonally present (Plate 6).

Uplands: sand bodies. Active, partially stabilized, stabilized dunes; partially stabilized, stabilized sand fields.

Distribution: i-SoCo, TraSN, MojD, ColD.
Elevation: 10 below sea level to 1200 m.

NDDB/Holland type and status: Desert dunes (22000).
   Active desert dunes (22100) G2 S2.2.
   Stabilized and partially stabilized dunes (22200) G 3 S2.2.
   Stabilized and partially stabilized desert sand fields (22300) G2 S2.2.

Other types:
Barry type: G7411323.
Cheatham & Haller type: Partially stabilized desert dunes, Stabilized desert dunes.
PSW-45 type: Croton series.
Stone & Sumida (1983): Sand plant community.
Thorne type: Desert dune sand plant.
WHR type: Desert scrub.

General references: Henry (1979), Thorne (1982), Paysen *et al.* (1980), Vasek & Barbour (1977).

Comments: Eighteen dune areas are scattered throughout the Trans-CA deserts, each with its own set of plant species (Pavlik 1985). Each area may be considered separately. Some are known floristically: Algodones dunes (Beauchamp 1977), Ballarat dunes (Pavlik 1985), Borrego dunes (Beauchamp 1986), Cadiza dunes, Chuckwalla Valley dunes,

Death Valley dunes (DeDecker 1979, 1984), (Norris 1982), Deep Springs dunes (Pavlik 1985), Dumont dunes, Eureka dunes (DeDecker 1984), Kelso dunes (Thorne *et al.* 1981), Mono dunes (Pavlik 1985), Olancha dunes (Pavlik 1985), Panamint Valley dunes, Rice Valley dunes, Saline Valley dunes (DeDecker 1984), Saratoga dunes (Pavlik 1985). Such Cis-CA areas as Colton dunes in Riverside Co. are included in this series.

Emergent shrubs are infrequent in this series in comparison to the Big galleta and Creosote bush series which also occur on sand.

Species mentioned in text	
Algodones buckwheat	*Eriogonum deserticola*
Big galleta	*Pleuraphis rigida*
Birdcage	*Oenothera deltoides*
California croton	*Croton californicus*
Creosote bush	*Larrea tridentata*
Desert sand-verbena	*Abronia villosa*
Desert gold	*Geraea canescens*
Desert dicoria	*Dicoria canescens*
Evening-primrose	*Oenothera* species
Fourwing saltbush	*Atriplex canescens*
Indian ricegrass	*Oryzopsis hymenoides*
Naked cleome	*Cleome sparsifolia*
White bursage	*Ambrosia dumosa*
Wild rhubarb	*Rumex hymenosepalus*

Plot-based descriptions: Bagley (1986), DeDecker (1982) describe Eureka Dunes in Inyo Co. MojD; Spolsky (1979) defines a Sonoran active dune forb association at Anza Borrego Desert State Park in San Diego Co. ColD.

Ditch-grass series

Ditch-grasses sole or dominant herb forming submerged beds in water; bladderwort, common water-nymph, horned-pondweed, hornworts, leafy pondweed, and/or milfoils may be present. Herbs < 0.2 m.

Wetlands: habitat permanently flooded, semipermanently flooded, seasonally flooded, irregularly exposed, temporarily exposed. Water chemistry: mixohaline, mixosaline, hypersaline. Channel, ditch, river, stream beds; estuaries; intertidal mud flats; ponds. Cowardin classes: Estuarine subtidal aquatic bed deepwater habitat, Estuarine subtidal aquatic bed wetland. The national list of wetland plants (Reed 1988) lists ditch-grasses as a OBL.

Distribution: o-NorCo, i-NorCo, CenCo, sj-CenV, s. m-SN, SoCo, ChaI, TraSN.
Elevation: sea level-2100 m.

NDDB/Holland type and status: **Marsh and swamp** (52000) .
>   Coastal brackish marsh (52200 *in part*) G2 S2.1.
>   Cismontane alkali marsh (52310 *in part*) G1 S1.1.
>   Transmontane alkali marsh (52320 *in part*) G3 S2.1.
>   Alkali seep (45320 in part) G3 S2.1.

Other types:
Barry type: G7412311.
Cheatham & Haller type: Alkali marshes.
Thorne type: Coastal salt marsh.
WHR type: Saline emergent wetland.

General references: Ferren (1989), Mason (1957), Thorne (1982).

Comments: Ditch-grass grows in environments which are often called alkali marshes. *The Jepson Manual* lists two species; *Ruppia cirrhosa* which is more inland and at higher elevations than *R. maritima*.

Species mentioned in text	
Bladderwort	*Utricularia vulgaris*
Common water-nymph	*Najas guadalupensis*
Ditch-grasses	*Ruppia* species
Horned-pondweed	*Zannichellia palustris*
Hornworts	*Ceratophyllum* species
Leafy pondweed	*Potamogeton foliosus*
Milfoils	*Myriophyllum* species

Plot-based descriptions: Odion *et al.* (1992) define a Ditch-grass-algae association at Fish Slough in Inyo Co. TraSN.

Duckweed series

Duckweeds sole or dominant plants floating on water surface; duckmeats, Mexican mosquito fern, mosquito fern, mud-midgets, and/or water-meals may be present. Emergent plants may be present. Plants 0.3-8 mm in size; cover continuous, intermittent, or open.

Wetlands: habitat permanently flooded, semipermanently flooded, seasonally flooded. Water chemistry: fresh. Ditch, river, stream channels; ponds. Cowardin class: Palustrine aquatic bed wetland. Riverine lower perennial aquatic bed wetland. The national list of wetland plants (Reed 1988) lists duckweeds as a OBL.

Distribution: Cis-CA, Trans-CA, worldwide.
Elevation: sea level-2300 m.

NDDB/Holland type and status: **Marsh and swamp** (52000).
> Coastal and valley freshwater marsh (52410 *in part*) G3 S2.1.
> Transmontane freshwater marsh (52420 *in part*) G3 S2.2.
> Montane freshwater marsh (52430 *in part*) G3 S3.

Other types:
Barry type: G7411221.
Cheatham & Haller type: Freshwater marshes.
Thorne type: Freshwater aquatic.
WHR type: Freshwater emergent wetland.

General references: Armstrong (1982), Armstrong & Thorne (1984), Mason (1957).

Comments: *The Jepson Manual* lists two species of duckmeat, seven of duckweed; two of mud-midget, and two of water-meal. Other aquatic plants can occur with these plants. For a stand to meet the series definition, duckweed species dominate over mosquito fern, floating-leaved, or emergent plants [*see* Mosquito fern series].

Species mentioned in text	
Duckmeats	*Spirodela* species
Duckweeds	*Lemna* species
Mexican mosquito fern	*Azolla mexicana*
Mosquito fern	*Azolla filiculoides*
Mud-midgets	*Wolffiella* species
Water-meals	*Wolffia* species

Plot-based descriptions: not available.

European beachgrass series

European beachgrass sole or dominant plant in the ground canopy; beach bursage, beach morning glory, beach pea, coyote brush, native dunegrass, yellow bush lupine, and/or yellow sand-verbena may be present. Scattered emergent shrubs may be present. Ground layer continuous.

Uplands: dunes of coastal bars, river mouths, spits along the immediate coastline. Soils sandy. The national list of wetland plants (Reed 1988) lists European beachgrass as a FAC.

Distribution: o-NorCo.
Elevation: sea level.

NDDB/Holland type and status: **Coastal dunes** (21000) G3 S2.2.
> Northern foredunes (21210 *in part*) G2 S2.1.
> Northern foredune grassland (21211 *in part*) G1 S1.1.

Other types:
Barry type: G7411331 BAMAR00.
Cheatham & Haller type: Northern coastal foredunes.
Thorne type: Coastal dune sand plant.
WHR type: Perennial grass.

General references: Barbour (1970), Barbour & Johnson (1977), Boyd (1992).

Comments: European beachgrass has been planted to stabilize dunes in o-NorCo. Changes have occurred as a result of its introduction (Parker 1974). Areas historically with stands of Native dunegrass and Sand-verbena - beach bursage series are now dominated by this and other exotic plants [*see* Native dunegrass series, Sand-verbena - beach bursage series, Yellow bush lupine series].

Much recent interest in dune ecology surrounds the role that this species plays in creating current vegetation patterns. At Humboldt Bay there is an attempt to restore native-dominated stands and enhance populations of rare Humboldt Bay wallflower (a CNPS List 1B plant) and beach layia (a CNPS List 1B plant) (Pickart *et al.* 1989, Skinner & Pavlik 1994).

Species mentioned in text

Australian fireweed	*Erechtites minima*
Beach bursage	*Ambrosia chamissonis*
Beach layia	*Layia carnosa*
Beach morning glory	*Calystegia soldanella*
Beach pea	*Lathyrus littoralis*
Coyote brush	*Baccharis pilularis*
European beachgrass	*Ammophila arenaria*
Humboldt Bay wallflower	*Erysimum menziesii*
	spp. *eurekense*
Native dunegrass	*Leymus mollis*
Yellow bush lupine	*Lupinus arboreus*
Yellow sand-verbena	*Abronia latifolia*

Plot-based descriptions: Barbour (1970) defines a stand at Bodega Head in Sonoma Co.; Parker (1974) in Barbour & Johnson (1977) and Duebendorfer (1989) each define one association at Clam Beach, one association along the North Spit of Humboldt Bay. LaBanca (1993) resampled the area studied by Parker, finding a rather different organization of dune plants. He recognizes one association in Humboldt Co. o-NorCo.

Associations:
Parker (1974):
> European beachgrass-Australian fireweed association [as *Ammophila-Erichtiches* association].

Duebendorfer (1989):
> Beachgrass association [= LaBanca (1993)].

Foothill needlegrass series

Foothill needlegrass sole or dominant grass in ground layer; California fescue, California melic, one-sided bluegrass, purple needlegrass, and/or tufted reedgrass may be present. Emergent shrubs and trees may be present. Grass < 1 m; cover open.

Uplands: all topographic locations. Soils sandstone or ultramafic-derived, deep with high clay content.

Distribution: NorCo, CenCo, l-KlaR, SoCo, m-PenR.
Elevation: sea level-1700 m.

NDDB/Holland type and status: Valley and foothill grasslands (42000).
   Serpentine bunchgrass (42130 *in part*) G2 S2.2.

Other types:
Barry type: G7411331 BSTLE30.
Cheatham & Haller type: Cismontane native grassland.
PSW-45 type: Needlegrass series.
Thorne type: Great Valley and Coast range grassland.
WHR type: Perennial grass.

General references: Heady (1977), Keeley (1993), Magney (1992), Paysen *et al.* (1980), Stoddart *et al.* (1975).

Comments: Stands of this series can have nonnative, annual species mixed with the perennial grasses and herbs. Soft chess is common. However, introduced species are not as important in this series as in other needlegrass series. If the stand occurs on ultramafic parent materials, serpentine species-adapted are common.

Foothill needlegrass, nodding needlegrass, and purple needlegrass grow sympatrically but do not typically mix, especially in SoCo. They tend to segregate based on substrate and slope factors [*see* Nodding needlegrass series and Purple needlegrass series].

In the ecological literature foothill needlegrass is referred to as *Stipa lepida. The Jepson Manual* places foothill needlegrass in the genus *Nassella.*

Species mentioned in text	
California fescue	*Festuca californica*
California melic	*Melica californica*
Foothill needlegrass	*Stipa lepida*
Nodding needlegrass	*Stipa cernua*
One-sided bluegrass	*Poa secunda*
Purple needlegrass	*Stipa pulchra*
Tufted reedgrass	*Calamagrostis koelerioides*
Soft chess	*Bromus hordeaceus*

Plot-based descriptions: Kellogg & Kellogg (1990, 1991) present unclassified plot data on areas dominated by foothill needlegrass, nodding needlegrass, or purple needlegrass at Pendleton Marine Corps Base in San Diego Co. o-SoCo; Boyd (1983) describes a stand in Gavilan Hills in Riverside Co. i. m-TraR; Keeler-Wolf (1990a) foothill needlegrass-dominated stands at King Creek RNA in San Diego Co. m-PenR.

Giant reed series

Giant reed sole or dominant in ground layer; few other species present. Emergent shrubs and trees may be present. Grass < 8 m; cover continuous.

Wetlands: habitat permanently saturated with shallow water table. Water chemistry: fresh. Cowardin class: Palustrine persistent emergent freshwater wetland. The national list of wetland plants (Reed 1988) lists giant reed as a FACW.

Distribution: o-NorCo, CenCo, f-SN, SoCo, m-TraR, MojD, ColD.
Elevation: sea level-500 m.

NDDB/Holland type and status: none.

Other types:
Barry type: G7411331 CARDO40.

General references: Reiger & Kreager (1989).

Comments: Giant reed, a very invasive grass, was introduced into CA in the 1880s. This species establishes and persists in riparian areas, and reduces or replaces native species. Today in many locations giant reed forms dense monocultures.

The treatment here of this series is broad to recognize the importance of this introduced species in the vegetation of California.

Species mentioned in text	
Giant reed	*Arundo donax*

Plot-based descriptions: Reiger & Kreager (1989) report giant reed density, frequency, and cover for two river systems in San Diego Co. PenR.

Green fescue series

Green fescue sole or dominant grass in ground layer; lupines, sedges, western needlegrass, and/or yarrow may be present. Emergent lodgepole pine or whitebark pine may be present. Grass < 1 m; cover open.

Uplands: slopes and ridges.

Distribution: su-KlaR, su-CasR, n. & c. su-SN, OR, WA.
Elevation: 2000-2700 m.

NDDB/Holland type and status: **Meadow and seep** (45000).
> Dry subalpine or alpine meadow (45220 *in part*) G3 S3.

Other types:
Cheatham & Haller type: Meadows and swamps.
PSW-45 type: Fescue series.
Rangeland type: SRM 103.

General references: Daubenmire & Daubenmire (1968), Franklin & Dyrness (1973), Johnson (1994), Johnson & Simon (1987), Paysen *et al.* (1980), Reid (1942), Weaver (1979).

Comments: This description follows Johnson (1994). Developed stands are nearly forb-free with continuous sod. Deteriorated stands have bare ground, hummocks of grass among many forbs and erosion pavement. This series may make a mosaic with stands of lodgepole pine or whitebark pine [*see* Lodgepole pine series, Whitebark pine series].

Species mentioned in text	
Green fescue	*Festuca viridula*
Lupines	*Lupinus* species
Sedges	*Carex* species
Western needlegrass	*Stipa occidentalis*
Yarrow	*Achillea millefolium*

Plot-based descriptions: not available.

Iceplant series

Iceplant sole or dominant herb in the ground canopy; beach bursage, dune buckwheat, seashore bluegrass, and/or yellow sand-verbena may be present. Emergent shrubs may be present. Herbs < 50 cm; canopy continuous.

Uplands: bluffs, disturbed land, sand dunes of immediate coastline.

Distribution: o-NorCo, o-CenCo, o-SoCo, ChaI.
Elevation: sea level-100 m.

NDDB/Holland type and status: none.

Other types:
Barry type: G7411321.

General references: Zedler & Scheid (1988).

Comments: The invasive character of several kinds of herbs in the Aizoaceae is well appreciated in California. Three genera are included in this series. *The Jepson Manual* recognizes two *Carpobrotus*, one *Malephora,* and two *Mesembryanthemum* in California. These five species have historically been placed in *Mesembryanthemum,* and referred to as crystalline iceplant, fig-marigold, Hottentot-fig, orange iceplant, sea-fig, or slenderleaf iceplant. All are invasive, and replace native species.

Iceplant stands are being removed to restore native stands and enhance populations of rare species. The treatment here is broad to recognize the importance of these introduced species in the vegetation of California.

Species mentioned in text	
Beach bursage	*Ambrosia chamissonis*
Crystalline iceplant	*Mesembryanthemum crystallinum*
Dune buckwheat	*Eriogonum latifolium*
Fig-marigold	*Carpobrotus edulis*
Hottentot-fig	*Carpobrotus edulis*
Orange iceplant	*Malephora crocea*
Sea-fig	*Carpobrotus chinensis*
Seashore bluegrass	*Poa douglasii*
Slenderleaf iceplant	*Mesembryanthemum nodiflorum*
Yellow sand-verbena	*Abronia latifolia*

Plot-based descriptions: not available.

Idaho fescue series

Idaho fescue sole or dominant grass in ground layer; bluebunch wheatgrass, bracken, California melic, California oatgrass, foothill sedge, one-sided bluegrass, purple needlegrass, red fescue, tall-oatgrass, velvet grass, and/or vernal grass may be present. Emergent shrubs and trees may be present. Grass < 1 m; cover open (Plate 5).

Uplands: slopes and ridges.

Distribution: NorCo, KlaR, m-CasR, n. & c. m-SN, o-CenCo, ModP, inter-West.
Elevation: 20-1800 m.

NDDB/Holland type and status: Coastal prairies (41000), **Great Basin grasslands** (43000) G1 S1.1.
    Bald hills prairie (41200 *in part*) G2 S2.1.
    Serpentine bushgrass (42130) G2 S2.2.

Other types:
Barry type: G741131 BFEID00.
Cheatham & Haller type: Cismontane native grassland.
PSW-45 type: Fescue series.
Rangeland types: SRM 102, SRM 214
Thorne type: Northern coastal prairie.
WHR type: Perennial grass.

General references: Daubenmire (1970), Heady *et al.* (1977b), Kruckeberg (1984), Paysen *et al.* (1980), Tisdale (1994).

Comments: Often considered part of the coastal prairie, this series extends from coastal terraces to inland bald hills where it can associate with other series on a fine scale [*see* California oatgrass series, Pacific reedgrass series, Tufted hairgrass series]. It often mixes with tree series at a coarser scale [*see* Douglas-fir - tanoak series, Oregon white oak series, Redwood series].

The series occurs in Nor-CA away from the coast with different associates. Idaho fescue is common on ultramafic-derived soils in KlaR, and extensive on basalt-derived soils in ModP.

Species mentioned in text	
Bluebunch wheatgrass	*Elymus spicatus*
California melic	*Melica californica*
California oatgrass	*Danthonia californica*
Foothill sedge	*Carex tumulicola*
Idaho fescue	*Festuca idahoensis*
One-sided bluegrass	*Poa secunda*
Purple needlegrass	*Stipa pulchra*
Red fescue	*Festuca rubra*
Tall-oatgrass	*Arrhenatherum elatius*
Velvet grass	*Holcus lanatus*
Vernal grass	*Anthoxanthum odoratum*
Yarrow	*Achillea millefolium*

Plot-based descriptions: Jimerson (1993) defines an Idaho fescue/yarrow association in Six Rivers National Forest l-KlaR. NDDB has plot data on file for stands in Modoc Co. su-WarR, Sheldon Antelope Refuge NV.

Indian ricegrass series

Indian ricegrass sole or dominant grass in ground layer; cheatgrass, Junegrass, needle-and-thread, one-sided bluegrass, squirreltail, and/or western wheatgrass may be present. Emergent shrubs may be present. Grass < 1.5 m; cover open.

Uplands: all topographic locations. Soils sandy.

Distribution: Trans-CA, inter-West.
Elevation: sea level-3400 m.

NDDB/Holland type and status: Mojavean desert scrub (34000), **Great Basin grasslands** (43000 *in part*) G2 S2.2.
   Mojave mixed steppe (34220) G1 S1.1.

Other types:
Barry type: G7411331 BORHY00.
Brown Lowe Pase type: 142.231.
Cheatham & Haller type: Great Basin native grassland.
PSW-45 type: Ricegrass series.
Thorne type: Great Basin sagebrush scrub.
WHR type: Perennial grass.

General references: Heady (1977), Paysen *et al.* (1980), Stoddart *et al.* (1975), Turner & Brown (1982).

Comments: This grass is a component of many Trans-CA series. Indian ricegrass stands prefer sandy substrates where they form open cover in mosaics with grass and shrub series. Only stands where it is dominant are included in this series; secondary species with Indian ricegrass vary regionally. Today most stands are small.

GB stands occur on Madeline Plain where it forms fine mosaics with stands of big sagebrush and bitterbrush [*see* Big sagebrush series, Bitterbrush series].

In the ecological literature Indian ricegrass is referred to as *Oryzopsis hymenoides*. *The Jepson Manual* places Indian ricegrass in the genus *Achnatherum*; other manuals place it in the genus *Stipa*.

Species mentioned in text	
Cheatgrass	*Bromus tectorum*
Granite-gilia	*Leptodactylon pungens*
Indian ricegrass	*Oryzopsis hymenoides*
Junegrass	*Koeleria macrantha*
Needle-and-thread	*Stipa comata*
One-sided bluegrass	*Poa secunda*
Squirreltail	*Elymus elymoides*
Western wheatgrass	*Elymus smithii*

Plot-based descriptions: NDDB has data from Madeline Plain in Modoc Co. and north of Honey Lake in Lassen Co. ModP; Major & Taylor (1977) define a Granitic-gilia - Indian ricegrass association in a-SN.

Introduced perennial grassland series

Introduced perennial grasses sole or dominant in ground layer; blue wildrye, bracken, California oatgrass, California brome, creeping bent, foothill sedge, green fescue, Kentucky bluegrass, one-sided bluegrass, orchard grass, red fescue, tall-oatgrass, velvet grass, and/or vernal grass may be present. Emergent shrubs and trees may be present. Grass < 1 m; cover open (Plate 5).

Wetlands: habitat seasonally, permanently saturated with shallow water table. Water chemistry: fresh. Cowardin class: Palustrine emergent wetland. **Uplands:** all topographic locations. Texture clay, loam, sand. The national list of wetland plants (Reed 1988) may list dominant as a FAC.

Distribution: Cis-CA, GB.
Elevation: sea level-3500 m.

NDDB/Holland type and status: **Coastal prairies** (41000), **Great Basin grasslands** (43000).

Other types:
Barry type: G741231.

General references: Foin & Platenkamp (1989), Peart (1982).

Comments: Often considered part of the coastal prairie, where introduced grasses, especially tall-oatgrass, velvet grass and vernal grass, dominate extensive areas. Areas dominated by an introduced grass in other regions of the state are included in this definition as well.

Plot-based descriptions: Heady *et al.* (1977b) and Foin & Hektner (1977) define associations in Sonoma Co., Grenier (1989), Saenz & Sawyer (1986) each one association in Redwood National Park, Humboldt Co. o-NorCo; Jimerson (1993) four

associations in Six Rivers National Forest o-NorCo, l-KlaR. NDDB has plot data on file for velvet grass-dominated stands in Sonoma Co. o-NorCo. Brown (1993) presents results of 40 transects, following the CNPS protocol. The purpose was to establish monitoring sites at Sonoma Coast State Beaches in Sonoma Co. o-NorCo. The most common species to dominate sites was velvet grass.

Species mentioned in text	
Blue wildrye	*Elymus glaucus*
Bracken	*Pteridium aquilinum*
California brome	*Bromus carinatus*
California oatgrass	*Danthonia californica*
Creeping bent	*Agrostis stolonifera*
Dandelion	*Taraxacum officinale*
Dogtail	*Cynosurus cristatus*
Foothill sedge	*Carex tumulicola*
Green fescue	*Festuca viridula*
Hairy oatgrass	*Danthonia pilosa*
Hareleaf	*Lagophylla glandulosa*
Kentucky bluegrass	*Poa pratensis*
One-sided bluegrass	*Poa secunda*
Orchard grass	*Dactylis glomerata*
Red fescue	*Festuca rubra*
Soft chess	*Bromus hordeaceus*
Tall-oatgrass	*Arrhenatherum elatius*
Tufted hairgrass	*Deschampsia cespitosa*
Velvet grass	*Holcus lanatus*
Vernal grass	*Anthoxanthum odoratum*

Associations:
Hektner & Foin (1977):
    Hairy oatgrass-vernal grass association.
Grenier (1989):
    Foothill sedge - tall-oatgrass association.
Heady *et al.* (1977b):
    Vernal grass-tufted hairgrass association
    [= Hektner & Foin (1977)].
Jimerson (1993):
    Dogtail-hareleaf association,
    Dogtail-soft chess-dandelion association,
    Tall-oatgrass - soft chess association.
Saenz & Sawyer (1986):
    Creeping bent - tall-oatgrass association.

Kentucky bluegrass series

Kentucky bluegrass sole or dominant in ground layer; blue wildrye, bracken, California brome, foothill sedge, one-sided bluegrass, orchard grass, sedges, velvet grass, and/or vernal grass may be present. Emergent shrubs and trees may be present. Grass < 1 m; cover open.

Wetlands: habitat seasonally, permanently saturated with shallow water table. Water chemistry: fresh, saline. Cowardin class: Palustrine emergent wetland. **Uplands:** all topographic locations. The national list of wetland plants (Reed 1988) lists Kentucky bluegrass as a FACU.

Distribution: Cis-CA, GB, inter-West.
Elevation: sea level-3500 m.

NDDB/Holland type and status: **Coastal prairies** (41000), **Valley and foothill grasslands** (42000), **Great Basin grasslands** (43000) G1 S1.1, **Meadows and seeps** (45000).

Other types:
Barry type: G7412311.

General references: Allen-Diaz (1991).

Comments: Kentucky bluegrass is probably a European introduction extensively naturalized in freshwater or saline, disturbed or stable meadows throughout California. It hybridizes with *Poa secunda* and its ecotypes [*P. ampla, P. canbyi, P. gracillima, P. incurva, P. limosa, P. nevadensis, P. sandbergii, P. scabrella; see* One-sided bluegrass series]. In the Sierra Nevada, Ratliff (1982, 1985) describes this series as containing many sedges.

Plot-based descriptions: Stuart *et al.* (1993) defines one association at Castle Crags State Park in Siskiyou Co. l-KlaR; Allen-Diaz (1991) three associations in Sierra Co. m-SN.

Species mentioned in text	
Blue wildrye	*Elymus glaucus*
Bracken	*Pteridium aquilinum*
California oatgrass	*Danthonia californica*
California brome	*Bromus carinatus*
Creeping bent	*Agrostis stolonifera*
Foothill sedge	*Carex tumulicola*
Green fescue	*Festuca viridula*
Hairy woodrush	*Luzula comosa*
Kentucky bluegrass	*Poa pratensis*
Narrower sedge	*Carex angustita*
One-sided bluegrass	*Poa secunda*
Orchard grass	*Dactylis glomerata*
Red fescue	*Festuca rubra*
Sedges	*Carex* species
Silver cinquefoil	*Potentilla gracilis*
Spreading rush	*Juncus patens*
Tall-oatgrass	*Arrhenatherum elatius*
Velvet grass	*Holcus lanatus*
Vernal grass	*Anthoxanthum odoratum*

Associations:
Allen-Diaz (1991):
    Kentucky bluegrass-sedge association,
    Kentucky bluegrass-silver cinquefoil association,
    Narrower sedge-Kentucky bluegrass association.
Stuart *et al.* (1993):
    Kentucky bluegrass-spreading rush-hairy woodrush association.

Mosquito fern series

Mosquito ferns sole or dominant plants floating on water surface; duckmeats, duckweeds, mud-midgets, water-meals, and/or woliffa may be present. Emergent plants may be present. Plants 0.3-8 mm in size; cover continuous, intermittent, or open (Plate 2).

Wetlands: habitat permanently flooded, semipermanently flooded, seasonally flooded. Water chemistry: fresh. Ditch, river, stream channels; ponds. Cowardin class: Palustrine aquatic bed wetland. Riverine lower perennial aquatic bed wetland. The national list of wetland plants (Reed 1988) lists mosquito ferns as a OBL.

Distribution: Cis-CA, Trans-CA, worldwide.
Elevation: sea level-2300 m.

NDDB/Holland type and status: **Marsh and swamp** (52000).
> Coastal and valley freshwater marsh (52410 *in part*) G3 S2.1.
> Transmontane freshwater marsh (52420 *in part*) G3 S2.2.
> Montane freshwater marsh (52430 *in part*) G3 S3.

Other types:
Barry type: G7412311.
Cheatham & Haller type: Freshwater marsh.
Thorne type: Freshwater aquatic.
WHR type: Freshwater emergent wetland.

General references: Mason (1957).

Comments: *The Jepson Manual* lists two species, Mexican mosquito fern and mosquito fern; the former is uncommon. This series can occur in the same area with other aquatic series [*see* especially the Duckweed series]. For a stand to meet this series definition the mosquito ferns dominate over duckweeds or floating-leaved or emergent plants.

Species mentioned in text	
Duckmeats	*Spirodela* species
Greater-duckweed	*Spirodela polyrrhiza*
Inflated duckweed	*Lemna gibba*
Lesser duckweed	*Lemna minor*
Mexican mosquito fern	*Azolla mexicana*
Mosquito fern	*Azolla filiculoides*
Mud-midgets	*Wolffiella* species
Water-meals	*Wolffia* species
Wolffia	*Wolffia borealis*

Plot-based descriptions: not available.

Native dunegrass series

Native dunegrass sole or dominant grass in a ground canopy; beach bursage, beach morning glory, European beachgrass, pink sand-verbena, sea-rocket, seashore bluegrass, and/or yellow sand-verbena may be present. Grass < 1m; cover sparse (Plate 4).

Uplands: dunes of coastal bars, river mouths, spits along the immediate coastline.

Distribution: o-NorCo, o-CenCo to AK, Asia.
Elevation: sea level.

NDDB/Holland type and status: **Coastal dunes** (21000).
    Northern foredune grassland (21211) G1 S1.1.

Other types:
Barry type: G7411311 FELMO30.
Cheatham & Haller type: Northern coastal foredunes.
Thorne type: Coastal dune sand plant.
WHR type: Perennial grass.

General references: Barbour *et al.* (1977d).

Comments: This series occurs along the coast on the seaward edge of coastal dunes (Barbour & Johnson 1977c). Stands of native dunegrass grow in the same habitat as European beachgrass which has reduced and replaced this series [*see* European beachgrass series]. Much recent interest in the dune ecology surrounds introduced (exotic) species, such as European beachgrass, that change vegetation patterns on the coastal dunes (Pickart *et al.* 1989).

Plot-based descriptions: Johnson (1963) and Parker (1974) in Barbour & Johnson (1977) each define one association in Humboldt Co. o-NorCo. LaBanca (1993) resampled the area studied by Parker, finding a rather different organization of dune plants. He recognizes two associations. Bluestone (1981) defines one association at Salinas River State Beach in Monterey Co. o-CenCo.

Species mentioned in text	
Beach bursage	*Ambrosia chamissonis*
Beach morning glory	*Calystegia soldanella*
European beachgrass	*Ammophila arenaria*
Hottentot-fig	*Carpobrotus edulis*
Native dunegrass	*Leymus mollis*
Pink sand-verbena	*Abronia maritima*
Sea-rocket	*Cakile* species
Seashore bluegrass	*Poa douglasii*
Yellow sand-verbena	*Abronia latifolia*

Associations:
Bluestone (1981):
    Native dunegrass-Hottentot-fig association.
LaBanca (1993):
    Native dunegrass-European beachgrass association.
Parker (1974):
    Native dunegrass-sea rocket association
    [as *Elymus-Cakile* and *Convolvulus-Elymus* communities, = Strand community (Johnson 1963), = *Elymus mollis* association (LaBanca 1993)].

Nebraska sedge series

Nebraska sedge sole or dominant herb in ground canopy; bulrushes, rushes, sedges, Sierra ricegrass, and/or spikerushes may be present. Herbs < 1 m; canopy continuous (Plate 3).

Wetlands: habitat seasonally flooded, semipermanently flooded, permanently saturated, seasonally saturated. Water chemistry: fresh. Margins of channels. Cowardin class: Palustrine nonpersistent emergent freshwater wetland. The national list of wetland plants (Reed 1988) lists Nebraska sedge as a OBL.

Distribution: m-KlaR, su-KlaR, m-CasR, su-CasR, m-SN, su-SN, GB, DesR, inter-West. **Elevation:** 900-2700 m.

NDDB/Holland type and status: Meadows and seeps (45000).
> Wet montane meadow (45110 *in part*) G3 S3.
> Freshwater seep (45400 *in part*) G4 S4.

Other types:
Barry type: G7411331.
Cheatham & Haller type: Meadows and swamps.
PSW-45 type: Sedge series.
Rangeland types: SRM 213, SRM 216.
Thorne type: Freshwater aquatic.
WHR type: Wet meadow.

General references: Allen-Diaz (1994), Hermann (1970), Mason (1957), Paysen *et al.* (1980).

Comments: This series occurs in inter-West mountains. If Nebraska sedge is present, but not dominant, the stand is a member of the Sedge series [*see* also Beaked sedge series, Rocky Mountain sedge series].

Plot-based descriptions: Halpern (1986) defines one association at Sequoia National Park in Fresno Co.

m-SN; Beguin & Major (1975), Burke (1987) in Keeler-Wolf (1990c) one association Grass Lake RNA in El Dorado Co. su-SN. Ratliff (1985) takes up montane and subalpine meadows for the central SN, especially Yosemite National Park and surrounding areas.

Species mentioned in text	
Bulrushes	*Scirpus* species
Beaked sedge	*Carex utriculata*
Nebraska sedge	*Carex nebrascensis*
Rushes	*Juncus* species
Rocky Mountain sedge	*Carex scopulorum*
Sedges	*Carex* species
Sierra ricegrass	*Oryzopsis kingii*
Spikerushes	*Eleocharis* species

Associations:
Beguin & Major (1975):
> Nebraska sedge association [as Cericetum nebraskensis, = Burke (1987) in Keeler-Wolf (1990e), included in Nebraska sedge vegetative series (Ratliff 1985)].

Halpern (1986):
> Nebraska sedge-Sierra ricegrass association [included in Nebraska sedge vegetative series (Ratliff 1985)].

Needle-and-thread series

Needle-and-thread sole or dominant grass in ground layer; ashy ryegrass, bluebunch wheatgrass, cheatgrass, Indian ricegrass, Junegrass, one-sided bluegrass, squirreltail, Thurber needlegrass, and/or western wheatgrass may be present. Emergent shrubs may be present. Grass < 1.5 m; cover open.

Uplands: all topographic locations.

Distribution: SN, PenR, GB, DesR, inter-West. **Elevation:** 200-3500 m.

NDDB/Holland type and status: **Great Basin grasslands** (43000 *in part*)

Other types:
Barry type: G7411331 BSTCO40.
Cheatham & Haller type: Great Basin native grassland.
PSW-45 type: Needlegrass series.
Thorne type: Great Basin sagebrush scrub.
WHR type: Perennial grass.

General references: Heady (1977), Paysen *et al.* (1980), Stoddart *et al.* (1975).

Comments: Stands of this series now typically include non-native, annual species mixed with native perennial grasses and herbs. Cheatgrass and medusa-head are common. NDDB finds no evidence of the presence of this series in GB today despite widespread presence of the species.

In the ecological literature needle-and-thread is referred to as *Stipa comata. The Jepson Manual* places needle-and-thread in the genus *Hesperostipa.*

Species mentioned in text	
Ashy ryegrass	*Leymus cinereus*
Baltic rush	*Juncus balticus*
Bluebunch wheatgrass	*Elymus spicatus*
Cheatgrass	*Bromus tectorum*
Indian ricegrass	*Oryzopsis hymenoides*
Junegrass	*Koeleria macrantha*
Medusa-head	*Taeniatherum caput-medusae*
Needle-and-thread	*Stipa comata*
One-sided bluegrass	*Poa secunda*
Squirreltail	*Elymus elymoides*
Thurber needlegrass	*Stipa thurberiana*
Western wheatgrass	*Elymus smithii*

Plot-based descriptions: Major & Taylor (1977) define a Needle-and-thread - Baltic rush association in Inyo Co. a-SN.

Nodding needlegrass series

Nodding needlegrass sole or dominant grass in ground layer; blue wildrye, California fescue, California melic, creeping wildrye, hook three-awn, Junegrass, and/or one-sided bluegrass may be present. Emergent shrubs and trees may be present. Grass < 1 m; cover open (Plate 5).

Uplands: all topographic locations. Soils deep with high clay content.

Distribution: i-NorCo, CenCo, m-TraR, i-SoCo, m-PenR, Baja CA.
Elevation: sea level-1400 m.

NDDB/Holland type and status: Valley and foothill grasslands (42000).
   Valley needlegrass grassland (42110 *in part*) G3 S3.1.

Other types:
Barry type: G7411331 BSICE00.
Cheatham & Haller type: Cismontane native grassland.
PSW-45 type: Needlegrass series.
Thorne type: Great Valley and Coast Range valley grassland.
WHR type: Perennial grass.

General references: Heady (1977), Keeley (1993), Magney (1992), Paysen *et al.* (1980), Stoddart *et al.* (1975).

Comments: Stands of this once extensive series now typically include non-native, annual species mixed with native perennial grasses and herbs. Ripgut, soft chess, and foxtail chess are common as are slender oats, wild oats, and Italian ryegrass.

Foothill needlegrass, nodding needlegrass, and purple needlegrass may occur sympatrically, but they do not typically mix, especially in SoCo. They tend to segregate based on substrate and slope factors [*see* Foothill needlegrass series, Purple needlegrass series].

In the ecological literature nodding needlegrass is referred to as *Stipa cernua. The Jepson Manual* places nodding needlegrass in the genus *Nassella.*

Species mentioned in text	
Blue oak	*Quercus douglasii*
Blue wildrye	*Elymus glaucus*
California fescue	*Festuca californica*
California melic	*Melica californica*
Creeping ryegrass	*Leymus triticoides*
Foothill needlegrass	*Stipa lepida*
Foxtail chess	*Bromus madritensis*
Hook three-awn	*Aristida ternipes*
Italian ryegrass	*Lolium multiflorum*
Junegrass	*Koeleria macrantha*
Nodding needlegrass	*Stipa cernua*
One-sided bluegrass	*Poa secunda*
Purple needlegrass	*Stipa pulchra*
Ripgut	*Bromus diandrus*
Slender oats	*Avena barbata*
Soft chess	*Bromus hordeaceus*
Wild oats	*Avena fatua*

Plot-based descriptions: Keeler-Wolf (1989a) describes blue oak stands with a high cover of nodding needlegrass at Wagon Caves candidate RNA in Monterey Co. o-CenCo; Kellogg & Kellogg (1990, 1991) present unclassified plot data for areas dominated by foothill needlegrass, nodding needlegrass, or purple needlegrass at Pendleton Marine Corps Base in San Diego Co. o-SoCo.

One-sided bluegrass series

One-sided bluegrass sole or dominant grass in ground layer; big squirreltail, creeping ryegrass, needle-and-thread, one-spike oatgrass, red three-awn, sand dropseed, and/ or squirreltail may be present. Emergent shrubs and trees may be present. Grass < 1 m; cover open.

Wetlands: habitat seasonally, permanently saturated with shallow water table. Water chemistry: fresh, saline. Valley bottoms, lower portions of alluvial slopes. Cowardin class: Palustrine emergent wetland. **Uplands:** all topographic locations. Soils may be ultramafic-derived. The national list of wetland plants (Reed 1988) lists one-sided bluegrass as a FAC.

Distribution: Cis-CA, Trans-CA, inter-West.
Elevation: sea level-3800 m.

Species mentioned in text	
Big squirreltail	*Elymus multisetus*
Creeping ryegrass	*Leymus triticoides*
Needle-and-thread	*Stipa comata*
One-sided bluegrass	*Poa secunda*
One-spike oatgrass	*Danthonia unispicata*
Red three-awn	*Aristida purpurea*
Sand dropseed	*Sporobolus cryptandrus*
Squirreltail	*Elymus elymoides*

Plot-based descriptions: NDDB has plot data on file for stands in Los Angeles, Kern, and Modoc Cos.

NDDB/Holland type and status: **Valley and foothill grasslands** (42000), **Great Basin grasslands** (43000).
    Valley grassland (42110 *in part*) G3 S3.1.
    Pine bunchgrass (42130 *in part*) G2 S2.1.
    Pine bluegrass grassland (421500) G3 S2.2.

Other types:
Barry type: G7411331 BPOSE00.
Cheatham & Haller type: Cismontane native grassland, Great Basin native grassland.
Thorne type: Great Basin sagebrush scrub.
WHR type: Perennial grass.

General references: Heady (1977).

Comments: *The Jepson Manual* includes two subspecies in *Poa secunda, P. s.* ssp. *juncifolia* and *P. s.* ssp. *s.,* and eight species of other authors (*P. ampla, P. canbyi, P. gracillima, P. incurva, P. limosa, P. nevadensis, P. sandbergii, P. scabrella*) in this species definition. These taxa and hybrids with *P. pratensis* are incorporated into this series.

Pacific reedgrass series

Pacific reedgrass sole or dominant grass in ground layer; alta fescue, bracken, California blackberry, coast mugwort, cow-parsnip, red fescue, velvet grass, and/or vernal grass may be present. Emergent shrubs, such as black huckleberry, and trees, such as Sitka spruce, may be present. Grass < 1.5 m; cover open (Plate 4).

Wetlands: habitat seasonally, permanently saturated with shallow water table. Water chemistry: fresh, saline. Valley bottoms, lower portions of alluvial slopes. Cowardin class: Palustrine emergent wetland. **Uplands:** coastal bluffs, terraces. The national list of wetland plants (Reed 1988) lists Pacific reedgrass as a FACW.

Distribution: o-NorCo, o-CenCo to AK.
Elevation: sea level-30 m.

NDDB/Holland type and status: Coastal prairies (41000).
    Coastal terrace prairie (41100 *in part*) G2 S2.2.

Other types:
Barry type: G7411331 BCANU30.
Cheatham & Haller type: Cismontane native grassland.
Rangeland type: SRM 214.
Thorne type: Northern coastal prairie.
WHR type: Perennial grass.

General references: Bartolome *et al.* (1994), Heady *et al.* (1977b).

Comments: Often considered part of the coastal prairie, this series is restricted to the coastal terraces where it mixes with other series at a fine scale [*see* California oatgrass series, Idaho fescue series, Introduced perennial grassland series, Tufted hairgrass series]. Stands of Pacific reedgrass series mix with woody series at a coarser scale, especially on headlands [*see* Coyote brush series, Hooker willow series, Red alder series, Salal-black huckleberry series, Sitka spruce series, Yellow bush lupine series].

Species mentioned in text	
Alta fescue	*Festuca arundinacea*
Black huckleberry	*Vaccinium ovatum*
Bracken	*Pteridium aquilinum*
California blackberry	*Rubus ursinus*
California oatgrass	*Danthonia californica*
Coast mugwort	*Artemisia suksdorfii*
Cow-parsnip	*Heracleum lanatum*
Coyote brush	*Baccharis pilularis*
Hooker willow	*Salix hookeriana*
Idaho fescue	*Festuca idahoensis*
Pacific reedgrass	*Calamagrostis nutkaensis*
Red alder	*Alnus rubra*
Red fescue	*Festuca rubra*
Salal	*Gaultheria shallon*
Sitka spruce	*Picea sitchensis*
Tufted hairgrass	*Deschampsia cespitosa*
Velvet grass	*Holcus lanatus*
Vernal grass	*Anthoxanthum odoratum*
Yellow bush lupine	*Lupinus arboreus*

Plot-based descriptions: Heady *et al.* (1977b) report cover and frequency for Pacific reedgrass type, Hektner & Foin (1977) define a Pacific reedgrass association in Sonoma Co., NDDB has plot data on file for Fort Ross State Park in Marin Co., Sonoma and Mendocino Cos., Table Bluff Ecological Area at Eel River Wildlife Area, and at Trinidad State Park in Humboldt Co. o-NorCo.

Pampas grass series

Pampas grasses sole or dominant in ground layer; California buckwheat, California sagebrush, and/or coyote brush may be present. Emergent trees may be present. Grass < 4 m; cover open.

Uplands: all topographic locations.

Distribution: o-NorCo, o-CenCo, SoCo, m-TraR.
Elevation: sea level-800 m.

NDDB/Holland type and status: none.

Other types:
Barry type: G7411331 BCOAT00.

General references: Cowan (1976), Kerbavaz (1985), Lippmann (1977).

Comments: Two species of this introduced, invasive grass from South America grow in disturbed stands of many series. White pampas grass is the less invasive species. Black pampas grass commonly invades Douglas-fir, redwood, and Sitka spruce forest sites after logging in NorCo. It dies out as the tree canopy develops and reduces light levels. This grass may be present, and persist, in stands of several shrub series. The treatment here is broad, recognizing the importance of this introduced genus in the vegetation of California.

Species mentioned in text	
Black pampas grass	*Cortaderia jubata*
California sagebrush	*Artemisia californica*
California buckwheat	*Eriogonum fasciculatum*
Coyote brush	*Baccharis pilularis*
Douglas-fir	*Pseudotsuga menziesii*
Pampas grasses	*Cortaderia* species
Redwood	*Sequoia sempervirens*
Sitka spruce	*Picea sitchensis*
White pampas grass	*Cortaderia selloana*

Plot-based descriptions: none available.

Pickleweed series

Pickleweed sole or dominant plant in ground canopy; alkali heath, arrow-grasses, dense-flowered cordgrass, dodder, fat-hen, jaumea, saltgrass, saltwort, sea-blite, and/ or sea-lavender may be present. Plants < 1.5 m; canopy continuous or intermittent (Plate 1).

Wetlands: habitat regularly flooded, irregularly flooded; permanently saturated with shallow water table. Water chemistry: mixohaline, euhaline, hyperhaline, saline. Estuaries: mud flats, banks, berms, and margins of bays, deltas, sandbars; valley bottoms, lower portions of alluvial slopes. Cowardin classes: Estuarine intertidal persistent emergent wetland, Palustrine persistent emergent saline wetland. The national list of wetland plants (Reed 1988) lists pickleweeds as a OBL.

Distribution: NorCo; o-CenCo; d-CenV, o-SoCo, ChaI, ModP, MojD, ColD.
Elevation: sea level-1200.

NDDB/Holland type and status: Marsh and swamp (32000).
Northern coastal salt marsh (52110 *in part*) G3 S3.2.
Southern coastal salt marsh (52120 *in part*) G2 S2.1.

Other types:
Barry type: G7412311 CSAVI00.
Cheatham & Haller type: Coastal salt marshes.
PSW-45 type: Pickleweed series.
Rangeland type: SRM 217.
Thorne type: Coastal salt marsh.
WHR type: Saline emergent wetland.

General references: Chapman (1977), Henrickson (1976), MacDonald (1977), MacDonald & Barbour (1974), Paysen *et al.* (1980), Vogl (1966), Zedler *et al.* (1992).

Comments: Five species of *Salicornia* grow in California. Bigelow pickleweed and common pickleweed, are frequently found in coastal salt marshes; *S. utahensis* occurs in alkali meadows. *S. europaea* and *S. subterminalis* are found in both habitats.

Bulrush series, Cordgrass series, Ditch-grass series, Pickleweed series, Saltgrass series, and Sedge series occur in CA salt marshes. Barnhart *et al.* (1992) describe Humboldt Bay marshes in Humboldt Co. o-NorCo; Josselyn (1983) details San Francisco Bay salt marshes o-CenCo; Zedler (1982) So-CA ones.

Species mentioned in text	
Alkali heath	*Frankenia salina*
Arrow-grasses	*Triglochin* species
Bigelow pickleweed	*Salicornia bigelovii*
Bulrushes	*Scirpus* species
Common pickleweed	*Salicornia virginica*
Cordgrasses	*Spartina* species
Dense-flowered cordgrass	*Spartina densiflora*
Ditch-grasses	*Ruppia* species
Dodder	*Cuscuta salina*
Fat-hen	*Atriplex patula*
Gumplant	*Grindelia stricta*
Jaumea	*Jaumea carnosa*
Pickleweeds	*Salicornia* species
Saltgrass	*Distichlis spicata*
Saltwort	*Batis maritima*
Sea-blite	*Suaeda californica*
Sea-lavender	*Limonium californicum*

Plot-based descriptions: The Barnhart *et al.* (1992) survey includes three associations defined by Eicher (1987) and Newton (1989), Eicher & Sawyer (1989) describe areas invaded by *Spartina alterniflora* for Humboldt Bay marshes in Humboldt Co., Cuneo (1987) San Francisco and San Pablo bay marshes o-NorCo; Atwater *et al.* (1979) define three associations in Contra Costa, Solano Co. d-CenV; Peinado *et al.* (1994) define one association for CA coastal salt marshes.

Associations:

Atwater *et al.* (1979):
> Common pickleweed association,
> Common pickleweed-gumplant association,
> Common pickleweed-saltgrass association
> [= Eicher (1987), = Newton (1989)].

Eicher (1987):
> Common pickleweed-jaumea-saltgrass association [= Common pickleweed-jaumea association (Newton 1989)].

Peinado et al. (1994):
> Bigelow pickleweed association
> [as Salicornietum bigelovii].

Pondweeds with floating leaves series

Pondweeds sole or dominant herbs with floating leaf blades; alpine pondweed, broadleaf pondweed, diverseleaf pondweed, floatingleaf pondweed, grassleaf pondweed, longleaf pondweed, Nuttall pondweed, and/ or shinning pondweed may be present. Emergent plants be may present. Plants 0.3-3 m in size; cover continuous, intermittent, or open (Plate 2).

Wetlands: habitat permanently flooded, intermittently exposed. Water chemistry: fresh. Ditches; lakes; ponds; slow streams. Cowardin classes: Lacustrine littoral aquatic bed wetland, Palustrine aquatic bed wetland, Riverine aquatic bed wetland. The national list of wetland plants (Reed 1988) lists pondweeds as a OBL.

Distribution: Cis-CA, Trans-CA.
Elevation: sea level-2750 m.

NDDB/Holland type and status: **Marsh and swamp** (52000).
　Coast and valley freshwater marsh (52410) G3 32.2.
　Transmontane freshwater marsh (52420) G. S2.2.
　Montane freshwater marsh (52430 *in part)* G3 S3.

Other types:
Barry type: G7422321 BPONA40.
Cheatham & Haller type: Freshwater marshes.
PSW-45 type: Pondweed series.
Thorne type: Freshwater aquatic.
WHR type: Freshwater emergent wetland.

General references: Mason (1957), Paysen *et al.* (1980).

Comments: Eight species of *Potamogeton* with floating leaves are included in this series. Other pondweeds lacking floating leaves are placed in the Pondweeds with submerged leaves series.

Species mentioned in text	
Alpine pondweed	*Potamogeton alpinus*
Broadleaf pondweed	*Potamogeton amplifolius*
Diverseleaf pondweed	*Potamogeton diversifolius*
Floatingleaf pondweed	*Potamogeton natans*
Grassleaf pondweed	*Potamogeton gramineus*
Longleaf pondweed	*Potamogeton nodosus*
Nuttall pondweed	*Potamogeton epihydrus*
Shinning pondweed	*Potamogeton illinoensis*

Plot-based descriptions: not available.

Pondweeds with submerged leaves series

Pondweeds sole or dominant herbs with submerged leaves; crispate pondweed, eelgrass pondweed, fennelleaf pondweed, leafy pondweed, Nevada pondweed, Richardson pondweed, Robbin pondweed, slenderleaf pondweed, small pondweed, and/or whitestem pondweed be may be present. Emergent plants may present. Plants 0.3-3 m in size; cover continuous, intermittent, or open.

Wetlands: habitat permanently flooded, intermittently exposed. Water chemistry: fresh. Ditches; lakes; ponds; slow streams. Cowardin classes: Lacustrine littoral aquatic bed wetland, Palustrine aquatic bed wetland, Riverine aquatic bed wetland. The national list of wetland plants (Reed 1988) lists pondweeds as a OBL.

Distribution: Cis-CA, Trans-CA.
Elevation: sea level-3300 m.

NDDB/Holland type and status: **Marsh and swamp** (52000).
>   Coast and valley freshwater marsh (52410) G3 S2.2.2
>   Transmontane freshwater marsh (52420) G3 S2.2.
>   Montane freshwater marsh (52430 *in part*) G3 S3.

Other types:
Barry type: G7422321.
Cheatham & Haller type: Freshwater marshes.
PSW-45 type: Pondweed series.
Thorne type: Freshwater aquatic.
WHR type: Freshwater emergent wetland.

General references: Mason (1957), Paysen *et al.* (1980).

Comments: The 10 species of *Potamogeton* in CA with submerged leaves are included in this series. Pondweeds with floating leaves are placed in the Pondweeds with floating leaves series.

Species mentioned in text	
Crispate pondweed	*Potamogeton crispus*
Eel-grass pondweed	*Potamogeton zosteriformis*
Fennelleaf pondweed	*Potamogeton pectinatus*
Leafy pondweed	*Potamogeton foliosus*
Nevada pondweed	*Potamogeton latifolius*
Richardson pondweed	*Potamogeton richardsonii*
Robbin pondweed	*Potamogeton robbinsii*
Slenderleaf pondweed	*Potamogeton filiformis*
Small pondweed	*Potamogeton pusillus*
Whitestem pondweed	*Potamogeton praelongus*

Plot-based descriptions: not available.

Purple needlegrass series

Purple needlegrass sole or dominant grass in ground layer; blue wildrye, California fescue, California melic, Junegrass, one-sided bluegrass, and/or oniongrass may be present. Annual grasses and flowers are common. Emergent shrubs and trees may be present. Grass < 1 m; cover open (Plate 6).

Uplands: all topographic locations. Soils deep with a high clay content.

Distribution: NorCo, CenCo, sac-CenV, SoCo, o. l-TraR, m-PenR. w. MojD, Baja CA.
Elevation: sea level-1300 m.

NDDB/Holland type and status: Valley and foothill grasslands (42000).
> Valley needlegrass grassland (42110 *in part*) G3 S3.1.

Other types:
Barry type: G7411331 BSTPU20.
Cheatham & Haller type: Cismontane native grassland.
PSW-45 type: Needlegrass series.
Thorne type: Great Valley and Coast Range grassland.
WHR type: Perennial grass.

General references: Barry (1972), Bartolome (1981), Bartolome & Gemmill (1981), Beetle (1947), Bittman (1985), Griggs (1980), Heady (1977), Hull & Muller (1977), Keeley (1993a, 1993), Magney (1992), Paysen *et al.* (1980), Stoddart *et al.* (1975), Turner & Brown (1982), White (1966b, 1967).

Comments: Stands of this once extensive series now typically include non-native annual species mixed with the perennial grasses and herbs. Ripgut, soft chess, and foxtail chess are common, as are slender oats, wild oats, and Italian ryegrass. Foothill needlegrass, nodding needlegrass, and purple needlegrass occur sympatrically, but do not typically mix, especially in SoCo. The species tend to segregate based on substrate and slope factors

[*see* Foothill needlegrass series, Nodding needlegrass series]. In the ecological literature purple needlegrass is referred to as *Stipa pulchra*. *The Jepson Manual* places purple needlegrass in the genus *Nassella*.

Species mentioned in text	
Blue wildrye	*Elymus glaucus*
California fescue	*Festuca californica*
California melic	*Melica californica*
Foothill needlegrass	*Stipa lepida*
Foxtail chess	*Bromus madritensis*
Italian ryegrass	*Lolium multiflorum*
Junegrass	*Koeleria micrantha*
Nodding needlegrass	*Stipa cernua*
One-sided bluegrass	*Poa secunda*
Oniongrass	*Melica imperfecta*
Purple needlegrass	*Stipa pulchra*
Purple sanicle	*Sanicula bipinnatifida*
Ripgut	*Bromus diandrus*
Slender oats	*Avena barbata*
Soft chess	*Bromus hordeaceus*
Wild oats	*Avena fatua*

Plot-based descriptions: Fiedler & Leidy (1987) define an association at Ring Mountain, Parker (1990) one association on Mount Tamalpais in Marin Co. o-NorCo; Stuart *et al.* (1993) an association at Castle Crags State Park in Siskiyou Co. l-KlaR; White (1966a, 1967) a native bunchgrass grassland dominated by purple needlegrass at Hasting Reservation in Monterey Co. o-CenCo; Kellogg & Kellogg (1990, 1991) present unclassified plot data on areas dominated by foothill needle-grass, nodding needlegrass, or purple needlegrass at Pendleton Marine Corps Base in San Diego Co. o-SoCo.
Associations:
Fiedler & Leidy (1987):
> Italian ryegrass-purple needlegrass association [as Serpentine bunchgrass grassland].

Parker (1990):
> Wild oats-purple needlegrass association.

Stuart *et al.* (1993):
> Purple needlegrass/purple sanicle association.

Quillwort series

Quillworts sole or dominant herb forming submerged beds in water. Herbs < 10 cm; cover intermittent or open.

Wetlands: habitat permanently flooded, seasonally flooded. Water chemistry: fresh. Lakes; ponds; stream margins; vernal pools. Cowardin classes: Lacustrine littoral aquatic bed wetland, Palustrine aquatic bed wetland. The national list of wetland plants (Reed 1988) lists quillworts as a OBL.

Distribution: m-NorCo, CenCo, CenV, m-KlaR, su-KlaR, m-CasR, f-SN, m-SN, SoCo, PenR, Baja CA, inter-West.
Elevation: sea level-1500 m.

NDDB/Holland type and status: **Marsh and swamp** (52000).
　　Freshwater marsh (52400 *in part*) G4 S4.

Other types:
Barry type: G7412311.
Cheatham & Haller type: Freshwater marshes.
Thorne type: Freshwater aquatic.
WHR type: Freshwater emergent wetland.

General references: Mason (1957).

Comments: *The Jepson Manual* recognizes six quillwort species. Three, *Isoetes bolanderi, I. echinospora,* and *I. occidentalis* are found in montane lakes and ponds; two, *I. howellii* and *I. orcuttii,* in vernal ponds, and *I. nuttallii* may occur along stream margins. Quillworts may dominate lake and pond bottoms, and also occur in vernal pools [*see* Vernal pools section].

Species mentioned in text	
Quillworts	*Isoetes* species
Torreyochloa moss	*Torreyochloa* species

Plot-based descriptions: Taylor (1984) defines a Torreyochloa moss-Bolander quillwort association at Harvey Monroe Hall RNA in Mono Co. a-SN.

Rocky Mountain sedge series

Rocky Mountain sedge sole or dominant herb in ground canopy; bulrushes, sedges, and/or spikerushes may be present. Herbs < 40 cm; canopy continuous.

Wetlands: habitat seasonally flooded, semi-permanently flooded, permanently saturated, seasonally saturated. Water chemistry: fresh. Margins of channels, lakes, ponds, overflow areas, streams; wet meadows. Cowardin classes: Lacustrine littoral unconsolidated shore wetland, Palustrine nonpersistent emergent freshwater wetland, Palustrine persistent emergent freshwater wetland, Palustrine unconsolidated freshwater shore wetland. The national list of wetland plants (Reed 1988) lists Rocky Mountain sedge as a FACW.

Distribution: m-NorCo, m-KlaR, su-KlaR, m-CasR, su-CasR, a-CasR, m-SN, su-SN, a-SN, WarR, WIS, inter-West.
Elevation: 1200-3400 m.

NDDB/Holland type and status: **Meadows and seeps** (45000).
    Wet montane meadow (45110 *in part*) G3 S3.
    Freshwater seep (45400 *in part*) G4 S4.

Other types:
Barry type: G7411331.
Cheatham & Haller type: Meadows and swamps.
PSW-45 type: Sedge series.
Thorne type: Freshwater aquatic.
Rangeland type: SRM 216.
WHR type: Wet meadow.

General references: Allen-Diaz (1994), Hermann (1970), Mason (1957).

Comments: This series occurs throughout in inter-West. If Rocky Mountain sedge is present in a stand, but not dominant, the stand is a member of the Sedge series [*see* also Beaked sedge series, Nebraska sedge series].

Plot-based descriptions: Major & Taylor (1977) define one association in su-SN. Helms & Ratliff (1987), Ratliff (1985) take up montane and subalpine meadows for the central SN, especially Yosemite National Park and surrounding areas. Taylor (1984) in Keeler-Wolf (1990e) defines two associations at Harvey Monroe Hall RNA in Mono Co. a-SN.

Species mentioned in text	
Beaked sedge	*Carex utriculata*
Bulrushes	*Scirpus* species
Cotton-grass	*Eriophorum criniger*
Elephant's ears	*Pedicularis groenlandica*
Nebraska sedge	*Carex nebrascensis*
Rocky Mountain sedge	*Carex scopulorum*
Sedges	*Carex* species
Spikerushes	*Eleocharis* species

Associations:
Major & Taylor (1977):
    Rocky Mountain sedge.
Taylor (1984):
    Rocky Mountain sedge-elephant's ears association,
    Rocky Mountain sedge - cotton-grass association.

Saltgrass series

Saltgrass sole or dominant grass in ground canopy; alkali cordgrass, alkali muhly, alkali sacaton, Baltic rush, common pickleweed, Cooper rush, one-sided bluegrass, saltgrass, sea-lavender, slender arrow-grass, and/or yerba mansa may be present. Emergent alkali rabbitbrush or iodine bush may be present. Plants < 1 m; canopy continuous, intermittent, or open (Plate 1).

Wetlands: habitat irregularly flooded, permanently saturated with shallow water table. Water chemistry: haline, saline. Estuaries: banks, berms, and margins of bays, deltas, sandbars. Valley bottoms, lower portions of alluvial slopes. Cowardin classes: Estuarine intertidal persistent emergent haline wetland, Palustrine persistent emergent saline wetland. The national list of wetland plants (Reed 1988) lists saltgrass as a FAC, FAC*.

Distribution: Cis-CA, Trans-CA, North America.
Elevation: sea level-1000 m.

NDDB/Holland type and status: **Meadow and seep** (45000), **Marsh and swamp** (52000).
 Alkali meadow (45310) G3 S2.1.
 Northern coastal salt marsh (52110 *in part*) G3 S3.3.
 Southern coastal salt marsh (52120 part) G2 S2.1.
Other types:
Barry type: G7411331 CDIST00.
Cheatham & Haller type: Alkali meadows,
Coastal salt marshes.
PSW-45 type: Saltgrass series.
Rangeland type: SRM 217.
Thorne type: Alkali meadow and aquatic,
Coastal salt marsh.
WHR type: Saline emergent wetland.

General references: Chapman (1977), MacDonald (1977), Paysen *et al.* (1980), Thorne (1982), Zedler *et al.* (1992).

Comments: Many regional descriptions include this series in an alkali meadows category. This series is also associated with coastal saltmarshes and grasslands [*see* Alkali sacaton series, Ashy ryegrass series, Creeping ryegrass series, Spikerush series].

Species mentioned in text	
Alkali cordgrass	*Spartina gracilis*
Alkali rabbitbrush	*Chrysothamnus albidus*
Alkali sacaton	*Sporobolus airoides*
Alkali muhly	*Muhlenbergia asperifolia*
Baltic rush	*Juncus balticus*
Common pickleweed	*Salicornia virginica*
Cooper rush	*Juncus cooperi*
Greasewood	*Sarcobatus vermiculatus*
Iodine bush	*Allenrolfea occidentalis*
Jaumea	*Jaumea carnosa*
One-sided bluegrass	*Poa secunda*
Saltgrass	*Distichlis spicata*
Sea-lavender	*Limonium californicum*
Slender arrow-grass	*Triglochin concinna*
Yerba mansa	*Anemopsis californica*

Plot-based descriptions: Newton (1989) defines one association in Humboldt Co. o-NorCo, Peinado *et al.* (1994) one association for CA coastal salt marshes; Atwater *et al.* (1979) one association in Contra Costa, Solano Cos. d-CenV; Boyd (1983) describes three stands in Gavilan Hills in Riverside Co. i. m-TraR; Bradley (1970) defines two associations in Death Valley National Park in Inyo Co. MojD; Ferren & Davis (1991) one association, Odion *et al.* (1992) two associations at Fish Slough TraSN.

Associations:
Atwater *et al.* (1979):
 Common pickleweed-saltgrass association [=(Newton 1989)].
Bradley (1970):
 Saltgrass-iodine bush association [as Eastern salt flat],

Saltgrass-Cooper rush association
[as Salt grass complex].
Ferren & Davis (1991):
Greasewood-saltgrass association.
Odion *et al.* (1992):
Saltgrass-alkali rabbitbrush association
[as *Distichlis-Chrysothamnus* association],
Baltic rush-saltgrass association
[as *Juncus-Distichlis* association].
Peinado et al. (1994):
Jaumea-saltgrass association
[as Jaumeo carnosae-Distichlidetum spicatae].

Sand-verbena - beach bursage series

Perennial forbs, grasses, and low shrubs form a ground canopy; beach bursage, beach morning glory, beach pea, California croton, dune buckwheat, dune lupine, dune sagebrush, pink sand-verbena, saltgrass, seashore bluegrass, sun cups, and/or yellow sand-verbena may be present. Individual emergent shrubs may be present. Ground layer open or continuous (Plate 6).

Uplands: sand dunes of coastal bars, river mouths, spits along the immediate coastline.

Distribution: o-NorCo, o-CenCo, o-SoCo, ChaI.
Elevation: sea level.

NDDB/Holland type and status: **Coastal dunes** (21000).
    Active coastal dunes (21100) G3 S2.2.
    Northern foredunes (21210) G2 S2.1.
    Southern foredunes (21320) G2 S2.1.

Other types:
Barry type: G7411321 CABOOO0.
Cheatham & Haller type: Coastal foredunes.
Thorne type: Coastal dune sand plant.
WHR type: Coastal scrub.

General references: Barbour (1970), Barbour & Johnson (1977), Breckon (1974), MacDonald & Barbour (1974), McBride & Stone (1976), Williams (1985).

Comments: The Sand-verbena - beach bursage series occurs along the coast at localized stretches of sand isolated by headlands and rocky tidepools. Davy (1902) presents an early account of Humboldt and Del Norte Cos. dunes.

Patches of European beachgrass, iceplant, coyote brush, and yellow bush lupine can occur as emergent plants.Much recent interest in the State's dune ecology focuses on introduced (exotic) species that have changed vegetation patterns and dynamics. European beachgrass, yellow bush lupine, and iceplant are being managed in some areas to restore native-dominated stands and enhance populations of rare Humboldt Bay wallflower (a CNPS List 1B plant) and beach layia (a CNPS List 1B plant) (Pickart *et al.* 1989, Skinner & Pavlik 1994). [*see* European beachgrass series, Yellow bush lupine series].

Other series that occur on coastal dunes are Dune lupine-heather goldenbush series, Coyote brush series, European beachgrass series, Hooker willow series, Iceplant series, Native dunegrass series, Sedge series, and Yellow bush lupine series.

Species mentioned in text	
Beach bursage	Ambrosia chamissonis
Beach layia	Layia carnosa
Beach morning glory	Calystegia soldanella
Beach pea	Lathyrus littoralis
California croton	Croton californicus
Chamise	Adenostoma fasciculatum
Coyote brush	Baccharis pilularis
Dune buckwheat	Eriogonum latifolium
Dune lupine	Lupinus chamissonis
Dune sagebrush	Artemisia pycnocephala
European beachgrass	Ammophila arenaria
Heather goldenbush	Ericameria ericoides
Hooker willow	Salix hookeriana
Humboldt wallflower	Erysimum menziesii var. eurekensis
Iceplant	Carpobrotus species
Native dunegrass	Leymus mollis
Pink sand-verbena	Abronia maritima
Sand-verbena	Abronia species
Saltgrass	Distichlis spicata
Seashore bluegrass	Poa douglasii
Seaside woolly-sunflower	Eriophyllum staechadifolium
Sedges	Carex species
Sun cups	Camissonia cheiranthifolia
Yellow bush lupine	Lupinus arboreus
Yellow sand-verbena	Abronia latifolia

Plot-based descriptions: Duebendorfer (1989), Johnson (1963), Parker (1974) in Barbour & Johnson (1977), Pickart (1989) define four associations from Humboldt Bay to Trinidad Head, Humboldt Co.; Holton & Johnson (1979) two associations at Point Reyes National Seashore in Marin Co. o-NorCo; Bluestone (1981) one association at Salinas River State Beach in Monterey Co., Williams & Potter (1972), Williams (1985) one association at Morro Bay in San Luis Obispo Co. o-CenCo. Couch (1914) presents counts for 7 plots at Manhattan Beach which he describes as "practically undisturbed" in Los Angeles Co. o-SoCo. The plots closest to the ocean are dominated by beach bursage, and the inland ones with chamise and dune plants. NDDB has data from Santa Catalina Island ChaI.

Associations:
Bluestone (1981):
    Beach morning glory-dune sagebrush
    [as Foredune hallow and middune].
Duebendorfer (1989):
    Seashore bluegrass-dune sagebrush association
    [as Southern dune mat, = *Artemisia* phase of
    dune mat (Pickart 1987)].
Holton & Johnson (1979):
    Beach bursage - seaside woolly-sunflower
    association,
    Seaside woolly-sunflower - yellow bush lupine
    association.

Johnson (1963):
    Plants on active sand
    [as Moving dunes
    community].
Parker (1974) in Barbour and Johnson (1977c):
    Seashore bluegrass-beach pea association [as
    Poa-Lathyrus community, = Stabilized ridge
    community, Foredune community (Johnson
    1963), Northern dune mat (Duebendorfer
    1989), *Poa-Lathyrus* phase of dune mat
    (Pickart 1987)].
Williams & Potter (1972):
    Strand.

Sedge series

Sedges sole, dominant, or important herbs in ground canopy; bulrushes, rushes, sedges, and/or spikerushes may be present. Herbs < 1 m; canopy continuous or intermittent (Plate 3).

Wetlands: habitat seasonally flooded, semipermanently flooded, permanently saturated, seasonally saturated, intermittently exposed. Water chemistry: fresh. Margins of channels, lakes, ponds, overflow areas, reservoirs, rivers, streams; depressions; seeps; swales. Cowardin classes: Lacustrine littoral unconsolidated shore wetland, Palustrine nonpersistent emergent freshwater wetland, Palustrine persistent emergent freshwater wetland, Palustrine unconsolidated freshwater shore wetland, Riverine emergent wetland. The national list of wetland plants (Reed 1988) lists sedges as a FACW, OBL.

Distribution: Cis-CA, Trans-CA.
Elevation: sea level-2900 m.

NDDB/Holland type and status: Meadows and seeps (45000), Alpine boulder and rock field (91000).
> Wet montane meadow (45110 *in part*) G3 S3.
> Freshwater seep (45400 *in part*) G4 S4.
> Wet subalpine or alpine meadow (45210) G3 S3.
> Dry subalpine or alpine meadow (45220) G3 S3.
> Vernal marsh (54500) G2 S2.1.
> Klamath Cascade fell field (91110) G4 S4.
> Sierra Nevada fell field (91120) G4 S4.
> Southern California fell field (91130) G2 S2.2.
> White Mountains fell field (91140) G2 S2.2.

Other types:
Barry type: G7411331.
Cheatham & Haller type: Meadows and swamps.
PSW-45 type: Sedge series.
Rangeland type: SRM 217.
Thorne type: Freshwater aquatic.
WHR type: Wet meadow.

General references: Hermann (1970), Mason (1957), Paysen *et al.* (1980).

Comments: This series contains areas dominated by sedges that have not been recognized as separate series at this time. Beaked sedge, Nebraska sedge series, Rocky Mountain sedge, and Shorthair sedge series are distinguished in this classification. Baker (1972) qualitatively describes a sedge stand at Ingelnook Fen in Mendocino Co. o-NorCo.

Species mentioned in text	
Alpine pussytoes	*Antennaria alpina*
Alpine shootingstar	*Dodecatheon alpinum*
Arrowleaf butterweed	*Senecio triangularis*
Beaked sedge	*Carex utriculata*
Bilberry	*Vaccinium caespitosum*
Blackish sedge	*Carex nigricans*
Brewer sedge	*Carex breweri*
Bulrushes	*Scirpus* species
Club-moss ivesia	*Ivesia lycopodioides*
Congdon sedge	*Carex congdonii*
Dewey sedge	*Carex deweyana*
Diego bentgrass	*Agrostis diegoensis*
Heller sedge	*Carex helleri*
Inflated sedge	*Carex vesicaria*
Junegrass	*Koeleria cristata*
Leafly sedge	*Carex ampilifolia*
Little elephant's head	*Pedicularis attollens*
Longbeak sedge	*Carex rostrata*
Luzulaleaf sedge	*Carex luzulifolia*
Many-nerved sedge	*Carex heteroneura*
Mount Dana sedge	*Carex subnigricans*
See next page	

Plot-based descriptions: Duebendorfer (1989) defines one association on sand in Humboldt Co., Fiedler & Leidy (1987) one association at Ring Mountain in Marin Co. o-NorCo; Beguin & Major (1975), Burke (1987) in Keeler-Wolf (1990e) four associations at Grass Lake RNA in El Dorado Co., Benedict (1983) one association at Rock Creek, Sequoia National Park, Major & Taylor (1977) four associations in a-SN and WIS; Taylor (1984) in Keeler-Wolf (1990e) 11 associations at Harvey Monroe Hall RNA in Mono Co. a-SN; Murray (1991) re-sampled meadows in the Marble Moun-

---

tains first sampled in 1977 by Stillman (1980). Only one higher-level category was present in both years in Siskiyou Co. m-KlaR. Of interest is an earlier qualitative description (Lewis 1966).

Species mentioned in text	
Mountain laurel	*Kalmia polifolia*
Nebraska sedge	*Carex nebrascensis*
Parry rush	*Juncus parryi*
Primrose monkeyflower	*Mimulus primuloides*
Ribbed sedge	*Carex multicostata*
Rocky Mountain sedge	*Carex scopulorum*
Rushes	*Juncus* species
Salt rush	*Juncus lesueurii*
Sedges	*Carex* species
Serrate sedge	*Carex subfusca*
Shore sedge	*Carex limosa*
Shortbeak sedge	*Carex simulata*
Shorthair sedge	*Carex filifolia*
Showy sedge	*Carex spectabilis*
Sibbaldia	*Sibbaldia procumbens*
Skyline bluegrass	*Poa cusickii*
Slough sedge	*Carex obnupta*
Spikerushes	*Eleocharis* species
Streambank arnica	*Arnica amplexicaulis*
Vernacular sedge	*Carex vernacula*
Water-plantain buttercup	*Ranunculus alismifolius*
Wheeler bluegrass	*Poa wheeleri*
Western yellow cress	*Rorippa curvisiliqua*
White-tipped sedge	*Carex albonigra*
Yarrow	*Achillea lanulosa*

See previous page

Associations:

Burke (1987):
> Longbeak sedge - shortbeak sedge association,
> Shortbeak-inflated sedge association.

Beguin & Major 1975):
> Primrose monkeyflower-shore sedge association [as Mimulo-Cericetum limosae, = category A (Rae 1970)],
> Skyline bluegrass-beaked sedge association [as Poo-Caricetum integrae]

Benedict (1983):
> Many-nerved sedge - yarrow association.

Duebendorfer (1989):
> Slough sedge-salt rush association [as Herbaceous hollows].

Fiedler & Leidy (1987):
> Serrate sedge-leafly sedge association.

Major & Taylor (1977):
> Alpine pussytoes-vernacular sedge association,
> Brewer sedge-Wheeler bluegrass association,
> Heller sedge-Parry rush association,
> Mount Dana sedge-little elephant's head association,
> Bilberry-blackish sedge association.

Stillman (1980):
> Diego bentgrass-ribbed sedge association [continuing *Ligusticum californicum-Erigeron aliceae, Calyptridium umbellatum-Castilleja arachnoidea* types (Murray 1991)].

Taylor (1984):
> Blackish sedge- mountain-laurel association,
> Brewer sedge association,
> Heller sedge - club-moss ivesia association,
> Luzulaleaf sedge - water-plantain buttercup association,
> Mount Dana sedge-alpine shootingstar association,
> Mount Dana sedge-alpine pussytoes association,
> Showy sedge-sibbaldia association,
> Streambank arnica-Congdon sedge association,
> Vernacular sedge-alpine pussytoes association,
> Western yellow cress-Dewey sedge association,
> White-tipped sedge-Junegrass association.

Shorthair reedgrass series

Shorthair reedgrass sole, dominant, or important with Sierra ricegrass in ground canopy; alpine aster, bilberry, Brewer heather, buckwheat, cinquefoils, heretic penstemon, Merten rush, shorthair sedge, tufted hairgrass, and/or yarrow may be present. Emergent shrubs may be present. Herbs < 0.15 m; canopy continuous or intermittent.

Uplands: flats, slopes.

Distribution: su-KlaR, su-SN, a-SN, OR.
Elevation: 1300-3800 m.

NDDB/Holland type and status: **Meadows and seeps** (45000).
  Wet montane meadow (45110 *in part*) G3 S3.2.
  Dry montane meadow (45120 *in part*) G3 S3.2.
  Wet subalpine or alpine meadow (45210 *in part*) G3 S3.3.
  Dry subalpine or alpine meadow (45220 *in part*) G3 S3.3.

Other types:
Barry type: G7411331 BCABR70.
Cheatham & Haller type: Subalpine meadow.
Rangeland type: SRM 216.
Thorne type: Subalpine meadow.
WHR type: Perennial grass.

General references: Allen-Diaz (1994), Major & Taylor (1977), Ratliff (1982, 1985).

Comments: Ratliff (1982, 1985) refers to stands dominated by shorthair reedgrass as members of the Shorthair reedgrass vegetative series. The series is extensive and may mix with Shorthair sedge series and Red-heather - bilberry series at a fine scale, and at a coarser scale with forest and woodland series. [*see* Alpine habitat, Subalpine meadow].

Plot-based descriptions: Benedict (1983) defines two associations at Rock Creek in Sequoia National Park in Tulare Co.; Burke (1982) two associations at Rae Lakes, King's Canyon National Park in Fresno Co.; Major & Taylor (1977) one association at Carson Pass in Alpine Co.; Taylor (1984) two associations at Harvey Monroe Hall RNA in Mono Co. su-SN.

Species mentioned in text	
Alpine aster	*Aster alpigenus*
Alpine-laurel	*Kalmia polifolia*
Bilberry	*Vaccinium caespitosum*
Brewer heather	*Phyllodoce breweri*
Buckwheats	*Eriogonum* species
Cinquefoils	*Potentilla* species
Heretic penstemon	*Penstemon heterodoxus*
Merten rush	*Juncus mertensianus*
Mountain heather	*Phyllodoce* species
Shorthair reedgrass	*Calamagrostis breweri*
Shorthair sedge	*Carex filifolia*
Sierra ricegrass	*Oryzopsis kingii*
Spike trisetum	*Trisetum spicatum*
Tufted hairgrass	*Deschampsia cespitosa*
Yarrow	*Achillea lanulosa*

Associations:
Benedict (1983):
  Shorthair reedgrass-alpine aster association,
  Shorthair reedgrass-spike trisetum association.
Major & Taylor (1977):
  Shorthair reedgrass-bilberry association
  [= Burke (1982), = Taylor (1984)].
Taylor (1984):
  Drummond sedge-shorthair reedgrass association,
  Shorthair reedgrass - mountain laurel association.

Shorthair sedge series

Shorthair sedge sole or dominant herb in ground canopy; alpine aster, buckwheats, cinquefoils, heretic penstemon, Merten rush, shorthair reedgrass, tufted hairgrass, and/or yarrow may be present. Emergent shrubs may be present. Herbs < 0.3 m; canopy continuous, intermittent, or open.

Uplands: slopes, ridges.

Distribution: su-SN, a-SN, su-TraR, ModP, WarR, WIS, inter-West.
Elevation: 1500-3700 m.

NDDB/Holland type and status: **Meadows and seeps** (45000).
>Dry subalpine or alpine meadow (45220 *in part*) G3 S3.2.

Other types:
Barry type: G7411331 BCAEX40.
Cheatham & Haller type: Subalpine meadow.
PSW-45 type: Sedge series.
Rangeland type: SRM 213, SRM 216.
Thorne type: Subalpine meadow.
WHR type: Perennial grass.

General references: Allen-Diaz (1994), Klikoff (1965), Major & Taylor (1977), Paysen *et al.* (1980), Ratliff (1979, 1982, 1985).

Comments: Ratliff (1982, 1985) defines to a Shorthair sedge vegetative series. Stands of this series are extensive and may mix with other meadow, forest and woodland series at a fine scale at subalpine and alpine elevations [*see* especially Alpine habitat, Subalpine meadow habitat].

The Jepson Manual uses the name *Carex filifolia* var. *erostrata* for shorthair sedge. The ecological literature refers to it as *Carex exserta*.

Plot-based descriptions: Benedict (1983) defines one association at Rock Creek in Sequoia National Park in Tulare Co.; Burke (1982) three associations at Rae Lakes, King's Canyon National Park in Fresno Co.; Major & Taylor (1977) one association at Carson Pass in Alpine Co.; (Taylor 1984) one at Harvey Monroe Hall RNA in Mono Co. su-SN.

Species mentioned in text	
Alpine aster	*Aster alpigenus*
Cinquefoils	*Potentilla* species
Buckwheats	*Eriogonum* species
Heretic penstemon	*Penstemon heterodoxus*
Merten rush	*Juncus mertensianus*
Nude buckwheat	*Eriogonum nudum*
Pussypaws	*Calyptridium umbellatum*
Saxifrage	*Saxifraga aprica*
Shorthair sedge	*Carex filifolia*
Shorthair reedgrass	*Calamagrostis breweri*
Sierra ricegrass	*Oryzopsis kingii*
Spike trisetum	*Trisetum spicatum*
Talus fleabane	*Erigeron algidus*
Tufted hairgrass	*Deschampsia cespitosa*
Yarrow	*Achillea lanulosa*

Associations:
Benedict (1983):
>Shorthair sedge-spike trisetum association, Shorthair sedge-Sierra ricegrass association.

Burke (1982):
>Pussypaws-shorthair sedge association, Nude buckwheat-shorthair sedge association, Talus fleabane-shorthair sedge association.

Major & Taylor (1977):
>Shorthair sedge association [= *Muhlenbergia richardsonii* association (Benedict 1983)].

Taylor (1984):
>Shorthair sedge-saxifrage association.

Spikerush series

Spikerushes sole or dominant herb in ground canopy; alpine aster, beaked sedge, bulrushes, creeping spikerush, few-flowered spikerush, littlebeak spikerush, marsh arrow-grass, mountain spikerush, mountain timothy, Nevada rush, primrose monkey flower, and/or sedges may be present. Herbs < 0.5 m; canopy continuous or intermittent (Plate 1).

Wetlands: habitat seasonally flooded, semi-permanently flooded, permanently saturated, seasonally saturated, intermittently exposed. Water chemistry: fresh. Margins of channels, lakes, ponds, overflow areas, reservoirs, rivers, streams; depressions; seeps; swales. Cowardin classes: Lacustrine littoral unconsolidated shore wetland, Palustrine nonpersistent emergent freshwater wetland, Palustrine persistent emergent freshwater wetland, Palustrine unconsolidated freshwater shore wetland, Riverine emergent wetland. The national list of wetland plants (Reed 1988) lists spikerushes as a FACW, OBL.

Distribution: Cis-CA, Trans-CA.
Elevation: sea level-2500 m.

NDDB/Holland type and status: **Meadows and seeps** (45000).
>   Wet montane meadow (45110 *in part*) G3 S3.
>   Freshwater seep (45400 *in part*) G4 S4.
>   Vernal marsh (52500) G2 S2.1.

Other types:
Barry type: G7411331.
Cheatham & Haller type: Meadows and seeps.
Thorne type: Freshwater aquatic.
WHR type: Wet meadow.

General references: Mason (1957).

Comments: This series contains both low and high elevation stands, but plot data are available only for montane and subalpine ones. Ratliff (1985) takes up montane and subalpine meadows for the central Sierra, especially Yosemite National Park and surrounding areas in terms of vegetative series.

Species mentioned in text	
Alkali muhly	*Muhlenbergia asperifolia*
Alpine aster	*Aster alpigenus*
Beaked sedge	*Carex utriculata*
Bulrushes	*Scirpus* species
Common three-square	*Scirpus americanus*
Cowbane	*Oxypolis occidentalis*
Creeping spikerush	*Eleocharis palustris*
Few-flowered spikerush	*Eleocharis pauciflora*
Littlebeak spikerush	*Eleocharis rostratata*
Pale spikerush	*Eleocharis macrostachya*
Marsh arrow-grass	*Triglochin palustris*
Mountain spikerush	*Eleocharis montevidensis*
Mountain timothy	*Phleum alpinum*
Nevada rush	*Juncus nevadensis*
Primrose monkeyflower	*Mimulus primuloides*
Sedges	*Carex* species
Spikerushes	*Eleocharis* species
Water pygmy	*Crassula aquatica*
Water-starwort	*Callitriche hermaphroditica*

Plot-based descriptions: Beguin & Major (1975) define one association at Grass Lake RNA in El Dorado Co., Benedict (1983) two associations at Rock Creek, Halpern (1986) two associations in Sequoia National Park in Fresno Co., Rae (1970) one association at Mason Bog in Nevada Co., Taylor (1984) one association at Harvey Monroe Hall RNA in Mono Co. m-SN, su-SN; Kepecko & Lathrop (1975) two associations at Santa Rosa Plateau in Riverside Co. o-SoCo; Odion *et al.* (1992) two associations at Fish Slough in Inyo Co. Tra-SN.

Associations:
Beguin & Major (1975):
>   Nevada rush-spikerush association [as Junco nevadensis-Eleochartetum quinqueflorae].
Benedict (1983):
>   Few-flowered spikerush association
>   [= Taylor (1984)],

Few-flowered spikerush - Primrose monkey
flower association [containing categories B, C,
D (Rae 1970)].

Halpern (1986):
Mountain spikerush-cowbane association
[containing 2 phases],
Mountain spikerush-moss association.

Kepecko & Lathrop (1975):
Pale spikerush - water-starwort association
[as Standing water zone],
Spikerush-water pygmy association
[as Muddy margin zone].

Odion *et al.* (1992):
Littlebeak spikerush - common three-square
association [as *Scirpus-Eleocharis* association],
Littlebeak spikerush-alkali muhly association
[as *Eleocharis-Muhlenbergia* association].

Tufted hairgrass series

Tufted hairgrass sole or dominant grass in ground layer; alta fescue, California oatgrass, colonial bent, Pacific reedgrass, red fescue, rushes, sedges, velvet grass, and/or vernal grass may be present. Emergent shrubs or trees may be present. Grass < 1 m; cover open (Plate 3).

Wetlands: habitat seasonally saturated. Water chemistry: fresh or hyperhaline. Banks, berms, and margins of bays; depressions; seeps; swales. Cowardin classes: Palustrine nonpersistent emergent freshwater wetland, Estuarine emergent nonpersistent wetland. **Uplands:** coastal bluffs, terraces, slopes and ridges. The national list of wetland plants (Reed 1988) lists tufted hairgrass as a FACW.

Distribution: NorCo, KlaR, CasR, SN, CenCo, TraR, WarR, WIS, North America, Eurasia. **Elevation:** sea level-3900 m.

NDDB/Holland type and status: **Coastal prairies** (41000), **Meadows and seeps** (45000).
> Coastal terrace prairie (41100 *in part*) G2 S2.1.
> Wet subalpine and alpine meadow (45210) G3 S3.2.

Other types:
Barry type: G7411331 BDECA00.
Cheatham & Haller type: Coastal prairie.
Rangeland types: SRM 213, SRM 214.
Thorne type: Northern coastal prairie.
WHR type: Perennial grass.

General references: Heady *et al.* (1977), Ratliff (1985).

Comments: *The Jepson Manual* recognizes two subspecies of tufted hairgrass, *Deschampsia cespitosa* ssp. *c.* and *D. c.* ssp. *holciformis. D. c.* ssp. *beringensis* has been submerged in the former subspecies.

The description of the series is based on o-NorCo studies of coastal prairie (Heady *et al.* 1977). Both ssp. of tufted hairgrass grow at low elevations in o-CenCo and o-NorCo. Tufted hairgrass also grows to alpine elevations.

Ratliff (1982, 1985) defines a Tufted hairgrass vegetative series in SN. Sawyer (1978a) in Keeler-Wolf (1990e) qualitatively describes tufted hairgrass meadows at Preacher Meadows candidate RNA in Trinity Co. m-KlaR.

Species mentioned in text	
Alta fescue	*Festuca arundinacea*
Brewer bitter-cress	*Cardamine breweri*
California oatgrass	*Danthonia californica*
Colonial bent	*Agrostis capillaris*
Coville ragwort	*Senecio scorzonella*
Mount Dana sedge	*Carex subnigricans*
Nebraska sedge	*Carex nebrascensis*
Northern goldenrod	*Solidago multiradiata*
Pacific reedgrass	*Calamagrostis nutkaensis*
Red fescue	*Festuca rubra*
Rushes	*Juncus* species
Sedges	*Carex* species
Tufted hairgrass	*Deschampsia cespitosa*
Velvet grass	*Holcus lanatus*
Vernal grass	*Anthoxanthum odoratum*
Yarrow	*Achillea lanulosa*

Plot-based descriptions: Elliott & Wehausen (1974) describe grazed and ungrazed plots of tufted hairgrass at Point Reyes National Seashore, NDDB has data for terraces at Audubon Canyon Ranch in Marin Co., Heady *et al.* (1977) define one association in Sonoma Co. o-NorCo.; Allen-Diaz (1991) one association in Sierra Co., Benedict (1983) three associations at Rock Creek in Sequoia National Park in Tulare Co., Taylor (1984) one association at Harvey Monroe Hall RNA in Mono Co. SN; Major & Taylor (1977) one association from White Mountains in Inyo Co. WIS.

Associations:
Allen-Diaz (1991):
> Tufted hairgrass-Nebraska sedge association.

Benedict (1983):
  Tufted hairgrass-Brewer bitter-cress association,
  Tufted hairgrass-Coville ragwort association,
  Tufted hairgrass-Coville ragwort-yarrow
  association.
Heady *et al.* (1977):
  Vernal grass-tufted hairgrass association.
Major & Taylor (1977):
  Mt. Dana sedge-tufted hairgrass association.
Taylor (1984):
  Tufted hairgrass-northern goldenrod
  association.

Yellow pond-lily series

Yellow pond-lily sole or dominant plant on water surface; bulrushes, bur-reeds, cattails, pondweeds, and/or water-shield may be present. Emergent plants may be present. Plants < 0.5 m in size; cover continuous, intermittent, or open (Plate 2).

Wetlands: habitat permanently flooded. Water chemistry: fresh. Lakes; ponds; slow streams. Cowardin classes: Lacustrine littoral aquatic bed wetland, Palustrine aquatic bed wetland. The national list of wetland plants (Reed 1988) lists yellow pond-lily as a OBL.

Distribution: NorCo, n. & c. CenCo, n. & c. SN, ModP, inter-West.
Elevation: sea level-2400 m.

NDDB/Holland type and status: Marsh and swamp (52000).
    Coast and valley freshwater marsh (52410) G3 S2.1.
    Transmontane freshwater marsh G3 S2.1.
    Montane freshwater marsh (52430 *in part*) G3 S3.

Other types:
Barry type: G7421321 BNUP00.
Cheatham & Haller type: Freshwater marshes.
Thorne type: Freshwater aquatic.
WHR type: Freshwater emergent wetland.

General references: Mason (1957).

Comments: Yellow pond-lily often dominates on surfaces of shallow ponds. It may also be occasional in stands of other aquatic series.

Species mentioned in text	
Bulrushes	*Scirpus* species
Bur-reeds	*Sparganium* species
Cattails	*Typha* species
Pondweeds	*Potamogeton* species
Water-shield	*Brasenia schreberi*
Yellow pond-lily	*Nuphar luteum*

Plot-based descriptions: Baker (1972) qualitatively defines stands as Open water at Ingelnook Fen in Mendocino Co. o-NorCo.

Key 2: Series Dominated by Shrubs

Key to shrublands dominated by subshrubs

Key to shrublands where one species dominates

1'   Emergent trees absent or if present over a shrub canopy then not in ColD, MojD 2

    2   Bitterbrush, sagebrush, or chenopod dominant 3

    2'   Bitterbrush, sagebrush, or chenopod not dominant 4

3   Bitterbrush dominant   *Bitterbrush series*

3'   Bitterbrush not dominant 5

    4   A sagebrush dominant   SAGEBRUSH SCRUBS
Big sagebrush series
Black sagebrush series
California sagebrush series
Low sagebrush series
Rothrock sagebrush series

    4'   A chenopod dominant   CHENOPOD SCRUBS
Allscale series
Bush seepweed series
Desert-holly series
Fourwing saltbush series
Greasewood series
Hop-sage series
Iodine bush series
Pickleweed series [see herbaceous plant series]
Shadscale series
Spinescale series
Winterfat series

5   Rabbitbrush dominant   RABBITBRUSH SCRUBS
Parry rabbitbrush series
Rubber rabbitbrush series

5'   Rabbitbrush not dominant 6

    6   A willow dominant   WILLOW THICKETS
Arroyo willow series [see tree series]
Black willow series [see tree series]
Hooker willow series [see tree series]
Narrowleaf willow series
Pacific willow series [see tree series]
Red willow series [see tree series]
Sandbar willow series
Sitka willow series [see tree series]

    6'   A willow not dominant 7

7   Alder dominant   *Mountain alder series*
Sitka alder series

7'   Alder not dominant 8

    8   Coastal scrub species dominant   COASTAL SCRUBS
Black sage series
Brittlebush series
Broom series
California buckwheat series
California encelia series

California sagebrush series
Coyote brush series
Purple sage series
Scalebroom series
Sumac series
White sage series

8'   Coast scrub species not dominant 9

9   Desert scrub species dominant; ColD, MojD   DESERT SCRUBS

Big galleta series
Black bush series
Brittlebush series
Catclaw acacia series
California buckwheat series
Creosote bush series
Iodine bush series
Mesquite series [see tree series]
Teddy-bear cholla series
Scalebroom series
White bursage series

9'   Desert scrub species not dominant 10

10   A *Ceanothus* species dominant   CEANOTHUS BRUSHFIELDS

Blue blossom series
Deerbrush series
Tobacco brush series

10'   A *Ceanothus* species not dominant 11

11   A *Holodiscus* species dominant   *Holodiscus series*

11'   A *Holodiscus* species not dominant 12

12   A *Quercus* species dominant   OAK BRUSHFIELDS

Brewer oak series
Huckleberry oak series
Sadler oak series

12'   A *Quercus* species not dominant 13

13   Wax currant or rock-gilia conspicuous   See HABITATS *section* p. 348

13'   Wax currant or rock-gilia not conspicuous 14

14   Exotic species dominant   NON-NATIVE BRUSHFIELDS

Broom series
Tamarisk series
Yellow bush lupine series

14'   Exotic species not dominant   NATIVE BRUSHFIELDS

Arrow weed series
Birchleaf mountain-mahogany series [see tree series]
Bush chinquapin series
Buttonbush series
Mexican elderberry series
Mountain heather-bilberry series
Mulefat series

McNab cypress series [see tree series]
Sumac series
Water birch series [see tree series]

Key to shrublands where two species are similarly important

1   Willows important   *Mixed willow series [see tree series]*
                                                     See HABITATS *section* p. 348
1'   Willows not important 2
      2   Mountain heather important   *Mountain heather-bilberry series*
      2'   Mountain heather not important 3
3   Goldenbush important   *Bladderpod-California ephedra-narrowleaf goldenbush series*
                                                     Dune lupine-goldenbush series
3'   Goldenbush not important 4
      4   Coastal scrub species important   COASTAL SCRUBS
                              Birchleaf mountain-mahogany-California buckwheat series
                                                     Brittlebush series
                                              Brittlebush-white bursage series
                                         California buckwheat-white sage series
                                    California sagebrush-California buckwheat series
                                              Chamise-black sage series
                                              Chamise-white sage series
                                                     Purple sage series
                                              Salal-black huckleberry series
                                                     White sage series
      4'   Coastal scrub species not important 5
5   Bush potentilla, rock-gilia or wax currant present   See HABITATS *section* p. 348
5'   Big sagebrush or creosote bush scrub species important   DESERT SCRUBS
                                                     Brittlebush series
                                                     Catclaw acacia series
                                         Creosote bush-white bursage series
                                         Foothill palo verde-saguaro series
                                                     Joshua tree series
                                         Mesquite series [see tree series]
                                                     Mixed saltbush series
                                                     Mojave yucca series
                                                     Nolina series
                                                     Ocotillo series
                                      See UNIQUE STANDS *section* p. 325

Key to shrublands where more than two species are similarly important

1   Willows important   *Mixed willow series*
                                                     See HABITATS *section* p. 348

1'		Willows not important 2	
	2	Creosote bush present	DESERT SCRUBS

Brittlebush-white bursage series
Catclaw acacia series
Foothill palo verde-saguaro series
Joshua tree series
Mesquite series [see tree series]
Mojave yucca series
Nolina series
Ocotillo series
See UNIQUE STANDS *section* p. 325

	2'	Creosote bush not present	MIXED SCRUBS

Bladderpod-California ephedra-narrowleaf goldenbush series
Coast prickly-pear series
Brittlebush-white bursage series
Dune lupine-goldenbush series
Mixed sage series
Mixed saltbush series
Sumac series
See HABITATS *section* p. 348

Key to chaparrals where one species dominates

1		An *Adenostoma* species dominant or important	*Chamise series*
			Red shank series
1'		An Adenostoma species not dominant, may be present 2	
	2	A *Ceanothus* species dominant	CEANOTHUS CHAPARRALS

Bigpod ceanothus series
Blue blossom series
Chaparral whitethorn series
Deerbrush series
Hairyleaf ceanothus series
Hoaryleaf ceanothus series
Mountain whitethorn series
Tobacco brush series
Wedgeleaf ceanothus series

	2'	A *Ceanothus* species not dominant 3	
3		A manzanita dominant	MANZANITA CHAPARRALS

Bigberry manzanita series
Eastwood manzanita series
Greenleaf manzanita series
Ione manzanita series
Whiteleaf manzanita series
Woollyleaf manzanita series

3'		A manzanita not dominant 4	
	4	Bush chinquapin dominant	*Bush chinquapin series*

4'   An oak dominant

OAK CHAPARRALS
Brewer oak series
Huckleberry oak series
Leather oak series
Scrub oak series
Sadler oak series
Canyon live oak shrub series
Interior live oak shrub series

Key to chaparrals where two species are similarly important

1   A *Salvia* species important

Chamise-black sage series
Chamise-white sage series

1'   A *Salvia* species not important 2

2   Chamise important

CHAMISE CHAPARRALS
Chamise-bigberry manzanita series
Chamise-back sage series
Chamise-cupleaf ceanothus series
Chamise-Eastwood manzanita series
Chamise-hoaryleaf ceanothus series
Chamise-wedgeleaf ceanothus series
Chamise-white sage series
Chamise-woollyleaf ceanothus series
Leather oak series
Mixed oak series
Red shank-chamise series
Scrub oak-chamise series
Sumac series

2'   Chamise not important 3

3   A live oak important

LIVE OAK CHAPARRALS
Interior live oak shrub series
Interior live oak-canyon live oak shrub series
Interior live oak-chaparral whitethorn shrub series
Leather oak series
Mixed scrub oak series
Scrub oak series
Scrub oak - birchleaf mountain-mahogany series
Scrub oak-chamise series
Scrub oak-chaparral whitethorn series

3'   A live oak not important 4

4   Bigpod ceanothus important

MIXED CHAPARRALS
Bigpod ceanothus - birchleaf mountain-mahogany series
Bigpod ceanothus-hollyleaf redberry series

4'   Birchleaf mountain-mahogany important 5

5   Red shank important

Red shank - birchleaf mountain-mahogany series

5'   Red shank not important 6
    6   Birchleaf mountain-mahogany important
            Birchleaf mountain-mahogany - California buckwheat series
    6'   Woollyleaf manzanita important   *Woollyleaf manzanita series*

Key to chaparrals where more than two species are similarly important

1   Chamise important   *Chamise - mission-manzanita - woollyleaf ceanothus series*
1'   Chamise not important 2
    2   Cupleaf ceanothus and fremontia important   *Cupleaf ceanothus-fremontia-oak series*
    2'   Scrub oak important   *Mixed scrub oak series*

Allscale series

Allscale sole or dominant shrub in canopy; bladderpod, bush buckwheat, cheesebush, California ephedra, paleleaf goldenbush, and/or saltgrass may be present. Emergent honey mesquite may be present. Shrubs < 3 m; canopy continuous or open. Ground layer variable (Plate 17).

Uplands: soil of old beach, lake deposits; dissected alluvial fans; rolling hills. Soils may be carbonate-rich, sandy. The national list of wetland plants (Reed 1988) lists allscale as a FACU.

Distribution: i-CenCo, sj-CenV, f-SN, i-SoCo, s. TraSN, MojD, ColD.
Elevation: 75 below sea level-1500 m.

NDDB/Holland type and status: **Interior dunes** (23000), **Chenopod scrubs** (36000).
   Relictual interior dunes (23200) G1 S1.1.
   Desert saltbush scrub (36110 *in part*) G3 S3.2.
   Valley saltbush scrub (36220 *in part*) G1 S1.1.
   Sierra-Tehachapi saltbush scrub (36310) G2 S2.1.
   Interior Coast Range saltbush scrub (36320) G2 S2.1.

Other types:
Barry type: G7411221 CATPO00.
Brown Lowe Pase type: 154.174.
Cheatham & Haller type: Saltbush scrub.
PSW-45 type: Saltbush series.
Thorne type: Shadscale scrub.
WHR type: Alkali sink.

General references: Bittman (1985), Burk (1977), Griggs (1980), Griggs & Zaninovich (1884), Johnson (1976), MacMahon (1988), MacMahon & Wagner (1985), McHargue (1973), Paysen *et al.* (1980), Vasek & Barbour (1977), Werschskull *et al.* (1984).

Comments: This series is often considered part of the chenopod or saltbush scrub. These scrubs are better thought of as a collection of series. If no dominant saltbush is present, the stand is assigned to Mixed saltbush series.

Allscale series occurs with different associates regionally as suggested by NDDB categories. Nor-CA stands have received the most attention. Many of these species occur in Bladderpod-California ephedra-narrowleaf goldenbush series. Small changes in topography and depth to water table create a fine-scale mosaic of allscale stands with those of Iodine bush and Bush seepweed series.

Species mentioned in text	
Allscale	*Atriplex polycarpa*
Bladderpod	*Isomeris arborea*
Bush buckwheat	*Eriogonum fasciculatum*
California ephedra	*Ephedra californica*
Cheesebush	*Hymenoclea salsola*
Honey mesquite	*Prosopis glandulosa*
Paleleaf goldenbush	*Isocoma acradenia*
Saltbush	*Atriplex* species
Saltgrass	*Distichlis spicata*

Plot-based descriptions: not available.

Arrow weed series

Arrow weed sole or dominant shrub in canopy; narrowleaf cattail, narrowleaf willow, and/or tamarisks may be present. Shrubs < 5 m; canopy continuous. Ground layer sparse.

Wetlands: habitats seasonally flooded, saturated. Water chemistry: fresh, mixohaline. Canyon bottoms; irrigation ditches, streamsides; around springs. Cowardin class: Palustrine shrub-scrub wetland. The national list of wetland plants (Reed 1988) lists arrow weed as a FACW.

Distribution: i-CenCo, sj-CenV, SoCo, MojD, ColD.
Elevation: sea level-600 m.

NDDB/Holland type and status: **Riparian scrubs** (63000).
Arrowweed scrub (63820) G3 S3.

Other types:
Barry type: G7411211.
Cheatham & Haller type: Bottomland woodlands and forest.
PSW-45 type: Arrow weed series.
Thorne type: Riparian woodland.
WHR type: Desert riparian.

General references: Paysen *et al.* (1980),

Comments: Arrow weed often forms pure stands. Secondary species, if present, vary regionally. Arrow weed stands may form a fine mosaic with other wetland series.

Species mentioned in text	
Arrow weed	*Pluchea sericea*
Narrowleaf cattail	*Typha angustifolia*
Narrowleaf willow	*Salix exigua*
Tamarisks	*Tamarix* species

Plot-based descriptions: not available.

Big sagebrush series

Big sagebrush sole or dominant shrub in canopy; bitterbrush, green ephedra, horsebrush, plateau gooseberry, rubber rabbitbrush, and/or yellow rabbitbrush may be present. Emergent junipers, Joshua tree, and/or pines may be present. Shrubs < 3 m; cover continuous, intermittent, or open. Ground layer sparse or grassy (Plate 13).

Uplands: bajadas, pediments, alluvium, valleys, dry washes. Soils well-drained, gravelly.

Distribution: i-CenCo, sj-CenV, m-CasR, su-CasR, m-SN, su-SN, TraR, PenR, GB, MojD, inter-West, Baja CA.
Elevation: 300-3000 m.

NDDB/Holland type and status: **Great Basin scrubs** (35000).
>   Great Basin mixed scrub (35100) G4 S4.
>   Big sagebrush (35210) G4 S4.
>   Sagebrush steppe 35300) G2 S2.1.

Other types:
Barry type: G7411211 CARTR20.
Brown Lowe Pase type: 142.213, 142.222, 152.111, 152.112.
Cheatham & Haller type: Great Basin sagebrush.
PSW-45 type: Sagebrush series.
Rangeland types: SRM 401, SRM 403.
Thorne type: Great Basin sagebrush scrub.
WHR type: Sagebrush.

General references: Paysen *et al.* (1980), Taylor (1976b), Tisdale (1994), Turner (1998a), Vale (1975), West (1988), Wolfram & Martin (1965), Young *et al.* (1977).

Comments: Some stands of this series have scattered Joshua trees, junipers, or pines. Big sagebrush occurs as an important understory shrub in stands of open woodland and forest series. If trees dominate the stand, then place it in a tree-domin-

ated series [*see* Jeffrey pine series, Joshua tree series, Singleleaf pinyon-Utah juniper series, Washoe pine series, Western juniper series].

Species mentioned in text	
Big sagebrush	*Artemisia tridentata*
Black bush	*Coleogyne ramosissima*
Bluebunch wheatgrass	*Elymus spicatus*
Bitterbrush	*Purshia tridentata*
Cheatgrass	*Bromus tectorum*
Crested wheatgrass	*Agropyron desertorum*
Desert snowberry	*Symphoricarpos longiflorus*
Green ephedra	*Ephedra viridis*
Horsebrush	*Tetradymia canescens*
Jeffrey pine	*Pinus jeffreyi*
Joshua tree	*Yucca brevifolia*
Junipers	*Juniperus* species
Needle-and-thread	*Stipa comata*
One-sided bluegrass	*Poa secunda*
Pines	*Pinus* species
Plateau gooseberry	*Ribes velutinum*
Rubber rabbitbrush	*Chrysothamnus nauseosus*
Singleleaf pinyon	*Pinus monophylla*
Utah juniper	*Juniperus osteosperma*
Washoe pine	*Pinus washoensis*
Western juniper	*Juniperus occidentalis* ssp. *occidentalis*
Yellow rabbitbrush	*Chrysothamnus viscidiflorus*

Another problem involves stands in grassland series which have a few big sagebrush. In this series shrubs are conspicuous, frequent, or dominant. If shrubs are infrequent, place the stand in a grassland series (*see* Bluebunch wheatgrass series, Cheatgrass series, Crested wheatgrass series, Needle-and-thread series, One-sided bluegrass series].

In California big sagebrush includes four subspecies. In inter-West classifications, subspecies define different series. There is some geographic separation of subspecies in California, but overlap is extensive, so subspecies are included in one series at this time.

Young *et al.* (1977) list 11 species as important

---

grasses in describing regional variation in the series, but unfortunately no plot data are presented to support the proposed types.

Keeler-Wolf (1990e) qualitatively describes a ridgetop stand at Mud Lake RNA in Plumas Co. m-CasR; TraSN; at Cahuilla Mountain RNA in Riverside Co. m-PenR; at Whippoorwill Flat RNA in Inyo Co.; Hanes (1976) describes vegetation types in the San Gabriel Mountains including the Big sagebrush series.

Plot-based descriptions: Taylor (1980) Keeler-Wolf (1990e) defines one association at Indiana Summit RNA in Mono Co., Ferren & Davis (1991) one association at Fish Slough TraSN; Franklin & Dyrness (1973) OR communities similar to those in ModP; Gordon & White (1994) one association in m-TraR, m-PenR, Spolsky (1979) one association in Anza Borrego State Park in San Diego Co. m-PenR.

Associations:
Ferren & Davis (1991):
    Big sagebrush-rubber rabbitbrush association.
Gordon & White (1994).
    Big sagebrush association.
Spolsky (1979):
    Desert slope sagebrush association.
Taylor (1980):
    Big sagebrush-desert snowberry association.

Bigberry manzanita series

Bigberry manzanita sole or dominant shrub in canopy; birchleaf mountain-mahogany, black sage, chamise, chaparral yucca, chaparral whitethorn, hollyleaf redberry, interior live oak, and/or scrub oak may be present. Emergent trees may be present. Shrubs < 8 m; canopy continuous. Ground layer sparse.

Uplands: outcrops, ridges, slopes north-facing, alluvial fans. Soils shallow, coarse-textured, may be ultramafic derived.

Distribution: i-SoCo, m-TraR, m-PenR, Baja CA.
Elevation: 600-1400 m.

NDDB/Holland type and status:: **Chaparral** (37000). Northern mixed chaparral (37110 *in part*) G3 S3. Upper Sonoran manzanita chaparral (37B00 *in part*) G4 S4.

Other types:
Barry type: G7411211 CARGL00.
Cheatham & Haller type: Mixed chaparral.
PSW-45 type: Manzanita series.
Rangeland type: SRM 208.
Thorne type: Mixed chaparral.
WHR type: Mixed chaparral.

General references: Hanes (1977, 1981), Horton (1960), Keeley (1987, 1992), Keeley & Keeley (1988), Pase (1982a), Patric & Hanes (1964), Paysen *et al.* (1980), Vasek & Clovis (1976), White (1994c), Wilson & Vogl (1965).

Comments: This definition follows Gordon & White (1994). In stands of this series, chamise is a minor canopy species. Bigberry manzanita plays a minor role in stands of Chamise series or Chamise-wedgeleaf ceanothus series. In the closely related Chamise-bigberry manzanita series both species are important. Bigberry manzanita can become very large in old stands.

Species mentioned in text	
Bigberry manzanita	*Arctostaphylos glauca*
Birchleaf mountain-mahogany	*Cercocarpus betuloides*
Black sage	*Salvia mellifera*
Chamise	*Adenostoma fasciculatum*
Chaparral yucca	*Yucca whipplei*
Chaparral whitethorn	*Ceanothus leucodermis*
Eastwood manzanita	*Arctostaphylos glandulosa*
Hollyleaf redberry	*Rhamnus ilicifolia*
Interior live oak	*Quercus wislizenii*
Scrub oak	*Quercus berberidifolia*
Wedgeleaf ceanothus	*Ceanothus cuneatus*

Plot-based descriptions: Gordon & White (1994) define a Bigberry manzanita association in m-PenR, Keeley & Keeley (1988) describes two stands in m-TraR.

Bigpod ceanothus series

Bigpod ceanothus sole or dominant shrub in canopy; birchleaf mountain-mahogany, black sage, chamise, hollyleaf redberry, scrub oak, and/or toyon may be present. Emergent trees may be present. Shrubs < 4 m; canopy continuous or intermittent. Ground layer sparse.

Uplands: slopes.

Distribution: o-CenCo, o-SoCo, o. l-TraR, w. m-PenR.
Elevation: 100-750 m.

NDDB/Holland type and status: **Chaparral** (37000). Ceanothus megacarpus chaparral (37840) G3 S3.2.

Other types:
Barry type: G7411211.
Cheatham & Haller type: Mixed chaparral.
PSW-45 type: Ceanothus series.
Rangeland type: SRM 208.
Thorne type: Mixed chaparral.
WHR type: Mixed chaparral.

General references: Keeley (1975), Keeley & Keeley (1988), Pase (1982a), Montygierd-Loba & Keeley (1987), Paysen *et al.* (1980), Schlesinger *et al.* (1982), White (1994c).

Comments: This definition follows Borchert *et al.* (1993b). If bigpod ceanothus has > 60% cover, the stand is a member of the Bigpod ceanothus series. If bigpod ceanothus covers 30-60% and another species covers 30-60%, then the stand is a member of a mixed series [*see* Bigpod ceanothus-hollyleaf redberry series, Bigpod ceanothus - birchleaf mountain-mahogany series].

Plot-based descriptions: Keeley & Keeley (1988) describe one stand, Borchert *et al.* (1993b) two associations in o-CenCo, w. m-TraR.

<table>
<tr><td colspan="2">***Species mentioned in text***</td></tr>
<tr><td>Bigpod ceanothus</td><td>*Ceanothus megacarpus*</td></tr>
<tr><td>Birchleaf mountain-mahogany</td><td>*Cercocarpus betuloides*</td></tr>
<tr><td>Black sage</td><td>*Salvia mellifera*</td></tr>
<tr><td>Chamise</td><td>*Adenostoma fasciculatum*</td></tr>
<tr><td>Hollyleaf redberry</td><td>*Rhamnus ilicifolia*</td></tr>
<tr><td>Scrub oak</td><td>*Quercus berberidifolia*</td></tr>
<tr><td>Toyon</td><td>*Heteromeles arbutifolia*</td></tr>
</table>

Associations:
Borchert *et al.* (1993b):
Bigpod ceanothus association,
Bigpod ceanothus-chamise/black sage association.

Bigpod ceanothus - birchleaf mountain-mahogany series

Bigpod ceanothus and birchleaf mountain-mahogany important shrubs in the canopy; black sage, chamise, hollyleaf redberry, scrub oak, and/or toyon may be present. Shrubs < 4 m; canopy continuous or intermittent. Ground layer sparse.

Uplands: slopes.

Distribution: o-CenCo, o. l-TraR, w. m-PenR.
Elevation: 100-750 m.

NDDB/Holland type and status: **Chaparral** (37000).
   Ceanothus megacarpus chaparral (37840) G3 S3.2.

Other types:
Barry type: G7411211.
Cheatham & Haller type: Mixed chaparral.
Rangeland type: SRM 208.
Thorne type: Mixed chaparral.
WHR type: Mixed chaparral.

General references: Keeley (1975), Keeley & Keeley (1988), Montygierd-Loba & Keeley (1987), Pase (1982a), Schlesinger *et al.* (1982), White (1994c).

Comments: This definition follows Borchert *et al.* (1993b). If bigpod ceanothus has > 60% cover, the stand is a member of the Bigpod ceanothus series. If bigpod ceanothus covers 30-60% and another species covers 30-60%, then the stand is a member of a mixed series [*see* Bigpod ceanothus series, Bigpod ceanothus-hollyleaf redberry series]. Stands are established after fire and develop into other kinds of chaparral.

Species mentioned in text	
Bigpod ceanothus	*Ceanothus megacarpus*
Birchleaf mountain-mahogany	*Cercocarpus betuloides*
Black sage	*Salvia mellifera*
Chamise	*Adenostoma fasciculatum*
Hollyleaf redberry	*Rhamnus ilicifolia*
Scrub oak	*Quercus berberidifolia*
Toyon	*Heteromeles arbutifolia*

Plot-based descriptions: Borchert *et al.* (1993b) defines a Bigpod ceanothus - birchleaf mountain-mahogany association in o-CenCo, w. m-TraR.

Bigpod ceanothus-hollyleaf redberry series

Bigpod ceanothus and hollyleaf redberry important shrubs in the canopy; birchleaf mountain-mahogany, black sage, chamise, scrub oak, and/or toyon may be present. Shrubs < 4 m; canopy continuous or intermittent. Ground layer sparse.

Uplands: slopes.

Distribution: o-CenCo, o. l-TraR, w. m-PenR.
Elevation: 100-750 m.

NDDB/Holland type and status: Chaparral (37000).
    Ceanothus megacarpus chaparral (37840) G3 S3.2.

Other types:
Barry type: G7411211.
Cheatham & Haller type: Mixed chaparral.
Rangeland type: SRM 208.
Thorne type: Mixed chaparral.
WHR type: Mixed chaparral.

General references: Keeley (1975), Keeley & Keeley (1988), Montygierd-Loba & Keeley (1987), Pase (1982a), Schlesinger *et al.* (1982), White (1994c).

Comments: This definition follows Borchert *et al.* (1993b). If bigpod ceanothus has > 60% cover, the stand is a member of the Bigpod ceanothus series. If bigpod ceanothus covers 30-60% and another species covers 30-60%, then the stand is a member of a mixed series [*see* Bigpod ceanothus series, Bigpod ceanothus - birchleaf mountain-mahogany series].

Species mentioned in text	
Bigpod ceanothus	*Ceanothus megacarpus*
Birchleaf mountain-mahogany	*Cercocarpus betuloides*
Black sage	*Salvia mellifera*
Chamise	*Adenostoma fasciculatum*
Hollyleaf redberry	*Rhamnus ilicifolia*
Scrub oak	*Quercus berberidifolia*
Toyon	*Heteromeles arbutifolia*

Plot-based descriptions: Borchert *et al.* (1993b) defines a Bigpod ceanothus-hollyleaf redberry association in o-CenCo, w m-TraR.

Birchleaf mountain-mahogany - California buckwheat series

Birchleaf mountain-mahogany and California buckwheat important shrubs in canopy; chamise, chaparral yucca, hollyleaf cherry, fremontia, hollyleaf redberry, pale silktassel, scrub oak, and/or white sage may be present. Shrubs < 4 m; canopy continuous. Ground layer sparse.

Uplands: all slopes, may be steep, alluvium. Soils deep or shallow. On seep rock, soil absent.

Distribution: SoCo, m-TraR, PenR, w. MojD, w. ColD.
Elevation: 900-1950 m.

NDDB/Holland type and status: **Chaparral** (37000).
Semi-desert chaparral (37400 *in part*) G3 S3.3.
Flannel bush chaparral (37J00 *in part*) G3 S3.3.

Other types:
Barry type: G7411211.
Cheatham & Haller type: Semi-desert chaparral.
Rangeland type: SRM 205.
Thorne type: Desert transitional chaparral.
WHR type: Mixed chaparral.

General references: Barbour (1994), Hanes (1976, 1977, 1981), Horton (1960), Keeley 1992, Keeley & Keeley (1988), O'Leary (1989), Pase (1982a).

Comments: This definition follows Gordon & White (1994). Birchleaf mountain-mahogany is found in stands of many chaparral series. It also can be a tree forming a forest [*see* Birchleaf mountain-mahogany series where it dominates, whether it is a tree or shrub]. In this series birchleaf mountain-mahogany mixes with coastal scrub and chaparral shrubs, the most important one being California buckwheat. Birchleaf mountain-mahogany shares importance in the Bigpod ceanothus - birchleaf mountain-mahogany series, Red shank - birchleaf

mountain-mahogany series, and Scrub oak -birchleaf mountain-mahogany series.

White (1994a) qualitatively describes stands at Cleghorn Canyon candidate RNA in San Bernardino Co. in m-TraR.

Species mentioned in text	
Bigpod ceanothus	*Ceanothus megacarpus*
Birchleaf mountain-mahogany	*Cercocarpus betuloides*
California buckwheat	*Eriogonum fasciculatum*
Chamise	*Adenostoma fasciculatum*
Chaparral yucca	*Yucca whipplei*
Fremontia	*Fremontodendron californicum*
Hollyleaf cherry	*Prunus ilicifolia* ssp. *ilicifolia*
Hollyleaf redberry	*Rhamnus ilicifolia*
Pale silktassel	*Garrya flavescens*
Red shank	*Adenostoma sparsifolium*
Scrub oak	*Quercus berberidifolia*
White sage	*Salvia apiana*

Plot-based descriptions: Gordon & White (1994) define a Birchleaf mountain-mahogany - California buckwheat association in m-TraR, m-PenR.

Bitterbrush series

Bitterbrush sole, dominant, or important shrub with big sagebrush or rubber rabbitbrush in canopy; curlleaf mountain-mahogany, green ephedra, desert peach, horsebrush, and/or yellow rabbitbrush may be present. Emergent Joshua tree, junipers, and/or pines may be present. Shrubs < 5 m; cover continuous, intermittent, or open. Ground layer sparse or grassy (Plate 13).

Uplands: slopes and flats. Soils well-drained, rapidly permeable.

Distribution: m-CasR, su-CasR, su-SN, n. i-TraR, e. PenR, GB, MojD, DesR, inter-West. **Elevation:** 1000-3400 m.

NDDB/Holland type and status: **Great Basin scrubs** (35000).
　　Great Basin mixed scrub (35100 *in part*) G4 S4.
　　Big sagebrush (35210 *in part*) G4 S4.
　　Sagebrush steppe (35300 *in part*) G2 S2.1.

Other types:
Barry type: G74.
Brown Lowe Pase type: 132.15.
Cheatham & Haller type: Great Basin sagebrush.
Rangeland type: SRM 210.
Thorne type: Great Basin sagebrush scrub.
WHR type: Sagebrush.

General references: Neal (1994), (Nord 1965), Young *et al.* (1977).

Comments: This series definition follows Neal (1994), where bitterbrush is at least an important species in the shrub layer. Some stands of this series may have scattered Joshua trees, junipers, or pines. Bitterbrush may be a component of other shrub series [*see* Big sagebrush series, Low sagebrush series, Rubber rabbitbrush series]. Bitterbrush also occurs as an important understory shrub in open woodland and forest series. If trees dominate the stand, place it in a tree-dominated series

[*see* Jeffrey pine series, Joshua tree series, Ponderosa pine series, Singleleaf pinyon-Utah juniper series, Washoe pine series, Western juniper series].

Stands may get as old as 125 years on deep, well drained sites, but stands commonly become decadent at 30 years of age and die at 40-50 years. Stands tend to be of one age, and appear to result from an either a disturbance event or a rare year when many seedlings survive.

Species mentioned in text	
Antelope bitterbrush	*Purshia tridentata* var. *tridentata*
Big sagebrush	*Artemisia tridentata*
Bitterbrush	*Purshia tridentata*
Curlleaf mountain-mahogany	*Cercocarpus ledifolius*
Desert peach	*Prunus andersonii*
Desert bitterbrush	*Purshia tridentata* var. *glandulosa*
Green ephedra	*Ephedra viridis*
Horsebrush	*Tetradymia canescens*
Jeffrey pine	*Pinus jeffreyi*
Joshua tree	*Yucca brevifolia*
Junipers	*Juniperus* species
Pines	*Pinus* species
Rubber rabbitbrush	*Chrysothamnus nauseosus*
Singleleaf pinyon	*Pinus monophylla*
Utah juniper	*Juniperus osteosperma*
Washoe pine	*Pinus washoensis*
Western juniper	*Juniperus occidentalis* ssp. *occidentalis*
Yellow rabbitbrush	*Chrysothamnus viscidiflorus*

In *The Jepson Manual* bitterbrush includes two varieties which are treated as species in many manuals. Antelope bitterbrush and desert bitterbrush are generally differentiated by range, but both grow in TraSN. Both are included in this series.

Plot-based descriptions: not available.

Black bush series

Black bush sole or dominant shrub in canopy; bladder-sage, budsage, California buckwheat, ephedras, hop-sage, greenfire, turpentine-broom, and/or winter fat may be present. Emergent California juniper, Joshua tree, or singleleaf pinyon may be present. Shrubs < 1 m; canopy continuous. Ground layer sparse (Plate 13).

Uplands: alluvial slopes; bajadas. Soils shallow often dolomitic limestone-derived.

Distribution: i. m-TraR, TraSN, MojD, AZ, NV, UT.
Elevation: 1200-1800 m.

NDDB/Holland type and status: Mojavean desert scrubs (34000).
    Black bush scrub (34300) G3 S3.2.

Other types:
Barry type: G7411222 BCORA00.
Brown Lowe Pase type: 153.121 153.131 153.122.
Cheatham & Haller type: Black bush scrub.
PSW-45 type: Black bush series.
Rangeland type: SRM 212.
Stone & Sumida (1983): Black bush scrub.
Thorne type: Black bush scrub.
WHR type: Sagebrush.

General references: Bates (1984), MacMahon (1988), Martin (1994), Paysen *et al.* (1980), Sampson & Jesperson (1983), Thorne (1982) Turner (1982b), Vasek & Barbour (1977).

Comments: This series occurs at transitional elevations between MojD and GB. Plot data are needed to clarify relationships between stands where black bush is a dominant and those where it is not [*see* California juniper series, Joshua tree series, and Singleleaf pinyon series].

Species mentioned in text	
Big sagebrush	*Artemisia tridentata*
Black bush	*Coleogyne ramosissima*
Bladder-sage	*Salazaria mexicana*
Budsage	*Artemisia spinescens*
California buckwheat	*Eriogonum fasciculatum*
California juniper	*Juniperus californica*
Ephedras	*Ephedra* species
Greenfire	*Menodora spinescens*
Hop-sage	*Grayia spinosa*
Joshua tree	*Yucca brevifolia*
Singleleaf pinyon	*Pinus monophylla*
Shadscale	*Atriplex confertifolia*
Turpentine-broom	*Thamnosma montana*
Winter fat	*Krascheninnikovia lanata*

Plot-based descriptions: Rundall (1972) recognizes black bush stands [as Group II of an ordination of 46 stands] in Saline Valley in Inyo Co. MojD; Spolsky (1979) a Sonoran black bush scrub association in Anza Borrego State Park in San Diego Co. m-PenR.

Black sage series

Black sage sole or dominant shrub in canopy; ash buckwheat, California buckwheat, California encelia, California sagebrush, chaparral mallow, chaparral yucca, coast prickly-pear, coyote brush, and/or white sage may be present. Emergent laurel sumac may be present. Shrubs < 2 m; canopy continuous or intermittent. Ground layer variable (Plate 12).

Uplands: slopes steep. Soils shallow.

Distribution: CenCo, SoCo, l-TraR, m-TraR, m-PenR, ChaI.
Elevation: sea level-1200 m.

NDDB/Holland type and status:Coastal bluff scrubs (31000), **Coastal scrubs** (32000).
 Southern coastal bluff scrub (31200) G1 S1.1.
 Central Lucian coastal scrub (32200 *in part*) G3 S3.3.
 Venturan coastal sage scrub (32300 *in part*) G3 S3.1.
 Diablan sage scrub (32600 *in part*) G3 S3.3.
 Riversidean upland sage scrub (32710 *in part*) G3 S3.1.

Other types:
Barry type: G7411221 CSAME30.
Cheatham & Haller type: Coastal sage scrub.
Jones & Stokes type: Black sage scrub, Bush mallow scrub, Coastal bluff.
PSW-45 type: Salvia series.
Rangeland type: SRM 205.
Thorne type: Southern coastal scrub.
Westman type: Venturan I.
WHR type: Coastal scrub.

General references: Axelrod (1978), Barbour (1994), Burk (1977), DeSimone & Burk (1992), Haidinger & Keeley (1993), Malanson (1984), Keeley & Keeley (1988), O'Leary (1989), Mooney (1977), Pase & Brown (1982), Paysen *et al.* (1980), Westman (1981a, 1981b, 1981c, 1983).

Comments: This series is often considered part of the coastal scrub, which is better thought of as a collection of series. This approach allows stands of comparable composition which can be considered regardless of geographic location. Coast bluff scrub is included here in part. Definitions differ in plant height and cover from coastal scrub, but contrast little in species composition. Keeler-Wolf (1990e) qualitatively describes black sage dominated stands at Limekiln Creek RNA (now part of the Cone Peak Gradient RNA) in Monterey Co. o-CenCo.

Species mentioned in text	
Ash buckwheat	*Eriogonum cinereum*
Black sage	*Salvia mellifera*
California buckwheat	*Eriogonum fasciculatum*
California encelia	*Encelia californica*
California sagebrush	*Artemisia californica*
Chaparral mallow	*Malacothamnus fasciculatum*
Chaparral yucca	*Yucca whipplei*
Coast prickly-pear	*Opuntia littoralis*
Coyote brush	*Baccharis pilularis*
Laurel sumac	*Malosma laurina*
White sage	*Salvia apiana*

Plot-based descriptions: Kirkpatrick & Hutchinson (1977) define two associations, Malanson (1984) two associations, Mooney (1977) one association in SoCo; White (1994b) recognizes this one among seven coastal scrub series in western Riverside Co. i-SoCo.

Associations:
Kirkpatrick & Hutchinson (1977):
 Black sage-laurel sumac association,
 Black sage-California buckwheat association
 [= Venturan 1a *in part* (Malanson 1984)].
Malanson (1984):
 Black sage association [as Venturan 1a *in part*],
 Black sage-California encelia association
 [as Venturan 1a *in part*].
Mooney (1977):
 Black sage-coast prickly-pear association
 [as coastal sage succulent scrub in Table 13-1].

Black sagebrush series

Black sagebrush sole or dominant shrub in canopy; green ephedra, shadscale, winter fat, and/or yellow rabbitbrush may be present. Emergent Jeffrey pine, singleleaf pinyon, and/or Utah juniper may be present. Shrub < 0.5 m; canopy continuous or open. Ground layer sparse or grassy.

Uplands: flats, depressions, slopes, ridges. Parent material limestone. Soils poorly drained, may be carbonate-rich.

Distribution: m-CasR, su-CasR, i. m-TraR, GB, DesR, inter-West.
Elevation: 1500-2300 m.

NDDB/Holland type and status: Great Basin scrubs (35000), **Pavement plain communities** (47000).
    Subalpine sagebrush scrub (35220 *in part*) G3 S3.2.
    Pebble plain scrub (35220 in part) G1 S1.1.

Other types:
Barry type: G7411211 CARNO00.
Brown Lowe Pase type: 152.113.
Cheatham & Haller type: Great Basin sagebrush.
PSW-45 type: Sagebrush series.
Rangeland type: SRM 405.
Stone & Sumida (1983): Calcareous community.
Thorne type: Great Basin sagebrush scrub.
WHR type: Low sagebrush.

General references: Derby & Wilson (1978, 1979), Krantz (1983, 1988), Paysen *et al.* (1980), Tisdale (1994), Turner (1982a), Young *et al.* (1977), West (1988).

Comments: Young *et al.* (1977) proposed four possible series in CA based on work done in the Lahontan Basin NV. Black sagebrush can be confused with low sagebrush unless plants are in flower or fruit.

The pebble plains at Big Bear Valley in San Bernardino Co. i. m-TraR are considered an association of this series (Derby & Wilson 1978, 1979, Krantz 1983, 1988).

Species mentioned in text	
Black sagebrush	*Artemisia nova*
Green ephedra	*Ephedra viridis*
Jeffrey pine	*Pinus jeffreyi*
Low sagebrush	*Artemisia arbuscula*
Shadscale	*Atriplex confertifolia*
Singleleaf pinyon	*Pinus monophylla*
Utah juniper	*Juniperus osteosperma*
Winter fat	*Krascheninnikovia lanata*
Yellow rabbitbrush	*Chrysothamnus viscidiflorus*

Plot-based descriptions: not available.

Bladderpod-California ephedra-narrowleaf goldenbush series

Bladderpod, California buckwheat, California ephedra, and/or narrowleaf goldenbush conspicuous shrubs in canopy; bush buckwheat, paleleaf goldenbush, slender buckwheat, and/or yellow mock aster present. Shrubs < 3 m; canopy open. Ground layer grassy.

Uplands: hilltops, slopes. Soils diatomaceous, sand, or shale-derived.

Distribution: i-CenCo, sj-CenV, f-SN, w. MojD, m-TraR.
Elevation: 400-1000 m.

NDDB/Holland type and status: **Interior dunes** (23000), **Upper Sonoran subshrub scrubs** (39000) G3 S3.2.
    Monvero residual dunes (23300) G1 S1.2.

Other types:
Barry type: G7411211.

General references: Twisselmann (1967).

Comments: The residual dunes in w. Fresno Co. on Monvero soils is a special case of a more widespread series. Study is needed throughout the range of vegetation to develop association-parent material relationships. This series shares species with the Allscale series in Nor-CA.

Species mentioned in text	
Bladderpod	*Isomeris arborea*
California buckwheat	*Eriogonum fasciculatum*
California ephedra	*Ephedra californica*
Narrowleaf goldenbush	*Ericameria linearifolia*
Paleleaf goldenbush	*Isocoma acradenia*
Slender buckwheat	*Eriogonum gracillimum*
Yellow mock aster	*Eastwoodia elegans*

Plot-based descriptions: not available.

Blue blossom series

Blue blossom sole or dominant shrub in canopy; California huckleberry, chamise, coast whitethorn, common manzanita, deerbrush, poison-oak, salal, scrub oak, tobacco brush, and/or toyon may be present. Emergent conifers may be present. Shrubs < 6 m; canopy continuous or intermittent. Ground layer sparse.

Uplands: ridges, upper slopes.

Distribution: o-NorCo, o-CenCo, w. l-KlaR, OR.
Elevation: 10-600 m.

NDDB/Holland type and status: Coastal bluff scrub (31000), **Chaparral** (37000).
    Northern coastal bluff scrub (31100) G2 S2.2.
    Blue brush chaparral (37820) G4 S4.
    Northern maritime chaparral (37C10 *in part*) G1 S1.2.
    Poison-oak chaparral (37F00 *in part*) G4 S4.

Other types:
Barry type: G7411211 CCETH00.
PSW-45 type: Ceanothus series
Rangeland type: SRM 204.

General references: Hanes (1977, 1981), Paysen *et al.* (1980).

Comments: Throughout its range, blue blossom occurs as scattered shrubs in forest understories or as the sole or dominant shrub forming shrublands. Stands establish after disturbance and are transitional to various kinds of forest.

The Jepson Manual recognizes two varieties of blue blossom [*Ceanothus thyrsiflorus* var. *t.*, *C. t.* var. *repens*]. The creeping variety grows in coastal headlands [*see* Coyote brush series, Salal-black huckleberry series, Yellow bush lupine series].

Keeler-Wolf (1990e) qualitatively describes a blue blossom chaparral at Limekiln Creek RNA (now part of the Cone Peak Gradient RNA) Monterey Co. m-CenCo.

Species mentioned in text	
Black huckleberry	*Vaccinium ovatum*
Blue blossom	*Ceanothus thyrsiflorus*
California huckleberry	*Vaccinium ovatum*
Chamise	*Adenostoma fasciculatum*
Coast whitethorn	*Ceanothus incanus*
Common manzanita	*Arctostaphylos manzanita*
Coyote brush	*Baccharis pilularis*
Deerbrush	*Ceanothus integerrimus*
Poison-oak	*Toxicodendron diversilobum*
Salal	*Gaultheria shallon*
Scrub oak	*Quercus berberidifolia*
Tobacco brush	*Ceanothus velutinus*
Toyon	*Heteromeles arbutifolia*
Yellow bush lupine	*Lupinus arboreus*

Plot-based descriptions: not available.

Brewer oak series

Brewer oak sole or dominant shrub in canopy; greenleaf manzanita, huckleberry oak, mountain whitethorn, pinemat manzanita, tobacco brush, and/or wedgeleaf ceanothus may be present. Emergent black oak, and/or conifers may be present. Shrubs < 5 m; canopy continuous or intermittent. Ground layer variable, may be grassy.

Uplands: upper slopes, may be steep, rocky.

Distribution: m-NorCo, l-KlaR, m-KlaR, f-CasR, f-SN, m-SN, m-TraR.
Elevation: 600-1800 m.

NDDB/Holland type and status: Chaparral (37000).
    Shin oak brush (37541) G3 S3.

Other types:
Barry type: G7411211.
Cheatham & Haller type: Lower montane chaparral.
Thorne type: Montane chaparral.
WHR type: Montane chaparral.

General references: Griffin & Critchfield (1972), Twisselmann (1967).

Comments: *The Jepson Manual* considers Brewer oak (*Quercus garryana* var. *breweri*) and shin oak (*Q. g.* var. *semota*) to be a single variety. These shrubby forms of Oregon white oak (*Q. g.* var. *g.*) grow in thin soils, commonly at higher elevations than the tree form. Brewer oak grows in NorCo, KlaR, f-CasR, n. f-SN, m-SN; shin oak in s. f-SN, m-SN, m-TraR. Stands of both varieties are included in this series.

Plot-based descriptions: Keeler-Wolf (1986c, 1988b) Keeler-Wolf (1990e) describes stands at Doll Basin candidate RNA in Mendocino Co., Thornburgh (1981), Keeler-Wolf (1989i) in Keeler-Wolf (1990e) stands at Ruth candidate RNA, Jimerson (1993) defines one association in Trinity Co. m-NorCo; Sawyer *et al.*

(1978b) in Keeler-Wolf (1990e) describe stands at Indian Creek, Keeler-Wolf & Keeler-Wolf (1975), Keeler-Wolf (1989e) in Keeler-Wolf (1990e) stands at Hosselkus Limestone candidate RNA in Shasta Co., Sawyer (1981b) in Keeler-Wolf (1990e) defines one association at North Trinity Mountain candidate RNA Humboldt Co. m-KlaR; Keeler-Wolf (1986b, 1990d) Keeler-Wolf (1990d) describes stands at Indian Creek proposed RNA in Tehama Co. f-CasR; Keeler-Wolf (1990c) stands at Long Canyon candidate RNA in Kern Co. m-SN.

Species mentioned in text	
Black oak	*Quercus kelloggii*
Brewer oak	*Quercus garryana* var. *breweri*
Brewer spruce	*Picea breweriana*
California fescue	*Festuca californica*
Greenleaf manzanita	*Arctostaphylos patula*
Huckleberry oak	*Quercus vaccinifolia*
Mountain whitethorn	*Ceanothus cordulatus*
Oregon white oak	*Quercus garryana* var. *garryana*
Pinemat manzanita	*Arctostaphylos nevadensis*
Tobacco brush	*Ceanothus velutinus*
Wedgeleaf ceanothus	*Ceanothus cuneatus*
White fir	*Abies concolor*

Associations:
Jimerson (1993):
    Oregon white oak-Brewer oak/California fescue association.
Sawyer (1981b):
    White fir/huckleberry oak association.

Brittlebush series

Brittlebush sole, dominant, or important with California buckwheat shrubs in canopy; Acton encelia, bebbia, California sagebrush, chamise, chollas, creosote bush, Engelmann hedgehog cactus, fourwing saltbush, prickly-pears, and/or white bursage may be present. Shrubs < 3 m; canopy continuous or intermittent. Ground layer open; annuals may be seasonally present (Plate 17).

Uplands: alluvial fans; bajadas; upland slopes. Soil well-drained, may have desert pavement surface.

Distribution: i-SoCo, m-PenR, MojD, ColD, AZ, NV, UT, Mexico.
Elevation: 75 below sea level-1000 m.

NDDB/Holland type and status: **Coastal scrubs** (32000), **Sonoran desert scrub** (33000), **Mojavean desert scrub** (34000).
  Riversidean desert scrub (32730) G3 S3.1.
  Mojave creosote bush scrub (34100 *in part*) G4 S4.
  Sonoran creosote bush scrub (33100 *in part*) G4 S4.

Other types:
Barry type: G7411221.
Brown Lowe Pase type: 154.126.
Cheatham & Haller type: Creosote bush scrub, Coastal sage scrub.
Jones & Stokes type: Brittlebush-buckwheat scrub.
PSW-45 type: Encelia series.
Rangeland type: SRM 205.
Thorne type: Creosote bush scrub, Southern coastal scrub.
WHR type: Coastal scrub, Desert scrub.

General references: Barbour (1994), Burk (1977), Hunt (1966), Keeley & Keeley (1988), MacMahon (1988), O'Leary (1989), Pase & Brown (1982), Paysen *et al.* (1980), Vasek & Barbour (1977).

Comments: There are several series in which brittlebush is a major component. This series is often considered part of the creosote bush scrub, coastal scrub, or inland sage scrub. These scrubs are better thought of as a collection of series. In Cis-CA this series is part of the coastal sage scrub or inland sage scrub where brittlebush is dominant [*see* California sagebrush series, California buckwheat series, Scalebroom series].

In Trans-CA this series is that part of the creosote bush scrub in which brittlebush is the dominant over any other shrubs which may be present [*see* Creosote bush series, Creosote bush-cactus series, Creosote bush-white bursage series, White bursage series]. The amount of shrub canopy varies among stands. Dense canopies are common on steep rocky slopes. Areas of desert pavement have occasional shrubs. It is difficult to consider such stands to be dominated by any one species, but brittlebush commonness places the stands here.

Species mentioned in text	
Acton encelia	*Encelia actoni*
Bebbia	*Bebbia juncea*
Brittlebush	*Encelia farinosa*
California buckwheat	*Eriogonum fasciculatum*
California sagebrush	*Artemisia californica*
Chamise	*Adenostoma fasciculatum*
Chollas	*Opuntia* species
Creosote bush	*Larrea tridentata*
Engelmann hedgehog cactus	*Echinocereus engelmannii*
Fourwing saltbush	*Atriplex canescens*
Prickly-pears	*Opuntia* species
Scalebroom	*Lepidospartum squamatum*
White bursage	*Ambrosia dumosa*
Wishbone bush	*Mirabilis californica*

Plot-based descriptions: Kirkpatrick & Hutchinson (1977) define a Brittlebush-wishbone bush association, White (1994b) recognizes this one among seven coastal scrub series in western Riverside Co. i-SoCo; Minnich *et al.* (1993) define a Brittlebush association [as Creosote bush scrub with brittlebush] at Marine Corps Air-ground Combat Center in San Bernardino Co. MojD.

Brittlebush-white bursage series

Shrubs and cacti conspicuous in shrub canopy; barrel cactus, brittlebush, chollas, clustered barrel cactus, desert agave, hedgehog cactus, jojoba, prickly-pears, pygmy-cedar, and/or white bursage may be present. Shrubs < 1.5 m; canopy intermittent or open. Ground layer open; annuals seasonally present.

Uplands: slopes rocky. Soils well-drained.

Distribution: ColD, Baja CA.
Elevation: 10-1000 m.

NDDB/Holland type and status: Sonoran desert scrubs (34000).
> Sonoran mixed woody scrub (33210 *in part*) G3 S3.
> Sonoran mixed woody and succulent scrub (33220 *in part*) G3 S2.2.

Other types:
Barry type: G74 G7411211.
Brown Lowe Pase type: 154.113.
Cheatham & Haller type: Desert cactus scrub.
Thorne type: Stem-succulent scrub.
WHR type: Desert succulent shrub.

General references: Burk (1977), MacMahon (1988), Thorne (1982).

Comments: This series has been described as cactus scrub or mixed scrub. Some scrub stands are dominated by brittlebush, others by white bursage, and others are mixed. The mixed stands are placed in this series. Cactus composition varies among stands as well. Ocotillo may be present, but not conspicuously [*see* Brittlebush series, Creosote bush series, Teddy-bear cholla series, Ocotillo series].

Species mentioned in text	
Barrel cactus	*Ferocactus cylindraceus*
Brittlebush	*Encelia farinosa*
Chollas	*Opuntia* species
Clustered barrel cactus	*Echinocactus polycephalus*
Desert agave	*Agave deserti*
Hedgehog cactus	*Echinocereus engelmannii*
Jojoba	*Simmondsia chinensis*
Ocotillo	*Fouquieria splendens*
Prickly-pears	*Opuntia* species
Pygmy-cedar	*Peucephyllum schottii*
Teddy-bear cholla	*Opuntia bigelovii*
White bursage	*Ambrosia dumosa*

Plot-based descriptions: Spolsky (1979) defines a Cholla-desert agave-stem succulent scrub association in Anza Borrego State Park in San Diego Co. ColD.

Broom series

Broom sole or dominant shrub in canopy; French broom, gorse, Spanish broom, Scotch broom and/or other species of *Cytisus* or *Genista* may be present. Emergent trees may be present. Shrubs < 6 m; canopy continuous. Ground layer sparse.

Uplands: all slopes.

Distribution: o-NorCo, CenCo, CenV, l-KlaR, f-CasR, m-CasR, f-SN, m-SN, SoCo, m-TraR, m-PenR, ChaI.
Elevation: sea level-1000 m.

NDDB/Holland type and status: none.

Other types:
Barry type: G7411221.

General references: Boyd (1985a, 1985b), McClintock (1985a, 1985b), Mountjoy (1979).

Comments: Invasiveness of broom is well appreciated. Ten species of broom or gorse from four legume genera are included in this series. *The Jepson Manual* recognizes three *Cytisus*, five *Genista*, one *Spartium*, and one *Ulex* in California. French broom, Spanish broom, and Scotch broom are the major invading brooms of disturbed areas in the state. The following are uncommon or locally common: *Cytisus multiflorus*, *Cytisus striatus*, *Genista canariensis*, *Genista linifolia*, *Genista maderensis*, *Genista stenopetala*, and *Ulex europaea*. In areas where trees form dense canopies with age, broom is killed; however a persistent seed bank remains. The treatment here is broad, recognizing the importance of these introduced species in the vegetation of California.

Species mentioned in text	
French broom	*Genista monspessulana*
Gorse	*Ulex europaea*
Spanish broom	*Spartium junceum*
Scotch broom	*Cytisus scoparius*

Plot-based descriptions: not available.

Bush chinquapin series

Bush chinquapin sole or dominant shrub in canopy; bitter cherry, creeping snowberry, Fremont silktassel, greenleaf manzanita, huckleberry oak, mountain whitethorn, pinemat manzanita, service berry, and/or tobacco brush may be present. Emergent conifers may be present. Shrubs < 2 m; canopy continuous or intermittent. Ground layer sparse (Plate 9).

Uplands: ridges, upper slopes, may be steep. Soils shallow, commonly granitic-derived, may be rocky.

Distribution: m-NorCo, m-KlaR, su-KlaR, m-CasR, su-CasR, m-SN, su-SN, o. su-TraR, su-PenR.
Elevation: 800-3300 m.

NDDB/Holland type and status: **Chaparral** (37000).
   Mixed montane chaparral (37510 *in part*) G4 S4.
   Bush chinquapin chaparral (37540) G3 S3.3.

Other types:
Barry type: G7411214.
Cheatham & Haller type: Upper montane chaparral.
PSW-45 type: Bush chinquapin series.
Thorne type: Montane chaparral.
WHR type: Montane chaparral.

General references: Pase (1982a), Paysen *et al.* (1980), Solinas *et al.* (1985), Vankat & Major (1978).

Comments: Bush chinquapin occurs as scattered shrubs in forest understories or is the sole or dominant shrub in what is often called montane chaparral. This series is found in montane and subalpine elevations, and is more extensive at higher elevations. Self-perpetuating stands grow on shallow soils; those transitional to forest occur on deeper soils. Huckleberry oak series stands, which primarily grow in Nor-CA, share shrub species and elevations with this series.

Stands of this series can be differentiated on the importance of huckleberry oak. In the Huckleberry oak series it is a major species, whereas here it is not an important species.

Keeler-Wolf (1990e) qualitatively describes bush chinquapin stands at Mount Shasta Mudflow RNA in Siskiyou Co. m-KlaR; at Mud Lake RNA in Plumas Co. m-CasR.

Species mentioned in text	
Bitter cherry	*Prunus emarginata*
Bush chinquapin	*Chrysolepis sempervirens*
Creeping snowberry	*Symphoricarpos mollis*
Fremont silktassel	*Garrya fremontii*
Greenleaf manzanita	*Arctostaphylos patula*
Huckleberry oak	*Quercus vaccinifolia*
Mountain whitethorn	*Ceanothus cordulatus*
Pinemat manzanita	*Arctostaphylos nevadensis*
Service berry	*Amelanchier alnifolia*
Tobacco brush	*Ceanothus velutinus*

Plot-based descriptions: Conard & Rodosevich (1982) describe stands in terms of cover in Sierra Co. m-SN.

Bush seepweed series

Bush seepweed sole or dominant shrub in canopy; alkali heath, greasewood, iodine bush, rusty molly, and/or samphire may be present. Shrubs < 1.5 m; canopy open. Ground layer absent.

Wetlands: habitats intermittently flooded, saturated. Water chemistry: mixosaline. Dry lake beds; plains; old lake beds perched above current drainages. Cowardin class: Palustrine shrub-scrub wetland. The national list of wetland plants (Reed 1988) lists bush seepweed as a FAC+.

Distribution: CenV, So-CA, Trans-CA.
Elevation: sea level-1600 m.

NDDB/Holland type and status: **Chenopod scrubs** (36000), **Alkali playa communities** (46000 *in part*) G3 S2.1.
> Desert sink scrub (36120 *in part*) G2 S2.1.
> Desert greasewood scrub (36130 *in part*) G3 S3.2.
> Valley sink scrub (36210 *in part*) G1 S1.1.

Other types:
Barry type: G7412321.
Brown Lowe Pase type: 153.171.
Cheatham & Haller type: Alkali sink scrub.
PSW-45 type: Suaeda series.
Thorne type: Alkali sink scrub.
WHR type: Alkali sink.

General references: Burk (1977), MacMahon (1988), Paysen *et al.* (1980), Vasek & Barbour (1977), West (1988).

Comments: *The Jepson Manual considers Suaeda fruticosa and S. torreyana, including S. t. var. ramosissima, misapplied to bush seepweed (S. moquinii).* Bush seepweed, iodine bush, and greasewood grow around margins of dry or wet lakes [*see* Greasewood series, Iodine bush series]. Whether a given stand is classed as a member of these series depends on which species dominates. Bush seepweed tolerates high salt concentrations. In CenV stands of this series form a fine scale with All-scale series and Alkali sacaton series, depending on depth to water table.

Plot-based descriptions: Bradley (1970) defines an association in Death Valley National Park MojD.

Species mentioned in text	
Alkali heath	*Frankenia salina*
Alkali sacaton	*Sporobolus airoides*
Allscale	*Atriplex polycarpa*
Bush seepweed	*Suaeda moquinii*
Greasewood	*Sarcobatus vermiculatus*
Iodine bush	*Allenrolfea occidentalis*
Rusty molly	*Kochia californica*
Samphire	*Salicornia subterminalis*

Associations:
Bradley (1970):
> Bush seepweed-iodine bush association [as Western salt flat].

Buttonbush series

Buttonbush sole or dominant shrub in canopy; red osier, narrowleaf willow, and/or Pacific willow may be present. Shrubs < 10 m tall; canopy continuous, intermittent, or open. Ground layer sparse or grassy (Plate 7).

Wetlands: habitat intermittently flooded, seasonally saturated. Water chemistry: fresh. Floodplains. Cowardin class: Palustrine shrub-scrub wetland. The national list of wetland plants (Reed 1988) lists buttonbush as a OBL.

Distribution: i-NorCo, i-CenCo, CenV, f-KlaR, f-CasR, f-SN, sw. U.S.
Elevation: 5-1000 m.

NDDB/Holland type and status: **Riparian scrubs** (63000).
    Buttonbush scrub (63430) G1 S1.1.

Other types:
Barry type: G7411211.
Cheatham & Haller type: Bottomland woodlands and forest.
Thorne type: Riparian woodland.
WHR type: Freshwater emergent wetland.

General references: Holland (1986).

Comments: Buttonbush, which is also called button-willow, is common as scattered shrubs in riparian settings. Where buttonbush occurs in stands, it forms small-grained patches in mosaic with other riparian series.

Species mentioned in text	
Buttonbush	*Cephalanthus occidentalis*
Narrowleaf willow	*Salix exigua*
Pacific willow	*Salix lucida* spp. *lasiandra*
Red osier	*Cornus sericea*

Plot-based descriptions: not available.

California buckwheat series

California buckwheat sole or dominant shrub in canopy; black sage, brittlebush, bush monkeyflower, California encelia, chaparral mallow, coast goldenbush, coyote brush, deer weed, and/or white sage may be present. Emergent lemonade berry may be present. Shrubs < 1 m; canopy continuous or intermittent. Ground layer variable, may be grassy.

Uplands: slopes, rarely flooded low-gradient deposits along streams. Soils shallow and rocky.

Distribution: CenCo, SoCo, m-TraR, m-PenR, ChaI, MojD, Baja CA.
Elevation: sea level-1200 m.

NDDB/Holland type and status: **Coastal bluff scrubs** (31000), **Coastal scrubs** (32000).
    Southern coastal bluff scrub (31200) G1 S1.1.
    Central Lucian coastal scrub (32200 *in part*) G3 S3.3.
    Venturan coastal sage scrub (32300 *in part*) G2 S3.1.
    Diegan coastal sage scrub (32500 *in part*) G2 S3.1.
    Diablan sage scrub (32600 *in part*) G3 S3.1.
    Riversidean sage scrub (32700 in part) G3 S3.1.
    Alluvial fan chaparral (37H00 *in part*) G3 S3.1.

Other types:
Barry type: G7411211 CERFA00.
Cheatham & Haller type: Central coastal scrub, Coastal sage scrub.
Jones & Stokes type: Buckwheat scrub, Coast bluff scrub.
PSW-45 type: California buckwheat series.
Rangeland type: SRM 205.
Thorne type: Southern coastal scrub.
WHR type: Coastal scrub.

General references: Axelrod (1978), Barbour (1994), DeSimone & Burk (1992), Keeley & Keeley (1988), Kirkpatrick & Hutchinson (1977), Malanson (1984),

Mooney (1977), O'Leary (1989), Paysen *et al.* (1980), Pase & Brown (1982), Vasek & Barbour (1977), Westman (1983).

Species mentioned in text	
Big sagebrush	*Artemisia tridentata*
Black sage	*Salvia mellifera*
Brittlebush	*Encelia farinosa*
Bush monkeyflower	*Mimulus aurantiacus*
California buckwheat	*Eriogonum fasciculatum*
California encelia	*Encelia californica*
California figwort	*Scrophularia californica*
California sagebrush	*Artemisia californica*
Chaparral mallow	*Malacothamnus fasciculatum*
Coast goldenbush	*Isocoma menziesii*
Coyote brush	*Baccharis pilularis*
Deer weed	*Lotus scoparius*
Lemonade berry	*Rhus integrifolia*
Phacelia	*Phacelia ramosissima*
White sage	*Salvia apiana*

Comments: This series is often considered part of the coastal scrub, which is better thought of as a collection of series. This approach allows stands of comparable composition to be considered regardless of geographic location. Coast bluff scrub is included here in part. Definitions differ in plant height and cover from coastal sage, but contrast little in species composition.

Three varieties of *Eriogonum fasciculatum* grow in the range of coastal scrub. There is some geographic separation between them: *E. f.* var. *f.* grows in o-CenCo, *E. f.* var. *polifolium* in i-SoCo, but *E. f.* var. *polifolium* is sympatric with them (Axelrod 1978). *E.* var. *flavoviride* occurs in MojD, ColD, as does *E. f.* var. *polifolium*. Stands dominated by *E. giganteum* on ChaI are included in this series. Stands of this series differ from the California sagebrush-California buckwheat series in that California buckwheat dominates here.

Plot-based descriptions: Hanes *et al.* (1989) report sites in six drainages [Big Tujunga Canyon, Cajon Creek, Cucamonga Canyon, Lytle Creek, San Savaine Canyon, Upper Santa Ana River] as domin-

ated by California buckwheat, Kirkpatrick &
Hutchinson (1977) define one association, Gordon
& White (1994) three associations in SoCo; White
(1994b) recognizes this one among seven coastal
scrub series in western Riverside Co. i-SoCo.

Associations:
Kirkpatrick & Hutchinson (1977):
>   California buckwheat-California figwort-
>   phacelia association.
Gordon & White (1994):
>   California buckwheat association,
>   California buckwheat-big sagebrush association,
>   California buckwheat alluvial fan association.

California buckwheat-white sage series

California buckwheat and white sage important shrubs in canopy; chamise, chaparral yucca, chaparral whitethorn, cupleaf ceanothus, deer weed, and/or scrub oak may be present. Emergent coast live oak may be present. Shrubs < 2 m; canopy intermittent. Ground layer variable, may be grassy.

Uplands: slopes south-facing, bouldery. Soils shallow.

Distribution: SoCo, Baja CA.
Elevation: 500-1500 m.

NDDB/Holland type and status: **Coastal scrubs** (32000), **Chaparral** (37000).
  Diegan coastal sage scrub (32500 *in part*) G3 S3.1.
  Riversidean sage scrub (32700 *in part*) G3 S3.1.
  Riversidean alluvial fan sage scrub (32720 *in part*) G1 S1.1.
  Alluvial fan chaparral (37H00 *in part*) G3 S3.2.

Other types:
Barry type: G7411221.
Cheatham & Haller types: Coastal sage scrub.
Jones & Stokes type: Coast bluff scrub.
Rangeland type: SRM 205.
Thorne type: Southern coastal scrub.
WHR type: Coastal scrub.

General references: Axelrod (1978), Barbour (1994), DeSimone & Burk (1992), Keeley & Keeley (1988), Kirkpatrick & Hutchinson (1977), Malanson (1984), O'Leary (1989), Pase (1982a), Pase & Brown (1982), Westman (1983).

Comments: This definition follows Gordon & White (1994). This series is often considered part of the coastal scrub, which is better thought of as a collection of series. This approach allows stands of comparable composition which can be considered regardless of geographic location.

Stands of this series differ from the California buckwheat series and White sage series in that California buckwheat and white sage are equally important. In addition, more chaparral species occur in stands of this series than in stands of the California buckwheat series or White sage series.

Three varieties of *Eriogonum fasciculatum* grow in the range of coastal scrub. There is some geographic separation of them: *E. f.* var. *f.* grows in o-CenCo, *E. f.* var. *polifolium* in i-SoCo, but *E. f.* var. *foliolosum* is sympatric with them (Axelrod 1978).

Species mentioned in text	
California buckwheat	*Eriogonum fasciculatum*
Chamise	*Adenostoma fasciculatum*
Chaparral yucca	*Yucca whipplei*
Chaparral whitethorn	*Ceanothus leucodermis*
Coast live oak	*Quercus agrifolia*
Cupleaf ceanothus	*Ceanothus greggii*
Deer weed	*Lotus scoparius*
Scrub oak	*Quercus berberidifolia*
White sage	*Salvia apiana*

Plot-based descriptions: Gordon & White (1994) define a California buckwheat-white sage association in SoCo; White (1994b) recognizes this one among seven coastal scrub series in western Riverside Co. in i-SoCo.

California encelia series

California encelia sole or dominant shrub in canopy; ash buckwheat, black sage, bladderpod, bush monkeyflower, California sagebrush, chaparral yucca, coast goldenbush, coyote brush, deer weed, purple sage, and/or wishbone bush may be present. Emergent lemonade berry or Mexican elderberry may be present. Shrubs < 2 m; canopy continuous or intermittent. Ground layer variable (Plate 13).

Uplands: slopes steep, south-facing. Soils colluvial derived.

Distribution: s. o-CenCo, o-SoCo, ChaI.
Elevation: sea level-1200 m.

NDDB/Holland type and status: Coastal bluff scrubs (3100), **Coastal scrubs** (32000).
>   Southern coastal bluff scrub (31200) G1 S1.1.
>   Venturan coastal sage scrub (32300 *in part*) G3 S3.1.
>   Diegan coastal sage scrub (32500 *in part*) G3 S3.1.

Other types:
Barry type: G7411221.
Cheatham & Haller types: Coastal sage scrub.
Jones & Stokes type: Coastal bluff scrub, Mixed scrub.
PSW-45 type: Encelia series.
Rangeland type: SRM 205.
Thorne type: Southern coastal scrub.
Westman type: Venturan I.
WHR type: Coastal scrub.

General references: Axelrod (1978), Barbour (1994), DeSimone & Burk (1992), Mooney (1977), Keeley & Keeley (1988), Kirkpatrick & Hutchinson (1977), Malanson (1984), O'Leary (1989), Pase & Brown (1982), Paysen *et al.* (1980), Westman (1981a, 1981b, 1981c).

Comments: This series is often considered part of the coastal scrub, which is better thought of as a collection of series. This approach allows stands of comparable composition which can be considered regardless of geographic location. Coast bluff scrub is included here in part. Definitions differ in plant height and cover from coastal scrub, but contrast little in species composition.

Species mentioned in text	
Ash buckwheat	*Eriogonum cinereum*
Black sage	*Salvia mellifera*
Bladderpod	*Isomeris arborea*
Bush monkeyflower	*Mimulus aurantiacus*
California encelia	*Encelia californica*
California sagebrush	*Artemisia californica*
Chaparral yucca	*Yucca whipplei*
Coast goldenbush	*Isocoma menziesii*
Coyote brush	*Baccharis pilularis*
Deer weed	*Lotus scoparius*
Lemonade berry	*Rhus integrifolia*
Mexican elderberry	*Sambucus mexicana*
Purple sage	*Salvia leucophylla*
Wishbone bush	*Mirabilis californica*

Plot-based descriptions: Kirkpatrick & Hutchinson (1977) define one association, Malanson (1984) one in SoCo.

Associations:
Kirkpatrick & Hutchinson (1977)
>   California encelia-California sagebrush association.
Malanson (1984)
>   California encelia association
>   [as Venturan Ib].

California sagebrush series

California sagebrush sole or dominant shrub; black sage, brittlebush, bush monkeyflower, California encelia, chamise, chaparral yucca, coast goldenbush, coyote brush, deer weed, poison-oak, purple sage, and/or white sage may be present. Emergent lemonade berry or Mexican elderberry may be present. Shrubs < 2 m; canopy continuous or intermittent. Ground layer variable (Plate 12).

Uplands: slopes steep, south-facing, rarely flooded low-gradient deposits along streams. Soils alluvial or colluvial-derived, shallow.

Distribution: CenCo, SoCo, ChaI, Baja CA.
Elevation: sea level-1200 m.

NDDB/Holland type and status: Coastal bluff scrub (31000), Coastal scrubs (32000).
    Southern coastal bluff scrub (31200) G1 S1.1.
    Northern coastal bluff scrub (31100 *in part*) G2 S2.2.
    Central Lucian coastal scrub (32200 *in part*) G3 S3.3.
    Venturan coastal sage scrub (32300 *in part*) G3 S3.1.
    Diablan sage scrub (32600 *in part*) G3 S3.3.
    Riversidean upland sage scrub (32700 *in part*) G3 S3.3.

Other types:
Barry type: G7411221 CARCA11.
Cheatham & Haller type: Northern coastal scrub, Central coastal scrub, Coastal sage scrub.
Jones & Stokes type: Coastal bluff scrub, sagebrush scrub, Sagebrush-monkeyflower scrub.
PSW-45 type: Coastal sagebrush series.
Rangeland type: SRM 205.
Thorne type: Southern coastal scrub.
WHR type: Coastal scrub.

General references: Axelrod (1978), Barbour (1994), DeSimone & Burk (1992), Keeley & Keeley (1988),

Kirkpatrick & Hutchinson (1977), Malanson (1984), Mooney (1977), O'Leary (1989), Pase & Brown (1982), Paysen *et al.* (1980), Westman (1981a).

Comments: This series is often considered part of the coastal scrub, which is better thought of as a collection of series. This approach allows stands of comparable composition which can be considered regardless of geographic location.

Species mentioned in text	
Black sage	*Salvia mellifera*
Brittlebush	*Encelia farinosa*
Bush monkeyflower	*Mimulus aurantiacus*
Bush-penstemon	*Keckiella cordifolia*
California buckwheat	*Eriogonum fasciculatum*
California encelia	*Encelia californica*
California sagebrush	*Artemisia californica*
Chamise	*Adenostoma fasciculatum*
Chaparral yucca	*Yucca whipplei*
Coast goldenbush	*Isocoma menziesii*
Coyote brush	*Baccharis pilularis*
Deer weed	*Lotus scoparius*
Lemonade berry	*Rhus integrifolia*
Mexican elderberry	*Sambucus mexicana*
Poison-oak	*Toxicodendron diversilobum*
Purple sage	*Salvia leucophylla*
White sage	*Salvia apiana*

Stands with California sagebrush and California buckwheat equally important are members of the California sagebrush-California buckwheat series. Stands with California sagebrush and black sage equally important are members of the California sagebrush-black sage series. Black sage or coyote brush may be present, but not important, in stands of this series. If these dominate, stands belong to the Black sage series or Coyote brush series.

Coast bluff scrub descriptions of most authors are included here in part. Definitions are similar in species composition to those for coastal scrub, but contrast little in species composition.

Keeler-Wolf (1990e) qualitatively describes California sagebrush stands at Limekiln Creek RNA (now part of the Cone Peak Gradient RNA) in Monterey Co. o-CenCo.

Plot-based descriptions: Cole (1980) reports cover for stands in the Purisima Hills in o-CenCo; Hanes *et al.* (1989) describe sites in Antonia Canyon with California buckwheat, Kirkpatrick & Hutchinson (1977) define one association, DeSimone & Burk (1992) one association, and Gordon & White (1994) three associations in SoCo; White (1994b) recognizes this one among seven coastal scrub series in western Riverside Co. i-SoCo; Boyd (1983) describes five stands in Gavilan Hills in Riverside Co. i. m-TraR.

Associations:
Kirkpatrick & Hutchinson (1977)
　　California sagebrush association.
DeSimone & Burk (1992):
　　California sagebrush-deer weed association
　　[as Group 3].
Gordon & White (1994):
　　California sagebrush association,
　　California sagebrush-bush penstemon association,
　　California sagebrush-purple sage association.

California sagebrush-black sage series

California sagebrush and black sage important shrubs in canopy; California buck-wheat, chamise, chaparral yucca, deer weed, sugar bush, and/or white sage may be present. Emergent lemonade berry may be present. Shrubs < 2 m; canopy continuous or intermittent. Ground layer variable.

Uplands: slopes steep, south-facing. Soils colluvial-derived.

Distribution: SoCo, m-TraR, m-PenR.
Elevation: 250-750 m.

NDDB/Holland type and status: **Coastal scrubs** (32000).
> Riversidean upland sage scrub (32700 *in part*) G2 S2.1.

Other types:
Barry type: G7411221.
Cheatham & Haller type: Coastal sage scrub.
Jones & Stokes type: Sagebrush-black sage scrub.
Rangeland type: SRM 205.
Thorne type: Southern coastal scrub.
WHR type: Coastal scrub.

General references: Axelrod (1978), Barbour (1994), DeSimone & Burk (1992), Kirkpatrick & Hutchinson (1977), Malanson (1984), O'Leary (1989), Pase & Brown (1982), Westman (1983).

Comments: This series is often considered part of the coastal scrub, which is better thought of as a collection of series. This approach allows stands of comparable composition which can be considered regardless of geographic location. Stands of this series differ from the California sagebrush series and Black sage series in that California sagebrush and black sage are equally important in this series.

Species mentioned in text	
Black sage	*Salvia mellifera*
California buckwheat	*Eriogonum fasciculatum*
California sagebrush	*Artemisia californica*
Chamise	*Adenostoma fasciculatum*
Chaparral yucca	*Yucca whipplei*
Deer weed	*Lotus scoparius*
Lemonade berry	*Rhus integrifolia*
Sugar bush	*Rhus ovata*
White sage	*Salvia apiana*

Plot-based descriptions: DeSimone & Burk (1992) define two associations, Gordon & White (1994) one association in SoCo.

Associations:
DeSimone & Burk (1992):
> California sagebrush-black sage association [as Group 4 = Gordon & White (1994)],
> Black sage-California sagebrush association [as Group 5].

California sagebrush-California buckwheat series

California sagebrush and California buckwheat important shrubs in canopy; chamise, bush monkeyflower, chaparral yucca, deer weed, sugar bush, and/or white sage may be present. Emergent laurel sumac or lemonade berry may be present. Shrubs < 2 m; canopy continuous or intermittent. Ground layer variable (Plate 12).

Uplands: slopes steep, south-facing. Soils colluvial-derived.

Distribution: i-SoCo, m-TraR, m-PenR, Baja CA.
Elevation: 250-950 m.

NDDB/Holland type and status: **Coastal scrubs** (32000).
> Diegan coastal sage scrub (32500 *in part*) G3 S3.1.
> Riversidean upland sage scrub (32700 *in part*) G3 S3.1.

Other types:
Cheatham & Haller type: Coastal sage scrub.
Jones & Stokes type: Sagebrush-California buckwheat scrub.
Rangeland type: SRM 205.
Thorne type: Southern coastal scrub.
WHR type: Coastal scrub.

General references: Axelrod (1978), Barbour (1994), DeSimone & Burk (1992), Keeley & Keeley (1988), Kirkpatrick & Hutchinson (1977), Malanson (1984), Mooney (1977), O'Leary (1989), Pase & Brown (1982), Westman (1983).

Comments: The series definition follows Gordon & White (1994). This series is often considered part of the coastal scrub, which is better thought of as a collection of series. This approach allows stands of comparable composition which can be considered regardless of geographic location. Stands of this series differ from the California sagebrush and California buckwheat series in that California sagebrush and California buckwheat are equally important in this series.

Species mentioned in text	
Bush monkeyflower	*Mimulus aurantiacus*
California buckwheat	*Eriogonum fasciculatum*
California sagebrush	*Artemisia californica*
Chamise	*Adenostoma fasciculatum*
Chaparral yucca	*Yucca whipplei*
Deer weed	*Lotus scoparius*
Laurel sumac	*Malosma laurina*
Sugar bush	*Rhus ovata*
White sage	*Salvia apiana*

Plot-based descriptions: Gordon & White (1994) define two associations, White (1994b) recognizes this as one among seven coastal scrub series in western Riverside Co. i-SoCo.

Associations:
Gordon & White (1994):
> California sagebrush-California buckwheat-sugar bush association,
> California sagebrush-California buckwheat-white sage association.

Canyon live oak shrub series

Canyon live oak sole or dominant shrub in canopy; California bay, chaparral whitethorn, deerbrush, interior live oak, and/or poison-oak may be present. Shrubs < 6 m; canopy continuous or intermittent. Ground layer variable.

Uplands: slopes commonly north-facing. Soil alluvial or bedrock-derived, may be rocky.

Distribution: m-TraR, m-PenR.
Elevation: 1000-2200 m.

NDDB/Holland type and status: **Chaparral** (37000).
   Interior live oak chaparral (37A00 *in part*) G3 S3.3.
   Southern north slope chaparral (37E20 *in part*) G3 S3.3.

Others types:
Cheatham & Haller type: Mixed chaparral.
Thorne type: Mixed chaparral.
WHR type: Mixed chaparral.

General references: Hanes (1977, 1981), Horton (1960), Griffin (1977), Keeley & Keeley (1988), Minnich (1976), Pase (1982a), Patric & Hanes (1964).

Comments: Canyon live oak is a widespread and common species in shrublands and forests. Stands where it occurs as a shrub, either because it is presumably distinct [*Quercus chrysolepis* var. *nana*] or because of age, are included in this series. Many of the So-CA shrub stands are believed to be the result of frequent sprouting after fires (White & Sawyer 1995). Stands old enough to form woodlands and forests are included in the Canyon live oak series.

This series definition follows Gordon & White (1994). If canyon live oak as a shrub has > 60% cover, the stand is a member of the Canyon live oak shrub series. If interior live oak as a shrub cover is 60-30% or canyon live oak as a shrub cover 30-

60%, then the stand is another series [*see* Interior live oak-chaparral whitethorn shrub series, Interior live oak-canyon live oak shrub series, Interior live oak-scrub oak shrub series].

Species mentioned in text	
California bay	*Umbellularia californica*
Canyon live oak	*Quercus chrysolepis*
Chaparral whitethorn	*Ceanothus leucodermis*
Deerbrush	*Ceanothus integerrimus*
Interior live oak	*Quercus wislizenii*
Poison-oak	*Toxicodendron diversilobum*

Plot-based descriptions: Gordon & White (1994) define two associations in m-TraR, m-PenR.

Associations:
Gordon & White (1994):
   Canyon live oak shrub association,
   Canyon live oak-(deerbrush-chaparral whitethorn) shrub association.

Catclaw acacia series

Catclaw acacia and creosote bush important shrubs in canopy; basket bush, bebbia, brittlebush, California ephedra, cheesebush, creosote bush, desert barberry, desert-lavender, desert-olive, fairy duster, Fremont box-thorn, and/or jojoba may be present. Emergent desert-willow or smoke tree may be present. Shrubs < 2 m; canopy continuous or open. Ground layer sparse; annuals seasonally present (Plate 15).

Uplands: Rarely flooded margins of arroyos and washes. The national list of wetland plants (Reed 1988) lists catclaw acacia as a FACU.

Distribution: ColD, MojD, Baja CA.
Elevation: 10-1550 m.

NDDB/Holland type and status: **Mojavean desert scrubs** (34000), **Riparian scrubs** (63000).
    Mojave wash scrub (34250) G3 S3.2.
    Mojave desert wash scrub (63700) G3 S2.1.

Other types:
Barry type: G7411124.
Brown Lowe Pase type: 143.153 153.141, 154.123, 153.161.
Cheatham & Haller type: Desert dry wash woodland.
PSW-45 type: Catclaw series.
Stone & Sumida (1983): Wash community.
Thorne type: Desert microphyll woodland.
WHR type: Desert wash.

General references: Johnson (1976), MacMahon (1988), Paysen *et al.* (1980), Thorne (1982), Turner & Brown (1982), Vasek & Barbour (1977).

Comments: Catclaw acacia stands occupy habitat similar to Blue palo verde-ironwood-smoke tree series at lower elevations. The Catclaw acacia series is mostly a MojD one that lacks the tree component of the Blue palo verde-ironwood-smoke tree series.

The pure species stands of desert-willow in MojD are included in this series at this time. Stands are transitional in character at intermediate elevations.

Cheesebush may be locally common, and these stands are included in this series. Upper reaches of washes in ColD lack the trees, but have catclaw acacia, fairy duster, and other shrubs. Such stands are included in this series until plot data are available to differentiate as a new series.

Species mentioned in text	
Basket bush	*Rhus trilobata*
Bebbia	*Bebbia juncea*
Blue palo verde	*Cercidium floridum*
Brittlebush	*Encelia farinosa*
California ephedra	*Ephedra californica*
Catclaw acacia	*Acacia greggii*
Cheesebush	*Hymenoclea salsola*
Creosote bush	*Larrea tridentata*
Desert barberry	*Berberis haematocarpa*
Desert lavender	*Hyptis emoryi*
Desert-olive	*Forestiera pubescens*
Desert-willow	*Chilopsis linearis*
Fairy duster	*Calliandra eriophylla*
Fremont box-thorn	*Lycium fremontii*
Ironwood	*Olneya tesota*
Jojoba	*Simmondsia chinensis*
Smoke tree	*Psorothamnus spinosa*

Plot-based descriptions: Vasek & Barbour (1977) report on wash composition [in Table 244] in the Whipple Mountains in San Bernardino Co. MojD; Spolsky (1979) defines Cheesebush scrub association in Anza Borrego State Park in San Diego Co. ColD.

Chamise series

Chamise sole or dominant shrub in canopy; black sage, California buckwheat, ceanothus, chaparral yucca, manzanita, poison-oak, interior live oak, red shank, scrub oak, toyon, and/or white sage may be present. Emergent trees may be present. Shrubs < 3 m; canopy continuous. Ground layer sparse (Plate 8).

Uplands: all slopes. Soils shallow, may be mafic-derived.

Distribution: i-NorCo, CenCo, f-KlaR, f-SN, m-SN, So-CA, ChaI, MojD, Baja CA. **Elevation:** 10-1800 m.

NDDB/Holland type and status: Chaparral (37000).
Gabbroic northern mixed chaparral (37111 *in part*) G2 S2.1.
Chamise chaparral (37200) G4 S4.
Upper Sonoran manzanita chaparral (37B00 *in part*) G4 S.
Northern maritime chaparral (37C10 *in part*) G1 S1.2.
Southern maritime chaparral (37C30 *in part*) G1 S1.1.
Northern north slope chaparral (37E10 *in part*) G3 S3.
Poison-oak chaparral (37F00 *in part*) G3 S3.3.

Other types:
Barry type: G7411212 BADFA00.
Cheatham & Haller type: Chamise chaparral.
PSW-45 type: Chamise series.
Rangeland type: SRM 206.
Thorne type: Chamise chaparral.
WHR type: Chamise-redshank chaparral.

General references: Davis & Hickson (1988), Hanes (1977, 1981, 1994), Horton (1960), Keeley (1992, 1993), Keeley & Keeley (1988), Hanes (1967, 1994), Pase (1982a), Patric & Hanes (1964), Paysen *et al.* (1980), Schmida & Whittaker (1981), Stohlgren *et al.* (1984), Vogl & Schorr (1972), Zammit & Zedler (1988).

Comments: This series definition follows Gordon & White (1994). Chamise occurs in many chaparral, coastal scrub, and tree series, but it does not dominate as it does in this series. If chamise has >

Species mentioned in text	
Bigberry manzanita	*Arctostaphylos glauca*
Black sage	*Salvia mellifera*
California buckwheat	*Eriogonum fasciculatum*
California sagebrush	*Artemisia californica*
Ceanothus	*Ceanothus* species
Chamise	*Adenostoma fasciculatum*
Chaparral yucca	*Yucca whipplei*
Cupleaf ceanothus	*Ceanothus greggii*
Eastwood manzanita	*Arctostaphylos glandulosa*
Hoaryleaf ceanothus	*Ceanothus crassifolius*
Interior live oak	*Quercus wislizenii*
Manzanita	*Arctostaphylos* species
Mission-manzanita	*Xylococcus bicolor*
Poison-oak	*Toxicodendron diversilobum*
Purple sage	*Salvia leucophylla*
Red shank	*Adenostoma sparsifolium*
Scrub oak	*Quercus berberidifolia*
Toyon	*Heteromeles arbutifolia*
Wedgeleaf ceanothus	*Ceanothus cuneatus*
White sage	*Salvia apiana*
Whiteleaf manzanita	*Arctostaphylos viscida*
Woollyleaf ceanothus	*Ceanothus tomentosus*

60% cover, the stand is a member of the Chamise series. If chamise cover is 60-30% and another species cover 30-60%, then the stand is a member of a mixed series [*see* Chamise-bigberry manzanita series, Chamise-cupleaf ceanothus series, Chamise-Eastwood manzanita series, Chamise-Eastwood manzanita-bigberry manzanita series, Chamise-hoaryleaf ceanothus series, Chamise - mission-manzanita - woollyleaf ceanothus series, Chamise-wedgeleaf ceanothus series, Chamise-woollyleaf ceanothus series]. Chamise may be important with coastal sage species [*see* Chamise-black sage series, Chamise-white sage series]. In some areas, stands of chamise mix with stands of coastal scrub forming a mosaic of series [*see* Black sage series, California buckwheat series, California sagebrush series, Mixed sage series, Purple sage series, White sage series].

Adenostoma fasciculatum var. *polifolium* in o-SoCo is included in this series. The Pine Hill area [Gabbroic northern mixed chaparral, *see* Whiteleaf manzanita series] with its several endemics mix in stands dominated with chamise form a local form of this series on Rescue loam soils in w. El Dorado Co. f-SN.

Keeler-Wolf (1990e) qualitatively describes chamise stands at Frenzel Creek RNA in Colusa Co., Hale Ridge candidate RNA in Lake Co. NorCo; at Limekiln Creek RNA (now part of the Cone Peak gradient RNA) in Monterey Co. o-CenCo; Fiedler (1986) at Jawbone Ridge candidate RNA in Tuolumne Co. f-SN; Keeler-Wolf (1990e) at Millard Canyon RNA in Riverside Co., Hanes (1976) describes vegetation types in the San Gabriel Mountains including a chamise chaparral in m-TraR.

Plot-based descriptions: Gray (1978) describes stands at Snow Mountain in Lake Co., Rundel (1991) stands at Wilder Ridge candidate RNA in Tehama Co. i-NorCo; Cole (1980) reports cover for stands at Purisima Hills in o-CenCo; Bauer (1936) describes stands in the Santa Monica Mountains, Meier (1979) Keeler-Wolf (1990e) stands at Fern Canyon RNA in Los Angeles Co. i-SoCo; Keeley & Keeley (1988) three stands, m-TraR, f-SN; Gordon & White (1994) defines 11 associations in m-TraR, SoCo.

Associations:
Gordon & White (1994):
> Chamise-(bigberry manzanita) association,
> Chamise-(black sage) association,
> Chamise-(California buckwheat-white sage) association,
> Chamise-(chaparral yucca) association,
> Chamise-(cupleaf ceanothus) association,
> Chamise-(cupleaf ceanothus-mafic soils) association,
> Chamise-(Eastwood manzanita) association,
> Chamise-(hoaryleaf ceanothus) association,
> Chamise-(scrub oak) association,
> Chamise-(wedgeleaf ceanothus) association,
> Chamise-(woollyleaf ceanothus) association.

Chamise-bigberry manzanita series

Chamise and bigberry manzanita important shrubs in canopy; ceanothus, chaparral yucca, interior live oak, manzanitas, poison-oak, scrub oak, and/or toyon may be present. Emergent trees may be present. Shrubs < 8 m; canopy continuous. Ground layer sparse or absent (Plate 9).

Uplands: slopes north-facing. Soils deep to shallow, may be mafic-derived.

Distribution: CenCo, SoCo, m-TraR, m-PenR, Baja CA.
Elevation: 450-1700 m.

NDDB/Holland type and status: Chaparral (37000).
    Chamise chaparral (37200) G4 S4.
    Upper Sonoran manzanita chaparral (37B00 *in part*) G4 S4.
    Northern maritime chaparral (37C10 *in part*) G1 S1.2.
    Northern north slope chaparral (37E10) G3 S3.
    Poison-oak chaparral (37F00 *in part*) G3 S3.3.

Other Types:
Barry type: G7411214 BAFA00.
Cheatham & Haller type: Chamise chaparral.
Rangeland type: SRM 206.
Thorne type: Chamisal.
WHR type: Chamise-redshank chaparral

General references: Hanes (1977, 1981, 1994), Horton 1960), Keeley (1993), Keeley & Keeley (1988), Pase (1982a), Patric & Hanes (1964), Vogl & Schorr (1972).

Comments: This series definition follows Gordon & White (1994). Chamise occurs in many chaparral, coastal scrub, and tree series. If chamise has < 60% cover, the stand is a member of the Chamise series. If chamise cover is 60-30% and bigberry manzanita cover 60-30%, then the stand is a member of this series [*see* Chamise series, Chamise-Eastwood manzanita series, Chamise-cupleaf ceanothus series, Chamise-woollyleaf ceanothus series, Chamise-

hoaryleaf ceanothus series, Chamise - mission-manzanita series, Chamise-wedgeleaf ceanothus series].

Burke (1985), Keeler-Wolf (1986d) in Keeler-Wolf (1990e) qualitatively defines Chamise-bigberry manzanita stands at Cahuilla Mountain RNA in Riverside Co., Organ Valley RNA in San Diego Co. m-PenR.

Species mentioned in text	
Bigberry manzanita	*Arctostaphylos glauca*
Bigpod ceanothus	*Ceanothus megacarpus*
Ceanothus	*Ceanothus* species
Chamise	*Adenostoma fasciculatum*
Chaparral whitethorn	*Ceanothus leucodermis*
Chaparral yucca	*Yucca whipplei*
Cupleaf ceanothus	*Ceanothus greggii*
Eastwood manzanita	*Arctostaphylos glandulosa*
Interior live oak	*Quercus wislizenii*
Hoaryleaf ceanothus	*Ceanothus crassifolius*
Manzanitas	*Arctostaphylos* species
Mission-manzanita	*Xylococcus bicolor*
Poison-oak	*Toxicodendron diversilobum*
Scrub oak	*Quercus berberidifolia*
Toyon	*Heteromeles arbutifolia*
Wedgeleaf ceanothus	*Ceanothus cuneatus*
Woollyleaf ceanothus	*Ceanothus tomentosus*

Plot-based descriptions: Gordon & White (1994) define seven associations in o-CenCo, SoCo, m-TraR.
Associations:
Gordon & White (1994):
    Chamise-bigberry manzanita association,
    Chamise-bigberry manzanita-(chaparral whitethorn) association,
    Chamise-bigberry manzanita-(chaparral yucca) association,
    Chamise-bigberry manzanita-(cupleaf ceanothus) association,
    Chamise-bigberry manzanita-(hoaryleaf ceanothus) association,
    Chamise-bigberry manzanita-(scrub oak) association,
    Chamise-bigberry manzanita-(wedgeleaf ceanothus) association.

Chamise-black sage series

Chamise and black sage important shrubs in canopy; California buckwheat, chaparral yucca, deerbrush, hoaryleaf ceanothus, wedgeleaf ceanothus, and/or white sage may be present. Emergent laurel sumac may be present. Shrubs < 3 m; canopy continuous. Ground layer sparse.

Uplands: slopes south-facing. Soils shallow, may be rocky.

Distribution: CenCo, So-CA.
Elevation: 10-1600 m.

NDDB/Holland type and status: **Chaparral** (37000).
 Chamise chaparral (37200) G4 S4.
 Coastal sage-chaparral scrub (37G00) G3 S3.2.

Other types:
Barry type: G7411214.
Cheatham & Haller type: Chamise chaparral.
Rangeland type: SRM 206.
Thorne type: Chamisal.
WHR type: Chamise-redshank chaparral.

General references: Gray (1983), Hanes (1977, 1981, 1994), Horton (1960), Keeley (1993), Keeley & Keeley (1988), Pase (1982a), Patric & Hanes (1964), Schmida & Whittaker (1981), Vogl & Schorr (1972).

Comments: This series definition follows Gordon & White (1994). Chamise occurs in many chaparral, coastal scrub, and tree series. If chamise has > 60% cover, the stand is a member of the Chamise series. If chamise cover is 60-30% and black sage cover 30-60%, then the stand is a member of this series. [*see* Chamise-white sage series]. In some areas stands of chamise mix with stands of coastal scrub forming a mosaic of different series [*see* Black sage series, California buckwheat series, California sagebrush series, Mixed sage series, Purple sage series, White sage series]. Further variation in stand composition can be handled at the association level.

Species mentioned in text	
Black sage	*Salvia mellifera*
California buckwheat	*Eriogonum fasciculatum*
California sagebrush	*Artemisia californica*
Chamise	*Adenostoma fasciculatum*
Chaparral yucca	*Yucca whipplei*
Deerbrush	*Ceanothus integerrimus*
Hoaryleaf ceanothus	*Ceanothus crassifolius*
Laurel sumac	*Malosma laurina*
Purple sage	*Salvia leucophylla*
Wedgeleaf ceanothus	*Ceanothus cuneatus*
White sage	*Salvia apiana*

Plot-based descriptions: Gordon & White (1994) define a Chamise-black sage association m-TraR, SoCo.

Chamise-cupleaf ceanothus series

Chamise and cupleaf ceanothus important shrubs in canopy; California buckwheat, chaparral yucca, chaparral whitethorn, red shank, scrub oak, sugar bush, and/or white sage may be present. Emergent trees may be present. Shrubs < 3 m; canopy continuous. Ground layer sparse.

Uplands: slopes north-facing. Soils shallow, may be rocky.

Distribution: CenCo, m-TraR, m-PenR, Baja CA.
Elevation: 600-1800 m.

NDDB/Holland type and status: Chaparral (37000).
> Chamise chaparral (37200) G4 S4.
> Upper Sonoran manzanita chaparral (37B00 *in part*) G4 S4.
> Northern north slope chaparral (37E10) G3 S3.3.

Other types:
Barry type: G7411214.
Cheatham & Haller type: Chamise chaparral.
Rangeland type: SRM 206.
Thorne type: Chamisal.
WHR type: Chamise-redshank chaparral.

General references: Bullock & Chamela (1991), Hanes (1977, 1981, 1994), Horton (1960), Keeley (1993), Keeley & Keeley (1988), Pase (1982a), Patric & Hanes (1964), Vogl & Schorr (1972), Zammit & Zedler (1992).

Comments: Chamise occurs in many chaparral, coastal scrub, and tree series. If chamise has > 60% cover, the stand is a member of the Chamise series. If chamise cover is 30-60% and cupleaf ceanothus cover 30-60%, then the stand is a member of this series [*see* Chamise-bigberry manzanita series, Chamise-Eastwood manzanita series, Chamise-woollyleaf ceanothus series, Chamise-hoaryleaf ceanothus series, Chamise - mission-manzanita series, Chamise-wedgeleaf ceanothus series].

Further variation in stand composition can be handled at the association level.

Species mentioned in text	
Birchleaf mountain-mahogany	*Cercocarpus betuloides*
Bigberry manzanita	*Arctostaphylos glauca*
California buckwheat	*Eriogonum fasciculatum*
Chamise	*Adenostoma fasciculatum*
Chaparral whitethorn	*Ceanothus leucodermis*
Chaparral yucca	*Yucca whipplei*
Cupleaf ceanothus	*Ceanothus greggii*
Eastwood manzanita	*Arctostaphylos glandulosa*
Hoaryleaf ceanothus	*Ceanothus crassifolius*
Mission-manzanita	*Xylococcus bicolor*
Red shank	*Adenostoma sparsifolium*
Scrub oak	*Quercus berberidifolia*
Sugar bush	*Rhus ovata*
Wedgeleaf ceanothus	*Ceanothus cuneatus*
White sage	*Salvia apiana*
Woollyleaf ceanothus	*Ceanothus tomentosus*

Plot-based descriptions: Gordon & White (1994) define two associations, Kummerow *et al.* (1985) present density data in San Diego Co. m-PenR.

Associations:
Gordon & White (1994):
> Chamise-cupleaf ceanothus association,
> Chamise/mafic soils association.

Chamise-Eastwood manzanita series

Chamise and Eastwood manzanita important shrubs in canopy; birchleaf mountain-mahogany, California buckwheat, ceanothus, chaparral yucca, manzanitas, scrub oak, and/or toyon may be present. Emergent trees may be present. Shrubs < 3 m; canopy continuous. Ground layer sparse (Plate 8).

Uplands: slopes south-facing. Soils shallow, may be mafic-derived.

Distribution: CenCo, SoCo, m-TraR, m-PenR.
Elevation: 450-1800 m.

NDDB/Holland type and status: Chaparral (37000).
 Chamise chaparral (37200) G4 S4.
 Mafic southern mixed chaparral (37122) G3 S3.2
 Upper Sonoran manzanita chaparral (37B00 *in part*) G4 S4.

Other types:
Barry type: G7411214.
Cheatham & Haller type: Chamise chaparral.
Rangeland type: SRM 206.
Thorne type: Chamisal.
WHR type: Chamise-redshank chaparral.
General references: Gray (1983), Hanes (1977, 1981, 1994), Horton (1960), Keeley (1993), Keeley & Keeley (1988), Pase (1982a), Patric & Hanes (1964), Vogl & Schorr (1972).

Comments: This series definition follows Gordon & White (1994). Chamise occurs in many chaparral, coastal scrub, and tree series. If chamise has > 60% cover, the stand is a member of the Chamise series. If chamise cover is 30-60% and Eastwood manzanita cover 30-60%, then the stand is a member of this series [see Chamise series, Chamise-woollyleaf ceanothus series, Chamise-hoaryleaf ceanothus series, Chamise - mission-manzanita - woollyleaf ceanothus series]. Keeler-Wolf (1990e) qualitatively describes Chamise-Eastwood manzanita stands at Millard Canyon RNA in Riverside Co. i. m-TraR; at Cahuilla Mountain RNA in Riverside Co., at Organ Valley RNA in San Diego Co. m-PenR.

Species mentioned in text	
Birchleaf mountain-mahogany	*Cercocarpus betuloides*
California buckwheat	*Eriogonum fasciculatum*
Ceanothus	*Ceanothus* species
Chamise	*Adenostoma fasciculatum*
Chaparral whitethorn	*Ceanothus leucodermis*
Chaparral yucca	*Yucca whipplei*
Cupleaf ceanothus	*Ceanothus greggii*
Eastwood manzanita	*Arctostaphylos glandulosa*
Hoaryleaf ceanothus	*Ceanothus crassifolius*
Manzanitas	*Arctostaphylos* species
Mission-manzanita	*Xylococcus bicolor*
Scrub oak	*Quercus berberidifolia*
Toyon	*Heteromeles arbutifolia*
Wedgeleaf ceanothus	*Ceanothus cuneatus*
Woollyleaf ceanothus	*Ceanothus tomentosus*

Plot-based descriptions: Parker (1990) defines an association at Mount Tamalpais in Marin Co. o-NorCo; Gordon & White (1994) seven associations in SoCo, m-TraR, White (1994a) an association at Cleghorn Canyon candidate RNA in San Bernardino Co. i. m-TraR; Keeler-Wolf (1990a) describes stands at King Creek RNA in San Diego Co. o-PenR.

Associations:
Gordon & White (1994):
 Chamise-Eastwood manzanita - (birchleaf mountain-mahogany) association,
 Chamise-Eastwood manzanita-(chaparral whitethorn) association [= White 1994],
 Chamise-Eastwood manzanita-(cupleaf

 Chamise-Eastwood manzanita-(hoaryleaf ceanothus) association,
 Chamise-Eastwood manzanita/mafic soils association,
 Chamise-Eastwood manzanita - (birchleaf mountain-mahogany) association,
 Chamise-Eastwood manzanita-(wedgeleaf ceanothus) association.
Parker (1990):
 Chamise-Eastwood manzanita association.

Chamise-hoaryleaf ceanothus series

Chamise and hoaryleaf ceanothus important shrubs in canopy; black sage, California buckwheat, chaparral yucca, manzanita, sugar bush, scrub oak, and/or toyon may be present. Emergent laurel sumac may be present. Shrubs < 3 m; canopy continuous. Ground layer sparse.

Uplands: slopes south-facing. Soils shallow, may be rocky.

Distribution: m-TraR, m-PenR.
Elevation: 400-1300 m.

NDDB/Holland type and status: Chaparral (37000).
  Granitic southern mixed chaparral (37121) G3 S3.3.
  Mafic southern mixed chaparral (37122) G3 S3.2.
  Chamise chaparral (37200) G4 S4.

Other types:
Barry type: G7411214.
Cheatham & Haller type: Chamise chaparral.
Rangeland type: SRM 206.
Thorne type: Chamisal.
WHR type: Chamise-redshank chaparral.

General references: Hanes (1977, 1981, 1994), Horton (1960), Keeley (1993), Keeley & Keeley (1988), Pase (1982a), Patric & Hanes (1964), Vogl & Schorr (1972).

Comments: This series definition follows Gordon & White (1994). Chamise occurs in many chaparral, coastal scrub, and tree series. If chamise has > 60% cover, the stand is a member of the Chamise series. If chamise cover is 60-30% and hoaryleaf ceanothus cover is 30-60%, then the stand is a member of this series. [*see* Chamise-cupleaf ceanothus series, Chamise-Eastwood manzanita series, Chamise-woollyleaf ceanothus series, Chamise - mission-manzanita - woolly ceanothus series].

Stands with localized species such as Cleveland monkeyflower (a CNPS List 4 plant), Dunn mariposa lily (a CNPS List 1B plant), or white coast ceanothus (a CNPS List 2 plant) may occur in this series (Skinner & Pavlik 1994).

Species mentioned in text	
Black sage	*Salvia mellifera*
California buckwheat	*Eriogonum fasciculatum*
Chamise	*Adenostoma fasciculatum*
Chaparral yucca	*Yucca whipplei*
Cleveland monkeyflower	*Mimulus clevelandii*
Cupleaf ceanothus	*Ceanothus greggii*
Dunn mariposa lily	*Calochortus dunnii*
Hoaryleaf ceanothus	*Ceanothus crassifolius*
Laurel sumac	*Malosma laurina*
Manzanita	*Arctostaphylos* species
Mission-manzanita	*Xylococcus bicolor*
Scrub oak	*Quercus berberidifolia*
Sugar bush	*Rhus ovata*
Toyon	*Heteromeles arbutifolia*
White coast ceanothus	*Ceanothus verrucosus*
Woollyleaf ceanothus	*Ceanothus tomentosus*

Plot-based descriptions: Gordon & White (1994) define two associations, White (1994a) one association at Cleghorn Canyon candidate RNA in San Bernardino Co. i. m-TraR.

Associations:
Gordon & White (1994):
  Chamise-hoaryleaf ceanothus association [= White 1994a],
  Chamise-hoaryleaf ceanothus-black sage association.

Chamise - mission-manzanita - woollyleaf ceanothus series

Chamise, mission-manzanita, and/or woolly-leaf ceanothus important shrubs in canopy; chaparral yucca, hoaryleaf ceanothus, hollyleaf redberry, scrub oak, sugar bush, toyon, and/or white sage may be present. Emergent laurel sumac may be present. Shrubs < 3 m; canopy continuous. Ground layer sparse.

Uplands: middle slopes. Soils shallow, may be rocky.

Distribution: o-SoCo, m-PenR.
Elevation: 300-950 m.

NDDB/Holland type and status: Chaparral (37000).
Granitic southern mixed chaparral (37121) G3 S3.3.
Mafic southern mixed chaparral (37122) G3 S3.2.
Chamise chaparral (37200 *in part*) G4 S4.
Southern maritime chaparral (37C30 *in part*) G1 S1.1.

Other types:
Barry type: G7411214 BAFA00.
Cheatham & Haller type: Chamise chaparral.
Rangeland type: SRM 206.
Thorne type: Chamisal.
WHR type: Chamise-redshank chaparral.

General references: Hanes (1977, 1981, 1994), Horton (1960), Keeley (1993), Keeley & Keeley (1988), Pase (1982a), Patric & Hanes (1964), Vogl & Schorr (1972).

Comments: This series definition follows Gordon & White (1994). Chamise occurs in many chaparral, coastal scrub, and tree series. If chamise has > 60% cover, the stand is a member of the Chamise series. If chamise cover is 30-60% and mission-manzanita or woollyleaf ceanothus cover together is 30-60%, then the stand is a member of this series [*see* Chamise-cupleaf ceanothus series, Chamise-East-

wood manzanita series, Chamise-hoaryleaf ceanothus series, Chamise-woollyleaf ceanothus series].

Stands with localized species as Cleveland monkeyflower (a CNPS List 4 plant), Dunn mariposa lily (a CNPS List 1B plant), or white coast ceanothus (a CNPS List 2 plant) occur in this series (Skinner & Pavlik 1994).

Species mentioned in text	
Chamise	*Adenostoma fasciculatum*
Chaparral yucca	*Yucca whipplei*
Cleveland monkeyflower	*Mimulus clevelandii*
Cupleaf ceanothus	*Ceanothus greggii*
Dunn mariposa lily	*Calochortus dunnii*
Eastwood manzanita	*Arctostaphylos glandulosa*
Hoaryleaf ceanothus	*Ceanothus crassifolius*
Hollyleaf redberry	*Rhamnus ilicifolia*
Laurel sumac	*Malosma laurina*
Mission-manzanita	*Xylococcus bicolor*
Scrub oak	*Quercus berberidifolia*
Sugar bush	*Rhus ovata*
Toyon	*Heteromeles arbutifolia*
White coast ceanothus	*Ceanothus verrucosus*
White sage	*Salvia apiana*
Woollyleaf ceanothus	*Ceanothus tomentosus*

Plot-based descriptions: Gordon & White (1994) define four associations in SoCo.

Associations:
Gordon & White (1994):
Chamise - mission-manzanita association,
Chamise - mission-manzanita - woollyleaf ceanothus association,
Chamise - mission-manzanita - woollyleaf ceanothus (mafic soils) association,
Chamise-woollyleaf ceanothus association.

Chamise-wedgeleaf ceanothus series

Chamise and wedgeleaf ceanothus important shrubs in canopy; black sage, California buckwheat, chaparral yucca, manzanitas, and/or scrub oak may be present. Emergent trees may be present. Shrubs < 3 m; canopy continuous. Ground layer sparse or absent.

Uplands: slopes variable. Soils shallow, may be rocky or fine-textured.

Distribution: i-NorCo, CenCo, f-SN, m-TraR, SoCo, m-PenR.
Elevation: 600-1500 m.

NDDB/Holland type and status: Chaparral (37000).
　Chamise chaparral (37200) G4 S4.
　Upper Sonoran manzanita chaparral (37B00 *in part*) G4 S4.

Other types:
Barry type: G7411214.
Cheatham & Haller type: Chamise chaparral.
Rangeland type: SRM 206.
Thorne type: Chamisal.
WHR type: Chamise-redshank chaparral.

General references: Gray (1982, 1983), Hanes (1977, 1981, 1994), Horton (1960), Keeley (1993), Keeley & Keeley (1988), Pase (1982a), Patric & Hanes (1964), Vogl & Schorr (1972).

Comments: This series definition follows Gordon & White (1994). Chamise occurs in many chaparral, coastal scrub, and tree series. If chamise has > 60% cover, the stand is a member of the Chamise series. If chamise cover is 60-30% and wedgeleaf ceanothus cover 30-60%, then the stand is a member of this series [*see* Chamise-bigberry manzanita series, Chamise-Eastwood manzanita series, Chamise-cupleaf ceanothus series]. Further variation in stand composition can be handled at the association level.

Species mentioned in text	
Bigberry manzanita	*Arctostaphylos glauca*
Black sage	*Salvia mellifera*
California buckwheat	*Eriogonum fasciculatum*
Chamise	*Adenostoma fasciculatum*
Chaparral yucca	*Yucca whipplei*
Cupleaf ceanothus	*Ceanothus greggii*
Eastwood manzanita	*Arctostaphylos glandulosa*
Manzanitas	*Arctostaphylos* species
Scrub oak	*Quercus berberidifolia*
Wedgeleaf ceanothus	*Ceanothus cuneatus*

Plot-based descriptions: Gordon & White (1994) define a Chamise-wedgeleaf ceanothus association in SoCo, m-TraR.

Chamise-white sage series

Chamise and white sage important shrubs in canopy; California buckwheat, California sagebrush, chaparral yucca, hollyleaf redberry, mission-manzanita, scrub oak, sugar bush, and/or yerba santa may be present. Emergent trees may be present. Shrubs < 3 m; canopy continuous. Ground layer sparse.

Uplands: slopes south or east-facing. Soils shallow, may be rocky.

Distribution: m-TraR, m-PenR.
Elevation: 300-1700 m.

NDDB/Holland type and status: Chaparral (37000).
   Chamise chaparral (37200) G4 S4.
   Coastal sage-chaparral scrub (37G00) G3 S3.2.

Other types:
Barry type: G7411214.
Cheatham & Haller type: Chamise chaparral.
Rangeland type: SRM 206.
Thorne type: Chamisal.
WHR type: Chamise-redshank chaparral.

General references: Gray (1982), Hanes (1977, 1981, 1994), Horton (1960), Keeley (1993), Keeley & Keeley (1988), Pase (1982a), Patric & Hanes (1964), Schmida & Whittaker (1981), Vogl & Schorr (1972).

Comments: This series definition follows Gordon & White (1994). Chamise occurs in many chaparral, coastal scrub, and tree series. If chamise has > 60% cover, the stand is a member of the Chamise series. If chamise cover is 60-30% and white sage cover 30-60%, then the stand is a member of this series [*see* Chamise-black sage series]. In some areas, stands of chamise mix with stands of coastal scrub forming a mosaic of different series [*see* Black sage series, California buckwheat series, California sagebrush series, Mixed sage series, Purple sage series, White sage series]. Further variation in stand composition can be handled at the association level.

Species mentioned in text	
Black sage	*Salvia mellifera*
California buckwheat	*Eriogonum fasciculatum*
California sagebrush	*Artemisia californica*
Chamise	*Adenostoma fasciculatum*
Chaparral yucca	*Yucca whipplei*
Hollyleaf redberry	*Rhamnus ilicifolia*
Mission-manzanita	*Xylococcus bicolor*
Purple sage	*Salvia leucophylla*
Scrub oak	*Quercus berberidifolia*
Sugar bush	*Rhus ovata*
Toyon	*Heteromeles arbutifolia*
Yerba santa	*Eriodictyon* species

Plot-based descriptions: Gordon & White (1994) define a Chamise-California buckwheat-white sage association in SoCo.

Chaparral whitethorn series

Chaparral whitethorn sole or dominant shrub in canopy; basket bush, bigberry manzanita, chaparral yucca, chamise, deerbrush, interior live oak, scrub oak, toyon, and/or yerba santa may be present. Emergent interior live oak trees may be present. Shrubs < 3 m; canopy continuous or intermittent. Ground layer sparse (Plate 11).

Uplands: slopes south-facing, may be steep. Soils alluvial or bedrock-derived, deep.

Distribution: f-SN, m-CenCo, TraR, PenR, Baja CA.
Elevation: 100-1900 m.

NDDB/Holland type and status: Chaparral (37000).
    Whitethorn chaparral (37532), G4 S4.
    Poison-oak chaparral (37F00 *in part*) G3 S3.3.

Other types:
Barry type: G7411214.
Cheatham & Haller type: Mixed chaparral.
PSW-45 type: Ceanothus series.
Rangeland type: SRM 208.
Thorne type: Mixed chaparral.
WHR type: Mixed chaparral.

General references: Hanes (1977, 1981), Keeley & Keeley (1988), Pase (1982a), Paysen *et al.* (1980), White (1994c).

Comments: Throughout its range, chaparral whitethorn occurs as scattered shrubs or as the sole or dominant shrub. In stands of this series chaparral whitethorn cover is > 60%; in stands of Scrub oak-chaparral whitethorn series its cover is between 30-60%. Stands of this series develop into other vegetation types.

Keeler-Wolf (1990e) qualitatively describes chaparral whitethorn stands at Fern Canyon RNA in Los Angeles Co. SoCo m-TraR; at Hall Canyon RNA in Riverside Co. m-PenR.

Species mentioned in text	
Basket bush	*Rhus trilobata*
Bigberry manzanita	*Arctostaphylos glauca*
Chaparral whitethorn	*Ceanothus leucodermis*
Chaparral yucca	*Yucca whipplei*
Chamise	*Adenostoma fasciculatum*
Deerbrush	*Ceanothus integerrimus*
Interior live oak	*Quercus wislizenii*
Scrub oak	*Quercus berberidifolia*
Toyon	*Heteromeles arbutifolia*
Yerba santa	*Eriodictyon* species

Plot-based descriptions: Gordon & White (1994) define a Chaparral whitethorn association in m-TraR, m-PenR.

Coast prickly-pear series

Bluff cholla, cane cholla, coast prickly-pear, or tree prickly-pear conspicuous with malacophyllous shrubs; bladderpod, black sage, box-thorns, bushrue, California buckwheat, California encelia, California sagebrush, cliff spurge, and/or wishbone bush may be present. Emergent lemonade berry or Mexican elderberry may be present. Shrubs < 2 m; canopy continuous or intermittent. Ground layer variable, may include *Dudleya* species (Plate 11).

Uplands: slopes steep. Soil shallow.

Distribution: o-SoCo, ChaI, Baja CA.
Elevation: sea level-1200 m.

NDDB/Holland type and status: Coastal bluff scrubs (31000), **Coastal scrubs** (32000).
    Southern coastal scrub (31200) G1 S1.1.
    Venturan coastal sage scrub (32300 *in part*) G3 S3.1.
    Maritime succulent scrub (32400) G2 S1.1.
    Diegan coastal sage scrub (32500) G3 S3.1.

Other types:
Barry type: G7411214.
Cheatham & Haller type: Maritime cactus scrub.
Jones & Stokes type: Coastal bluff scrub, Maritime succulent cactus scrub, Southern cactus scrub.
Thorne type: Southern coastal scrub.
WHR type: Coastal scrub.

General references: Axelrod (1978), Keeley & Keeley (1988), Magney (1992), O'Leary (1989), Pase & Brown (1982).

Comments: This series is often considered part of the coastal scrub, which is better thought of as a collection of series. This series differs from the other coastal scrub series in having common cacti and Diegan division species (Axelrod 1978). Coast bluff scrub is included here in part. Definitions differ in plant height and cover from coastal scrub, but contrast little in species composition.

Stands of localized species as goldenspine cereus (a CNPS List 2 plant), San Diego barrel cactus (a CNPS List 2 plant), Shaw agave (a CNPS List 2 plant) occur in this series (Skinner & Pavlik 1994).

Species mentioned in text	
Black sage	*Salvia mellifera*
Bladderpod	*Isomeris arborea*
Bluff cholla	*Opuntia prolifera*
Box-thorns	*Lycium* species
Bushrue	*Cneoridium dumosum*
California buckwheat	*Eriogonum fasciculatum*
California encelia	*Encelia californica*
California sagebrush	*Artemisia californica*
Cane cholla	*Opuntia parryi*
Cliff spurge	*Euphorbia misera*
Coast prickly-pear	*Opuntia littoralis* and hybrids
Dudleya	*Dudleya* species
Goldenspine cereus	*Bergerocactus emoryi*
Lemonade berry	*Rhus integrifolia*
Mexican elderberry	*Sambucus mexicana*
San Diego barrel cactus	*Ferocactus viridescens*
Shaw agave	*Agave shawii*
Tree prickly-pear	*Opuntia oricola*
Wishbone bush	*Mirabilis californica*

Plot-based descriptions: not available.

Coyote brush series

Coyote brush sole or dominant shrub; black sage, California blackberry, California buckwheat, California coffeeberry, California sagebrush, wax-myrtle, poison-oak, salal, white sage, and/or yellow bush lupine may be present. Shrubs < 2 m; canopy continuous or intermittent. Ground layer variable (Plate 12).

Uplands: stabilized dunes of coastal bars, river mouths, spits along coastline; coastal bluffs, open slopes, terraces.

Distribution: o-NorCo, o-CenCo, o-SoCo.
Elevation: sea level-1000 m.

NDDB/Holland type and status: Coast dunes (21000), **Coastal bluff scrubs** (31000), **Coastal scrubs** (32000), **Chaparral** (37000).
> Northern dune scrub (21321 *in part*) G2 S1.2.
> Northern (Franciscan) coastal bluff scrub (31100 *in part*) G2 S2.2.
> Northern coyotebrush scrub (321100 G3 S3.
> Central Lucian coastal scrub (32200 *in part*) G3 S3.3.
> Diablan sage scrub (32600 *in part*) G3 S3.3.

Other types:
Barry type: G7411211 CBAPIC0.
Cheatham & Haller type: Northern coastal scrub, Central coastal scrub, Coastal sage scrub.
Jones & Stokes type: Coastal bluff scrub, Coyote brush scrub, Sagebrush-coyote brush scrub.
PSW-45 type: Baccharis series.
Rangeland type: SRM 204.
Thorne type: Northern coastal scrub, Southern coastal scrub.
WHR type: Coastal scrub.

General references: Barbour (1994), DaSilva & Bartolome (1984), Heady *et al.* (1977b), Grams *et al.* (1977), McBride (1974), Paysen *et al.* (1980).

Comments: Heady *et al.* (1977b) describe coyote brush shrublands on o-NorCo bluffs, slopes, and terraces, but stands occur in CenCo and SoCo as well. Axelrod (1978) notes that coyote brush is important in all divisions of his coastal scrub except the Diegan. Coyote brush invades recently logged land in NorCo well away from the coast. Coyote brush stands are considered ones that develop into forest (McBride 1974), or are permanent (Grams *et al.* 1977). In Humboldt Co. coyote brush grows on coastal dunes, where it invades areas stabilized by European beachgrass and yellow bush lupine (Parker 1974) which were originally Sand-verbena - beach bursage series vegetation.

Coast bluff scrub descriptions are included in part. Definitions differ in plant height and cover from coastal scrub, but contrast little in species composition.

Keeler-Wolf (1990e) qualitatively describes coyote brush stands at Limekiln Creek RNA (now part of the Cone Peak Gradient RNA) in Monterey Co. o-CenCo.

Species mentioned in text	
Beach bursage	*Ambrosia chamissonis*
Black sage	*Salvia mellifera*
California blackberry	*Rubus ursinus*
California buckwheat	*Eriogonum fasciculatum*
California coffeeberry	*Rhamnus californica*
California figwort	*Scrophularia californica*
California sagebrush	*Artemisia californica*
Coyote brush	*Baccharis pilularis*
Creeping ryegrass	*Leymus triticoides*
European beachgrass	*Ammophila arenaria*
Poison-oak	*Toxicodendron diversilobum*
Salal	*Gaultheria shallon*
Seaside woolly-sunflower	*Eriophyllum staechadifolium*
Sword fern	*Polystichum munitum*
Tufted hairgrass	*Deschampsia cespitosa*
Wax-myrtle	*Myrica californica*
White sage	*Salvia apiana*
Yellow bush lupine	*Lupinus arboreus*
Yellow sand-verbena	*Abronia latifolia*

Plot-based descriptions: Elliott & Wehausen (1974) define one association at Point Reyes National Seashore, Fiedler & Leidy (1987) one association in Marin Co., Grams *et al.* (1977) one association, which is comparable to (Heady *et al.* 1977b) one association in Sonoma Co., Parker (1974) in Barbour & Johnson (1977) three associations, LaBanca (1993) two associations on sand Humboldt Co. o-NorCo; Baxter (1992) defines one association, McBride (1974) compares scrub to forest in Contra Costa Co. o-CenCo.

Associations:
Baxter (1992):
> Coyote brush/seaside woolly-sunflower association.

Elliott & Wehausen (1974):
> Coyote brush/tufted hairgrass association.

Fiedler & Leidy (1987):
> Coyote brush/creeping ryegrass association.

Grams *et al.* (1977):
> Coyote brush/sword fern association
> [as Coastal scrub; Mt. Vision, Jenner, Stuarts Pt. locations in Table 21-7 (Heady *et al.* 1977b)].

Heady *et al.* (1977b):
> Coyote brush-California sagebrush association
> [as Coastal scrub; Monterey, Santa Cruz, San Mateo Co. locations in Table 21-7 (Barbour & Johnson 1977); = *Artemisia californica-Baccharis pilularis-Elymus condensatus* association (Kirkpatrick & Hutchinson 1977)].

Parker (1974) in (Barbour & Johnson 1977):
> Coyote brush-yellow bush lupine association,
> Coyote brush/European beachgrass association
> [as *Ammophila-Baccharis* community],
> Coyote brush/California figwort association
> [= LaBanca (1993)].

Creosote bush series

Creosote bush sole or dominant shrub in canopy; desert-holly, prickly-pears, brittle-bush, chollas, ephedras, hop-sage, indigo bush, saltbushes, and/or white bursage may be present. Shrubs < 3 m; canopy open. Emergent Joshua tree or mesquites may be present. Ground layer open; annuals seasonally present (Plate 15).

Uplands: alluvial fans; bajadas; upland slopes. Soils well-drained, may have pavement surface.

Distribution: MojD, ColD, AZ, NM, Mexico.
Elevation: 75 below sea level-1000 m.

NDDB/Holland type and status: Sonoran desert scrubs (33000), **Mojavean desert scrubs** (34000).
Sonoran creosote bush scrub (33100 *in part*) G4 S4.
Mojave creosote bush scrub (34100 *in part*) G4 S4.

Other types:
Barry type: G74 G7411211 CLADI20.
Brown Lowe Pase type: 153.111, 153.113, 154.111.
Cheatham & Haller type: Mojave creosote bush scrub, Sonoran creosote bush scrub.
PSW-45 type: Creosote bush series.
Rangeland type: SRM 211.
Stone & Sumida (1983): Creosote bush scrub.
Thorne type: Creosote bush scrub.
WHR type: Desert scrub.

General references: Burk (1977), Holton (1994), Hunt (1966), MacMahon (1988), O'Leary & Minnich (1981), Paysen *et al.* (1980), Phillips & MacMahon (1981), Turner (1982b), Turner & Brown (1982), Vasek (1980), Vasek & Barbour (1977), Vasek *et al.* (1975).

Comments: This series is often considered part of the creosote bush scrub, which is better thought of as a collection of series. In this series creosote bush dominates; white bursage may be present, but not

as an important shrub [*see* Creosote bush-white bursage series]. Big galleta may be present, but is not important in the ground layer [*see* Big galleta series]. Turner (1982a) presents qualitative descriptions of several areas in CA which describe vegetation variation in the series.

Species mentioned in text	
Big galleta	*Pleuraphis rigida*
Brittlebush	*Encelia farinosa*
Chollas	*Opuntia* species
Creosote bush	*Larrea tridentata*
Desert-holly	*Atriplex hymenelytra*
Ephedras	*Ephedra* species
Hop-sage	*Grayia spinosa*
Joshua tree	*Yucca brevifolia*
Indigo bush	*Psorothamnus schottii*
Mesquites	*Prosopis* species
Prickly-pears	*Opuntia* species
Saltbushes	*Atriplex* species
White bursage	*Ambrosia dumosa*

Plot-based descriptions: Davidson & Fox (1974) describe off-road-vehicle effects in Kern Co., McHargue (1973) stands on bajadas, hillsides, and dunelands in Coachella Valley in Riverside Co., Phillips & MacMahon (1981) in MacMahon (1988) report shrub density and cover, Randell (1972) describes stands [as Group III] in Saline Valley in Inyo Co., Vasek & Barbour (1977) stands, Minnich *et al.* (1993) define one association at Marine Corps Air-ground Combat Center in San Bernardino Co. MojD.; Spolsky (1979) defines four associations at Anza Borrego State Park in San Diego Co. ColD.

Associations:
Minnich *et al.* (1993):
Creosote bush with disturbance association.
Spolsky (1979):
Sonoran dune scrub association,
High diversity creosote scrub association,
Creosote bush-brittlebush association,
Saltbush-creosote bush association.

Creosote bush-white bursage series

Creosote bush and white bursage important or conspicuous shrubs in canopy; box-thorn, brittlebush, chollas, desert-holly, ephedras, indigo bush, prickly-pears, and/or saltbushes may be present. Shrubs < 3 m. Shrub canopy is two-tiered. Emergent Joshua tree or ocotillo may be present. Ground layer open; annuals seasonally present (Plate 15).

Uplands: alluvial fans; bajadas; upland slopes. Soils well-drained, may have pavement surface.

Distribution: MojD, ColD, Baja CA.
Elevation: 75 below sea level-1200 m.

NDDB/Holland type and status: **Sonoran desert scrubs** (33000), **Mojavean desert scrub** (34000).
　　Mojave creosote bush scrub (34100 *in part*) G4 S4.
　　Sonoran creosote bush scrub (33100 *in part*) G4 S4.

Other types:
Barry type: G74 G7411211.
Brown Lowe Pase type: 153.112, 153.141, 154.112.
Cheatham & Haller type: Mojave creosote bush scrub, Sonoran creosote bush scrub.
Rangeland types: SRM 211, SRM 506.
Thorne type: Creosote bush scrub.
WHR type: Desert scrub.

General references: Burk (1977), Holzman (1994), Hunt (1966), MacMahon (1988), Thorne (1982), Turner (1982b), Turner & Brown (1982), Vasek & Barbour (1977).

Comments: This series is often considered part of the creosote bush scrub, which is best to consider as a collection of series. In this series creosote bush and white bursage are equally important, and brittlebush can be a third common species. If only one species dominates the stand, it is part of a different series [*see* Brittlebush series, Creosote bush series, or White bursage series].

The amount of shrub canopy varies among stands. Dense stands are as common as those with sparse shrubs in areas of desert pavement.

Species mentioned in text	
Box-thorn	*Lycium* species
Brittlebush	*Encelia farinosa*
Chollas	*Opuntia* species
Creosote bush	*Larrea tridentata*
Desert-holly	*Atriplex hymenelytra*
Ephedras	*Ephedra* species
Indigo bush	*Psorothamnus schottii*
Joshua tree	*Yucca brevifolia*
Ocotillo	*Fouquieria splendens*
Prickly-pears	*Opuntia* species
Saltbushes	*Atriplex* species
White bursage	*Ambrosia dumosa*

Plot-based descriptions: Randell (1972) describes stands [as Group III of an ordination of 46 stands] in Saline Valley in Inyo Co., Phillips & MacMahon (1981) in MacMahon (1988) stands in terms of shrub cover and density, Vasek *et al.* (1975) stands in San Bernardino Co. MojD; McHargue (1973) defines three habitats in Coachella Valley in Riverside Co., Spolsky (1979) defines two associations at Anza Borrego State Park in San Diego Co. ColD.

Associations:
McHargue (1973):
　　Rocky sides of hills and mountains,
　　Rocky bajadas,
　　Gravelly bajadas.
Spolsky (1979):
　　Sonoran creosote bush scrub association,
　　Uniform creosote scrub association.

Cupleaf ceanothus-fremontia-oak series

Cupleaf ceanothus, fremontia and/or oaks important shrubs in canopy; desert bitterbrush, Apache plume, bush poppy, California buckwheat, chaparral yucca, chamise, cliffrose, desert-almond, desert-apricot, manzanita, and/or Veatch silktassel may be present. Emergent birchleaf mountain-mahogany, California juniper, and/or singleleaf pinyon may be present. Shrubs < 3 m; canopy intermittent. Ground layer sparse.

Uplands: slopes north-facing. Soils shallow, may be rocky.

Distribution: s. i-CenCo, s. sj-CenV, m-TraR, m-PenR, w. MojD, w. ColD, Baja CA.
Elevation: 700-1700 m.

NDDB/Holland type and status: **Chaparral** (37000).
    Semi-desert chaparral (37400) G3 S3.2.
    Flannel bush chaparral (37J00) G3 S3.3.

Other types:
Cheatham & Haller type: Semi-desert chaparral.
Rangeland type: 208.
Thorne type: Desert transitional chaparral.
WHR type: Mixed chaparral.

General references: Hanes (1976, 1977, 1981), Horton (1960), Keeley & Keeley (1988), Zammit & Zedler (1992).

Comments: The various desert or semi-desert chaparral descriptions stress the open shrub canopy and mix of Cis-CA and Trans-CA species, but differ in species composition. Some allow a significant tree component. The series builds on the original desert chaparral definition (Horton 1960). If trees such as California juniper or singleleaf pinyon are common, the stand is not a member of this series [*see* California juniper series, Singleleaf pinyon series]. If California buckwheat or chamise are dominant in a stand, they are assigned to those series even though the stands are in Trans-CA [*see* California buckwheat series, Chamise series].

Species mentioned in text	
Apache plume	*Fallugia paradoxa*
Birchleaf mountain-mahogany	*Cercocarpus betuloides*
Bush poppy	*Dendromecon rigida*
California juniper	*Juniperus californica*
California buckwheat	*Eriogonum fasciculatum*
Chamise	*Adenostoma fasciculatum*
Chaparral yucca	*Yucca whipplei*
Cliffrose	*Purshia mexicana*
Cupleaf ceanothus	*Ceanothus greggii*
Desert-almond	*Prunus fasciculatum*
Desert-apricot	*Prunus fremontii*
Desert bitterbrush	*Purshia tridentata* var. *glandulosa*
Desert scrub oak	*Quercus turbinella*
Fremontia	*Fremontodendron californicum*
Interior live oak	*Quercus wislizenii*
Manzanita	*Arctostaphylos* species
Muller oak	*Quercus cornelius-mulleri*
Palmer oak	*Quercus palmeri*
Scrub oak	*Quercus berberidifolia*
Singleleaf pinyon	*Pinus monophylla*
Tucker oak	*Quercus john-tuckeri*
Veatch silktassel	*Garrya veatchii*

The term "scrub oak" is used for several shrubby live oaks. *The Jepson Manual* applies the name *Quercus dumosa*, "Nuttall scrub oak" (a CNPS List 1B plant) to a SoCo species (Skinner & Pavlik 1994). References in the ecological literature to *Q. dumosa* are more appropriately *Q. berberidifolia* which ranges through much of Cis-CA. Interior live oak as a shrub, desert scrub oak, and the rarer Muller oak, Tucker oak [=*Q. turbinella* var. *californica*], and Palmer oak are called scrub oaks as well. According to *The Jepson Manual*, desert scrub oak is restricted to e. MojD. This series is in the range of Muller oak, Palmer oak, and Tucker oak. Careful identification of the oaks in a stand will be necessary to correctly assign the stand to a series. Keeler-Wolf (1990c) qualitatively describes Cupleaf ceanothus-fremontia-oak stands in the Long Canyon proposed RNA in Kern Co. m-SN.

Plot-based descriptions: Gordon & White (1994) describe four associations in m-PenR.

Associations:
Gordon & White (1994):
    Cupleaf ceanothus association,
    Birchleaf mountain-mahogany - chamise association,
    Birchleaf mountain-mahogany - fremontia association,
    Tucker oak or Muller oak association.

Deerbrush series

Deerbrush sole or dominant shrub in canopy; bitter cherry, canyon live oak, common manzanita, creeping snowberry, greenleaf manzanita, interior live oak, mountain whitethorn, ocean spray, scrub oak, tobacco brush, and/or wedgeleaf ceanothus may be present. Emergent conifers may be present. Shrubs < 3 m; canopy continuous or intermittent. Ground layer sparse (Plate 10).

Uplands: ridges, upper slopes.

Distribution: NorCo, CenCo, KlaR, CasR, SN, TraR, PenR.
Elevation: 300-2100 m.

NDDB/Holland type and status: **Chaparral** (37000).
  Deer brush chaparral (37531) G4 S4.
  Mixed montane chaparral (37510 *in part*) G4 S4.
  Montane ceanothus chaparral (37520) G4 S4.
  Poison-oak chaparral (37F00 *in part*) G3 S3.3.

Other types:
Barry type: G7411214.
Cheatham & Haller type: Lower montane chaparral.
PSW-45 type: Ceanothus series.
Rangeland type: 208.
Thorne type: Mixed chaparral.
WHR type: Mixed chaparral.

General references; Cronemiller (1959), Paysen *et al.* (1980).

Comments: Deerbrush occurs as scattered shrubs in forest understories or as the sole or dominant shrub. This series is found in several elevation zones in many regions. Self-perpetuating stands are unusual; instead, most stands are established after fire or logging and are transitional to various kinds of forest. Four varieties of deerbrush [*Ceanothus integerrimus* var. *i.*, *C. i.* var. *californicus*, *C. i.* var. *puberulus*, *C. i.* var. *macrothyrsus*] are often recog-

nized, but *The Jepson Manual* considers them poorly defined.

Keeler-Wolf (1990e) qualitatively describes deerbrush stands at Hale Ridge candidate RNA in Lake Co. o-NorCo; at Fern Canyon RNA in Los Angeles Co., White (1994a) at Cleghorn Canyon candidate RNA in San Bernardino Co. i. m-TraR; Keeler-Wolf (1990e) at Hall Canyon RNA in Riverside Co. m-PenR.

Species mentioned in text	
Blue wildrye	*Elymus glaucus*
Bitter cherry	*Prunus emarginata*
Canyon live oak	*Quercus chrysolepis*
Common manzanita	*Arctostaphylos manzanita*
Creeping snowberry	*Symphoricarpos mollis*
Deerbrush	*Ceanothus integerrimus*
Douglas-fir	*Pseudotsuga menziesii*
Interior live oak	*Quercus wislizenii*
Greenleaf manzanita	*Arctostaphylos patula*
Madrone	*Arbutus menziesii*
Mountain whitethorn	*Ceanothus cordulatus*
Ocean spray	*Holodiscus discolor*
Scrub oak	*Quercus berberidifolia*
Tobacco brush	*Ceanothus velutinus*
Tanoak	*Lithocarpus densiflora*
Wedgeleaf ceanothus	*Ceanothus cuneatus*

Plot-based descriptions: Stuart *et al.* (1993) defines two associations following tree harvest in Douglas-fir - tanoak series stands in Siskiyou Co. l-KlaR; Gordon & White (1994) define one association in m-PenR.

Associations:
Gordon & White (1994):
  Deerbrush association.
Stuart *et al.* (1993):
  Deerbrush-canyon live oak-blue wildrye association,
  Tanoak-madrone-deerbrush association.

Desert-holly series

Desert-holly sole or conspicuous shrub in canopy; brittlebush, creosote bush, bush seepweed, tidestromia, and/or white bursage may be present. Shrubs < 1 m; canopy open. Ground layer sparse; annuals seasonally present.

Uplands: dissected alluvial fans, along washes. Soils may be carbonate-rich.

Distribution: MojD, ColD.
Elevation: 75 below sea level-1400 m.

NDDB/Holland type and status: **Chenopod scrubs** (36000).
    Desert saltbush scrub (36110 *in part*) G3 S3.2.

Other types:
Barry type: G7411221.
Cheatham & Haller type: Creosote bush scrub.
PSW-45 type: Saltbush series.
Thorne type: Creosote bush scrub.
WHR type: Desert scrub.

General references: Hunt (1966), Johnson (1976), MacMahon (1988), Paysen *et al.* (1980), Thorne (1982).

Comments: This series is often considered part of the chenopod or saltbush scrub. These scrubs are better thought of as a collection of series. This series shares species with the Creosote bush-white bursage series. Stands of both series have low shrub count.

Species mentioned in text	
Brittlebush	*Encelia farinosa*
Bush seepweed	*Suaeda moquinii*
Creosote bush	*Larrea tridentata*
Desert-holly	*Atriplex hymenelytra*
Tidestromia	*Tidestromia oblongifolia*
White bursage	*Ambrosia dumosa*

Plot-based descriptions: not available.

Dune lupine-goldenbush series

California ephedra, California sagebrush, coast goldenbush, coast prickly-pear, dune lupine, dune sagebrush, and/or heather goldenbush may be important shrubs in canopy. Emergent lemonade berry may be present. Shrubs < 1 m; canopy continuous, intermittent, or open. Ground layer sparse or abundant (Plate 12).

Uplands: stabilized backdune slopes of bars, river mouths, spits along coastline. Soils sandy.

Distribution: s. o-NorCo, o-CenCo, o-SoCo.
Elevation: sea level.

NDDB/Holland type and status: **Coastal dunes** (21000).
    Central dune scrub (21320) G2 S2.2.
    Southern dune scrub (21330) G1 S1.1.

Other types:
Barry type: G7411221.
Cheatham & Haller type: Coastal dune scrub.
Thorne type: Coastal scrub.
WHR type: Coastal scrub.

General references: Barbour (1970), Barbour & Johnson (1977c), Barbour *et al.* (1981a), Breckon (1974), Jones (1984), Russell (1983).

Comments: This series occurs as separate patches of sand along o-CenCo and o-SoCo coast, and shares species with the Sand-verbena - beach bursage series and coastal scrub series.

Plot-based descriptions: Holton & Johnson (1979) define two associations at Point Reyes National Seashore in Marin Co. o-NorCo; Barbour & Johnson (1977) describe a stand at Vandenberg Air Base in Santa Barbara Co., Bluestone (1981) defines one association at Salinas River State Beach in Monterey Co. o-CenCo.

Species mentioned in text	
California ephedra	*Ephedra californica*
California sagebrush	*Artemisia californica*
Coast goldenbush	*Isocoma menziesii*
Coast prickly-pear	*Opuntia littoralis*
Dune lupine	*Lupinus chamissonis*
Dune sagebrush	*Artemisia pycnocephala*
Heather goldenbush	*Ericameria ericoides*
Lemonade berry	*Rhus integrifolia*

Associations:
Bluestone (1981):
    Heather goldenbush association [as Mid-dune hallow, Reardune foreslope, Rear dune crown, Rear leeslope].
Holton & Johnson (1979):
    Dune lupine association,
    Dune lupine-heather goldenbush association.

Eastwood manzanita series

Forms of Eastwood manzanita sole or dominant shrub in canopy; chamise, chaparral yucca, chaparral whitethorn, coyote brush, interior live oak, scrub oak, sugar bush, and/or toyon may be present. Emergent Coulter pine may be present. Shrubs < 3 m; canopy continuous. Ground layer sparse or absent (Plate 8).

Uplands: outcrops, ridges, slopes north-facing. Soils shallow, may be rocky.

Distribution: o-NorCo, o-CenCo, m-TraR, m-PenR.
Elevation: 300-2200 m.

NDDB/Holland type and status:: Chaparral (37000).
Northern mixed chaparral (37110 *in part*) G3 S3.3.
Upper Sonoran manzanita chaparral (37B00 *in part*) G4 S4.

Other types:
Barry type: G7411211 CARGL30.
Cheatham & Haller type: Mixed chaparral.
PSW-45 type: Manzanita series.
Thorne type: Mixed chaparral.
WHR type: Mixed chaparral.

General references: Hanes (1977, 1981), Horton (1960), Keeley (1987a, 1993), Keeley & Keeley (1988), Pase (1982a), Patric & Hanes (1964), Paysen *et al.* (1980), Vogl & Schorr (1972).

Comments: This series definition follows Gordon & White (1994). Eastwood manzanita can be a dominant or minor species throughout its range. Stands where it dominates (> 60%) are included in this series. Eastwood manzanita is important in stands of the Chamise-Eastwood manzanita series. *The Jepson Manual* recognizes 6 subspecies of Eastwood manzanita, some local or rare. The wide

ranging *A. glandulosa* ssp. *g.* in Nor-CA; *A. g.* ssp. *adamsii, A. g.* ssp. *crassifolia, A. g.* ssp. *glaucomollis, A. g.* ssp. *mollis,* and *A. g.* ssp. *zacaensis* in So-CA.

Species mentioned in text	
Bigberry manzanita	*Arctostaphylos glauca*
Chamise	*Adenostoma fasciculatum*
Chaparral whitethorn	*Ceanothus leucodermis*
Chaparral yucca	*Yucca whipplei*
Coulter pine	*Pinus coulteri*
Coyote brush	*Baccharis pilularis*
Eastwood manzanita	*Arctostaphylos glandulosa*
Interior live oak	*Quercus wislizenii*
Scrub oak	*Quercus berberidifolia*
Sugar bush	*Rhus ovata*
Toyon	*Heteromeles arbutifolia*

Plot-based descriptions: Gordon & White (1994) define an Eastwood manzanita association, Keeley & Keeley (1988) describe a stand in m-TraR.

Foothill palo verde-saguaro series

Saguaro and foothill palo verde emergent over a shrub canopy; brittlebush, Anderson box-thorn, catclaw acacia, creosote bush, desert-lavender, and/or white bursage may be present. Trees < 3 m; canopy intermittent. Shrubs common. Ground layer rich in cacti and perennial grasses; annuals seasonally present (Plate 17).

Uplands: slopes, valleys. Soils metamorphic volcanic-derived; alluvial or colluvial.

Distribution: e. ColD, AZ.
Elevation: 150-500 m in Trans-CA.

NDDB/Holland type and status: **Sonoran thorn woodlands** (75000).
     Arizonan woodland (75400) G1.1 S3.

Other types:
Brown Lowe Pase type 154.12.
Cheatham & Haller type: Low desert scrub.
Thorne type: Creosote bush scrub.
WHR habitat type: Desert scrub.

General references: Brum (1973), MacMahon (1988), Turner & Brown (1982).

Comments: Stands of the Foothill palo verde-saguaro series are known from Whipple Mountains San Bernardino Co. Saguaro is a rare plant (a CNPS List 2 plant) in CA (Skinner & Pavlik 1994). Foothill palo verde is a common plant in AZ [where it is highly safeguarded] and Mexico, but is only known in CA from these mountains.

Saguaro is known from other California locations along the Colorado River in Imperial and Riverside Cos. This is a case where one of the defining species is rare and it has a greater range in CA than does the series. This series is extensive in the Arizona upland region of the Sonoran Desert (Shreve & Wiggins 1964).

Species mentioned in text	
Anderson box-thorn	*Lycium andersonii*
Brittlebush	*Encelia farinosa*
Catclaw acacia	*Acacia greggii*
Creosote bush	*Larrea tridentata*
Desert-lavender	*Hyptis emoryi*
Foothill palo verde	*Cercidium microphyllum*
Saguaro	*Carnegiea gigantea*
White bursage	*Ambrosia dumosa*

Plot-based descriptions: NDDB has plot data on file for the Copper Basin area of the Whipple Mountains in San Bernardino Co. ColD.

Fourwing saltbush series

Fourwing saltbush sole or dominant shrub in canopy; cheesebush, green ephedra, hopsage, and/or yellow rabbitbrush may be present. Emergent honey mesquite may be present. Shrubs < 2 m; canopy continuous, intermittent, or open. Ground layer sparse; annuals seasonally present.

Uplands: bluffs; dunes; lower, rocky slopes; washes. Soils may be carbonate-rich. The national list of wetland plants (Reed 1988) lists fourwing saltbush as a FACU.

Distribution: i-CenCo, SoCo, m-TraR, m-PenR, GB, MojD, ColD, inter-West.
Elevation: 75 below sea level-2200 m.

NDDB/Holland type and status: Chenopod scrubs (36000).
> Desert saltbush scrub (36110 *in part*) G3 S3.2.
> Desert sink scrub (36120 *in part*) G3 S2.1.

Other types:
Barry type: G7411221.
Brown Lowe Pase type: 142.121, 152.172.
Cheatham & Haller type: Saltbush scrub.
PSW-45 type: Saltbush series.
Rangeland type: SRM 414.
Thorne type: Alkali sink scrub.
WHR type: Alkali sink.

General references: Burk (1977), MacMahon (1988), McHargue (1973), Paysen *et al.* (1980), Thorne (1982), Tueller (1994).

Comments: This series is often considered part of the chenopod or saltbush scrub. These scrubs are better thought of as a collection of series [*see* Allscale series, Desert-holly series, Fourwing saltbush series, Shadscale series, Spinescale series]. Stands with fourwing saltbush alone or mixed with other shrubs occur in many environments from the coast to interior deserts. This species can occur in wetland habitats. Vegetation descriptions suggest the series is an upland one.

Species mentioned in text	
Cheesebush	*Hymenoclea salsola*
Creosote bush	*Larrea tridentata*
Fourwing saltbush	*Atriplex canescens*
Green ephedra	*Ephedra viridis*
Honey mesquite	*Prosopis glandulosa*
Hop-sage	*Grayia spinosa*
Yellow rabbitbrush	*Chrysothamnus viscidiflorus*

Plot-based descriptions: Spolsky (1979) defines a Saltbush-creosote bush scrub association in Anza Borrego Desert State Park in San Diego Co. ColD.

Greasewood series

Greasewood sole or dominant shrub in canopy; alkali heath, big sagebrush, fourwing saltbush, iodine bush, rubber rabbitbrush, rusty molly, bush seepweed, spinescale, shadscale, and/or yellow rabbitbrush may be present. Shrubs < 3 m; canopy continuous or open. Ground layer variable (Plate 17).

Wetlands: habitats intermittently flooded, saturated. Water chemistry: mixosaline. Barrier beaches; dry lake beds; plains; lagoon bars; old lake beds perched above current drainages; stable dunes. Cowardin class: Palustrine shrub-scrub wetland. The national list of wetland plants (Reed 1988) lists greasewood as a FACU.

Distribution: CenV, ModP, TraSN, MojD, ColD.
Elevation: 100-2000 m.

NDDB/Holland type and status: **Chenopod scrubs** (36000), **Alkali playa community** (46000 *in part*) G3 S2.1.
    Desert sink scrub (36120 *in part*) G2 S2.1.
    Desert greasewood scrub (36130) G3 S3.2.

Other types:
Barry type: G7411221.
Brown Lowe Pase type: 152.171, 153.171, 154.171.
Cheatham & Haller type: Alkali sink scrub, Saltbush scrub.
PSW-45 type: Greasewood series.
Rangeland types: SRM 414, SRM 501.
Thorne type: Alkali sink scrub.
WHR type: Alkali sink.

General references: Burk (1977), MacMahon (1988), Paysen *et al.* (1980), Thorne (1982), Tueller (1994), Vasek & Barbour (1977b), Young *et al.* (1977), West (1988).

Comments: This series is often considered part of the alkali sink scrub, which is better thought of as a collection of series [*see* Bush seepweed series, Iodine bush series].

Whether a given stand is classed as a member of either series depends on which species dominates. Iodine bush tolerates higher salt concentrations than greasewood (MacMahon & Wagner 1985).

Species mentioned in text	
Alkali heath	*Frankenia salina*
Big sagebrush	*Artemisia tridentata*
Bush seepweed	*Suaeda moquinii*
Fourwing saltbush	*Atriplex canescens*
Greasewood	*Sarcobatus vermiculatus*
Iodine bush	*Allenrolfea occidentalis*
Rubber rabbitbrush	*Chrysothamnus nauseosus*
Rusty molly	*Kochia californica*
Shadscale	*Atriplex confertifolia*
Spinescale	*Atriplex spinifera*
Yellow rabbitbrush	*Chrysothamnus viscidiflorus*

Plot-based descriptions: Young *et al.* (1986) describe stands in w. NV. NDDB has plot data on file for Surprise Valley in Modoc Co. ModP; Ferren & Davis (1991) define a Greasewood-shadscale association at Fish Slough in TraSN.

Greenleaf manzanita series

Greenleaf manzanita sole or dominant shrub in canopy; bitter cherry, bush chinquapin, Fremont silktassel, huckleberry oak, mountain whitethorn, ocean spray, pinemat manzanita, service berry, and/or tobacco brush may be present. Emergent conifers may be present. Shrubs < 2 m; canopy continuous or intermittent. Ground layer sparse (Plate 8).

Uplands: ridges, upper slopes, may be steep. Soils shallow, commonly granitic or volcanic-derived.

Distribution: m-NorCo, m-KlaR, su-KlaR, m-CasR, su-CasR, m-SN, su-SN, i. m-TraR, PenR, inter-West, Baja CA.
Elevation: 750-3350 m.

NDDB/Holland type and status:: **Chaparral** (37000).
  Mixed montane chaparral (37510 *in part*) G4 S4.
  Montane manzanita chaparral (37520 *in part*) G4 S4.
  Upper Sonoran manzanita chaparral (37B00 *in part*) G4 S4.

Other types:
Barry type: G7411211 CARPA60.
Cheatham & Haller type: Montane chaparral.
PSW-45 type: Manzanita series.
Rangeland type: SRM 209.
Thorne type: Montane chaparral.
WHR type: Montane chaparral.

General references: Barbour (1988), Paysen *et al.* (1980), Riser & Fry (1994), Solinas *et al.* (1985).

Comments: Greenleaf manzanita occurs as scattered shrubs in a forest understory or as a sole or dominant shrub in what is often called montane chaparral. Self-perpetuating stands grow on shallow soils; those transitional to forest occur on deeper soils. This series is found in the montane and subalpine zones, where it is extensive.

Greenleaf manzanita series stands share shrub species and elevations with stands of the Huckleberry oak series and Tobacco brush series. Stands of these series can be differentiated on the importance of huckleberry oak. In this series, huckleberry oak is a minor species.

Parry manzanita replaces greenleaf manzanita in TraR. These stands will be included in this series.

Keeler-Wolf (1990e) qualitatively describes greenleaf manzanita stands as montane chaparral at Pearch Creek in Humboldt Co. m-KlaR; at Antelope Creek Lakes candidate RNA, at Shasta Mudflow RNA, at Shasta Red Fir candidate RNA in Siskiyou Co., Feidler *et al.* (1986) at Green Island Lake candidate RNA in Butte Co. m-CasR, su-CasR; Keeler-Wolf (1990e) at Millard Canyon RNA in Riverside Co. i. m-TraR.

Species mentioned in text	
Bitter cherry	*Prunus emarginata*
Bush chinquapin	*Chrysolepis sempervirens*
Fremont silktassel	*Garrya fremontii*
Greenleaf manzanita	*Arctostaphylos patula*
Huckleberry oak	*Quercus vaccinifolia*
Mountain whitethorn	*Ceanothus cordulatus*
Ocean spray	*Holodiscus discolor*
Parry manzanita	*Arctostaphylos parryana*
Pinemat manzanita	*Arctostaphylos nevadensis*
Service berry	*Amelanchier alnifolia*
Tobacco brush	*Ceanothus velutinus*

Plot-based descriptions: Sawyer & Thornburgh (1977b) define a Huckleberry oak-greenleaf manzanita association on granites in Siskiyou and Trinity Cos. m-KlaR, su-KlaR.

Hairyleaf ceanothus series

Hairyleaf ceanothus sole or dominant shrub in canopy; canyon live oak, chamise, poison-oak, scrub oak, sugar bush, and/or toyon may be present. Emergent trees may be present. Shrubs < 3.5 m; canopy continuous or intermittent. Ground layer sparse (Plate 10).

Uplands: slopes north-facing, steep. Soils deep.

Distribution: NorCo, CenCo, o. l-TraR, w. m-PenR.
Elevation: 600-950 m.

NDDB/Holland type and status: **Chaparral** (37000). Upper Sonoran ceanothus chaparral (37800) G3 S3.3.

Other types:
Barry type: G7411211 CCEOL00.
Cheatham & Haller type: Mixed chaparral.
PSW-45 type: Ceanothus series.
Rangeland type: SRM 208.
Thorne type: Mixed chaparral.
WHR type: Mixed chaparral.

General references: Hanes (1977, 1981), Keeley & Keeley 1988), Pase (1982a), Paysen *et al.* (1980), White (1994c).

Comments: This series definition follows Gordon & White (1994). Throughout its range, hairyleaf ceanothus occurs as scattered shrubs or as the sole or dominant shrub. The differences between pure and mixed stands can be handled at the association level.

The Jepson Manual recognizes two varieties of *Ceanothus oliganthus*, the typical *C. o.* var. *o.* [hairyleaf ceanothus] and *C. o.* var. *sorediatus* [Jim brush]. Both varieties are included in this series.

Keeler-Wolf (1990e) qualitatively describes a hairyleaf ceanothus chaparral at Limekiln RNA (now part of the Cone Peak Gradient RNA) in Monterey Co. m-CenCo.

Species mentioned in text	
Canyon live oak	*Quercus chrysolepis*
Chamise	*Adenostoma fasciculatum*
Hairyleaf ceanothus	*Ceanothus oliganthus*
Poison-oak	*Toxicodendron diversilobum*
Scrub oak	*Quercus berberidifolia*
Sugar bush	*Rhus ovata*
Toyon	*Heteromeles arbutifolia*

Plot-based descriptions: Gordon & White (1994) define a Hairyleaf ceanothus association m-TraR, m-PenR.

Hoaryleaf ceanothus series

Hoaryleaf ceanothus sole or dominant shrub in canopy; bigberry manzanita, birchleaf mountain-mahogany, black sage, California buckwheat, chamise, chaparral whitethorn, hollyleaf redberry, scrub oak, and/or toyon may be present. Emergent trees may be present. Shrubs < 3.5 m; canopy continuous or intermittent. Ground layer sparse (Plate 10).

Uplands: slopes south-facing at highest elevations. Soils deep or shallow, usually coarse-textured.

Distribution: s. CenCo, i-SoCo, TraR, PenR, Baja CA.
Elevation: 100-1350 m.

Species mentioned in text	
Bigberry manzanita	*Arctostaphylos glauca*
Birchleaf mountain-mahogany	*Cercocarpus betuloides*
Black sage	*Salvia mellifera*
California buckwheat	*Eriogonum fasciculatum*
Chamise	*Adenostoma fasciculatum*
Chaparral whitethorn	*Ceanothus leucodermis*
Hoaryleaf ceanothus	*Ceanothus crassifolius*
Hollyleaf redberry	*Rhamnus ilicifolia*
Scrub oak	*Quercus berberidifolia*
Toyon	*Heteromeles arbutifolia*

Plot-based descriptions: Gordon & White (1994) define a Hoaryleaf ceanothus association in PenR.

NDDB/Holland type and status: **Chaparral** (37000).
    Ceanothus crassifolius chaparral (37830) G3 S3.2.

Other types:
Barry type: G7411211 CCECR00.
Cheatham & Haller type: Mixed chaparral.
PSW-45 type: Ceanothus series.
Rangeland type: SRM 208.
Thorne type: Mixed chaparral.
WHR type: Mixed chaparral.

General references: Hanes (1977, 1981), Keeley (1987b), Keeley & Keeley (1988), Montygierd-Loba & Keeley (1987), Pase (1982a), Paysen *et al.* (1980), White (1994c).

Comments: This series definition follows Gordon & White (1994). Hoaryleaf ceanothus occurs as scattered shrubs or as the sole or dominant shrub in chaparral. Stands where hoaryleaf ceanothus and chamise are equally important are members of the Chamise-hoaryleaf ceanothus series.

Holodiscus series

Ocean spray or rock-spiraea sole, dominant, or important shrub in canopy; bitter cherry, bush chinquapin, creeping snowberry, Fremont silktassel, greenleaf manzanita, huckleberry oak, rubber rabbitbrush, Sadler oak, service berry, and/or tobacco brush may be present. Emergent conifers may be present. Shrubs < 1 m; canopy intermittent. Ground layer sparse (Plate 10).

Uplands: ridges, upper slopes, may be steep, talus. Soils skeletal. Texture loam or sand, may be rocky.

Distribution: m-NorCo, m-CenCo, m-KlaR, su-KlaR, m-CasR, su-CasR, a-CasR, m-SN, su-SN, a-SN, m-TraR, su-TraR, su-PenR.
Elevation: 700-2800 m.

NDDB/Holland type and status: none.

Other types:
Barry type: G7411321 EHOMI00.

General references: Sawyer & Thornburgh (1977b).

Comments: This series is common on mountain ridges and slopes with thin soils. Ocean spray and rock-spiraea do not occur in the same stands as they are geographically segregated. This series is found in the montane, subalpine, and alpine zones. Unlike many montane chaparral, these stands are self-perpetuating. This series does not include areas where shrubs are scattered in forest understories.

The treatment of *Holodiscus* varies among manuals. *The Jepson Manual* recognizes two species, ocean spray [*H. discolor*; not recognizing three varieties or *H. boursieri*] and rock-spiraea [*H. microphyllus*]. Two varieties are recognized for rock-spiraea, one *H. m.* var. *glabrescens* m-NorCo, ModP, WarR and *H. m.* var. *microphyllus* [including *H. m.* var. *sericeus*] SN, TraR, PenR, SN. These species are commonly misidentified [*see The Jepson Manual*].

Keeler-Wolf (1990e) qualitatively describes rock-spiraea stands at Cedar Basin candidate RNA in Siskiyou Co. su-KlaR; at Antelope Creek Lakes candidate RNA in Siskiyou Co. su-CasR.

Species mentioned in text	
Bitter cherry	*Prunus emarginata*
Bush chinquapin	*Chrysolepis sempervirens*
Creeping snowberry	*Symphoricarpos mollis*
Fremont silktassel	*Garrya fremontii*
Greenleaf manzanita	*Arctostaphylos patula*
Huckleberry oak	*Quercus vaccinifolia*
Nude buckwheat	*Eriogonum nudum*
Ocean spray	*Holodiscus discolor*
Rock-spiraea	*Holodiscus microphyllus*
Rubber rabbitbrush	*Chrysothamnus nauseosus*
Sadler oak	*Quercus sadleriana*
Service berry	*Amelanchier alnifolia*
Suksdorf monkey flower	*Mimulus suksdorfii*
Tobacco brush	*Ceanothus velutinus*
Wedgeleaf keckiella	*Keckiella corymbosa*
Western needlegrass	*Stipa occidentalis*

Plot-based descriptions: Taylor & Teare (1979a), Keeler-Wolf (1990e) define one association at Manzanita Creek candidate RNA in Trinity Co., Sawyer & Thornburgh (1977b) one association on granite in Siskiyou and Trinity Cos. su-KlaR; Burke (1982) two associations at Rae Lakes area, King's Canyon National Park in Fresno Co. su-SN.

Associations:
Burke (1982):
    Rock-spiraea/Suksdorf monkey flower association,
    Western needlegrass-nude buckwheat association.
Sawyer & Thornburgh (1977b):
    Ocean spray/greenleaf manzanita association [as *Pinus albicaulis/Holodiscus microphyllus* type].
Taylor & Teare (1979a):
    Ocean spray/wedgeleaf keckiella association [as *Holodiscus microphyllus/Penstemon corymbosus* alliance].

Hop-sage series

Hop-sage sole or dominant shrub in canopy; big sagebrush, black bush, box-thorns, creosote bush, fourwing saltbush, shadscale, spinescale, white bursage, and/or winter fat may be present. Emergent Joshua trees may be present. Shrubs < 1 m; canopy continuous, intermittent, or open. Ground layer sparse (Plate 16).

Uplands: basins. Soils alluvial-derived.

Distribution: i-CenCo, s. m-SN, m-TraR, s. TraSN, MojD, inter West.
Elevation: 500-2800 m.

NDDB/Holland type and status: Chenopod scrubs (36000).
    Shadscale scrub (36140) G4 S3.2.

Other types:
Barry type: G7411221.
Cheatham & Haller type: Saltbush scrub.
Thorne type: Shadscale scrub.
WHR type: Desert scrub.

General references: Beatley (1975), Johnson (1976), Young *at al.* (1977), Thorne (1982).

Comments: This series is often considered part of the chenopod or saltbush scrub. These scrubs are better considered as a collection of series. Hop-sage is a common, but not a dominant species in several series [*see* Black sagebrush series, Creosote bush series, Fourwing saltbush series, Joshua tree series, Shadscale series].

Species mentioned in text	
Big sagebrush	*Artemisia tridentata*
Black bush	*Coleogyne ramosissima*
Black sagebrush	*Artemisia nova*
Box-thorns	*Lycium* species
Creosote bush	*Larrea tridentata*
Fourwing saltbush	*Atriplex canescens*
Hop-sage	*Grayia spinosa*
Joshua tree	*Yucca brevifolia*
Shadscale	*Atriplex confertifolia*
Spinescale	*Atriplex spinifera*
White bursage	*Ambrosia dumosa*
Winter fat	*Krascheninnikovia lanata*

Plot-based descriptions: Beatley (1965) describes stands in NV [*see* Yoder *et al.* (1983)].

Huckleberry oak series

Huckleberry oak sole or dominant shrub in canopy; bitter cherry, bush chinquapin, creeping snowberry, Fremont silktassel, greenleaf manzanita, mountain whitethorn, pinemat manzanita, ocean spray, Sadler oak, service berry, and/or tobacco brush may be present. Emergent conifers may be present. Shrubs < 2 m; canopy continuous or intermittent. Ground layer sparse.

Uplands: ridges, upper slopes, may be steep. Soils shallow, commonly granitic-derived, may be rocky.

Distribution: m-NorCo, m-KlaR, su-KlaR, m-CasR, su-CasR, m-SN, su-SN.
Elevation: 700-2800 m.

NDDB/Holland type and status: Chaparral (37000).
Mixed montane chaparral (37510 *in part*) G4 S4.
Huckleberry oak chaparral (37542) G3 S3.

Other types:
Barry type: G7411214.
Cheatham & Haller type: Upper montane chaparral.
Rangeland type: SRM 209.
Thorne type: Montane chaparral.
WHR type: Montane chaparral.

General references: Riser & Fry (1994), Sawyer & Thornburgh (1977b), Vankat & Major (1978).

Comments: Huckleberry oak occurs as scattered shrubs in forest understories or as the sole or dominant shrub in what is often called montane chaparral. This series is montane and subalpine, and is more extensive at the higher elevations. Self-perpetuating stands grow on shallow soils; those transitional to forest occur on deeper soils. Bush chinquapin, primarily So-CA, shares shrub species and elevations with this series. Stands of these series can be differentiated on the importance of huckleberry oak. Huckleberry oak may only be a minor species in the Bush chinquapin series, where here it is the dominant species.

Keeler-Wolf (1990e) qualitatively describes huckleberry oak stands as montane chaparral at Cedar Basin candidate RNA, Rock Creek Butte candidate RNA in Siskiyou Co., at Pearch Creek in Humboldt Co. m-KlaR; at Soda Ridge candidate RNA, Green Island candidate RNA in Butte Co., Cub Creek RNA in Tehama Co. m-CasR; Mount Pleasant RNA in Plumas Co., Sugar Pine Point RNA in Nevada Co., Onion Creek in Placer Co., Bell Meadow RNA in Tuolumne Co. m-SN.

Species mentioned in text	
Bitter cherry	*Prunus emarginata*
Bush chinquapin	*Chrysolepis sempervirens*
Creeping snowberry	*Symphoricarpos mollis*
Fremont silktassel	*Garrya fremontii*
Greenleaf manzanita	*Arctostaphylos patula*
Huckleberry oak	*Quercus vaccinifolia*
Mountain whitethorn	*Ceanothus cordulatus*
Ocean spray	*Holodiscus discolor*
Pinemat manzanita	*Arctostaphylos nevadensis*
Sadler oak	*Quercus sadleriana*
Service berry	*Amelanchier alnifolia*
Tobacco brush	*Ceanothus velutinus*

Plot-based descriptions: Sawyer & Thornburgh (1977b) define a Huckleberry oak-greenleaf manzanita association on granites in Siskiyou and Trinity Cos. m-KlaR, su-KlaR.

Interior live oak shrub series

Interior live oak sole or dominant shrub in canopy; bigberry manzanita, birchleaf mountain-mahogany, California coffeeberry, canyon live oak, chamise, chaparral whitethorn, Eastwood manzanita, hollyleaf cherry, hollyleaf redberry, manzanitas, poison-oak, scrub oak, and/or wedgeleaf ceanothus may be present. Emergent California buckeye, California juniper, Coulter pine, foothill pine, and/or knobcone pine may be present. Shrubs < 6 m; canopy continuous or intermittent. Ground layer variable.

Uplands: all slopes, may be steep. Soils alluvial or bedrock-derived, may be rocky.

Distribution: i-NorCo, f-CasR, f-SN, m-SN m-TraR, m-PenR.
Elevation: 300-1850 m.

NDDB/Holland type and status: **Chaparral** (37000).
Interior live oak chaparral (37A00) G3 S3.3,
Northern north slope chaparral (37E10 *in part*) G3 S3.3.
Southern north slope chaparral (37E20 *in part*) G3 S3.3.
Poison-oak chaparral (37F00 *in part*) G3 S3.3.

Other types:
Barry type: G7411211 CQUWI20.
Cheatham & Haller type: Mixed chaparral.
Thorne type: Mixed chaparral.
WHR type: Mixed chaparral

General references: Hanes (1977, 1981), Horton (1960), Griffin (1977), Keeley & Keeley (1988), Minnich (1976), Pase (1982a), Patric & Hanes (1964), (White 1992).

Comments: Interior live oak is a widespread and common species in shrublands and forests. Stands where it occurs as a shrub, either because it is presumably distinct taxon [*Quercus wislizenii* var.

frutescens] or because of age, are included in this series. Many of the So-CA shrub stands are believed to be the result of frequent sprouting after fires (White & Sawyer 1995). Stands old enough to form woodlands and forests are included in the Interior live oak series.

This series definition follows Gordon & White (1994). If interior live oak as a shrub has > 60% cover, the stand is a member of this series. If interior live oak as a shrub cover is 60-30% and another species cover 30-60%, then the stand is a member of a mixed series [*see* Interior live oak-chaparral whitethorn shrub series, Interior live oak-scrub oak shrub series, Interior live oak-canyon live oak shrub series].

Species mentioned in text	
Bigberry manzanita	*Arctostaphylos glauca*
Birchleaf mountain-mahogany	*Cercocarpus betuloides*
California buckeye	*Aesculus californica*
California coffeeberry	*Rhamnus californica*
California juniper	*Juniperus californica*
Canyon live oak	*Quercus chrysolepis*
Ceanothus	*Ceanothus* species
Chamise	*Adenostoma fasciculatum*
Chaparral whitethorn	*Ceanothus leucodermis*
Coulter pine	*Pinus coulteri*
Eastwood manzanita	*Arctostaphylos glandulosa*
Foothill pine	*Pinus sabiniana*
Hollyleaf cherry	*Prunus ilicifolia* ssp. *ilicifolia*
Hollyleaf redberry	*Rhamnus ilicifolia*
Interior live oak	*Quercus wislizenii*
Knobcone pine	*Pinus attenuata*
Manzanitas	*Arctostaphylos* species
Poison-oak	*Toxicodendron diversilobum*
Scrub oak	*Quercus berberidifolia*
Wedgeleaf ceanothus	*Ceanothus cuneatus*

Keeler-Wolf (1990d) qualitatively describes interior live oak shrub stands at Indian Creek candidate RNA in Tehama Co. f-CasR; at Fall Canyon RNA in Los Angeles Co., at Millard Canyon RNA in Riverside Co. m-TraR; at Cahuilla Mountain RNA in Riverside Co. m-PenR.

Plot-based descriptions: Keeler-Wolf (1983) Keeler-Wolf (1990e) describes stands as Non-serpentinite chaparral at Frenzel Creek RNA in Colusa Co. i-NorCo. White & Sawyer (1995) show that oak chaparral is distinct from interior live oak forest [*see* Interior live oak series] and oak-ceanothus shrubland [*see* Chaparral whitethorn series] in San Bernardino Co. i. m-TraR. Oak chaparral is equivalent to the interior live oak phase of north-slope chaparral (Patric & Hanes 1964) and the interior live oak phase of oak chaparral (Minnich 1976). Gordon & White (1994) define a Shrub interior live oak association in So-CA.

Interior live oak-canyon live oak shrub series

Interior live oak and canyon live oak important shrubs in canopy; California coffeeberry, chamise, chaparral whitethorn, deerbrush, hollyleaf redberry, manzanitas, scrub oak, and/or toyon may be present. Shrubs < 6 m; canopy continuous or intermittent. Ground layer variable.

Uplands: slopes north-facing, may be steep. Soil alluvial or bedrock-derived, may be rocky.

Distribution: m-TraR, m-PenR.
Elevation: 700-1900 m.

NDDB/Holland type and status: **Chaparral** (37000).
Interior live oak chaparral (37A00 *in part*) G3 S3.3.
Southern north slope chaparral (37E20 *in part*) G3 S3.3.

Others types:
Barry type: G7411211.
Cheatham & Haller type: Mixed chaparral.
Thorne type: Mixed chaparral.
WHR type: Mixed chaparral.

General references: Hanes (1977, 1981), Horton (1960), Griffin (1977), Keeley & Keeley (1988), Minnich (1976), Pase (1982a), Patric & Hanes (1964), White (1991).

Comments: Interior live oak is a widespread and common species in shrublands and forests. Stands where it occurs as a shrub, either because it is presumably distinct [*Quercus wislizenii* var. *frutescens*] or because of age, are included in this series. Many of the So-CA shrub stands are believed to be the result of frequent sprouting after fires (White & Sawyer 1995). Stands old enough to form woodlands and forests are included in the Interior live oak series.

This series definition follows Gordon & White (1994). If interior live oak as a shrub has > 60% cover, the stand is a member of the Interior live oak shrub series. If interior live oak as a shrub cover is 60-30% and canyon live oak as a shrub cover is 30-60%, then the stand is a member of this series [*see* Interior live oak-chaparral whitethorn shrub series, Interior live oak-scrub oak shrub series].

Keeler-Wolf (1990e) qualitatively describes interior live oak shrub stands at Fall Canyon RNA in Los Angeles Co., White (1994a) at Cleghorn Canyon candidate RNA in San Bernardino Co. i. m-TraR; Keeler-Wolf (1990e) at Cahuilla Mountain RNA in Riverside Co. m-PenR.

Species mentioned in text

California coffeeberry	*Rhamnus californica*
Canyon live oak	*Quercus chrysolepis*
Chamise	*Adenostoma fasciculatum*
Chaparral whitethorn	*Ceanothus leucodermis*
Deerbrush	*Ceanothus integerrimus*
Hollyleaf redberry	*Rhamnus ilicifolia*
Interior live oak	*Quercus wislizenii*
Manzanitas	*Arctostaphylos* species
Scrub oak	*Quercus berberidifolia*
Toyon	*Heteromeles arbutifolia*

Plot-based descriptions: Gordon & White (1994) define an Interior live oak-canyon live oak shrub association in m-TraR, m-PenR. White & Sawyer (1995) describe unclassified plots.

Interior live oak-chaparral whitethorn shrub series

Interior live oak and chaparral whitethorn important shrubs in canopy; birchleaf mountain-mahogany, California coffeeberry, chamise, chaparral yucca, Eastwood manzanita, basket bush, and/or yerba santa may be present. Emergent Coulter pine may be present. Shrubs < 6 m; canopy continuous or intermittent. Ground layer variable.

Uplands: slopes south-facing, may be steep. Soils shallow, may be rocky.

Distribution: m-TraR, m-PenR.
Elevation: 900-1950 m.

NDDB/Holland type and status: Chaparral (37000).
   Interior live oak chaparral (37A00 *in part*) G3 S3.3.
   Southern north slope chaparral (37E20 *in part*) G3 S3.

Others types:
Barry type: G7411214.
Cheatham & Haller type: Mixed chaparral.
Thorne type: Mixed chaparral.
WHR type: Mixed chaparral.

General references: Hanes (1977, 1981), Horton (1960), Griffin (1977), Keeley & Keeley (1988), Minnich (1976), Pase (1982a), Patric & Hanes (1964), White (1991).

Comments: Interior live oak is a widespread and common species in shrublands and forests. Stands where it occurs as a shrub, either because it is presumably distinct [*Quercus wislizenii* var. *frutescens*] or because of age, are included in this series. Many of the So-CA shrub stands are believed to be the result of frequent sprouting after fires (White & Sawyer 1995). Stands old enough to form woodlands and forests are included in the Interior live oak series.

This series definition follows Gordon & White (1994). If interior live oak as a shrub has > 60% cover, the stand is a member of the Interior live oak shrub series. If interior live oak as a shrub cover is 60-30% and chaparral whitethorn cover 30-60%, then the stand is a member of this series [*see* Interior live oak-scrub oak shrub series, Interior live oak-canyon live oak shrub series].

Keeler-Wolf (1990e) qualitatively describes interior live oak shrub stands at Fall Canyon RNA in Los Angeles Co. i. m-TraR; at Cahuilla Mountain RNA in Riverside Co. m-PenR.

Plot-based descriptions: Gordon & White (1994) define two associations, White & Sawyer (1995) one association in m-TraR, m-PenR.

Species mentioned in text	
Basket bush	*Rhus trilobata*
Birchleaf mountain-mahogany	*Cercocarpus betuloides*
California coffeeberry	*Rhamnus californica*
Canyon live oak	*Quercus chrysolepis*
Chamise	*Adenostoma fasciculatum*
Chaparral whitethorn	*Ceanothus leucodermis*
Chaparral yucca	*Yucca whipplei*
Coulter pine	*Pinus coulteri*
Eastwood manzanita	*Arctostaphylos glandulosa*
Interior live oak	*Quercus wislizenii*
Scrub oak	*Quercus berberidifolia*
Yerba santa	*Eriodictyon* species

Associations:
Gordon & White (1994):
   Shrub interior oak-chaparral whitethorn association [= Oak-ceanothus shrubland (White & Sawyer 1995)],
   Shrub interior oak-chaparral whitethorn-(Eastwood manzanita) association.

Interior live oak-scrub oak shrub series

Interior live oak and scrub oak important shrubs in canopy; bigberry manzanita, birchleaf mountain-mahogany, California coffeeberry, canyon live oak, chamise, hollyleaf cherry, hollyleaf redberry, poison-oak, and/or wedgeleaf ceanothus may be present. Emergent trees may be present. Shrubs < 6 m; canopy continuous or intermittent. Ground layer variable.

Uplands: slopes often north-facing, may be steep. Soils residual, may be rocky.

Distribution: m-TraR, m-PenR.
Elevation: 700-1850 m.

NDDB/Holland type and status: **Chaparral** (37000).
Interior live oak chaparral (37A00) G3 S3.3.
Southern north slope chaparral (37E20 *in part*) G3 S3.3.

Other types:
Barry type: G7411211.
Cheatham & Haller type: Mixed chaparral.
Thorne type: Mixed chaparral.
WHR type: Mixed chaparral.

General references: Hanes (1977, 1981), Horton (1960), Griffin (1977), Keeley & Keeley (1988), Minnich (1976), Pase (1982a), Patric & Hanes (1964), White (1991).

Comments: Interior live oak is a widespread and common species in shrublands and forests. Stands where it occurs as a shrub, either because it is presumably distinct [*Quercus wislizenii* var. *frutescens*] or because of age (White & Sawyer 1995), are included in this series. Stands are old enough to form woodlands and forests are included in the Interior live oak series.

his series definition follows Gordon & White (1994). If interior live oak as a shrub has > 60% cover, the stand is a member of the Interior live oak shrub series. If interior live oak as a shrub cover is 60-30% and scrub oak cover 30-60%, then the stand is a member of this series [*see* Interior live oak-chaparral whitethorn shrub series, Interior live oak-canyon live oak shrub series].

Keeler-Wolf (1990e) qualitatively describes interior live oak shrub stands at Fall Canyon RNA in Los Angeles Co. i. m-TraR; at Cahuilla Mountain RNA in Riverside Co. m-PenR.

Species mentioned in text	
Bigberry manzanita	*Arctostaphylos glauca*
Birchleaf mountain-mahogany	*Cercocarpus betuloides*
California coffeeberry	*Rhamnus californica*
Canyon live oak	*Quercus chrysolepis*
Chamise	*Adenostoma fasciculatum*
Hollyleaf cherry	*Prunus ilicifolia* ssp. *ilicifolia*
Hollyleaf redberry	*Rhamnus ilicifolia*
Interior live oak	*Quercus wislizenii*
Poison-oak	*Toxicodendron diversilobum*
Scrub oak	*Quercus berberidifolia*
Wedgeleaf ceanothus	*Ceanothus cuneatus*

Plot-based descriptions: Gordon & White (1994) define an Interior live oak-scrub oak shrub association in m-TraR, m-PenR.

Iodine bush series

Iodine bush sole or dominant shrub in canopy; alkali heath, greasewood, rusty molly, saltgrass, samphire, bush seepweed, and/or shadscale may be present. Shrubs < 2 m; canopy continuous or open. Ground layer variable.

Wetlands: habitats intermittently flooded, saturated. Water chemistry: hypersaline. Dry lake beds, margins; hummocks; lagoon bars; old lake beds perched above current drainages; seeps. Cowardin class: Palustrine shrub-scrub wetland. The national list of wetland plants (Reed 1988) lists iodine bush as a FACW+.

Distribution: sj-CenV, ModP, TraSN, MojD, ColD.
Elevation: sea level-1800 m.

NDDB/Holland type and status: **Chenopod scrubs** (36000), **Alkali playa communities** (46000 *in part*) G3 S2.1.
　　Desert sink scrub (36120 *in part*) G3 S2.1.
　　Desert greasewood scrub (36130 *in part*) G3 S2.2.
　　Valley sink scrub (36210) G1 S1.1.

Other types:
Barry type: G7412213 BALOC20.
Brown Lowe Pase type: 153.172, 154.171, 154.171.
Cheatham & Haller type: Alkali sink scrub, Saltbush scrub.
PSW-45 type: Iodine bush series.
Thorne type: Alkali sink scrub.
WHR type: Alkali sink.

General references: Bittman (1985), Burk (1977), Griggs (1980), MacMahon (1988), MacMahon & Wagner (1985), Paysen *et al.* (1980), Thorne (1982), Vasek & Barbour (1977b), Werschskull *et al.* (1984), Young *et al.* (1977).

Comments: This is one of several series included in alkali scrub or sink vegetation. These categories are better considered a collection of series. Whether a given stand is classed as a member of the Alkali sacaton series, Bush seepweed series, Iodine bush series, or Greasewood series depends on which species dominates. Stands of these series can form a fine mosaic in response to microtopograhy.

Species mentioned in text

Alkali heath	*Frankenia salina*
Alkali sacaton	*Sporobolus airoides*
Bush seepweed	*Suaeda moquinii*
Greasewood	*Sarcobatus vermiculatus*
Iodine bush	*Allenrolfea occidentalis*
Rusty molly	*Kochia californica*
Saltgrass	*Distichlis spicata*
Samphire	*Salicornia subterminalis*
Shadscale	*Atriplex confertifolia*

Plot-based descriptions: Bradley (1970) defines two associations in Death Valley National Park MojD; Odion *et al.* (1992) one association at Fish Slough in Inyo Co. TraSN [Some plots in Odion *et al.* associations have low values of iodine bush; these associations might be considered as belonging to the Alkali sacaton series or Saltgrass series]; McHargue (1973) describes stands in Coachella Valley in Riverside Co. ColD.

Associations:
Bradley (1970):
　　Saltgrass-iodine bush association
　　[as Eastern salt flat],
　　Bush seepweed-Iodine bush association
　　[as Western salt flat].
Odion *et al.* (1992):
　　Alkali sacaton-iodine bush association
　　[as *Sporobolus-Allenrolfea* association].
McHargue (1973):
　　Iodine bush association
　　[as Level-wet sites].

Ione manzanita series

Ione manzanita sole or dominant shrub in canopy; chamise, deer weed, hoary coffeeberry, interior live oak, scrub oak, toyon, whiteleaf manzanita, woollyleaf ceanothus, and/or yerba santa may be present. Emergent canyon live oak, foothill pine, or knobcone pine may be present. Shrubs < 2 m; canopy continuous. Ground layer sparse (Plate 10).

Uplands: slopes. Soils derived from Eocene Ione formation.

Distribution: f-SN.
Elevation: 100-300 m.

NDDB/Holland type and status:: **Chaparral** (37000). Ione chaparral (37D00) G1 S1.1.

Other types:
Barry type: G7411214.
PSW-45 type: Manzanita series.

General references: Gankin & Major (1964), Paysen *et al.* (1980), Singer (1978), Wood & Parker (1988).

Comments: Unlike many localized manzanitas, this one dominates where it grows in Amador and Calaveras Cos. Five rare species are associated with this series: Parry horkelia (a CNPS List 1B plant), Ione manzanita (a CNPS List 1B plant), Ione buckwheat (a CNPS List 1B plant), Irish Hill buckwheat (a CNPS List 1B plant), Bisbee Peak rush-rose (a CNPS List 1B plant) (Skinner & Pavlik 1994).

Species mentioned in text	
Bisbee Peak rush-rose	*Helianthemum suffrutescens*
Canyon live oak	*Quercus chrysolepis*
Chamise	*Adenostoma fasciculatum*
Deer weed	*Lotus scoparius*
Foothill pine	*Pinus sabiniana*
Hoary coffeeberry	*Rhamnus tomentella*
Interior live oak	*Quercus wislizenii*
Ione buckwheat	*Eriogonum apricum* var. *apricum*
Ione manzanita	*Arctostaphylos myrtifolia*
Irish Hill buckwheat	*Eriogonum apricum* var. *prostratum*
Knobcone pine	*Pinus attenuata*
Parry horkelia	*Horkelia parryi*
Scrub oak	*Quercus berberidifolia*
Toyon	*Heteromeles arbutifolia*
Whiteleaf manzanita	*Arctostaphylos viscida*
Woollyleaf ceanothus	*Ceanothus tomentosus*
Yerba santa	*Eriodictyon californicum*

Plot-based descriptions: Wood & Parker (1988) present shrub density and cover stands at Apricum Hill Ecological Reserve in Amador Co. m-SN.

Joshua tree series

Joshua tree emergent and abundant over a shrub canopy; big sagebrush, black bush, bladderpod, bush buckwheat, box-thorns, cheesebush, creosote bush, ephedras, Mojave yucca, rabbitbrushes, and/or Spanish bayonet may be present. Shrubs <3 m; intermittent or open. Emergent Joshua tree < 12 m. Ground layer with cacti and perennial grass; annuals seasonally present (Plate 16).

Uplands: gentle alluvial fans. Soils colluvial, alluvial-derived.

Distribution: m-SN, m-TraR, MojD, NV, UT.
Elevation: 700-1800 m.

NDDB/Holland type and status: **Mojavean desert scrub** (34000), **Joshua tree woodland** (73000).
   Mojave mixed steppe (34220) G3 S2.2
   Mojave mixed woody scrub (34210) G3 S3.2
   Joshua tree woodland (73000) G 4 S3.2

Other types:
Brown Lowe Pase types 153.151, 153.152, 153.153.
Cheatham & Haller type: Joshua tree woodland.
PSW-45 type: Joshua tree series.
Stone & Sumida (1983): Joshua tree community.
Thorne type: Joshua tree woodland.
WHR habitat type: Joshua tree.

General references: Johnson (1976), MacMahon (1988), Paysen *et al.* (1980), Thorne (1982), Turner (1982a, 1982b), Vasek & Thorne (1977).

Comments: Joshua tree has a large range and grows with many species, so much so that Rowlands (1978) questions the existence of a plant community. He suggests that it is better to consider all Joshua tree populations as a component of other grasslands and scrubs no matter the Joshua tree density. Because Joshua tree is common and conspicuous in some areas and uncommon in others,

this approach defines a series to include stands of dense populations. Trees other than Joshua tree can be present in this series, and Joshua tree can be a component of other series as well. Hanes (1976) describes vegetation types in the San Gabriel Mountains including a Joshua tree community.

Phillips *et al.* (1980) present data on two stands in Joshua Tree National Monument which they call Joshua tree woodland. In both stands California juniper dominates over Joshua tree, and in one, singleleaf pinyon and desert scrub oak have more cover than Joshua tree. These stands are better considered as belonging to the California juniper series.

Three varieties of Joshua tree are proposed: the arboreal *Yucca brevifolia* var. *brevifolia* from a single trunk and *Y. b.* var. *jaegeriana* with several trunks, and shrubby *Y. b.* var. *herbertii*. NDDB information suggests that variety shrub is usually associated with stands better treated as belonging to the California juniper series. *The Jepson Manual* does not recognize these varieties.

Species mentioned in text	
Big sagebrush	*Artemisia tridentata*
Black bush	*Coleogyne ramosissima*
Bladderpod	*Isomeris arborea*
Box-thorns	*Lycium* species
Bush buckwheat	*Eriogonum fasciculatum*
Cheesebush	*Hymenoclea salsola*
Creosote bush	*Larrea tridentata*
Desert scrub oak	*Quercus turbinella*
Ephedras	*Ephedra* species
Joshua tree	*Yucca brevifolia*
Mojave yucca	*Yucca schidigera*
Rabbitbrushes	*Chrysothamnus* species
Singleleaf pinyon	*Pinus monophylla*
Spanish bayonet	*Yucca baccata*

Plot-based descriptions: Vasek & Barbour (1977) describe a stand north of Cima in San Bernardino Co. MojD.

Leather oak series

Leather oak sole, dominant or important shrub with chamise and toyon in canopy; musk bush, silktassels, scrub oak, whiteleaf manzanita, and/or yerba santa may be present. Emergent California juniper, foothill pine, knobcone pine, McNab cypress, or Sargent cypress may be present. Shrubs < 1.5 m; canopy intermittent. Ground sparse (Plate 8).

Uplands: all slopes. Soils shallow, serpentinite-derived, may be rocky.

Distribution: s.o-NorCo, i-NorCo, CenCo.
Elevation: 150-1500 m.

NDDB/Holland type and status: Chaparral (37000).
   Mixed serpentine chaparral (37610) G2 S2.1.
   Leather oak chaparral (37620) G3 S3.2.

Other types:
Barry type: G7411211.
Cheatham & Haller type: Serpentine chaparral.
Thorne type: Serpentine chaparral.
WHR type: Mixed chaparral.

General references: Hanes (1977, 1981), Keeley & Keeley (1988), Kruckeberg (1984).

Comments: Habitats of this series are famous for restricted species. Locally important shrubs as coyote ceanothus (a CNPS List 1B plant), Raiche manzanita (a CNPS List 1B plant), Tamalpais manzanita (a CNPS List 1B plant), and The Cedars manzanita (a CNPS List 1B plant) are found in this series (Skinner & Pavlik 1994). Several tree-dominated series share species and habitat with this series [*see* California juniper series, Foothill pine series, Knobcone pine series, McNab cypress series, Sargent cypress series]. There can be a few trees present in the stands of this series.

Species mentioned in text	
California juniper	*Juniperus californica*
Chamise	*Adenostoma fasciculatum*
Coyote ceanothus	*Ceanothus ferrisae*
Foothill pine	*Pinus sabiniana*
Knobcone pine	*Pinus attenuata*
Leather oak	*Quercus durata*
McNab cypress	*Cupressus macnabiana*
Musk bush	*Ceanothus jepsonii*
Raiche manzanita	*Arctostaphylos stanfordiana* ssp. *raichei*
Sargent cypress	*Cupressus sargentii*
Scrub oak	*Quercus berberidifolia*
Silktassels	*Garrya* species
Tamalpais manzanita	*Arctostaphylos hookeri* ssp. *montana*
The Cedars manzanita	*Arctostaphylos bakeri* ssp. *sublaevis*
Toyon	*Heteromeles arbutifolia*
Whiteleaf manzanita	*Arctostaphylos viscida*
Yerba santa	*Eriodictyon californicum*

Plot-based descriptions: Keeler-Wolf (1983) Keeler-Wolf (1990e) defines a Serpentinite chaparral at Frenzel Creek RNA in Colusa Co. i-NorCo.

Low sagebrush series

Low sagebrush sole or dominant shrub in canopy; bitterbrush, horsebrush, rubber rabbitbrush, and/or yellow rabbitbrush may be present. Emergent mountain juniper may be present. Shrub < 0.5 m; canopy continuous or open. Ground layer sparse or grassy (Plate 14).

Uplands: flats, depressions, slopes, ridges. Soils may be poorly drained.

Distribution: m-NorCo, su-KlaR, m-CasR, su-CasR, m-SN, su-SN, ModP, inter-West. **Elevation:** 1500-3800 m.

NDDB/Holland type and status: **Great Basin scrubs (35000), Alpine boulder and rock field (91000).**
Subalpine sagebrush scrub (35220) G3 S3.2.
White Mountain fell-field (91140 *in part*) G2 S2.2.

Other types:
Barry type: G7411211 CARAR80.
Cheatham & Haller type: Subalpine sagebrush.
PSW-45 type: Sagebrush series.
Rangeland type: SRM 406.
Thorne type: Mountain sagebrush scrub.
WHR type: Low sagebrush.

General references: Major & Taylor (1977), Paysen *et al.* (1980), Taylor (1976a), Tisdale (1994), West (1988), Young *et al.* (1977).

Comments: Two general conditions exist in this series. Nearly pure stands of low sagebrush occur at high elevations and in GB depressions with perched watertables. At alpine and subalpine elevations low sagebrush covers upper slopes and ridges. Ecologically related Black sagebrush series occurs in inter-West, MojD, and PenR.

Plot-based descriptions: Stillman (1980) defines

one association in the Marble Mountains that was resampled (Murray 1991) in Siskiyou Co. su-KlaR; Major & Taylor (1977) one association at Carson Pass area in Alpine Co. su-SN, a-SN, and associations in White Mountains in Inyo Co. a-WIS.

The KlaR studies are important in that the same area was studied twice [1980 & 1991] using the same methods; series composition was similar at both times, but subtypes were rather different in composition.

Species mentioned in text	
Bitterbrush	*Purshia tridentata*
Black sagebrush	*Artemisia nova*
Horsebrush	*Tetradymia canescens*
Idaho fescue	*Festuca idahoensis*
Low sagebrush	*Artemisia arbuscula*
Mono clover	*Trifolium monoense*
Mountain juniper	*Juniperus occidentalis* ssp. *australis*
Rubber rabbitbrush	*Chrysothamnus nauseosus*
Stemless haplopappus	*Stenotus acaulis*
Yellow rabbitbrush	*Chrysothamnus viscidiflorus*

Associations:
Major & Taylor (1977):
Low sagebrush/Mono clover association [as *Artemisia arbuscula* association],
Low sagebrush/stemless haplopappus association [containing *Haplopappus acaulis - Leptodactylon pungens*, *Haplopappus acaulis - Tetradymia canescens*, *Haplopappus acaulis - Geum canescens* associations].
Stillman (1980):
Low sagebrush/Idaho fescue association [as *Artemisia arbuscula* series, containing *Artemisia arbuscula/Castilleja schizotricha*, *Artemisia arbuscula/Castilleja applegatei*, *Eriogonum nudum/Monardella odoratissima* types, = Murray (1991) *Sitanion hystrix-Eriogonum douglasii*, *Castilleja applegatei-Lomatium dissectum* types].

Mexican elderberry series

Mexican elderberry sole or dominant shrub in canopy; California wild grape, narrowleaf willow, poison-oak, and/or Oregon ash may be present. Emergent Fremont cottonwood or valley oak may be present. Shrubs < 8 m tall; canopy continuous, intermittent, or open. Ground layer grassy.

Wetlands: soil intermittently flooded, seasonally saturated. Water chemistry: fresh. Floodplains. Cowardin class: Palustrine forested wetland. The national list of wetland plants (Reed 1988) lists Mexican elderberry as a FAC.

Distribution: i-CenCo, CenV, SoCo.
Elevation: sea level-300 m.

NDDB/Holland type and status: **Riparian scrubs** (63000).
Elderberry savanna (63430) G2 S2.1.

Other types:
Barry type: G7411221.
Cheatham & Haller type: Bottomland woodlands and forest.
Thorne type: Riparian woodland.
WHR type: Freshwater emergent wetland.

General references: Holland (1986).

Comments: Mexican elderberry is common in many series, often as an emergent tree over coastal scrub, chaparral, or as an understory shrub in forests. This series includes only those stands with large populations of elderberry.

NDDB has qualitative descriptions for four occurrences in Sacramento and San Joaquin Cos. CenV.

Species mentioned in text	
California wild grape	*Vitis californica*
Fremont cottonwood	*Populus fremontii*
Mexican elderberry	*Sambucus mexicana*
Narrowleaf willow	*Salix exigua*
Oregon ash	*Fraxinus latifolia*
Poison-oak	*Toxicodendron diversilobum*
Valley oak	*Quercus lobata*

Plot-based descriptions: not available.

Mixed sage series

Black sage, brittlebush, bush monkeyflower, California sagebrush, California encelia, California buckwheat, prickly-pears, purple sage, and/or white sage equally important shrubs in canopy. Emergent laurel sumac, lemonade berry, Mexican elderberry may be present. Shrubs < 2 m; canopy continuous or intermittent. Ground layer variable.

Uplands: slopes. Soil shallow.

Distribution: CenCo, SoCo, ChaI, Baja CA. **Elevation:** sea level-1200 m.

NDDB/Holland type and status: **Coastal bluff scrubs** (31000), **Coastal scrubs** (32000).
   Southern coastal bluff scrub (31200) G1 S1.1.
   Central Lucian coastal scrub (32200 *in part*) G3 S3.3.
   Venturan coastal sage scrub (32300 *in part*) G3 S3.1.
   Diablan sage scrub (32600 *in part*)) G3 S3.3.
   Riversidean upland sage scrub (32700 *in part*) G3 S3.1.
   Diegan coastal sage scrub (32500 *in part*) G3 S3.1.

Other Types:
Barry type: G7411221.
Cheatham & Haller type: Central coastal scrub, Coastal sage scrub.
Jones & Stokes type: Mixed scrub.
Thorne type: Southern coastal scrub.
WHR type: Coastal scrub.

General references: Axelrod (1978), DeSimone & Burk (1992), Keeley & Keeley (1988), Malanson (1984), Mooney (1977), O'Leary (1989), Pase & Brown (1982), Westman (1983).

Comments: No single species or pair of species can dominate stands of this series. Instead three or more species equally share commonness and cover. If there is a single dominant or two important species [*see* Black sage series, California buckwheat series, Cali-

fornia buckwheat-white sage series, California sagebrush series, California sagebrush-black sage series, California sagebrush-California buckwheat series, Coyote brush series, Purple sage series, or White sage series].

This series is often considered part of the coastal scrub, which is better thought of as a collection of series. This approach has advantages over the regional one (Axelrod 1978); stands of comparable composition can be considered regardless of geographic location. Mixed sage series is more narrowly defined than is coastal scrub. Coast bluff scrub is included here, in part. Definitions differ in plant height and cover from coastal scrub, but contrast little in species composition.

Species mentioned in text

Black sage	*Salvia mellifera*
Brittlebush	*Encelia farinosa*
Bush monkeyflower	*Mimulus aurantiacus*
California buckwheat	*Eriogonum fasciculatum*
California encelia	*Encelia californica*
California sagebrush	*Artemisia californica*
Coyote brush	*Baccharis pilularis*
Laurel sumac	*Malosma laurina*
Lemonade berry	*Rhus integrifolia*
Mexican elderberry	*Sambucus mexicana*
Prickly-pears	*Opuntia* species
Purple sage	*Salvia leucophylla*
White sage	*Salvia apiana*

Plot-based descriptions: DeSimone & Burk (1992) define a White sage-California sagebrush-California buckwheat association [= *Artemisia californica-Eriogonum fasciculatum-Salvia apiana* Kirkpatrick & Hutchinson (1977)]. The DeSimone & Burke description is based on cover, the Kirkpatrick & Hutchinson on frequency, which is less useful in definition.

Mixed saltbush series

No sole or dominant saltbush species in canopy; allscale, big saltbush, brittlescale, fourwing saltbush, and/or shadscale may be present. Shrubs < 3 m; canopy continuous, intermittent, or open. Ground layer sparse.

Uplands: bajadas; flats; lower slopes; playas; valleys. Soils may be carbonate-rich.

Distribution: i-CenCo, SoCo, m-TraR, m-PenR, GB, MojD, ColD, inter-West.
Elevation: 75 below sea level-2500 m.

NDDB/Holland type and status: **Chenopod scrubs** (36000).
    Desert saltbush scrub (36110 *in part*) G3 S3.2.
    Desert sink scrub (36120 *in part*) G3 S2.1.

Other types:
Barry type: G7411221.
Brown Lowe Pase type: 152.121, 152.172.
Cheatham & Haller type: Saltbush scrub.
PSW-45 type: Saltbush series.
Rangeland type: SRM 414.
Thorne type: Shadscale scrub.
WHR type: Alkali sink.

General references: Burk (1977), MacMahon (1988), McHargue (1973), Paysen *et al.* (1980), Thorne (1982), Tueller (1994).

Comments: This series is often considered part of the chenopod or saltbush scrub. These scrubs are better thought of as a collection of series. In the Mixed saltbush series no saltbush dominates [*see* Allscale series, Desert-holly series, Fourwing saltbush series, Shadscale series, Spinescale series].

Species mentioned in text	
Allscale	*Atriplex polycarpa*
Big saltbush	*Atriplex lentiformis*
Brittlescale	*Atriplex depressa*
Fourwing saltbush	*Atriplex canescens*
Shadscale	*Atriplex confertifolia*

Plot-based descriptions: Yoder *et al.* (1983) defines one association in Alabama Hills in Inyo Co. TraSN; McHargue (1973) one association in Coachella Valley in Riverside Co. ColD.

Associations:
McHargue (1973): Mixed saltbush association
    [as Level-moist sites].
Yoder *et al.* (1983):
    Allscale-shadscale association.

Mixed scrub oak series

Scrub oak and bigberry manzanita, ceanothus, or toyon important shrubs in canopy; chamise, deerbrush, hollyleaf redberry, hairyleaf ceanothus, interior live oak, wedgeleaf ceanothus, and/or woollyleaf ceanothus may be present. Shrubs < 4 m; canopy continuous. Ground layer sparse.

Uplands: all slopes. Soils deep or shallow, may be rocky.

Distribution: So-CA.
Elevation: 300-1500 m.

NDDB/Holland type and status: Chaparral (37000).
   Granitic southern mixed chaparral (37121) G3 S3.1
   Mafic southern mixed chaparral (37122) G3 S3.1
   Scrub oak chaparral (37900 *in part*) G3 S3.
   Southern north slope chaparral (37E20 *in part*) G3 S3.1.
Other types:
Barry type: G7411214.
Cheatham & Haller type: Mixed chaparral.
Thorne type: Mixed chaparral.
WHR type: Mixed chaparral.

General references: Hanes (1977, 1981), (Horton 1960), Keeley & Keeley (1988), Pase (1982a).

Comments: The term scrub oak has been used for several shrubby live oaks. *The Jepson Manual* applies the name *Quercus dumosa*, Nuttall scrub oak, (a CNPS List 1B plant) to a rare SoCo species (Skinner & Pavlik 1994). References in the ecological literature to *Q. dumosa* are more appropriately *Q. berberidifolia* which ranges through much of Cis-CA. Interior live oak, desert scrub oak, and the rarer Muller oak, Tucker oak, and Palmer oak are also called scrub oaks. Careful identification of the oaks in a stand will be necessary to correctly assign them to a series. [*see* California juniper series, Interior live oak shrub series]. Gordon & White (1994) define four series in So-CA in which scrub oak shares importance with one other shrub.

In these types scrub oak cover is 60-30% and another species cover 30-60% [*see* Scrub oak-birchleaf mountain-mahogany series, Scrub oak - chamise series, Scrub oak-chaparral whitethorn series]. If stand composition does not meet these definitions, then place it here.

Species mentioned in text	
Birchleaf mountain-mahogany	*Cercocarpus betuloides*
Bigberry manzanita	*Arctostaphylos glauca*
Ceanothus	*Ceanothus* species
Chamise	*Adenostoma fasciculatum*
Chaparral whitethorn	*Ceanothus leucodermis*
Cleveland monkeyflower	*Mimulus clevelandii*
Deerbrush	*Ceanothus integerrimus*
Desert ceanothus	*Ceanothus greggii*
Desert scrub oak	*Quercus turbinella*
Dunn mariposa lily	*Calochortus dunnii*
Hairyleaf ceanothus	*Ceanothus oliganthus*
Hollyleaf redberry	*Rhamnus ilicifolia*
Interior live oak	*Quercus wislizenii*
Muller oak	*Quercus cornelius-mulleri*
Palmer oak	*Quercus palmeri*
Scrub oak	*Quercus berberidifolia*
Toyon	*Heteromeles arbutifolia*
Tucker oak	*Quercus john-tuckeri*
Wedgeleaf ceanothus	*Ceanothus cuneatus*
White coast ceanothus	*Ceanothus verrucosus*
Woollyleaf ceanothus	*Ceanothus tomentosus*

Stands with localized species as Cleveland monkeyflower (a CNPS List 4 plant), Dunn mariposa lily (a CNPS List 1B plant), or white coast ceanothus (a CNPS List 2 plant) can occur in this series (Skinner & Pavlik 1994). Keeler-Wolf (1990e) qualitatively describes mixed scrub oak stands at Cahuilla Mountain RNA in Riverside Co. m-PenR.

Plot-based descriptions: Gordon & White (1994) define six associations in So-CA.
Associations:
Gordon & White (1994):
   Scrub oak-bigberry manzanita association,
   Scrub oak-deerbrush association,
   Scrub oak-hairyleaf ceanothus association,
   Scrub oak-toyon association,
   Scrub oak-wedgeleaf ceanothus association,
   Scrub oak-woollyleaf ceanothus association.

Mojave yucca series

Mojave yucca emergent tree over a shrub canopy; big galleta, brittlebush, cholla, creosote bush, jojoba, prickly-pears, and/or white bursage may be present. Trees < 5 m. Ground layer open; annuals seasonally present (Plate 16).

Uplands: rocky slopes. Soils well-drained.

Distribution: MojD, ColD.
Elevation: 50-2500m.

NDDB/Holland type and status: **Mojavean desert scrub** (34000).
> Mojave mixed steppe (34220) G3 S2.2,
> Mojave yucca scrub and steppe (34230) G3 S3.2.

Other types:
Cheatham & Haller type: Low desert scrub.
Thorne type: Semi-succulent scrub.
WHR habitat type: Desert succulent scrub.

General references: Burk (1977), MacMahon (1988), Thorne (1982).

Comments: This series is often considered part of the Creosote bush scrub, which is better thought of a as collection of series. Areas of dominance by Mojave yucca can also occur through its range as small populations in other series.

Species mentioned in text	
Barrel cactus	*Ferocactus cylindraceus*
Big galleta	*Pleuraphis rigida*
Black bush	*Coleogyne ramosissima*
Brittlebush	*Encelia farinosa*
Cholla	*Opuntia* species
Clustered barrel cactus	*Echinocactus polycephalus*
Desert agave	*Agave deserti*
Hedgehog cactus	*Echinocereus engelmannii*
Jojoba	*Simmondsia chinensis*
Mojave yucca	*Yucca schidigera*
Prickly-pears	*Opuntia* species
Pygmy-cedar	*Peucephyllum schottii*
White bursage	*Ambrosia dumosa*

Plot-based descriptions: Minnich *et al.* (1993) define a Creosote bush with black bush/Mojave yucca association at Marine Corps Air-ground Combat Center in San Bernardino Co. MojD.; Spolsky (1979) a Yucca-big galleta association in Anza Borrego State Park in San Diego Co. ColD.

Mountain alder series

Mountain alder sole or dominant shrub in canopy; red osier, Jepson willow, Lemmon willow, Mackenzie willow, mountain maple, and/or Scouler willow may be present. Emergent trees may be present. Shrubs < 10 m; canopy continuous. Ground layer variable (Plate 7).

Wetlands: habitats seasonally flooded, saturated. Water chemistry: fresh. Floodplains; low-gradient depositions along streams. Cowardin class: Palustrine shrub/scrub wetland. The national list of wetland plants (Reed 1988) lists mountain alder as a NI.

Distribution: m-NorCo, m-KlaR, su-KlaR, m-CasR, m-SN, su-SN, inter-West, Canada, AK. **Elevation:** 1200-2400 m.

NDDB/Holland type and status: **Riparian scrubs** (63000).
  Montane riparian scrub (63500 *in part*) G4 S4.

Other types:
Barry type: G7412121.
Cheatham & Haller type: Willow thickets.
Thorne type: Mountain meadow.
WHR type: Montane riparian.

General references: Brayshaw (1976).

Comments: In *The Jepson Manual*, mountain alder is treated as a subspecies (*Alnus incana* ssp. *tenuifolia*) of a widespread North America species. The elevational and regional ranges of the two montane alders, mountain alder and Sitka alder (*Alnus viridis*), differ in California. They rarely occur in the same stand [*see* Sitka alder series]. Stands of the Montane wetland shrub habitat and Subalpine wetland habitat occur in similar conditions. Mountain alder tolerates cooler and shadier riparian situations than most willows.

Keeler-Wolf (1990e) presents qualitative descriptions at Soda Ridge candidate RNA, Feidler *et al.* (1986) at Green Island Lake candidate RNA in Butte Co., Keeler-Wolf (1990e) at Mud Lake RNA in Plumas Co. m-CasR; at Onion Creek in Placer Co. m-SN.

Species mentioned in text	
Jepson willow	*Salix jepsonii*
Lemmon willow	*Salix lemmonii*
Mackenzie willow	*Salix prolixa*
Mountain alder	*Alnus incana*
Mountain maple	*Acer glabrum*
Red osier	*Cornus sericea*
Scouler willow	*Salix scouleriana*

Plot-based descriptions: not available.

Mountain heather-bilberry series

Mountain heathers (Brewer heather or Cascade heather) and bilberry dominant or important shrubs in canopy; alpine-laurel, blackish sedge, Brewer cinquefoil, Drummond cinquefoil, Jeffrey shooting star, Merten rush, shorthair reedgrass, sibbaldia, and/or Sierra primrose may be present. Shrubs < 1 m; canopy continuous, intermittent, or open. Ground layer variable.

Wetlands: habitats seasonally flooded, saturated. Water chemistry: fresh. Floodplains; low gradient depositions along streams. Cowardin class: Palustrine shrubscrub wetland.

Distribution: su-KlaR, su-CasR, su-SN, su-PenR.
Elevation 1800-3600 m.

NDDB/Holland type and status: Meadows and seeps (45000), **Riparian scrubs** (63000).
    Wet montane meadow (45110) G3 S3.2.
    Wet alpine and subalpine meadow (45210) G3 S3.2.
    Montane riparian scrub (63500 *in part*) G4 S4.
Other types:
Barry type: G7411211 CPHBR40, G7411211, CPHEM00.
Cheatham & Haller type: Alpine meadow.
Thorne type: Mountain meadow.
WHR type: Montane riparian.
General references: Major & Taylor (1977), Thorne (1982).

Comments: Cascade heather grows in su-KlaR; Brewer heather in su-CasR, su-SN, su-PenR. This series is closely related ecologically to the Subalpine wetland habitat and Shorthair reedgrass series. Stands of this series mix with Shorthair sedge series and Sedge series at fine scale, and with forest and woodland series at a coarse scale.

Species mentioned in text	
Alpine-laurel	*Kalmia polifolia*
Bilberry	*Vaccinium caespitosum*
Blackish sedge	*Carex nigricans*
Brewer cinquefoil	*Potentilla breweri*
Brewer heather	*Phyllodoce breweri*
Cascade heather	*Phyllodoce empetriformis*
Drummond cinquefoil	*Potentilla drummondii*
Jeffrey shooting star	*Dodecatheon jeffreyi*
Merten rush	*Juncus mertensianus*
Mountain heathers	*Phyllodoce* species
Shorthair reedgrass	*Calamagrostis breweri*
Shorthair sedge	*Carex filifolia*
Sibbaldia	*Sibbaldia procumbens*
Sierra primrose	*Primula suffrutescens*

Plot-based descriptions: Burke (1982) defines three associations at Rae Lakes area, King Canyon National Park in Fresno Co.; Benedict (1983) one association at Rock Creek in Sequoia National Park in Tulare Co; Major & Taylor (1977) three associations at Carson Pass area in Alpine Co.; Taylor (1984) one association at Harvey Monroe Hall RNA in Mono Co. su-SN.

Associations:
Burke (1982):
    Sierra primrose association
    [= *Primula suffrutescens-Silene sargentii* association (Major & Taylor 1977d)].
Major & Taylor (1977):
    Bilberry-blackish sedge association;
    Drummond cinquefoil-Brewer cinquefoil association [containing *Carex helleri-Juncus parryi* association];
    Shorthair reedgrass-bilberry association
    [= Benedict (1983), = Sibbaldia-Merten rush association (Burke 1982)].
Palmer (1979):
    Jeffrey shooting star-Merten rush association.
Taylor (1984):
    Shorthair reedgrass - alpine-laurel association.

Mountain whitethorn series

Mountain whitethorn sole or dominant shrub in canopy; bitter cherry, bush chinquapin, creeping snowberry, greenleaf manzanita, huckleberry oak, ocean spray, tobacco brush, and/or wedgeleaf ceanothus may be present. Emergent conifers may be present. Shrubs < 2 m; canopy continuous or intermittent. Ground layer sparse (Plate 9).

Uplands: ridges, upper slopes.

Distribution: m-NorCo, m-KlaR, m-CasR, m-SN, su-SN, i. m-TraR, i. su-TraR, m-PenR, su-PenR.
Elevation: 900-2900 m.

NDDB/Holland type and status: **Chaparral** (37000).
    Mixed montane chaparral (37510 *in part*) G4 S4.
    Montane ceanothus chaparrals (37530 *in part*) G3 S3.
    Whitethorn chaparral (37532) G4 S4.
    Upper Sonoran manzanita chaparral (37B00 *in part*) G4 S4.

Other types:
Barry type: G7411214.
Cheatham & Haller type: Upper montane chaparral.
PSW-45 type: Ceanothus series.
Rangeland type: SRM 209.
Thorne type: Montane chaparral.
WHR type: Montane chaparral.

General references: Paysen *et al.* (1980), Riser & Fry (1994).

Comments: Mountain whitethorn occurs as scattered individual shrubs in forest understories or as the sole or dominant shrub in what is often called montane chaparral. Self-perpetuating stands are unusual; instead, most stands are established after fire or logging and are transitional to various kinds of forest.

Species mentioned in text	
Bitter cherry	*Prunus emarginata*
Bush chinquapin	*Chrysolepis sempervirens*
Creeping snowberry	*Symphoricarpos mollis*
Greenleaf manzanita	*Arctostaphylos patula*
Huckleberry oak	*Quercus vaccinifolia*
Mountain whitethorn	*Ceanothus cordulatus*
Ocean spray	*Holodiscus discolor*
Tobacco brush	*Ceanothus velutinus*
Wedgeleaf ceanothus	*Ceanothus cuneatus*

Plot-based descriptions: none available.

Mulefat series

Mulefat sole or dominant shrub in canopy; arroyo willow, and/or narrowleaf willow may be present. Shrubs < 4 m; canopy continuous. Ground layer sparse.

Wetlands: habitats seasonally flooded, saturated. Water chemistry: fresh. Canyon bottoms; irrigation ditches, stream channels. Cowardin class: Palustrine shrub-scrub wetland. The national list of wetland plants (Reed 1988) lists mulefat as a FACU*.

Distribution: i-NorCo, CenCo, CenV, f-KlaR, f-CasR, s-SN, SoCo, m-TraR, m-PenR, MojD, ColD.
Elevation: sea level-1250 m.

NDDB/Holland type and status: Riparian scrubs (63000).
   Mulefat scrub (63310) G4 S4.

Other types:
Barry type: G7411211.
Cheatham & Haller type: Bottomland woodlands and forest.
Thorne type: Riparian woodland.
WHR type: Freshwater emergent wetland.

General references: Shanfield (1984).

Comments: Mulefat often occurs in pure stands. Secondary species vary regionally. Stands of this series may mix at a fine scale with other wetland series [*see* California sycamore series, Coast live oak series, Fremont cottonwood series].

Species mentioned in text	
Arroyo willow	*Salix lasiolepis*
Mulefat	*Baccharis salicifolia*
Narrowleaf willow	*Salix exigua*

Plot-based descriptions: not available.

Narrowleaf willow series

Narrowleaf willow sole or dominant shrub in canopy; Fremont cottonwood, white alder, and/or willows may be present. Emergent trees may be present. Shrubs < 7 m; canopy continuous. Ground layer variable (Plate 8).

Wetlands: habitats seasonally flooded, saturated. Water chemistry: fresh. Floodplains; depositions along rivers, streams. Cowardin class: Palustrine shrub-scrub wetland. The national list of wetland plants (Reed 1988) lists narrowleaf willow as a OBL.

Distribution: Cis-CA, Trans-CA, North America.
Elevation: sea level-2700 m.

NDDB/Holland type and status: **Riparian forests** (61000), **Riparian scrubs** (63000).
    Southern cottonwood-willow riparian forest (61330 *in part*) G3 S3.
    Great Valley cottonwood riparian forest (61410 *in part*) G2 S2.1.
    Modoc-Great Basin cottonwood-willow riparian forest (61610 *in part*) G3 S2.1.
    Mojave riparian forest (61700 *in part*) G1 S1.1.
    Sonoran cottonwood-willow riparian forest (61810 *in part*) G2 S1.1.
    North Coast riparian scrub (63100 *in part*) G3 S3.2.
    Central Coast riparian scrub (63200 *in part*) G3 S3.2.
    Southern willow scrub (63320 *in part*) G3 S2.1.
    Great Valley willow scrub (63410 *in part*) G3 S3.2.

Other types:
Barry type: G7411221.
Cheatham & Haller type: Bottomland woodlands and forest.
PSW-45 type: Willow series
Thorne type: Riparian woodland.
WHR type: Desert riparian, Freshwater emergent wetland.

General references: Bowler (1989), Brayshaw (1976), Holstein (1984), Paysen *et al.* (1980), Shanfield (1984).

Comments: Many willow stands are composed of a single species and easily assigned to a series using the dominance rule. If no dominant willow is present at low elevations, then place the stands in the Mixed willow series. Montane and subalpine willow stands are placed in separate classes since other willow species are restricted to those elevations [*see* Montane wetland shrub habitat, Subalpine wetland shrub habitat].

Stands of this series have similar environmental conditions as alder, cottonwood, and other willow series [*see* Black cottonwood series, Black willow series, Fremont cottonwood series, Hooker willow series, Pacific willow series, Red alder series, Red willow series, Sitka willow series, White alder series].

This treatment follows *The Jepson Manual* by including *Salix exigua, S. hindsiana,* and hybrids with *S. sessilifolia* in *S. exigua*. Narrowleaf willow thickets occur along margins of streams and rivers that are continually disturbed by point-bar deposition.

Species mentioned in text	
Black cottonwood	*Populus balsamifera*
Black willow	*Salix gooddingii*
Fremont cottonwood	*Populus fremontii*
Hooker willow	*Salix hookeriana*
Narrowleaf willow	*Salix exigua*
Pacific willow	*Salix lucida* spp. *lasiandra*
Red alder	*Alnus rubra*
Red willow	*Salix laevigata*
Sitka willow	*Salix sitchensis*
White alder	*Alnus rhombifolia*
Willows	*Salix* species

Plot-based descriptions: not available.

Nolina series

Bigelow or Parry nolina emergent over a shrub canopy; desert agave, barrel cactus, black bush, brittlebush, bush buckwheat, catclaw acacia, desert-lavender, jojoba, narrowleaf goldenbush, and/or white bursage may be present. Emergent California juniper or ocotillo may present. Shrubs < 5 m; canopy open. Ground layer sparse (Plate 16).

Uplands: steep slopes, valleys. Soils rocky, thin, calcareous or granitic-derived.

Distribution: s. m-SN, m-TraR, m-PenR, MojD, ColD, AZ, NV.
Elevation: 250-2250 m.

NDDB/Holland type and status: Mojavean desert scrubs (34000).
    Nolina scrub (34400) G3 S1.3.

Other types:
Barry type: G7411111.
Cheatham & Haller type: Enriched desert scrub.
Stone & Sumida (1983): Nolina woodland.
Thorne type: Semi-succulent scrub.
WHR type: Desert succulent scrub.

General references: Stone and Sumida (1983).

Comments: Bigelow or Parry nolina form local scattered stands which are especially conspicuous when in flower. Plot data are available for Trans-CA stands even though the series goes beyond the deserts. Individual nolina may occur in other series.

Species mentioned in text	
Barrel cactus	*Ferocactus cylindraceus*
Bigelow nolina	*Nolina bigelovii*
Black bush	*Coleogyne ramosissima*
Brittlebush	*Encelia farinosa*
Bush buckwheat	*Eriogonum fasciculatum*
California juniper	*Juniperus californica*
Catclaw acacia	*Acacia greggii*
Desert agave	*Agave deserti*
Desert-lavender	*Hyptis emoryi*
Jojoba	*Simmondsia chinensis*
Narrowleaf goldenbush	*Ericameria linearifolia*
Ocotillo	*Fouquieria splendens*
Parry nolina	*Nolina parryi*
White bursage	*Ambrosia dumosa*

Plot-based descriptions: NDDB has plot data for Parry nolina stands in the Kingston Mountains in San Bernardino Co. MojD; Spolsky (1979) defines Bigelow nolina stands as Sonoran nolina scrub association in Anza Borrego Desert State Park in San Diego Co. ColD.

Ocotillo series

Ocotillo emergent and abundant over a shrub canopy; barrel cactus, box-thorns, brittlebush, chollas, creosote bush, desert agave, and/or white bursage may be present. Shrubs < 1 m; intermittent or open. Emergent ocotillo < 3 m. Ground layer open; annuals seasonally present (Plate 16).

Uplands: alluvial fans; bajadas; rocky slopes. Soils well-drained.

Distribution: ColD.
Elevation: sea level-800 m.

NDDB/Holland type and status: Sonoran desert scrubs (33000).
> Sonoran mixed woody scrub (33210 *in part*) G3 S3.
> Sonoran mixed woody and succulent scrub (33220 in part) G4 S3.2.

Other types:
Barry type: G74 G7411221 BFOSP20.
Brown Lowe Pase type: 154.116.
Cheatham & Haller type: Enriched desert scrub.
PSW-45 type: Ocotillo series.
Thorne type: Semi-succulent scrub.
WHR type: Desert succulent scrub.

General references: Burk (1977), MacMahon (1988), Paysen *et al.* (1980), Thorne (1982), Turner & Brown (1982).

Comments: Populations of ocotillo are often encountered in ColD. There they have been generally treated as a part of creosote bush scrub, which is better considered a collection of series.

What distinguishes this series is presence of visually conspicuous populations of ocotillo. It can also occur in the Creosote bush-cactus series, but there ocotillo is not abundant. Scattered ocotillo can occur in other ColD series as well [*see* Blue palo verde-ironwood-smoke tree series, Brittlebush series, Creosote bush series, White bursage series].

Species mentioned in text	
Barrel cactus	*Ferocactus cylindraceus*
Blue palo verde	*Cercidium floridum*
Box-thorns	*Lycium* species
Brittlebush	*Encelia farinosa*
Chollas	*Opuntia* species
Creosote bush	*Larrea tridentata*
Desert agave	*Agave deserti*
Ironwood	*Olneya tesota*
Ocotillo	*Fouquieria splendens*
Smoke tree	*Psorothamnus spinosa*
White bursage	*Ambrosia dumosa*

Plot-based descriptions: Spolsky (1979) defines a Ocotillo fan scrub association in Anza Borrego State Park in San Diego Co. ColD.

Parry rabbitbrush series

Parry rabbitbrush sole or dominant shrub in canopy; bitterbrush, big sagebrush, ephedras, and/or yellow rabbitbrush may be present. Emergent junipers or pines may be present; or emergent shrubs may occur over a ground layer of grass. Trees scattered, if present. Shrubs < 1 m; canopy continuous, intermittent, or open. Ground layer sparse or grassy (Plate 14).

Uplands: bajadas, pediments, valleys. Soils well-drained, gravelly.

Distribution: m-NorCo, su-KlaR, su-CasR, m-Sn, su-SN, TraR, GB, DesR, OR.
Elevation: 700-3700 m.

NDDB/Holland type and status: Great Basin scrubs (35000).
    Mono pumice flat (35410) G1 S1.1.

Other types:
Barry type: G7411221.
Cheatham & Haller type: Great Basin sagebrush.
PSW-45 type: Rabbitbrush series.
Thorne type: Great Basin sagebrush scrub.
WHR type: Low sagebrush.

General references: Paysen *et al.* (1980), West (1988), Young *et al.* (1977).

Comments: Stands dominated by any of the 6 ssp. of Parry rabbitbrush growing in CA are included in this series. Some subspecies are local, others are more wide ranging. The species sometimes grows in stands dominated by trees; such stands are referred to a tree-dominated series.

Species mentioned in text	
Big sagebrush	*Artemisia tridentata*
Bitterbrush	*Purshia tridentata*
Ephedras	*Ephedra* species
Gayophytum	*Gayophytum diffusum*
Junipers	*Juniperus* species
Parry rabbitbrush	*Chrysothamnus parryi*
Pines	*Pinus* species
Yellow rabbitbrush	*Chrysothamnus viscidiflorus*

Plot-based descriptions: Taylor (1980c) Keeler-Wolf (1990e) define a Parry rabbitbrush-gayophytum association at Indiana Summit RNA in Mono Co. TraSN.

Purple sage series

Purple sage sole, dominant, or important shrub with California sagebrush in canopy; black sage, bush monkeyflower, California buckwheat, coast goldenbush, and/or white sage may be present. Emergent coast live oak, California walnut, laurel sumac, lemonade berry, or Mexican elderberry may be present. Shrubs < 1.5 m; canopy continuous or intermittent. Ground layer variable, may be grassy (Plate 11).

Uplands: slopes north-facing, steep. Soils colluvial-derived, may be rocky.

Distribution: o-CenCo, o-SoCo, m-TraR.
Elevation: 50-800 m.

NDDB/Holland type and status: Coastal bluff scrubs (31000), **Coastal scrubs** (32000).
    Southern coastal bluff scrub (31200) G1 S1.1.
    Venturan coastal sage scrub (32300 *in part*) G3 S3.1.
    Diegan coastal sage scrub (32500 *in part*) G 3 S3.1.
    Riversidean sage scrub (32700) G3 S3.1.

Other types:
Barry type: G7411221 CSALE30.
Cheatham & Haller type: Coastal bluff scrub, Coastal sage scrub.
Jones & Stokes type: Coastal bluff scrub, Purple sage scrub.
PSW-45 type: Salvia series.
Thorne type: Southern coastal scrub.
Westman type: Venturan II.
WHR type: Coastal scrub.

General references: Axelrod (1978), Keeley & Keeley (1988), Kirkpatrick & Hutchinson (1977), Malanson (1984), Mooney (1977), O'Leary (1989), Pase & Brown (1982), Paysen *et al.* (1980), Westman (1981a, 1981b, 1981c, 1982).

Comments: This series is often considered part of the coastal scrub, which is better thought of as a collection of series. This approach allows stands of similar composition can be considered regardless of geographic location. Stands of this series make mosaics with Coast live oak and California walnut series. Coast bluff scrub is included here. Descriptions differ in plant height and cover from coastal scrub, but contrast little in species composition.

Species mentioned in text	
Black sage	*Salvia mellifera*
Bush monkeyflower	*Mimulus aurantiacus*
California buckwheat	*Eriogonum fasciculatum*
California sagebrush	*Artemisia californica*
California walnut	*Juglans californica* var. *californica*
Coast goldenbush	*Isocoma menziesii*
Coast live oak	*Quercus agrifolia*
Laurel sumac	*Malosma laurina*
Lemonade berry	*Rhus integrifolia*
Mexican elderberry	*Sambucus mexicana*
Purple sage	*Salvia leucophylla*
White sage	*Salvia apiana*

Plot-based descriptions: Kirkpatrick & Hutchinson (1977) define two associations in s. CenCo, SoCo.

Associations:
Kirkpatrick & Hutchinson (1977):
    Purple sage-California sagebrush association [= Venturan II (Malanson 1984)],
    Purple sage-laurel sumac association [= White *et al.* (1993)].

Red shank series

Red shank sole or dominant shrub in canopy; bigberry manzanita, birchleaf mountain-mahogany, chamise, cupleaf ceanothus, hollyleaf redberry, Muller oak, and/or Veatch silktassel may be present. Shrubs < 5 m; canopy continuous. Ground layer sparse (Plate 9).

Uplands: all slopes. Soils deep, alluvial or bedrock-derived.

Distribution: s. o-CenCo, o. l-TraR, m-PenR. **Elevation:** 300 2000 m.

NDDB/Holland type and status: Chaparral (37000).
Red shank chaparral (37300) G3 S3.2.

Other types:
Barry type: G7411212 BADSP00.
Cheatham & Haller type: Red shank chaparral.
PSW-45 type: Redshank series.
Thorne type: Red-shanks chaparral.
WHR type: Chamise-redshank chaparral.

General references: Hanes (1965, 1977, 1981), Keeley & Keeley (1988), Pase (1982a), Paysen *et al.* (1980).

Comments: This series definition follows Gordon & White (1994). If red shank has > 60% cover, the stand is a member of this series. If red shank cover is 60-30% and another species cover 30-60%, then the stand is a member of a mixed series. Stands where red shank is the dominant are less often encountered than those in which it is important with birchleaf mountain-mahogany or chamise [*see* Red shank - birchleaf mountain-mahogany series, Red shank-chamise series].

Species mentioned in text	
Bigberry manzanita	*Arctostaphylos glauca*
Birchleaf mountain-mahogany	*Cercocarpus betuloides*
Chamise	*Adenostoma fasciculatum*
Cupleaf ceanothus	*Ceanothus greggii*
Hollyleaf redberry	*Rhamnus ilicifolia*
Muller oak	*Quercus cornelius-mulleri*
Red shank	*Adenostoma sparsifolium*
Veatch silktassel	*Garrya veatchii*

Plot-based descriptions: Gordon & White (1994) define a Red shank association in m-PenR.

Red shank - birchleaf mountain-mahogany series

Red shank and birchleaf mountain-mahogany important shrubs in canopy; cupleaf ceanothus, chaparral yucca, Eastwood manzanita, interior live oak, and/or scrub oak may be present. Shrubs < 5 m; canopy continuous. Ground layer sparse.

Uplands: all slopes. Soils alluvial or bedrock-derived, may be rocky.

Distribution: s. o-CenCo, o. l-TraR, m-PenR, Baja CA.
Elevation: 1300-1800 m.

NDDB/Holland type and status: **Chaparral** (37000).
　　Red shank chaparral (37300) G3 S3.2.

Other types:
Barry type: G7411212.
Cheatham & Haller type: Red shank chaparral.
Thorne type: Red-shanks chaparral.
WHR type: Chamise-redshank chaparral.

General references: Hanes (1965, 1977, 1981), Keeley & Keeley (1988), Pase (1982a).

Comments: This series definition follows Gordon & White (1994). If red shank has > 60% cover, the stand is a member of the Red shank series. If red shank cover is 60-30% and birchleaf mountain-mahogany cover 30-60%, then the stand is a member of this series. Stands of the Red shank-birchleaf mountain-mahogany series have little or no chamise. If chamise is important with red shank, the stands belong to the Red shank-chamise series.

Species mentioned in text	
Birchleaf mountain-mahogany	*Cercocarpus betuloides*
Cupleaf ceanothus	*Ceanothus greggii*
Chamise	*Adenostoma fasciculatum*
Chaparral yucca	*Yucca whipplei*
Eastwood manzanita	*Arctostaphylos glandulosa*
Interior live oak	*Quercus wislizenii*
Red shank	*Adenostoma sparsifolium*
Scrub oak	*Quercus berberidifolia*

Plot-based descriptions: Gordon & White (1994) define a Red shank - birchleaf mountain-mahogany association in m-PenR.

Red shank-chamise series

Red shank and chamise important shrubs in canopy; bigberry manzanita, birchleaf mountain-mahogany, California buckwheat, chaparral yucca, cupleaf ceanothus, Eastwood manzanita, hollyleaf redberry, pointleaf manzanita, and/or scrub oak may be present. Shrubs < 5 m; canopy continuous. Ground layer sparse (Plate 9).

Uplands: all slopes. Soils alluvial or bedrock-derived, may be rocky.

Distribution: s. o-CenCo, o. l-TraR, m-PenR.
Elevation: 600-2000 m.

NDDB/Holland type and status: Chaparral (37000).
Red shank chaparral (37300) G3 S3.3.

Others types:
Barry type: G7411212.
Cheatham & Haller type: Red shank chaparral.
Thorne type: Red-shanks chaparral.
WHR type: Chamise-redshank chaparral.

General references: Hanes (1965, 1977, 1981), Keeley & Keeley (1988), Pase (1982a).

Comments: This series definition follows Gordon & White (1994). If red shank has > 60% cover, the stand is a member of the Red shank series. If red shank cover is 60-30% and chamise cover 30-60%, then the stand is a member of this series. Stands where red shank is the dominant are less often encountered than those in which it is important with birchleaf mountain-mahogany or chamise [*see* Red shank - birchleaf mountain-mahogany series].

Plot-based descriptions: Gordon & White (1994) define three associations in m-PenR.

Species mentioned in text	
Bigberry manzanita	*Arctostaphylos glauca*
Birchleaf mountain-mahogany	*Cercocarpus betuloides*
California buckwheat	*Eriogonum fasciculatum*
Chamise	*Adenostoma fasciculatum*
Chaparral yucca	*Yucca whipplei*
Cupleaf ceanothus	*Ceanothus greggii*
Eastwood manzanita	*Arctostaphylos glandulosa*
Hollyleaf redberry	*Rhamnus ilicifolia*
Pointleaf manzanita	*Arctostaphylos pungens*
Red shank	*Adenostoma sparsifolium*
Scrub oak	*Quercus berberidifolia*

Associations:
Gordon & White (1994):
Red shank-chamise-(birchleaf mountain-mahogany) association,
Red shank-chamise-(cupleaf ceanothus) association,
Red shank-chamise-(pointleaf manzanita) association.

Rothrock sagebrush series

Rothrock sagebrush sole or dominant shrub in canopy; alpine goldenbush, heretic penstemon, pennyroyal, and/or small haplopappus may be present. Emergent foxtail pine, lodgepole pine, or whitebark pine may be present. Shrub < 1 m; canopy intermittent. Ground layer sparse (Plate 14).

Uplands: slopes, ridges. Soils well-drained, gravelly.

Distribution: su-SN, a-SN, TraR, PenR.
Elevation: 2000-3100 m.

NDDB/Holland type and status: **Great Basin scrubs** (35000).
    Subalpine sagebrush scrub (35220 *in part*) G3 S3.2.

Other types:
Barry type: G7411211.
Cheatham & Haller type: Subalpine sagebrush.
PSW-45 type: Sagebrush series.
Thorne type: Mountain sagebrush scrub.
WHR type: Low sagebrush.

General references: Major & Taylor (1977), Paysen *et al.* (1980), West (1988).

Comments: Ratliff (1985) proposed a Rothrock sagebrush vegetative series in SN. This series is more restricted than other sagebrush series. It is generally present between meadow and forest environments. On the Kern Plateau, the series is common, especially at Big Meadows, Monache Meadows, Ramshaw Meadows, and Templeton Meadows.

Plot-based descriptions: Benedict (1983) defines one association in Rock Creek area in Sequoia National Park in Tulare Co. su-SN; Taylor (1984) Keeler-Wolf (1990e) one association in Harvey Monroe Hall RNA in Mono Co. su-SN.

Species mentioned in text	
Alpine goldenbush	*Ericameria discoidea*
Foxtail pine	*Pinus balfouriana*
Heretic penstemon	*Penstemon heterodoxus*
Lodgepole pine	*Pinus contorta* ssp. *murrayana*
Pennyroyal	*Monardella odoratissima*
Rothrock sagebrush	*Artemisia rothrockii*
Small haplopappus	*Ericameria suffruticosa*
Whitebark pine	*Pinus albicaulis*

Associations:
Benedict (1983):
    Rothrock sagebrush/heretic penstemon association
    [as Rothrock sagebrush association].
Taylor (1984):
    Rothrock sagebrush/pennyroyal association.

Rubber rabbitbrush series

Rubber rabbitbrush sole or dominant shrub in canopy; big sagebrush, bitterbrush, ephedras, and/or yellow rabbitbrush may be present. Emergent junipers or pines may be present; or emergent shrubs may occur over a ground layer of grass. Trees scattered, if present. Shrubs < 3 m; canopy continuous, intermittent, or open. Ground layer sparse or grassy (Plate 14).

Uplands: bajadas, pediments, valleys. Soils well-drained, gravelly.

Distribution: i-NorCo, i-CenCo, m-CenCo, m-CasR, su-CasR, m-SN, su-SN, TraR, PenR, GB, MojD, inter-West.
Elevation: 50-3300 m.

Species mentioned in text	
Alkali sacaton	*Sporobolus airoides*
Big sagebrush	*Artemisia tridentata*
Bitterbrush	*Purshia tridentata*
Ephedras	*Ephedra* species
Junipers	*Juniperus* species
Pines	*Pinus* species
Rubber rabbitbrush	*Chrysothamnus nauseosus*
Yellow rabbitbrush	*Chrysothamnus viscidiflorus*

Plot-based descriptions: Ferren & Davis (1991) defines a Rubber rabbitbrush-alkali sacaton association at Fish Slough in Inyo Co. TraSN.

NDDB/Holland type and status: **Great Basin scrubs** (35000).
    Rabbitbrush scrub (35400) G5 S5.

Other types:
Barry type: G7411221 BCHNA20.
Brown Lowe Pase type: 142.141.
Cheatham & Haller type: Great Basin sagebrush.
PSW-45 type: Rabbitbrush series.
Thorne type: Great Basin sagebrush scrub.
WHR type: Sagebrush.

General references: Paysen *et al.* (1980), West (1988), Young *et al.* (1977).

Comments: California stands are dominated by any of 8 ssp. of rubber rabbitbrush. Some ssp. are local, others have extensive ranges including disturbed areas as abandoned agricultural land and overgrazed pastures. The species grow in series dominated by trees, shrubs, and even grasses.

Sadler oak series

Sadler oak sole or dominant shrub in canopy; bitter cherry, wedgeleaf ceanothus, bush chinquapin, creeping snowberry, greenleaf manzanita, huckleberry oak, mountain whitethorn, ocean spray, and/or tobacco brush may be present. Emergent conifers may be present. Shrubs < 2 m; canopy continuous or intermittent. Ground layer sparse.

Uplands: ridges, upper slopes.

Distribution: m-KlaR, su-KlaR, OR.
Elevation: 600-2200 m.

NDDB/Holland type and status: **Chaparral** (37000). Mixed montane chaparral (37510 *in part*) G4 S4.

Other types:
Barry type: G7411214.
Cheatham & Haller type: Upper montane chaparral.
Thorne type: Montane chaparral.
WHR type: Montane chaparral.

General references: Holland (1986).

Comments: Sadler oak occurs as scattered shrubs in forest understories or as the sole or dominant shrub in what is often called montane chaparral. Self-perpetuating stands are unusual; most stands are established after fire or logging and are transitional to forest.

Species mentioned in text	
Bitter cherry	*Prunus emarginata*
Wedgeleaf ceanothus	*Ceanothus cuneatus*
Bush chinquapin	*Chrysolepis sempervirens*
Creeping snowberry	*Symphoricarpos mollis*
Greenleaf manzanita	*Arctostaphylos patula*
Huckleberry oak	*Quercus vaccinifolia*
Mountain whitethorn	*Ceanothus cordulatus*
Ocean spray	*Holodiscus discolor*
Sadler oak	*Quercus sadleriana*
Tobacco brush	*Ceanothus velutinus*

Plot-based descriptions: Sawyer *et al.* (1978b) in Keeler-Wolf (1990e) describe stands at Indian Creek in Siskiyou Co. m-KlaR.

Salal-black huckleberry series

Salal and black huckleberry important shrubs in canopy; bearberry, bush monkeyflower, California blackberry, coast silktassel, coast mugwort, coyote brush, red flowering currant, and/or seaside woolly-sunflower may be present. Emergent red alder, Sitka spruce, and/or willows may present. Shrubs < 1 m; canopy continuous or intermittent. Ground layer may include grasses, such as Pacific reedgrass, and herbs, such as cow-parsnip, as tall as shrubs (Plate 7).

Uplands: coastal bluffs, terraces.

Distribution: o-NorCo, OR.
Elevation: sea level-30 m.

NDDB/Holland type and status: Coastal bluff scrubs (31000), **Coastal scrubs** (32000), **Chaparral** (37000).
Northern (Franciscan) coastal bluff scrub (32110 *in part*) G2 2.2.
Northern salal scrub (32120) G3 S3.2.
Northgern silk-tassel scrub (32130) G3 S3.2.
Poison-oak chaparral (37F00 *in part*) G3 S3.3.

Other types:
Barry type: G7411221.
Cheatham & Haller type: Northern coastal bluff scrub, Northern coastal scrub.
Rangeland type: SRM 204.
Thorne type: Northern coastal scrub.
WHR type: Coastal scrub.

General references: Barbour (1994), Heady *et al.* (1977).

Comments: This series is often considered part of the coastal scrub, which is better considered a collection of series. Axelrod (1978) referred to it, at least in part, as Franciscan scrub. His description is more broadly based and includes stands dominated by coyote brush or yellow bush lupine [*see* Coyote brush series, Yellow bush lupine series]. Stands of this series are many times mix with other series at a fine scale, especially on coastal

headlands [*see* Hooker willow series, Pacific reedgrass series, Red alder series, Sitka spruce series, Tufted hairgrass series].

Species mentioned in text	
Bearberry	*Arctostaphylos uva-ursi*
Black huckleberry	*Vaccinium ovatum*
Bush monkeyflower	*Mimulus aurantiacus*
California blackberry	*Rubus ursinus*
Coast silktassel	*Garrya elliptica*
Coast mugwort	*Artemisia suksdorfii*
Cow-parsnip	*Heracleum lanatum*
Coyote brush	*Baccharis pilularis*
Pacific reedgrass	*Calamagrostis nutkaensis*
Poison-oak	*Toxicodendron diversilobum*
Red alder	*Alnus rubra*
Red flowering current	*Ribes sanguineum*
Salal	*Gaultheria shallon*
Seaside woolly-sunflower	*Eriophyllum staechadifolium*
Sitka spruce	*Picea sitchensis*
Tufted hairgrass	*Deschampsia caespitosa*
Willows	*Salix* species
Yellow bush lupine	*Lupinus arboreus*

Plot-based descriptions: Stuart *et al.* (1986) defines a Salal-coyote brush-blue blossom association at Stagecoach Hill Azalea Preserve in Humboldt Co., Baker (1972) describes a stand at Ingelnook fen as a Fen carr, NDDB has data for a bearberry-dominated stand at Point Arena in Mendocino Co., and a poison-oak - salal-dominated stand at Table Bluff, Eel River Wildlife Area in Humboldt Co. o-NorCo.

Sandbar willow series

Sandbar willow sole or dominant shrub in canopy; bigleaf maple, black cottonwood, red alder, red osier, white alder, and/or willows may be present. Emergent trees may be present. Shrubs < 5 m; canopy continuous. Ground layer variable.

Wetlands: habitats seasonally flooded, saturated. Water chemistry: fresh. Flood-plains; depositions along rivers, streams. Cowardin class: Palustrine shrub-scrub wetland.

Distribution: o-NorCo, l-KlaR.
Elevation: sea level-200 m.

NDDB/Holland type and status: **Riparian scrubs** (63000).
   North Coast riparian scrub (63100 *in part*) G3 S3.2.

Other types:
Barry type: G7411211.
Cheatham & Haller type: Bottomland woodlands and forest.
PSW-45 type: Willow series.
Thorne type: Riparian woodland.
WHR type: Freshwater emergent wetland.

General references: Brayshaw (1976), Paysen *et al.* (1980), Read & Sprackling (1980).

Comments: Many willow stands are composed of a single species and easily assigned to a series using the dominance rule. If no dominant willow is present at low elevations, then place the stands in the Mixed willow series. Montane and subalpine willow stands are placed in separate classes since different willow species are restricted to those elevations [*see* Montane wetland shrub habitat, Subalpine wetland shrub habitat].

Stands of this series have similar environmental conditions to alder, cottonwood, and other willow series [*see* Black cottonwood series, Black willow series, Fremont cottonwood series, Hooker willow series, Pacific willow series, Red alder series, Red willow series, Sitka willow series, White alder series].

Species mentioned in text	
Bigleaf maple	*Acer macrophyllum*
Black cottonwood	*Populus balsamifera*
Black willow	*Salix gooddingii*
Fremont cottonwood	*Populus fremontii*
Hooker willow	*Salix hookeriana*
Pacific willow	*Salix lucida* spp. *lasiandra*
Red alder	*Alnus rubra*
Red osier	*Cornus sericea*
Red willow	*Salix laevigata*
Sandbar willow	*Salix sessilifolia*
Sitka willow	*Salix sitchensis*
White alder	*Alnus rhombifolia*
Willows	*Salix* species

Plot-based descriptions: not available.

Scalebroom series

Scalebroom sole, dominant, or important with other shrubs in canopy; basket bush, big sagebrush, bladderpod, brittlebush, California buckwheat, California sagebrush, catclaw acacia, chaparral yucca, cheesebush, deer weed, hairy yerba santa, mulefat, prickly-pears, poison-oak, and/or sugar bush, may be present. Emergent birchleaf mountain-mahogany, California juniper, California sycamore, California walnut, Fremont cottonwood, laurel sumac, lemonade berry, or Mexican elderberry may be present. Shrubs < 1.5 m, canopy continuous or intermittent. Ground layer variable, may be grassy (Plate 17).

Uplands: Rarely flooded, low gradient deposits along streams.

Distribution: s. f-SN, i-CenCo, SoCo, MojD, ColD.
Elevation: sea level-1800 m.

NDDB/Holland type and status: Coastal scrubs (32000), **Chaparral** (37000), **Riparian scrub** (63000).
    Riversidean alluvial fan sage scrub (32720) G1 S1.1.
    Alluvial fan chaparral (37H00) G3 S3.1.
    Mojave desert wash scrub (63700) G3 S2.1.

Other types:
Barry type: G7411221.
Jones & Stokes type: Scalebroom scrub.

General references: Hanes *et al.* (1989), Ingles (1929), Kirkpatrick & Hutchinson (1977), Magney (1992).

Comments: Species composition differs greatly among stands; disturbance may account for the high variation. Some stands in this habitat may have sufficient emergent trees to be placed in tree-dominated series [*see* Birchleaf mountain-mahogany series]. The federal and state listed Slender-horned spineflower (a CNPS list 1B plant) and Santa Ana river woollystar (a CNPS list 1B plant) grow in some stands of this series (Skinner & Pavlik 1994).

Species mentioned in text	
Basket bush	*Rhus trilobata*
Big sagebrush	*Artemisia tridentata*
Birchleaf mountain-mahogany	*Cercocarpus betuloides*
Bladderpod	*Isomeris arborea*
Brittlebush	*Encelia farinosa*
California buckwheat	*Eriogonum fasciculatum*
California juniper	*Juniperus californica*
California sagebrush	*Artemisia californica*
California sycamore	*Platanus racemosa*
California walnut	*Juglans californica* var. *californica*
Catclaw acacia	*Acacia greggii*
Cheesebush	*Hymenoclea salsola*
Chaparral yucca	*Yucca whipplei*
Deer weed	*Lotus scoparius*
Fremont cottonwood	*Populus fremontii*
Hairy yerba santa	*Eriodictyon crassifolium*
Laurel sumac	*Malosma laurina*
Lemonade berry	*Rhus integrifolia*
Mexican elderberry	*Sambucus mexicana*
Mulefat	*Baccharis salicifolia*
Poison-oak	*Toxicodendron diversilobum*
Prickly-pears	*Opuntia* species
Santa Ana River woollystar	*Eriastrum densifolium* ssp. *sanctorum*
Scalebroom	*Lepidospartum squamatum*
Slender-horned spineflower	*Dodecahema leptoceras*
Sugar bush	*Rhus ovata*

Plot-based descriptions: Hanes *et al.* (1989) report sites in three drainages [Cucamonga Canyon, San Gabriel Canyon, San Jacinto River] as dominated by scalebroom, Kirkpatrick & Hutchinson (1977) define one association in SoCo. Smith (1980) describes pioneer, intermediate, and mature stands along the San Gabriel River floodplain in Los Angles Co. SoCo. Gordon & White (1994) define one association in m-TraR, Boyd (1983) describes one stand in Gavilan Hills in Riverside Co. i. m-TraR.

Associations:
Gordon & White (1994):
    California buckwheat-scalebroom association.
Kirkpatrick & Hutchinson (1977):
    Scalebroom-hairy yerba santa-chaparral yucca association.

Scrub oak series

Scrub oak sole or dominant shrub in canopy; blue blossom, California coffeeberry, chamise, chaparral pea, chaparral whitethorn, hollyleaf redberry, interior live oak, manzanita, poison-oak, red shank, and/or toyon may be present. Emergent birchleaf mountain-mahogany, California buckeye, and/or foothill pine may be present. Shrubs < 3 m; canopy continuous. Ground layer sparse.

Uplands: all slopes. Soils deep or shallow, may be rocky.

Distribution: i-NorCo, CenCo, f-CasR, f-SN, So-CA, Baja CA.
Elevation: 300-1500 m.

NDDB/Holland type and status: Chaparral (37000).
 Granitic southern mixed chaparral (37121) G3 S3.3.
 Mafic southern mixed chaparral (37122) G3 S3.1.
 Island chaparral (37700) G3 S3.
 Scrub oak chaparral (37900 *in part*) G3 S3.3.
 Northern maritime chaparral (37C10 *in part*) G1 S1.2.
 Northern north slope chaparral (37E10 *in part*) G3 S3.3.
 Southern north slope chaparral (37E20 *in part*) G3 S3.
 Poison-oak chaparral (37F00 *in part*) G3 S3.3.

Other types:
Barry type: G7411211 CQUDU00.
Cheatham & Haller type: Mixed chaparral.
PSW-45 type: Scrub oak series.
Rangeland type: SRM 207.
Thorne type: Mixed chaparral.
WHR type: Mixed chaparral.

General references: Hanes (1977, 1981), Horton (1960), Keeley & Keeley (1988), Paysen *et al.* (1980), White (1994b).

Comments: The term scrub oak has been used for

several shrubby live oaks. *The Jepson Manual* applies the name *Quercus dumosa*, Nuttall oak, to a rare (a CNPS List 1B plant) SoCo species (Skinner & Pavlik 1994). References in the ecological literature to *Q. dumosa* are more appropriately *Q. berberidifolia* which ranges through much of Cis-CA. Desert scrub oak, interior live oak, and the rarer Muller oak, Tucker oak, and Palmer oak are also called scrub oaks. Careful identification of the oaks in a stand will be necessary to correctly assign them to a series. [*see* California juniper series, Interior live oak shrub series].

Species mentioned in text	
Birchleaf mountain-mahogany	*Cercocarpus betuloides*
Blue blossom	*Ceanothus thyrsiflorus*
California buckeye	*Aesculus californica*
California coffeeberry	*Rhamnus californica*
Chamise	*Adenostoma fasciculatum*
Chaparral pea	*Pickeringia montana*
Chaparral whitethorn	*Ceanothus leucodermis*
Cleveland monkeyflower	*Mimulus clevelandii*
Dense reedgrass	*Calamagrostis densa*
Desert scrub oak	*Quercus turbinella*
Dunn mariposa lily	*Calochortus dunnii*
Eastwood manzanita	*Arctostaphylos glandulosa*
Foothill pine	*Pinus sabiniana*
Hollyleaf redberry	*Rhamnus ilicifolia*
Interior live oak	*Quercus wislizenii*
Manzanita	*Arctostaphylos* species
Poison-oak	*Toxicodendron diversilobum*
Muller oak	*Quercus cornelius-mulleri*
Palmer oak	*Quercus palmeri*
Red shank	*Adenostoma sparsifolium*
Scrub oak	*Quercus berberidifolia*
Toyon	*Heteromeles arbutifolia*
Tucker oak	*Quercus john-tuckeri*
Wart-stemmed ceanothus	*Ceanothus verrucosus*

This series definition follows Gordon & White (1994). If scrub oak cover is > 60% then the stand is in the Scrub oak series. If scrub oak is 30-60% and the other species cover is 30-60%, then the stand is a member of a mixed series [*see* Scrub oak - birchleaf mountain-mahogany series, Scrub oak-chamise series, Scrub oak-chaparral series]. Stands with localized species as Cleveland monkeyflower (a CNPS List 4 plant), Dunn mariposa lily (a CNPS

List 1B plant), or white coast ceanothus (a CNPS List 2 plant) may occur in this series (Skinner & Pavlik 1994).

Hanes (1976) qualitatively describes vegetation types in the San Gabriel Mountains including a Scrub oak chaparral; Keeler-Wolf (1990e) scrub oak stands at Cahuilla Mountain RNA in Riverside Co., at King Creek RNA in San Diego Co. m-PenR.

Plot-based descriptions: Newton (1987) in Keeler-Wolf (1990e) defines an association at Devil's Basin candidate RNA in Tehama Co. i-NorCo, Gordon & White (1994) define five associations in So-CA; Boyd (1983) describes three stands in Gavilan Hills in Riverside Co. i. m-TraR.

Associations:
Newton (1987):
    Scrub oak/California buckeye association
    [as *Quercus dumosa*/California buckeye association].
Gordon & White (1994):
    Scrub oak association,
    Scrub oak-(chamise-Eastwood manzanita) association,
    Scrub oak-(toyon) association,
    Scrub oak-(chaparral whitethorn) association,
    Scrub oak-(birchleaf mountain-mahogany) association.

Scrub oak - birchleaf mountain-mahogany series

Scrub oak and birchleaf mountain-mahogany important shrubs in canopy; chamise, chaparral whitethorn, hollyleaf redberry, interior live oak, manzanita, poison-oak, red shank, and/or toyon may be present. Shrubs < 4 m; canopy continuous. Ground layer sparse or absent.

Uplands: all slopes. Soils deep or shallow. Texture loam, may be rocky.

Distribution: SoCo, m-TraR, m-PenR.
Elevation: 300-1500 m.

NDDB/Holland type and status: Chaparral (37000).
    Granitic southern mixed chaparral (37121) G3 S3.3.
    Mafic southern mixed chaparral (37122) G3 S3.2.
    Scrub oak chaparral (37900) G3 S3.
    Southern north slope chaparral (37E20 *in part*) G3 S3.3.
Other types:
Barry type: G7411214.
Cheatham & Haller type: Mixed chaparral.
Rangeland type: SRM 207.
Thorne type: Mixed chaparral.
WHR type: Mixed chaparral.
General references: Hanes (1977, 1981), Horton (1960), Keeley & Keeley (1988), Pase (1982a), White (1994b).

Comments: The term scrub oak has been used for several shrubby live oaks. *The Jepson Manual* applies the name *Quercus dumosa*, Nuttall scrub oak, to a rare (a CNPS List 1B plant) SoCo species (Skinner & Pavlik 1994). References in the ecological literature to *Q. dumosa* are more appropriately *Q. berberidifolia* which ranges through much of Cis-CA. Desert scrub oak, interior live oak, and the rarer Muller oak, Tucker oak, and Palmer oak are also called scrub oaks. Careful identification of the oaks in a stand will be necessary to correctly assign them to a series. [*see* California juniper series, Interior live oak shrub series].

This series definition follows Gordon & White (1994). If either scrub oak or birchleaf mountain-mahogany cover is > 60%, then the stand is a member of the Scrub oak series or Birchleaf mountain-mahogany series. Stands with localized species as Cleveland monkeyflower (a CNPS List 4 plant), Dunn mariposa lily (a CNPS List 1B plant), or white coast ceanothus (a CNPS List 2 plant) may occur in this series (Skinner & Pavlik 1994). Keeler-Wolf (1990e) qualitatively describes stands at Cahuilla Mountain RNA in Riverside Co. m-PenR.

Species mentioned in text	
Birchleaf mountain-mahogany	*Cercocarpus betuloides*
California buckeye	*Aesculus californica*
Chamise	*Adenostoma fasciculatum*
Chaparral whitethorn	*Ceanothus leucodermis*
Cleveland monkeyflower	*Mimulus clevelandii*
Cupleaf ceanothus	*Ceanothus greggii*
Desert scrub oak	*Quercus turbinella*
Dunn mariposa lily	*Calochortus dunnii*
Eastwood manzanita	*Arctostaphylos glandulosa*
Foothill ash	*Fraxinus dipetala*
Hollyleaf redberry	*Rhamnus ilicifolia*
Interior live oak	*Quercus wislizenii*
Manzanita	*Arctostaphylos* species
Muller oak	*Quercus cornelius-mulleri*
Palmer ceanothus	*Ceanothus palmeri*
Palmer oak	*Quercus palmeri*
Red shank	*Adenostoma sparsifolium*
Scrub oak	*Quercus berberidifolia*
Toyon	*Heteromeles arbutifolia*
Tucker oak	*Quercus john-tuckeri*
White coast ceanothus	*Ceanothus verrucosus*

Plot-based descriptions: Gordon & White (1994) define four associations in So-CA.
Associations:
Gordon & White (1994):
    Scrub oak - birchleaf mountain-mahogany association,
    Scrub oak - birchleaf mountain-mahogany - (cupleaf ceanothus) association,
    Foothill ash - birchleaf mountain-mahogany - scrub oak association,
    Scrub oak - birchleaf mountain-mahogany - (Palmer ceanothus) association.

Scrub oak-chamise series

Scrub oak and chamise important shrubs in canopy; birchleaf mountain-mahogany, chaparral whitethorn, hoaryleaf ceanothus, hollyleaf redberry, manzanita, mission-manzanita, and/or toyon may be present. Emergent knobcone pine may be present. Shrubs < 4 m; canopy continuous. Ground layer sparse.

Uplands: all slopes. Soils deep or shallow, may be rocky.

Distribution: SoCo, m-TraR, PenR.
Elevation: 400-1650 m.

NDDB/Holland type and status: **Chaparral** (37000).
Granitic southern mixed chaparral (37121) G3 S3.3.
Mafic southern mixed chaparral (37122) G3 S3.2.
Scrub oak chaparral (37900) G3 S3.
Southern north slope chaparral (37E20 *in part*) G3 S3.3.

Other types:
Barry type: G7411214.
Cheatham & Haller type: Chamise chaparral.
Rangeland type: SRM 207.
Thorne type: Chamisal.
WHR type: Mixed chaparral.

General references: Hanes (1977, 1981), Horton (1960), Keeley & Keeley (1988), Pase (1982a), White (1994b).

Comments: The term scrub oak has been used for several shrubby live oaks. *The Jepson Manual* applies the name *Quercus dumosa,* Nuttall scrub oak, to a rare (a CNPS List 1B plant) SoCo species (Skinner & Pavlik 1994). References in the ecological literature to *Q. dumosa* are more appropriately *Q. berberidifolia* which ranges through much of Cis-CA. Desert scrub oak, interior live oak, and the rarer Muller oak, Tucker oak, and Palmer oak are also called scrub oaks. Careful identification

of the oaks in a stand will be necessary to correctly assign them to a series. [*see* California juniper series, Interior live oak shrub series]. This series definition follows Gordon & White (1994). If scrub oak or chamise cover is > 60% then the stand is a member of the Scrub oak series or Chamise series. Stands with localized species as Cleveland monkey-flower (a CNPS List 4 plant), Dunn mariposa lily (a CNPS List 1B plant), or white coast ceanothus (a CNPS List 2 plant) may occur in this series (Skinner & Pavlik 1994).

Keeler-Wolf (1990e) qualitatively describes scrub oak stands at Cahuilla Mountain RNA in Riverside Co., King Creek RNA in San Diego Co. m-PenR.

Species mentioned in text	
Birchleaf mountain-mahogany	*Cercocarpus betuloides*
Chamise	*Adenostoma fasciculatum*
Chaparral whitethorn	*Ceanothus leucodermis*
Cleveland monkeyflower	*Mimulus clevelandii*
Cupleaf ceanothus	*Ceanothus greggii*
Dense reedgrass	*Calamagrostis densa*
Desert scrub oak	*Quercus turbinella*
Hollyleaf redberry	*Rhamnus ilicifolia*
Hoaryleaf ceanothus	*Ceanothus crassifolius*
Interior live oak	*Quercus wislizenii*
Knobcone pine	*Pinus attenuata*
Manzanita	*Arctostaphylos* species
Mission manzanita	*Xylococcus bicolor*
Muller oak	*Quercus cornelius-mulleri*
Palmer oak	*Quercus palmeri*
Scrub oak	*Quercus berberidifolia*
Toyon	*Heteromeles arbutifolia*
Tucker oak	*Quercus john-tuckeri*
White coast ceanothus	*Ceanothus verrucosus*

Plot-based descriptions: Gordon & White (1994) define two associations in So-CA.

Associations:
Gordon & White (1994):
Scrub oak-chamise-(hoaryleaf ceanothus). association,
Scrub oak-chamise-(cupleaf ceanothus) association.

Scrub oak-chaparral whitethorn series

Scrub oak and chaparral whitethorn important shrubs in canopy; chamise, hollyleaf redberry, interior live oak, manzanita, and/or toyon may be present. Emergent trees may be present. Shrubs < 3 m; canopy continuous. Ground layer sparse (Plate 11).

Uplands: all slopes. Soils deep or shallow, may be rocky.

Distribution: SoCo, Baja CA.
Elevation: 300-1600 m.

NDDB/Holland type and status: **Chaparral** (37000).
   Granitic southern mixed chaparral (37121) G3 S3.
   Mafic southern mixed chaparral (37122) G2 S2.1
   Scrub oak chaparral (37900 *in part*) G3 S3.
   Southern north slope chaparral (37E20 *in part*) G3 S3.

Other types:
Barry type: G7411214.
Cheatham & Haller type: Mixed chaparral.
Rangeland type: SRM 207.
Thorne type: Mixed chaparral.
WHR type: Mixed chaparral.

General references: Hanes (1977, 1981), Horton (1960), Keeley & Keeley (1988), White (1994b).

Comments: The term scrub oak has been used for several shrubby live oaks. *The Jepson Manual* applies the name *Quercus dumosa,* Nuttall scrub oak, to a rare (a CNPS List 1B plant) SoCo species (Skinner & Pavlik 1994). References in the ecological literature to *Q. dumosa* are more appropriately *Q. berberidifolia* which ranges through much of Cis-CA. Desert scrub oak, interior live oak, and the rarer Muller oak, Tucker oak, and Palmer oak are also called scrub oaks. Careful identification of the oaks in a stand will be necessary to correctly assign them to a series. [*see* California juniper series, Interior live oak shrub series].

This series definition follows Gordon & White (1994). If scrub oak or chaparral whitethorn cover is > 60% then the stand is a member of the Scrub oak series or Chaparral whitethorn series. Stands with localized species as Cleveland monkeyflower (a CNPS List 4 plant), Dunn mariposa lily (a CNPS List 1B plant), or white coast ceanothus (a CNPS List 2 plant) may occur in this series (Skinner & Pavlik 1994).

Keeler-Wolf (1990e) qualitatively describes scrub oak stands at Cahuilla Mountain RNA in Riverside Co. m-PenR.

Species mentioned in text	
Chamise	*Adenostoma fasciculatum*
Chaparral whitethorn	*Ceanothus leucodermis*
Cleveland monkeyflower	*Mimulus clevelandii*
Dunn mariposa lily	*Calochortus dunnii*
Desert scrub oak	*Quercus turbinella*
Hollyleaf redberry	*Rhamnus ilicifolia*
Interior live oak	*Quercus wislizenii*
Manzanita	*Arctostaphylos* species
Muller oak	*Quercus cornelius-mulleri*
Palmer oak	*Quercus palmeri*
Scrub oak	*Quercus berberidifolia*
Toyon	*Heteromeles arbutifolia*
Tucker oak	*Quercus john-tuckeri*
White coast ceanothus	*Ceanothus verrucosus*

Plot-based descriptions: Gordon & White (1994) define a Scrub oak-chaparral whitethorn association in So-CA.

Shadscale series

Shadscale sole or dominant shrub in canopy; budsage, creosote bush, greasewood, hopsage, horsebrush, Nevada ephedra, spinescale, yellow rabbitbrush, and/or winter fat may be present. Shrubs < 1 m; canopy continuous, intermittent, or open. Ground layer sparse (Plate 14).

Uplands: bajadas; flats; lower slopes; playas; valleys; washes. Soils carbonate-rich, may have desert pavement.

Distribution: i-CenCo, sj-CenV, m-TraR, m-PenR, ModP, s. TraSN, MojD, inter-West. **Elevation:** 450-2500 m.

NDDB/Holland type and status: **Chenopod scrubs** (36000).
  Shadscale scrub (36140) G4 S3.2.

Other types:
Barry type: G7411221.
Brown Lowe Pase type: 152.122.
Cheatham & Haller type: Shadscale scrub.
PSW-45 type: Saltbrush series.
Rangeland types: SRM 414, SRM 501.
Thorne type: Shadscale scrub.
WHR type: Alkali sink.

General references: Burk (1977), MacMahon (1988), MacMahon & Wagner (1985), McHargue (1973), Paysen *et al.* (1980), Thorne (1982), Turner (1982b), Vasek & Barbour (1977), Young *et al.* (1977).

Comments: This series is often considered part of the chenopod or saltbush scrub. These scrubs are better thought of as a collection of series. Stands of shadscale alone or mixed with spinescale are common and extensive. If no dominant saltbush is present, the stand is assigned to Mixed saltbush series. If one dominates then the stand is a member if the Shadscale series or Spinescale series. Young *et al.* (1977) qualitatively define associations that illustrate the variation within this extensive series.

Species mentioned in text	
Budsage	*Artemisia spinescens*
Creosote bush	*Larrea tridentata*
Greasewood	*Sarcobatus vermiculatus*
Hop-sage	*Grayia spinosa*
Horsebrush	*Tetradymia canescens*
Nevada ephedra	*Ephedra nevadensis*
Shadscale	*Atriplex confertifolia*
Spinescale	*Atriplex spinifera*
Winter fat	*Krascheninnikovia lanata*
Yellow rabbitbrush	*Chrysothamnus viscidiflorus*

Plot-based descriptions: Billings (1949) describes 9 widely distributed stands in e. San Bernardino Co. MojD and w. NV. NDDB has plot data on file for Surprise Valley in Modoc Co. ModP; Ferren & Davis (1991) define two associations at Fish Slough in Inyo Co. TraSN.

Associations:
Ferren & Davis (1991):
  Greasewood-shadscale association,
  Shadscale-Nevada ephedra association.

Sitka alder series

Sitka alder sole or dominant shrub in canopy; red osier, Jepson willow, Lemmon willow, Mackenzie willow, mountain maple, and/or Scouler willow may be present. Emergent trees may be present. Shrubs < 10 m; canopy continuous. Ground layer variable (Plate 7).

Wetlands: habitats seasonally flooded, saturated. Water chemistry: fresh. Floodplains; low-gradient depositions along streams. Cowardin class: Palustrine shrub/scrub wetland. The national list of wetland plants (Reed 1988) lists Sitka alder as a FACW*.

Distribution: m-NorCo, m-KlaR, su-KlaR to AK.
Elevation: 1000-2700 m.

NDDB/Holland type and status: **Riparian scrubs** (63000).
    Montane riparian scrub (63500 *in part*) G4 S4.

Other types:
Barry type: G7412121.
Cheatham & Haller type: Willow thickets.
Thorne type: Mountain meadow.
WHR type: Montane riparian.

General references: Brayshaw (1976).

Comments: In *The Jepson Manual*, Sitka alder is treated as a subspecies (*Alnus viridis* ssp. *sinuata*) of a widespread North America and Eurasia species. The elevational and regional ranges of the montane alders, mountain alder (*Alnus incana*), and Sitka alder, differ in California. They rarely occur in the same stand [*see* Mountain alder series]. This series is best developed along m-KlaR streams. Stands of

the Montane wetland habitat and Subalpine wetland habitat occur in similar environmental conditions. Sitka alder tolerates cooler and shadier riparian situations than most willows.

Species mentioned in text	
Jepson willow	*Salix jepsonii*
Lemmon willow	*Salix lemmonii*
Mackenzie willow	*Salix prolixa*
Mountain alder	*Alnus incana*
Mountain maple	*Acer glabrum*
Red osier	*Cornus sericea*
Scouler willow	*Salix scouleriana*
Sitka alder	*Alnus viridis*

Plot-based descriptions: not available.

Spinescale series

Spinescale sole or dominant shrub in canopy; alkali heath, California ephedra, cheesebush, and/ or saltgrass may be present. Shrubs < 2 m; canopy open. Ground layer variable, annuals seasonally present (Plate 15).

Wetlands: habitats intermittently flooded, saturated. Water chemistry: mixosaline. Dry lake beds; plains. Cowardin class: Palustrine shrub-scrub wetland. **Uplands:** alluvial fans; old lake beds perched above current drainages. Soils may be carbonate-rich. The national list of wetland plants (Reed 1988) lists spinescale as a FAC.

Distribution: i-CenCo, sj-CenV, MojD.
Elevation: 50-800 m.

NDDB/Holland type and status: **Chenopod scrubs** (36000).
>   Desert saltbush scrub (36110 *in part*) G3 S3.2.
>   Valley saltbush scrub (36220 *in part*) G1 S1.1.

Other types;
Barry type: G7411221.
Cheatham & Haller type: Saltbush scrub.
PSW-45 type: Saltbush series.
Thorne type: Shadscale scrub.
WHR type: Alkali sink.

General references: Bittman (1985), Griggs (1980), Griggs & Zaninovich (1984), Paysen *et al.* (1980), Vasek & Barbour (1977), Werschkull (1984).

Comments: This is a series often considered part of the chenopod or saltbush scrub. These scrubs are better thought of as a collection of series. Stands of spinescale alone or mixed with shadscale are common and extensive. If no dominant saltbush is present, the stand is assigned to Mixed saltbush series. If one dominates then the stand is a member if the Shadscale series or Spinescale series [*see* Shadscale series].

Species mentioned in text	
Allscale	*Atriplex polycarpa*
Alkali heath	*Frankenia salina*
California ephedra	*Ephedra californica*
Cheesebush	*Hymenoclea salsola*
Fourwing saltbush	*Atriplex canescens*
Saltgrass	*Distichlis spicata*
Shadscale	*Atriplex confertifolia*
Spinescale	*Atriplex spinifera*

Plot-based descriptions: Phillips & MacMahon (1981) in MacMahon (1988) report density and cover for stands in San Bernardino Co. MojD.

Sumac series

Laurel sumac or lemonade berry sole, dominant, or important shrubs with black sage, hollyleaf redberry, or toyon in canopy; bush monkeyflower, California encelia, California sagebrush, hollyleaf cherry, poison-oak, purple sage, sugar bush, and or yellow bush penstemon may be present. Emergent California walnut may occur. Shrubs < 4 m; canopy open or continuous. Ground layer sparse.

Uplands: slopes often steep. Soils shallow, coarse.

Distribution: o-SoCo, w. m-TraR, m-PenR, Baja CA.
Elevation: near sea level-400 m.

NDDB/Holland type and status: Chaparral (37000).
    Granitic southern mixed chaparral (37121 *in part*)
    G3 S3.
    Mafic southern mixed chaparral (37122 *in part*)
    G2 S2.1.

Other types:
Brown Lowe Pase type: Sumac series.
Cheatham & Haller type: South mixed chaparral.
Jones & Stokes: Mixed chaparral (coastal form).
PSW-45 type: Sumac series.
Thorne type: Mixed chaparral.
WHR type: Mixed chaparral.

General references: Keeley & Keeley (1988), Paysen *et al.* (1980).

Comments: This series is often overlooked by combining it with mixed chaparral, but many characteristic chaparral genera (*Adenostoma, Arctostaphylos, Ceanothus, Quercus*) are absent or occur only uncommonly. Structure is continuous in the northern part of its range, but increasingly open the south. Lloret & Zedler (1991) describe lemonade berry recruitment; Miller & Pole (1979) describe water use by sumacs and other shrubs.

Peruvian pepper and Brazilian pepper are naturalized and increasingly common in sumac stands.

White (1994b) qualitatively describes sumac stands in the Puente Hills in Los Angeles Co. o-SoCo.

Species mentioned in text	
Black sage	*Salvia mellifera*
Brazilian pepper	*Schinus terebinthifolius*
Bush monkeyflower	*Mimulus aurantiacus*
California encelia	*Encelia californica*
California sagebrush	*Artemisia californica*
California walnut	*Juglans californica* var. *californica*
Hollyleaf cherry	*Prunus ilicifolia* ssp. *ilicifolia*
Laurel sumac	*Malosma laurina*
Lemonade berry	*Rhus integrifolia*
Mexican elderberry	*Sambucus mexicana*
Peruvian pepper	*Schinus molle*
Poison-oak	*Toxicodendron diversilobum*
Toyon	*Heteromeles arbutifolia*
Purple sage	*Salvia leucophylla*
Yellow bush penstemon	*Keckiella antirrhinoides*

Plot-based descriptions: not available.

Tamarisk series

Tamarisks sole or dominant shrub; catclaw acacia, cheesebush, saltbushes, and/or willows may be present. Shrubs < 5 m. Shrubs continuous or open. Emergent trees may be present. Ground layer sparse (Plate 7).

Wetlands: habitats intermittently flooded, saturated. Water chemistry: fresh. Arroyo margins, ditches; washes, watercourses. Cowardin class: Temporarily flooded palustrine shrub-scrub wetland. The national list of wetland plants (Reed 1988) lists tamarisks as a FACW.

Distribution: o-NorCo, i-NorCo, i-CenCo, CenV, s. f-SN, s. m-SN, SoCo, TraSN, MojD, ColD.
Elevation: 75 below sea level-800 m.

NDDB/Holland type and status: **Riparian scrub** (63000).
  Tamarisk scrub (63810) G5 S4.

Other types:
Barry type: G7411212 BTACH00.
Cheatham & Haller type: Alluvial woodlands.
PSW-45 type: Salt-cedar series.
Thorne type: Desert riparian woodland.
WHR type: Desert riparian.

General references: Johnson (1987), Neill (1985), MacMahon (1988), Paysen *et al.* (1980).

Comments: Five introduced species of tamarisk are known to grow in CA; *Tamarix parviflora* and *T. ramosissima* are the most common species (*The Jepson Manual*). Tamarisk supplants native plants and reduces water for wildlife. Active programs to remove tamarisk are ongoing in the state.

Species mentioned in text	
Catclaw acacia	*Acacia greggii*
Cheesebush	*Hymenoclea salsola*
Saltbushes	*Atriplex* species
Tamarisks	*Tamarix* species
Willows	*Salix* species

Plot-based descriptions: not available.

Teddy-bear cholla series

Teddy-bear cholla sole or dominant shrub in canopy; brittlebush, creosote bush, and/or white bursage may be present. Shrubs < 2 m; canopy continuous or intermittent. Ground layer open; annuals seasonally present (Plate 15).

Uplands: bajadas; upland slopes. Soils well-drained.

Distribution: MojD, ColD.
Elevation: sea level-1000 m.

NDDB/Holland type and status: **Sonoran desert scrubs** (33000).
  Sonoran creosote bush scrub (33100 *in part*) G4 S4.

Other types:
Barry type: G74 G7411213 BOPBI00.
Brown Lowe Pase type: 154.117.
Cheatham & Haller type: Desert cactus scrub.
PSW-45 type: Opuntia series.
Thorne type: Stem-succulent scrub.
WHR type: Desert succulent shrub.

General references: Burk (1977), MacMahon (1988), Paysen *et al.* (1980), Thorne (1982), Turner & Brown (1982).

Comments: Regional descriptions mention areas with dense populations of Teddy-bear cholla as part of a mosaic with creosote bush scrub. These stands of Teddy-bear cholla are separated here as distinctive series, since the cactus dominates the landscape. Teddy-bear cholla can occur as a minor component in other series [*see* Creosote bush series, Creosote bush-big galleta series, Ocotillo series]. Collecting data on this series will be a challenge.

Species mentioned in text	
Big galleta	*Pleuraphis rigida*
Brittlebush	*Encelia farinosa*
Creosote bush	*Larrea tridentata*
Ocotillo	*Fouquieria splendens*
Teddy-bear cholla	*Opuntia bigelovii*
White bursage	*Ambrosia dumosa*

Plot-based descriptions: not available.

Tobacco brush series

Tobacco brush sole or dominant shrub in canopy; bitter cherry, bush chinquapin, creeping snowberry, greenleaf manzanita, huckleberry oak, mountain whitethorn, ocean spray, Sadler oak, wedgeleaf ceanothus, and/or whiteleaf manzanita may be present. Emergent conifers may be present. Shrubs < 2 m; canopy continuous or intermittent. Ground layer sparse.

Uplands: ridges, upper slopes.

Distribution: o-NorCo, m-NorCo, m-KlaR, m-CasR, su-CasR, m-SN, su-SN, WarR, TraSN.
Elevation: 50-3000 m.

NDDB/Holland type and status: **Chaparral** (37000).
   Mixed montane chaparral (37510 *in part*) G4 S4.
   Tobacco brush chaparral (37533) G3 S3.3.

Other types:
Barry type: G7411214.
Cheatham & Haller type: Upper montane chaparral.
PSW-45 type: Ceanothus series.
Rangeland type: SRM 209, SRM 420.
Thorne type: Montane chaparral.
WHR type: Montane chaparral.

General references: Paysen *et al.* (1980), Riser & Fry (1994), Solinas *et al.* (1985), Winward (1994).

Comments: Tobacco brush occurs as scattered shrubs in forest understories or as the sole or dominant shrub in what is often called montane chaparral. Self-perpetuating stands are unusual; instead, most stands are established after fire or logging and are transitional to various kinds of forest.
Keeler-Wolf (1990e) qualitatively describes tobacco brush stands at Upper Goose Creek candidate RNA in Del Norte Co. l-KlaR; at Antelope Creek Lakes candidate RNA su-CasR, Shasta Mudflow RNA in Siskiyou Co. m-CasR; at Mud Lake RNA in Plumas Co.; Babbitt Peak RNA in Sierra Co. su-SN.

Species mentioned in text	
Bitter cherry	*Prunus emarginata*
Bush chinquapin	*Chrysolepis sempervirens*
Creeping snowberry	*Symphoricarpos mollis*
Greenleaf manzanita	*Arctostaphylos patula*
Huckleberry oak	*Quercus vaccinifolia*
Mountain whitethorn	*Ceanothus cordulatus*
Ocean spray	*Holodiscus discolor*
Sadler oak	*Quercus sadleriana*
Tobacco brush	*Ceanothus velutinus*
Wedgeleaf ceanothus	*Ceanothus cuneatus*
Whiteleaf manzanita	*Arctostaphylos viscida*

Plot-based descriptions: Imper (1988a) in Keeler-Wolf (1990e) defines a Tobacco brush-greenleaf manzanita association at Haypress Meadows candidate RNA in Siskiyou Co. m-KlaR; Conard & Radoevich (1982) report cover for Sierra Co. m-SN stands.

Wedgeleaf ceanothus series

Wedgeleaf ceanothus sole or dominant shrub in canopy; chamise, common manzanita, deerbrush, Fremont silktassel, greenleaf manzanita, scrub oak, toyon, and/or woollyleaf manzanita may be present. Emergent conifers may be present. Shrubs < 3 m; canopy continuous or intermittent. Ground layer sparse (Plate 11).

Uplands: ridges, upper slopes.

Distribution: NorCo, m-CenCo, KlaR, CasR, SN, o. m-TraR, PenR, OR, Baja CA.
Elevation: 15-1800 m.

NDDB/Holland type and status: **Chaparral** (37000).
Mixed montane chaparral (37510 *in part*) G4 S4.
Buck brush chaparral (37810) G4 S4.
Poison-oak chaparral (37F00 *in part*) G4 S4.

Other types:
Barry type: G7411211 CCECU00.
Cheatham & Haller type: Lower montane chaparral.
PSW-45 type: Ceanothus series.
Rangeland type: SRM 208.
Thorne type: Mixed chaparral.
WHR type: Mixed chaparral.

General references: Hanes (1977, 1981), Keeley & Keeley (1988), Paysen *et al.* (1980).

Comments: Throughout its range, wedgeleaf ceanothus may occur as scattered shrubs in a forest understory or the sole or dominant shrub in shrublands. This series is found in several elevation zones in all regions. Self-perpetuating stands are unusual; instead, most stands are established after disturbance and are transitional to various kinds of forest and chaparral.

The Jepson Manual recognizes three varieties of *Ceanothus cuneatus*. Wedgeleaf ceanothus is *C. c.* var. *c.*, coast ceanothus is *C. c.* var. *fascicularis*, and Monterey ceanothus is *C. c.* var. *rigidus*. The last two

varieties are found in the Woollyleaf manzanita series.

Keeler-Wolf (1990e) qualitatively describes wedgeleaf ceanothus stands at Hale Ridge candidate RNA in Lake Co. o-NorCo; Keeler-Wolf (1990d) at Indian Creek candidate RNA in Tehama Co. f-CasR; Keeler-Wolf (1990e) at Big Grizzly Mountain RNA in Mariposa Co. m-SN.

Species mentioned in text	
Black sage	*Salvia mellifera*
Buck bush	*Ceanothus cuneatus*
Chamise	*Adenostoma fasciculatum*
Common manzanita	*Arctostaphylos manzanita*
Deerbrush	*Ceanothus integerrimus*
Fremont silktassel	*Garrya fremontii*
Greenleaf manzanita	*Arctostaphylos patula*
Hollyleaf redberry	*Rhamnus ilicifolia*
Incense-cedar	*Calocedrus decurrens*
Scrub oak	*Quercus berberidifolia*
Squirreltail	*Elymus elymoides*
Toyon	*Heteromeles arbutifolia*
Wedgeleaf ceanothus	*Ceanothus cuneatus*
Woollyleaf manzanita	*Arctostaphylos tomentosa*
	spp. *tomentosa*

Plot-based descriptions: Stuart *et al.* (1992) define two associations in Siskiyou Co. l-KlaR; Taylor & Teare (1979b) Keeler-Wolf (1990e) one association at Smoky Creek candidate RNA in Trinity Co. m-KlaR.

Associations:
tuart *et al.* (1992):
Wedgeleaf ceanothus association,
Wedgeleaf ceanothus - incense-cedar association.
Taylor & Teare (1979b):
Wedgeleaf ceanothus/squirreltail association [as *Ceanothus cuneatus/Sitanion hystrix* association].

White bursage series

White bursage sole or dominant shrub in canopy; brittlebush, chollas, creosote bush, and/or saltbushes may be present. Shrubs < 3 m. Shrub canopy is two-tiered, few creosote bushes in the upper tier over the lower one of white bursage. Ground layer open; annuals seasonally present.

Uplands: alluvial fans; bajadas; partially stabilized, stabilized sand fields; upland slopes. Soils well-drained, may have pavement surface.

Distribution: MojD, ColD, Baja CA.
Elevation: 75 below sea level-1200 m.

NDDB/Holland type and status: **Sonoran desert scrubs** (33000), **Mojavean desert scrubs** (34000).
    Sonoran creosote bush scrub (33100 *in part*) G4 S4.
    Mojave creosote bush scrub (34100 *in part*) G4 S4.

Other types:
Brown Lowe Pase type: 154.113.
Cheatham & Haller type: Mojave creosote bush scrub, Sonoran creosote bush scrub.
Thorne type: Desert dune sand plant community.
WHR type: Desert scrub.

General references: Burk (1977), Hunt (1966), MacMahon (1988), Thorne (1982), Turner (1982b), Turner & Brown (1982), Vasek & Barbour (1977).

Comments: This series is often considered part of the creosote bush scrub, which is better thought of as a collection of series. In this series white bursage dominates, and in the Creosote bush-white bursage series it is important [*see* Brittlebush series, Creosote bush series, Creosote bush-white bursage series].

The shrub cover varies among stands. Dense ones are as common as are areas of desert pavement with occasional shrubs. This variation is included in this series.

Species mentioned in text	
Brittlebush	*Encelia farinosa*
Chollas	*Opuntia* species
Creosote bush	*Larrea tridentata*
Saltbushes	*Atriplex* species
White bursage	*Ambrosia dumosa*

Plot based descriptions: Zedler (1981) reports shrub densities at a site near Ocotillo in Imperial Co. ColD.

White sage series

White sage sole, dominant, or important shrub with California sagebrush in canopy; bush monkeyflower, California buckwheat, chaparral mallow, chaparral yucca, and/or coast goldenbush may be present. Emergent trees may be present. Shrubs < 1 m; canopy continuous or intermittent. Ground layer variable (Plate 12).

Uplands: slopes, rarely flooded low-gradient deposits along streams. Soil shallow.

Distribution: SoCo, m-TraR, m-PenR, ChaI, Baja CA.
Elevation: sea level-1600 m.

NDDB/Holland type and status: Coastal scrubs (32000), **Chaparral** (37000).
    Venturan coastal sage scrub (32300 *in part*) G3 S3.1.
    Diegan coastal sage scrub (32500 *in part*) G3 S3.1.
    Riversidean upland sage scrub (32700 *in part*) G3 S3.1.
    Alluvial fan chaparral (37H00 *in part*) G3 S3.1.

Other types:
Barry type: G7411231 CSAME30.
Cheatham & Haller type: Central coastal scrub, Coastal sage scrub.
Jones & Stoke type: White sage scrub.
PSW-45 type: Salvia series.
Rangeland type: SRM 205.
Thorne type: Southern coastal scrub.
WHR type: Coastal scrub.

General references: Axelrod (1978), Barbour (1994), DeSimone & Burk (1992), Keeley & Keeley (1988), Kirkpatrick & Hutchinson (1977), Malanson (1984), Mooney (1977), O'Leary (1989), Pase & Brown (1982), Paysen *et al.* (1980), Westman (1983).

Comments: This series is often considered part of the coastal scrub, which is better thought of as a collection of series. This approach allows stands of comparable composition to be considered regardless of geographic location.

Keeler-Wolf (1990e) qualitatively describes white sage stands as Riversidean sage scrub at Millard Canyon RNA in Riverside Co. m-TraR; as *Salvia* herbland at Cahuilla Mountain RNA, King Creek RNA in Riverside Co. m-PenR.

If stands have white sage equally important with California buckwheat, they are members of the California buckwheat-white sage series.

Species mentioned in text	
Bush monkeyflower	*Mimulus aurantiacus*
California sagebrush	*Artemisia californica*
California buckwheat	*Eriogonum fasciculatum*
Chaparral mallow	*Malacothamnus fasciculatum*
Chaparral yucca	*Yucca whipplei*
Coast goldenbush	*Isocoma menziesii*
Laurel sumac	*Malosma laurina*
White sage	*Salvia apiana*

Plot-based descriptions: DeSimone & Burk (1992) define a White sage-California sagebrush association [as Group 2], Hanes *et al.* (1989) report sites in two drainages [Elwanda Canyon, upper Santa Ana River] as dominated by white sage, in Orange Co. o-SoCo.

Whiteleaf manzanita series

Whiteleaf manzanita sole or dominant shrub in canopy; chamise, common manzanita, Fremont silktassel, greenleaf manzanita, mountain whitethorn, ocean spray, scrub oak, service berry, tobacco brush, and/or wedgeleaf ceanothus may be present. Emergent conifers may be present. Shrubs < 4 m; canopy continuous or intermittent. Ground layer sparse (Plate 10).

Uplands: ridges, upper slopes, may be steep. Soil shallow, may be ultra-mafic derived.

Distribution: NorCo, m-KlaR, su-KlaR, f-CasR, f-SN, m-SN, OR.
Elevation: 150-1850 m.

NDDB/Holland type and status: **Chaparral** (37000).
  Gabbroic northern mixed chaparral (37111 *in part*) G2 S2.1.
  Mixed montane chaparral (37510 *in part*) G4 S4.
  Montane manzanita chaparral (37520 *in part*) G4 S4.
  Serpentine chaparral (37600 *in part*) G2 S2.1.
  Upper Sonoran manzanita chaparral (37800 *in part*) G4 S4.

Other types:
Barry type: G7411214.
Cheatham & Haller type: Mixed chaparral.
PSW-45 type: Manzanita series.
Thorne type: Mixed chaparral.
WHR type: Mixed chaparral.

General references: Hanes (1977, 1981), Horton (1960), Keeley & Keeley (1988), Patric & Hanes (1964), Pase (1982a), Paysen *et al.* (1980).

Comments: Whiteleaf manzanita occurs as scattered shrubs as a forest understory or the sole or dominant shrub in what is often called montane chaparral. Self-perpetuating stands grow on shallow soils; those transitional to forest occur on deeper soils. This series is found at both low and montane elevations. Whiteleaf manzanita series stands share shrub species and elevations with several shrub-dominated series.

The Pine Hill area with its several endemics mix in stands dominated by chamise [*see* Chamise series] or by whiteleaf manzanita locally forming this series [Gabbroic northern mixed chaparral] on Rescue loam soils in w. El Dorado Co. f-SN (Wilson 1986).

The Jepson Manual recognizes three subspecies of whiteleaf manzanita. *Arctostaphylos viscida* ssp. *mariposa*) grows in SN, *A. v.* ssp. *pulchella* in NorCo, KlaR, and *A. v.* ssp. *viscida* throughout the range of the other subspecies. All subspecies are included in this series.

Keeler-Wolf (1990e) qualitatively describes whiteleaf manzanita stands at Frenzel Creek RNA called serpentine chaparral in Colusa Co. i-NorCo.; Grizzly Mountain RNA in Mariposa Co. m-SN.

Species mentioned in text	
Canyon live oak	*Quercus chrysolepis*
Chamise	*Adenostoma fasciculatum*
Common manzanita	*Arctostaphylos manzanita*
Fremont silktassel	*Garrya fremontii*
Greenleaf manzanita	*Arctostaphylos patula*
Mountain whitethorn	*Ceanothus cordulatus*
Ocean spray	*Holodiscus discolor*
Service berry	*Amelanchier alnifolia*
Scrub oak	*Quercus berberidifolia*
Tobacco brush	*Ceanothus velutinus*
Wedgeleaf ceanothus	*Ceanothus cuneatus*
Whiteleaf manzanita	*Arctostaphylos viscida*

Plot-based descriptions: Stuart *et al.* (1992) defines a Whiteleaf manzanita-greenleaf manzanita-canyon live oak association at Castle Crags State Park in Siskiyou Co. l-KlaR; Keeler-Wolf (1983) in Keeler-Wolf (1990e) describes stands at Frenzel Creek RNA called serpentine chaparral in Colusa Co. i-NorCo.

Winter fat series

Winter fat sole or dominant shrub in canopy; black sagebrush, black bush, shadscale, and/or yellow rabbitbrush may be present. Shrubs < 1 m; canopy continuous, intermittent, or open. Ground layer sparse, grassy (Plate 13).

Wetlands: habitats intermittently flooded, saturated. Water chemistry: mixosaline. Plains; old lake beds perched above current drainages. Cowardin class: Palustrine shrub-scrub wetland. **Uplands:** flats; lower slopes. Soils may be rocky.

Distribution: sj-CenV, s. m-SN, GB, MojD, inter-West.
Elevation: 100-2700 m.

NDDB/Holland type and status: **Chenopod scrubs** (36000).
　　Shadscale scrub (36140 *in part*) G4 S3.

Other types:
Barry type: G7411221.
Brown Lowe Pase type: 152.152, 152.152.
Cheatham & Haller type: Shadscale scrub.
Rangeland type: SRM 414.
Thorne type: Alkali sink scrub.
WHR type: Alkali sink.

General references: Turner (1982a), Young *at al.* (1977), West (1988).

Species mentioned in text	
Big sagebrush	*Artemisia tridentata*
Black sagebrush	*Artemisia nova*
Black bush	*Coleogyne ramosissima*
Shadscale	*Atriplex confertifolia*
Winter fat	*Krascheninnikovia lanata*
Yellow rabbitbrush	*Chrysothamnus viscidiflorus*

Comments: Winter fat generally occurs here as pure stands. It can be a minor component in several series [*see* Black bush series, Black sagebrush series, Big sagebrush series, Shadscale series].

Plot-based descriptions: not available.

Woollyleaf manzanita series

Forms of woollyleaf manzanita dominant or important shrub with one or more rare ceanothus or manzanita in canopy; black sage, California buckwheat, California coffeeberry, California sagebrush, chamise, coyote brush, poison-oak, and/or toyon may be present. Emergent birchleaf mountain-mahogany, and /or coast live oak may be present. Shrubs < 3 m; canopy continuous. Ground layer sparse.

Uplands: bluffs, dunes, mesas, outcrops, slopes, terraces. Soils sand, sandstone, shale, or volcanic-derived.

Distribution: i-CenCo, m-CenCo.
Elevation: sea level-1500 m.

NDDB/Holland type and status:: **Chaparral** (37000).
 Northern maritime chaparral (37C10) G1 S1.2.
 Central maritime chaparral (37C20) G2 S2.1.
 Poison-oak chaparral (37F00 *in part*) G4 S4.

Other types:
Barry type: G7411214.
Cheatham & Haller type: Mixed chaparral.
PSW-45 type: Manzanita series.
Rangeland type: SRM 208.
Thorne type: Mixed chaparral.
WHR type: Mixed chaparral.

General references: Davis & Hickson (1988), Keeley & Keeley (1988), Pase (1982a), Paysen *et al.* (1980), Wells (1962).

Comments: Many areas of chaparral on the o-CenCo and m-CenCo have concentrations of local ceanothus and manzanita species. Such areas are often called maritime chaparral. In this series, forms of woollyleaf manzanita are a common component along with familiar members of chaparral and coastal scrub.

Species mentioned in text	
Birchleaf mountain-mahogany	*Cercocarpus betuloides*
Bishop manzanita	*Arctostaphylos obispoensis*
Black sage	*Salvia mellifera*
Bonnie Doon manzanita	*Arctostaphylos silvicola*
Bracted manzanita	*Arctostaphylos tomentosa* ssp. *bracteosa*
Brittleleaf manzanita	*Arctostaphylos tomentosa* ssp. *crustacea*
California buckwheat	*Eriogonum fasciculatum*
California coffeeberry	*Rhamnus californica*
California sagebrush	*Artemisia californica*
Chamise	*Adenostoma fasciculatum*
Coast ceanothus	*Ceanothus cuneatus* var. *fascicularis*
Coast live oak	*Quercus agrifolia*
Coyote brush	*Baccharis pilularis*
Cropleaf ceanothus	*Ceanothus dentatus*
Dacite manzanita	*Arctostaphylos tomentosa* ssp. *daciticola*
Hearst ceanothus	*Ceanothus hearstiorum*
Hearst manzanita	*Arctostaphylos hookeri* ssp. *hearstiorum*
Hooker manzanita	*Arctostaphylos hookeri* ssp. *hookeri*
Hoover manzanita	*Arctostaphylos hooveri*
La Cruz manzanita	*Arctostaphylos cruzensis*
La Purisima manzanita	*Arctostaphylos purissima*
Little Sur manzanita	*Arctostaphylos edmundsii*
Maritime ceanothus	*Ceanothus maritimus*
Montara manzanita	*Arctostaphylos montaraensis*
Monterey ceanothus	*Ceanothus cuneatus* var. *rigidus*
Morro manzanita	*Arctostaphylos morroensis*
Nipomo Mesa ceanothus	*Ceanothus impressus*
Oso manzanita	*Arctostaphylos osoensis*
Pajaro manzanita	*Arctostaphylos pajaroensis*
Pecho manzanita	*Arctostaphylos pechoensis*
Poison-oak	*Toxicodendron diversilobum*
San Bruno Mountain manzanita	*Arctostaphylos imbricata*
Sand mesa manzanita	*Arctostaphylos rudis*
Sandmat manzanita	*Arctostaphylos pumila*
Santa Barbara manzanita	*Arctostaphylos tomentosa* ssp. *eastwoodiana*
Santa Cruz manzanita	*Arctostaphylos nummularia* var. *sensitiva*
Santa Lucia manzanita	*Arctostaphylos luciana*
Santa Margarita manzanita	*Arctostaphylos pilosula*
Schreiber manzanita	*Arctostaphylos glutinosa*
Toyon	*Heteromeles arbutifolia*
Wells manzanita	*Arctostaphylos wellsii*
Woollyleaf manzanita	*Arctostaphylos tomentosa* ssp. *tomentosa*

The following local manzanitas are found along CenCo: bishop manzanita (a CNPS List 4 plant), Bonnie Doon manzanita (a CNPS List 1B plant), bracted manzanita, brittleleaf manzanita, dacite manzanita (a CNPS List 1B plant), Hearst manzanita (a CNPS List 1B plant), Hooker manzanita (a CNPS List 1B plant), Hoover manzanita (a CNPS List 4 plant), La Cruz manzanita (a CNPS List 1B plant), La Purisima manzanita (a CNPS List 1B plant), Little Sur manzanita (a CNPS List 1B plant), Montara manzanita (a CNPS List 1B plant), Morro manzanita (a CNPS List 1B plant), Oso manzanita (a CNPS List 1B plant), Pajaro manzanita (a CNPS List 1B plant), Pecho manzanita (a CNPS List 1B plant), San Bruno Mountain manzanita (a CNPS List 1B plant), Santa Barbara manzanita (a CNPS List 1B plant), Santa Cruz manzanita, sand mesa manzanita (a CNPS List 1B plant), sandmat manzanita (a CNPS List 1B plant), Santa Lucia manzanita (a CNPS List 1B plant), Santa Margarita manzanita (a CNPS List 1B plant), Schreiber manzanita (a CNPS List 1B plant), and Wells manzanita (a CNPS List 1B plant) (Skinner & Pavlik 1994).

Local ceanothus species are also well represented along CenCo: coast ceanothus, cropleaf ceanothus, Hearst ceanothus, maritime ceanothus (a CNPS List 1B plant), Monterey ceanothus, and Nipomo Mesa ceanothus (Skinner & Pavlik 1994).

Plot-based descriptions: Griffin (1978) describes stands near Fort Ord in Monterey Co.; NDDB has plot data on file for Contra Costa Co., Fort Ord Military Reservation in Monterey Co., and Burton Mesa in Santa Barbara Co. o-CenCo.

Yellow bush lupine series

Yellow bush lupine sole or dominant shrub in canopy; coyote brush, red alder, and/or wax-myrtle present. Shrubs < 2 m; canopy continuous or intermittent. Ground layer variable.

Uplands: stabilized dunes of coastal bars, river mouths, spits along coastline; coastal bluffs, terraces.

Distribution: o-NorCo.
Elevation: sea level-30 m.

NDDB/Holland type and status: Coastal dunes (21000), **Coastal bluff scrubs** (31000), **Coastal scrubs** (32000).
> Northern dune scrub (21310 *in part*) G2 S1.2.
> Northern coastal bluff scrub (31100 *in part*) G2 S2.2.
> Venturan coastal sage scrub (32300 *in part*) G3 S3.1.

Other types:
Barry type: G7411211 CLUAR00.
Cheatham & Haller type: Partially stabilized coastal dunes, Coastal bluff scrub, Northern coastal scrub.
PSW-45 type: Lupine series.
Rangeland type: SRM 204.
Thorne type: Northern coastal scrub.
WHR type: Coastal scrub.

General references: Barbour (1994), Barbour & Johnson (1977c), Davidson & Barbour (1977), Foin & Hektner (1977), Miller (1988), Paysen *et al.* (1980).

Comments: The species range covers most of o-CenCo and o-NorCo. It grows on coastal dunes, headlands, and terraces. This CA native is considered an "exotic" north of Manchester Beach in Mendocino Co. In Humboldt Co. it has a history of being planted to stabilize sand (Miller 1988). It is very successful at colonizing and enriching the sand, and as a result is changing vegetation patterns. Yellow bush lupine stands in Humboldt Co. are being managed by annual removal to restore Sand-verbena - beach bursage series stands and enhance populations of rare species (Pickart *et al.* 1989).

Species mentioned in text	
Beach bursage	*Ambrosia chamissonis*
California figwort	*Scrophularia californica*
Coyote brush	*Baccharis pilularis*
Heather goldenbush	*Ericameria ericoides*
Red alder	*Alnus rubra*
Ripgut	*Bromus diandrus*
Vernal grass	*Anthoxanthum odoratum*
Wax-myrtle	*Myrica californica*
Yellow bush lupine	*Lupinus arboreus*
Yellow sand-verbena	*Abronia latifolia*

Plot-based descriptions: Hektner & Foin (1977) define one association at Sea Ranch on terraces in Sonoma Co.; Holton & Johnson (1979) two associations at Point Reyes National Seashore on sand in Marin Co.; Duebendorfer (1989), LaBanca (1993), Parker (1974) in Major and Barbour (1977a) three associations at Humboldt Bay on sand in Humboldt Co. o-NorCo.

Associations:
Duebendorfer (1989):
> Yellow bush lupine-ripgut association
> [as Lupine/grass type].
Holton & Johnson (1979):
> Yellow bush lupine association
> [= LaBanca (1994)],
> Yellow bush lupine-heather goldenbush association.
Hektner & Foin (1977):
> Yellow bush lupine-vernal grass association.
Parker (1974):
> Yellow bush lupine-California figwort association.

Key 3: Series Dominated by Trees

1 One species dominates tree canopy 2
1' No single species dominates tree canopy 3
   2   A conifer species dominates tree canopy
                               Key to forests where one conifer species dominates p. 214
   2' A single non-conifer species dominates tree canopy
                                 Key to forests where one non-conifer dominates p. 216
3 Two species similarly important in tree canopy
                             Key to forests where two species are similarly important p. 217
3' More than two species similarly important in tree canopy
                   Key to forests where more than two species are similarly important p. 218

Key to forests where one conifer species dominates

1 A *Chamaecyparis* species dominant   *Port Orford-cedar series*
                               See UNIQUE STANDS *section* p. 325
1' A *Chamaecyparis* species not dominant 2
   2   A *Pseudotsuga* species dominant   *Bigcone Douglas-fir series*
                                     Douglas-fir series
   2' A *Pseudotsuga* species not dominant 3
3 A *Picea* species dominant   *Engelmann spruce series*
                                     Sitka spruce series
3' A *Picea* species not dominant 4
   4   A *Tsuga* species dominant   *Mountain hemlock series*
                               Western hemlock series
   4' A *Tsuga* species not dominant 5
5 Incense-cedar dominant   *Incense-cedar series*
5' Incense-cedar not important 6
   6   Giant sequoia dominant   *Giant sequoia series*
   6' Giant sequoia not dominant 7
7 Redwood dominant   *Redwood series*
7' Redwood not dominant 8
   8   Cypress or juniper dominant 9
   8' A fir or pine dominant 10
9 A cypress dominant   CYPRESS WOODLANDS
                                   McNab cypress series
                                   Pygmy cypress series
                                 Sargent cypress series
                            See UNIQUE STANDS *section* p. 325

9' A juniper dominant

JUNIPER WOODLANDS
California juniper series
Mountain juniper series
Utah juniper series
Western juniper series

10 A fir dominant

FIR FORESTS
Grand fir series
Red fir series
Santa Lucia fir series
Subalpine fir series
White fir series
See UNIQUE STANDS *section p. 325*

10' A pine dominant 11

11 A closed-cone pine present

CLOSED-CONE PINE FORESTS
Beach pine series
Bishop pine series
Coulter pine series
Knobcone pine series
Monterey pine series
See UNIQUE STANDS *section p. 325*

11' A closed-cone pine not present 12

12   A pinyon dominant

PINYON WOODLANDS
Parry pinyon series
Singleleaf pinyon series
See UNIQUE STANDS *section p. 325*

12'   A pinyon not dominant 13

13 2- leaved pines dominant

TWO-LEAVED PINE FORESTS
Beach pine series
Bishop pine series
Lodgepole pine series
See UNIQUE STANDS *section p. 325*

13' 3 or 5-leaved pines dominant 14

14 3-leaved pines dominant

THREE-LEAVED PINE FORESTS
Coulter pine series
Foothill pine series
Jeffrey pine series
Knobcone pine series
Monterey pine series
Ponderosa pine series
Washoe pine series
See UNIQUE STANDS *section p. 325*

14' 5-leaved pines dominant

FIVE-LEAVED PINE FORESTS
Bristlecone pine series
Foxtail pine series
Limber pine series

Western white pine series
Whitebark pine series
See UNIQUE STANDS *section p. 325*

Key to forests where one non-conifer dominates

1 Emergent trees present over desert scrub species   DESERT WOODLANDS
Nolina series [see shrub series]
Fan palm series
Joshua tree series [see shrub series]
Mesquite series
Mojave yucca series [see shrub series]
Ocotillo series [see shrub series]
Foothill palo verde-saguaro series [see shrub series]
See UNIQUE STANDS *section p. 325*

1' Desert scrub species not present 2
   2   Tree leaves deciduous 3
   2'   Tree leaves evergreen 8
3   An oak dominant   DECIDUOUS FORESTS AND WOODLANDS
Black oak series
Blue oak series
Oregon white oak series
Valley oak series

3' An oak not dominant 4
   4   An *Alnus* species dominant   *Red alder series*
White alder series

   4'   An *Alnus* species not dominant 5
5   A *Populus* species dominant   *Aspen series*
Black cottonwood series
Fremont cottonwood series

5' A *Populus* species not dominant 6
   6   A *Juglans* species dominant   *California walnut series*
See UNIQUE STANDS *section p. 325*

   6'   A *Juglans* species not dominant 7
7   A *Salix* species dominant   WILLOW THICKETS
Arroyo willow series
Black willow series
Hooker willow series
Pacific willow series
Red willow series
Sitka willow series

7' A *Salix* species not dominant   DECIDUOUS WOODLANDS
California buckeye series
California sycamore series
Mesquite series

8   An oak dominant   EVERGREEN OAK FORESTS AND WOODLANDS
Blue oak series
Canyon live oak series
Coast live oak series
Engelmann oak series
Interior live oak series
Island oak series
Tanoak series

8' An oak not dominant 9

9 A *Cercocarpus* species important   MOUNTAIN-MAHOGANY WOODLANDS
Birchleaf mountain-mahogany series
Curlleaf mountain-mahogany series

9 A *Cercocarpus* species not important   EVERGREEN FORESTS
California bay series
Eucalyptus series
Fan palm series
Tanoak series
See UNIQUE STANDS *section* p. 325

Key to forests where two species are similarly important

1   A *Populus* species important   COTTONWOOD FORESTS
Aspen series
Black cottonwood series
Fan palm series
Fremont cottonwood series

1' A *Populus* species not important 2
  2   California buckeye important   *California buckeye series*
  2' California buckeye and hollyleaf cherry not important 3
3 Blue oak important   *Blue oak series*
3' Blue oak and foothill pine not important 4
  4   A mountain-mahogany and conifer species important   MOUNTAIN-MAHOGANY WOODLANDS
Birchleaf mountain-mahogany series
Curlleaf mountain-mahogany series

  4' A mountain-mahogany and conifer species not important 5
5 A live oak important   MIXED LIVE OAK FORESTS
Bigcone Douglas-fir - canyon live oak series
Black oak series
Canyon live oak series
Coast live oak series
Coulter pine-canyon live oak series
Engelmann oak series
Santa Lucia fir series

5' Canyon live oak not important 6

6   McNab cypress and Sargent cypress important   *Sargent cypress series*
6'   McNab and Sargent cypress not important 7
7 Douglas-fir or white fir important   MIXED FORESTS AND WOODLANDS
                                                                    Black oak series
                                                                    Douglas-fir series
                                                    Douglas-fir - ponderosa pine series
                                                        Douglas-fir - tanoak series
                                                            Port Orford-cedar series
                                                                    Redwood series

7' Douglas-fir or white fir not important 8
    8   Ponderosa pine important   *Black oak series*
                                                    Jeffrey pine-ponderosa pine series
                                                            Ponderosa pine series

    8'   Ponderosa pine not important 9
9 Singleleaf pinyon important   *Singleleaf pinyon-Utah juniper series*
9' Singleleaf pinyon not important   *Giant sequoia series*
                                                            Port Orford-cedar series
                                                                Sitka spruce series

Key to forests where more than two species are similarly important

1 Non-conifer species important   MIXED FORESTS AND WOODLANDS
                                            Blue palo verde-ironwood-smoke tree series
                                                                    Mixed oak series
                                                                    Mixed willow series

1' Conifer species important 2
    2   Giant sequoia important   *Giant sequoia series*
    2'   Giant sequoia not important 3
3 Douglas-fir, incense-cedar, Jeffrey pine, ponderosa pine, sugar pine, or white fir similarly
    important   *Mixed conifer series*
3' Douglas-fir, incense-cedar, Jeffrey pine, ponderosa pine, sugar pine, or white fir not similarly
    important 4
    4   Brewer spruce present   *See* UNIQUE STANDS *section p. 235*
    4'   Brewer spruce absent   *Mixed subalpine forest series*

Arroyo willow series

Arroyo willow sole or dominant shrub or tree in canopy; bigleaf maple, black cottonwood, buttonbush, California sycamore, coyote brush, Fremont cottonwood, Mexican elderberry, mulefat, red osier, wax-myrtle, white alder, and/or willows may be present. If shrubland, emergent trees may be present. Trees < 10 m; canopy continuous. Shrubs sparse under tree canopy. Ground layer sparse or abundant (Plate 18).
Wetlands: habitats seasonally flooded, saturated. Water chemistry: fresh. Floodplains; low gradient depositions along rivers, streams. Cowardin classes: Palustrine forested or shrub-scrub wetland, Palustrine forested wetland. The national inventory of wetland plants (Reed 1988) lists arroyo willow as a FACW.
Distribution: NorCo, CenCo, CenV, f-KlaR, f-CasR, f-SN, So-CA, GB, Baja CA.
Elevation: sea level-1800 m.

NDDB/Holland type and status: Riparian forests (61000), **Riparian woodlands** (62000), **Riparian scrubs** (63000).

Central Coast arroyo willow riparian forest (61230) G3 S3.2.
Southern arroyo willow riparian forest (61320) G2 S2.1.
Southern cottonwood-willow riparian forest (61330 *in part*) G3 S3.
Great Valley willow scrub (63410 *in part*) G3 S3.2.
North Coast riparian scrub (63100 *in part*) G3 S3.2.
Central Coast riparian scrub (63200 *in part*) G3 S3.2.
Southern willow scrub (63320 *in part*) G3 S2.1.

Other types:
Barry type: G7411221 BSALA60.
Cheatham & Haller type: Bottomland woodlands and forest.
PSW-45 type: Willow series.

Thorne type: Riparian woodland.
WHR type: Freshwater emergent wetland.
General references: Bowler (1989), Capelli & Stanley (1984), Faber *et al.* (1989), Holstein (1984), Minnich (1976), Paysen *et al.* (1980), Read & Sprackling (1980), Zedler & Scheid (1988).

Comments: Willow stands may or may not be dominated by a single species. If no dominant willow is present at low elevations, then place the stand in the Mixed willow series. Montane and subalpine willow stands are placed in separate classes since different willow species are restricted to those elevations [*see* Montane wetland shrub habitat, Subalpine wetland shrub habitat]. Stands of the Arroyo willow series have environmental conditions similar to alder, cottonwood, and other willow series [*see* Black cottonwood series, Black willow series, Fremont cottonwood series, Hooker willow series, Pacific willow series, Red alder series, Red willow series, Sitka willow series, White alder series].

Species mentioned in text	
Arroyo willow	*Salix lasiolepis*
Bigleaf maple	*Acer macrophyllum*
Black cottonwood	*Populus balsamifera*
Buttonbush	*Cephalanthus occidentalis*
California sycamore	*Platanus racemosa*
Coyote brush	*Baccharis pilularis*
Fremont cottonwood	*Populus fremontii*
Mexican elderberry	*Sambucus mexicana*
Mulefat	*Baccharis salicifolia*
Mugwort	*Artemisia douglasiana*
Red osier	*Cornus sericea*
Wax-myrtle	*Myrica californica*
Willows	*Salix* species

Plot-based descriptions: Zembal (1989) reports tree density and cover along Santa Margarita and Santa Ana Rivers in Orange Co. o-SoCo; NDDB has data from Santa Catalina Island ChaI; White (1994a) defines a Arroyo willow/mugwort association at Cleghorn Canyon candidate RNA in San Bernardino Co. i. m-TraR; Stromberg & Patten (1992) report willow density for stands in TraSN.

Aspen series

Aspen sole, dominant or important tree with red fir or white fir in canopy. Trees < 35 m; canopy continuous, intermittent, or open. Shrubs common or infrequent. Ground layer abundant, if not grazed (Plate 19).

Wetlands: soils seasonally saturated, permanently saturated. Water chemistry: fresh. Streamsides; springs. Cowardin class: Palustrine forested wetland. **Uplands:** soils shallow, may be rocky. The national inventory of wetland plants (Reed 1988) lists aspen as a FAC+.

Distribution: m-KlaR, m-CasR, su-CasR, m-SN, su-SN, GB, i. m-TraR, Baja CA.
Elevation: 1800-3000 m.

NDDB/Holland type and status: **Riparian forests** (61000), **Broadleaved upland forests** (81000).
    Aspen riparian forest (61520) G4 S3.2.
    Aspen forest (81B00) G5 S3.2.

Other types:
Barry type: G7422212 BPOTR00.
Cheatham & Haller type: Aspen groves.
PSW-45 type: Aspen series.
Rangeland type: SRM 411.
Thorne type: Aspen woodland.
WHR type: Aspen.

General references: Aspen range in CA (Griffin & Critchfield 1972); silvics (Perala 1990). Barbour (1988), Barry (1971), DeByle (1980), DeByle & Zasada (1980), Minnich (1976), Mueggler (1994), Paysen *et al.* (1980), Thorne (1977).

Comments: Aspen stands may be transitional or self-perpetuating (Perala 1990). Most stands are altered by grazing. Mueggler (1988) summarizes intermountain information on this series.

Species mentioned in text	
Aspen	*Populus tremuloides*
Leafy aster	*Aster foliaceus*
Red fir	*Abies magnifica*
White fir	*Abies concolor*
White corn-lily	*Veratrum californicum*

Plot-based descriptions: Talley (1977), Keeler-Wolf (1989j) in Keeler-Wolf (1990e) defines a riparian stand at Babbitt Peak RNA in Sierra Co., Potter (1994) two associations in c. SN, Keeler-Wolf (1985c, 1991) in Keeler-Wolf (1990e) one association at Bell Meadow RNA in Tuolumne Co. in SN; Riegel *et al.* (1990) two upland associations in Modoc Co. WarR.

Associations:
Talley (1977):
    Riparian aspen forest
    [= Keeler-Wolf (1985c, 1989l)].
Potter (1994):
    Aspen association,
    Aspen/upland association.
Riegel *et al.* (1990):
    Aspen/white corn-lily association,
    Aspen/leafy aster association.

Beach pine series

Beach pine sole or dominant tree in canopy; Bishop pine, grand fir, madrone, Sitka spruce, and/or western hemlock may be present. Trees < 25 m; canopy continuous. Shrubs common. Ground layer abundant (Plate 23).

Wetlands: soils intermittently flooded. Water chemistry: fresh. Cowardin class: Palustrine forested wetland. **Uplands:** maritime terraces or sand dunes. Soils excessively or poorly drained. The national inventory of wetland plants (Reed 1988) lists lodgepole pine as a FAC.

Distribution: o-NorCo, OR, WA, Canada.
Elevation: sea level-150 m.

NDDB/Holland type and status: Closed-cone coniferous forests (83000).
Beach pine forest (83110) G4 S2.1.

Other types:
Barry type: G7411112 BPICOCO.
Cheatham & Haller type: Closed-cone coniferous forest.
Thorne type: Shore pine woodland.
WHR type: Closed-cone pine-cypress.

General references: Beach pine range in CA (Griffin & Critchfield 1972); silvics (Lotan & Critchfield 1990). Vogl *et al.* (1977).

Comments: *Pinus contorta* includes four subspecies which are best treated as ecological races (Lotan & Critchfield 1990). The Rocky Mountain race (*P. c.* ssp. *latifolia*) grows outside CA. The Sierra-Cascade race (*P. c.* ssp. *murrayana*) is included in the Lodgepole series. The Mendocino white plains race (*P. c.* ssp. *bolanderi*) is a component of the Pygmy cypress series. The coastal race (*P. c.* ssp. *contorta*), also called shore pine, is the dominant tree in this series. Since the varieties have very different

ecology, each is placed in its own separate series.

The coastal race's cones tend to open at maturity, but do remain attached to the branches. Some stands are self-replacing; others are not. Stands of this series occur along the coast to Alaska.

Species mentioned in text	
Beach pine	*Pinus contorta* ssp. *contorta*
Bishop pine	*Pinus muricata*
Douglas-fir	*Pseudotsuga menziesii*
Grand fir	*Abies grandis*
Madrone	*Arbutus menziesii*
Sitka spruce	*Picea sitchensis*
Western hemlock	*Tsuga heterophylla*

Plot-based descriptions: not available.

Bigcone Douglas-fir series

Bigcone Douglas-fir sole or dominant tree in canopy; black oak, canyon live oak, Coulter pine, incense-cedar, ponderosa pine, single-leaf pinyon, sugar pine, and/or white fir may be present. Trees < 30 m; canopy continuous. Shrubs infrequent. Ground layer sparse.

Uplands: aspect primarily north-facing. Soils sandstone, schist-derived, shallow, well-drained.

Distribution: m-CenCo, m-TraR, m-PenR.
Elevation: 275-2400 m.

NDDB/Holland type and status: Lower montane coniferous forests (84000).
   Bigcone spruce-canyon oak forest (84150 *in part*) G3 S3.1.

Other types:
Barry type: G7411112 BPSMA00.
Cheatham & Haller type: Bigcone spruce-canyon live oak forest.
PSW-45 type: Bigcone Douglas-fir series.
Thorne type: Southern mixed evergreen forest.
WHR type: Montane hardwood conifer.

General references: Bigcone Douglas-fir range in CA (Griffin & Critchfield 1972); silvics (McDonald 1990a). Barbour (1988), McDonald & Littrell (1976), Minnich (1976, 1980b, 1982, 1987), Paysen *et al.* (1980), Sawyer *et al.* (1977).

Comments: In this series bigcone Douglas-fir is the dominant canopy tree whereas in the Bigcone Douglas-fir - canyon live oak series it is important. The canopy is one-tiered here, lacking the dense lower tier of canyon live oak characteristic of the Bigcone Douglas-fir - canyon live oak series.

Species mentioned in text	
Bigcone Douglas-fir	*Pseudotsuga macrocarpa*
Black oak	*Quercus kelloggii*
Canyon live oak	*Quercus chrysolepis*
Coulter pine	*Pinus coulteri*
Incense-cedar	*Calocedrus decurrens*
Ponderosa pine	*Pinus ponderosa*
Singleleaf pinyon	*Pinus monophylla*
Sugar pine	*Pinus lambertiana*
White fir	*Abies concolor*

Plot-based descriptions: McDonald & Littrell (1976) define stands in So-CA; Sproul (1981) in Keeler-Wolf (1990e) in Falls Canyon RNA in Los Angeles Co. i. m-TraR.

Bigcone Douglas-fir - canyon live oak series

Bigcone Douglas-fir and canyon live oak important trees in canopy; California bay, California walnut, black oak, Coulter pine, foothill pine, knobcone pine, and/or single-leaf pinyon may be present. Trees < 30 m; canopy continuous and two-tiered in older stands. Shrubs infrequent. Ground layer sparse.

Uplands: all aspects. Soils shallow, well-drained.

Distribution: m-CenCo, m-TraR, m-PenR. **Elevation:** 275-2400 m.

NDDB/Holland type and status: Lower montane coniferous forests (84000).
　　Bigcone spruce-canyon oak forest (84150 *in part*) G3 S3.1.

Other types:
Barry type: G74 G7411114 BQUCH20.
Cheatham & Haller type: Mixed evergreen.
Thorne type: Mixed evergreen.
WHR type: Montane hardwood conifer forest.

General references: Bigcone Douglas-fir and canyon live oak ranges in CA (Griffin & Critchfield 1972), silvics McDonald (1990a), Thornburgh (1990). Barbour (1988), Cooper (1922), McDonald & Littrell (1976), Minnich (1976, 1980, 1982), Pavlik *et al.* (1991), Sawyer *et al.* (1977), Shreve (1927).

Comments: Bigcone Douglas-fir is dominant in the Bigcone Douglas-fir series and canyon live oak is dominant in the Canyon live oak series. Here they share dominance in the tree canopy. In this series the canopy is two-tiered, with a upper tier of conifer trees above a lower tier of canyon live oak.

Species mentioned in text	
Bigcone Douglas-fir	*Pseudotsuga macrocarpa*
Black oak	*Quercus kelloggii*
California bay	*Umbellularia californica*
California walnut	*Juglans californica*
Canyon live oak	*Quercus chrysolepis*
Coulter pine	*Pinus coulteri*
Foothill pine	*Pinus sabiniana*
Knobcone pine	*Pinus attenuata*
Singleleaf pinyon	*Pinus monophylla*

Plot-based descriptions: McDonald & Littrell (1976), Minnich (1976, 1980a, 1982) describe stands in So-CA; White (1994a) stands at Cleghorn Canyon candidate RNA in San Bernardino Co., Keeler-Wolf (1988d) in Keeler-Wolf (1990e) stands at Millard Canyon RNA in Riverside Co. i. m-TraR.

Birchleaf mountain-mahogany series

Birchleaf mountain-mahogany sole, dominant, or important tree with California juniper, foothill pine, and/or singleleaf pinyon in canopy; California buckwheat, chamise, Eastwood manzanita, and/or holly-leaf redberry. Trees < 8 m. Shrubs common. Ground layer sparse or grassy.

Uplands: ridges, upper slopes, rarely flooded low-gradient deposits along streams. Soils shallow and rocky.

Distribution: i-NorCo, m-NorCo, i-CenCo, f-KlaR, m-KlaR, f-CasR, f-SN, m-SN, So-CA, ChaI, ModP, OR, AZ.
Elevation: 100-2400 m.

NDDB/Holland type and status: **Chaparral** (37000), **Cismontane woodlands** (71000), **Broadleaved upland forests** (81000).

Other types:
Barry type: G74 G7411211.
Cheatham & Haller type: Semi-desert chaparral.
PSW-45 type: Mountain mahogany series.
Thorne type: Desert transitional chaparral.
WHR type: Mixed chaparral.

General references: Paysen *et al.* (1980).

Comments: Birchleaf mountain-mahogany is a wide ranging species. It grows on rocky ridges and steep slopes with thin soil, or on depositions along streams. This plant can be the dominant as a tall shrub or tree. Other trees may be present in these areas. Trees, if present, are from series of the region. Birchleaf mountain-mahogany can occur in other series as well, especially chaparral series where it is important [*see* Bigpod ceanothus - birchleaf mountain-mahogany series, Red shank - birchleaf mountain-mahogany series, Scrub oak - birchleaf mountain-mahogany series]. Coastal scrub

[*see* Birchleaf mountain-mahogany series] and scalebroom stands [*see* Scalebroom series] are closely related to this series.

Keeler-Wolf (1990e) qualitatively describes a Birchleaf mountain-mahogany chaparral at Millard Canyon RNA in Riverside Co. PenR, White (1994a) stands at Cleghorn Canyon candidate RNA in San Bernardino Co. i. m-TraR.

The Jepson Manual recognizes three varieties of birchleaf mountain-mahogany. The typical one (*C. betuloides* var. *b.*) grows throughout CA. *C. b.* var. *macrourus* is a KlaR, CasR, ModP form. *C. b.* var. *blancheae* [Island mountain-mahogany] (a CNPS List 4 plant) is a local ChaI form. *C. minutiflorus* in San Diego Co. s. PenR is included in this series. Catalina Island mountain-mahogany (*C. traskiae*, a CNPS List 1B plant) is a separate species (Skinner & Pavlik 1994).

+---+
| ***Species mentioned in text*** |
| Bigpod ceanothus   *Ceanothus megacarpus* |
| Birchleaf mountain-mahogany *Cercocarpus betuloides* |
| California buckwheat   *Eriogonum fasciculatum* |
| California juniper   *Juniperus californica* |
| Chamise   *Adenostoma fasciculatum* |
| Eastwood manzanita   *Arctostaphylos glandulosa* |
| Foothill pine   *Pinus sabiniana* |
| Hollyleaf redberry   *Rhamnus ilicifolia* |
| Scrub oak   *Quercus berberidifolia* |
| Singleleaf pinyon   *Pinus monophylla* |
+---+

Plot-based descriptions: Taylor & Teare (1979a) in Keeler-Wolf (1990e) define one association on ridges and steep slopes at Manzanita Creek candidate RNA in Trinity Co. m-KlaR; Gordon & White (1994) one association in m-TraR, m-PenR.

Associations:
Taylor & Teare (1979a):
    Birchleaf mountain-mahogany association.
Gordon & White (1994):
    Birchleaf mountain-mahogany association.

Bishop pine series

Bishop pine sole or dominant tree in canopy; beach pine, Bolander pine, Douglas-fir, madrone, Monterey pine, pygmy cypress, and/or redwood may be present. Trees < 25 m; canopy continuous. Shrubs absent, infrequent, or common. Ground layer sparse or abundant.

Uplands: maritime terraces, headlands, rocky ridges. Soils shallow, acid, may be inadequately drained.

Distribution: o-NorCo, o-CenCo, ChaI, Baja CA.
Elevation: sea level-400 m.

NDDB/Holland type and status: Closed-cone coniferous forests (83000).
>   Northern Bishop pine forest (83121) G2 S2.2.
>   Southern Bishop pine forest (83122) G1 S1.1.

Other types:
Barry type: G7411112 BPIMU00.
Cheatham & Haller type: Closed-cone coniferous forest.
PSW-45 type: Bishop pine series.
Thorne type: Closed-cone conifer woodland.
WHR type: Closed pine-cypress.

General references: Bishop pine range in CA (Griffin & Critchfield 1972). Axelrod (1980a), Millar (1986), Minnich (1987), Paysen *et al.* (1980), Philbrick & Haller (1977), Vogl *et al.* (1977).

Comments: Populations of Bishop pine have been treated taxonomically several ways (Haller 1967). Following Vogl *et al.* (1977), *Pinus remorata* and *P. r.* var. *borealis* are included in this series. The degree of cone persistence varies among the stands (Millar 1986).

Only in Mendocino and Humboldt Cos. does Bishop pine commonly mix with other conifers. Elsewhere it forms stands distinct from local chaparral and forest of several series in the mainland. Island stands are particularly different in species composition from one another (Philbrick & Haller 1977). Individuals or small populations of Bishop pine can occur in other series. The extensive Gray Creek stand at its northern extent is logged.

Species mentioned in text	
Beach pine	*Pinus contorta* ssp. *contorta*
Bear-grass	*Xerophyllum tenax*
Bishop pine	*Pinus muricata*
Bolander pine	*Pinus contorta* ssp. *bolanderi*
Douglas-fir	*Pseudotsuga menziesii*
Labrador-tea	*Ledum glandulosum*
Madrone	*Arbutus menziesii*
Monterey pine	*Pinus radiata*
Pygmy cypress	*Cupressus goveniana* ssp. *pygmaea*
Rayless arnica	*Arnica discoidea*
Redwood	*Sequoia sempervirens*

Plot-based descriptions: Cole (1980), Vogl *et al.* (1977) report species of the La Purisima Hills stands on diatomaceous mudstone in Santa Barbara Co. o-CenCo; Westman & Whittaker (1975) defines four associations as part of ordination analysis of gradients from pygmy cypress to redwood-dominated stands in Mendocino Co., NDDB has plot data in the Gary Creek stand before it was logged in Humboldt Co. o-NorCo.

Associations:
Westman & Whittaker (1975):
>   Bishop pine/bear-grass association
>   [as Bishop pine type],
>   Bishop pine-Bolander pine/Labrador-tea association
>   [as Bishop pine-Bolander pine type],
>   Bishop pine-Bolander pine/rayless arnica association [as Xeric Bishop pine type],
>   Bishop pine - Douglas-fir association.

Black cottonwood series

Black cottonwood sole, dominant, or important tree in canopy; aspen, bigleaf maple, box elder, Fremont cottonwood, Hooker willow, Jeffrey pine, lodgepole pine, narrowleaf willow, Oregon ash, red alder, Scouler willow, wax-myrtle, western juniper, white fir, and/or yellow willow may be present. Trees < 25 m; canopy continuous or intermittent. Shrubs common. Ground layer abundant or sparse.

Wetlands: soils seasonally flooded, permanently saturated. Water chemistry: fresh. Canyon slopes; margins of lakes, meadows, floodplains, deltas; river mouths and terraces; stream gravel bars. Cowardin class: Palustrine forested wetland. The national inventory of wetland plants (Reed 1988) lists balsam cottonwood as a FACW.

Distribution: o-NorCo, m-NorCo, CenCo, l-KlaR, m-KlaR, m-CasR, SN, o-SoCo, m-TraR, m-PenR, TraSN, OR, WA, Canada.
Elevation: sea level-2800 m.

NDDB/Holland type and status: **Riparian forests** (61000).
    North Coast black cottonwood riparian forest (61110) S1.1.
    Montane black cottonwood riparian forest (61530) G4 S3.2.

Other types:
Barry type: G7411121 BPOTR60.
Cheatham & Haller type: Northern riparian woodland, Montane riparian woodland.
PSW-45 type: Cottonwood series.
Rangeland type: SRM 203.
Thorne type: Riparian woodlands.
WHR type: Montane riparian, Valley foothill riparian.

General references: Black cottonwood range in CA

(Griffin & Critchfield 1972); silvics (DeBell 1990) DeBell (1980), Holstein (1984), McBride (1994), Paysen *et al.* (1980), Roberts (1984), Roberts *et al.* (1980).

Comments: Low elevation and montane elevation stands are included in this series. Largest low elevation stands occur along the Eel River o-NorCo. Montane stands are small, but widely dispersed.

Species mentioned in text	
Aspen	*Populus tremuloides*
Bigleaf maple	*Acer macrophyllum*
Black cottonwood	*Populus balsamifera*
Box elder	*Acer negundo*
Hooker willow	*Salix hookeriana*
Fremont cottonwood	*Populus fremontii*
Jeffrey pine	*Pinus jeffreyi*
Lodgepole pine	*Pinus contorta* ssp. *murrayana*
Narrowleaf willow	*Salix exigua*
Oregon ash	*Fraxinus latifolia*
Red alder	*Alnus rubra*
Scouler willow	*Salix scouleriana*
Wax-myrtle	*Myrica californica*
Western juniper	*Juniperus occidentalis*
White fir	*Abies concolor*
Yellow willow	*Salix lutea*

Plot-based descriptions: Zembal (1989) reports tree density and cover along Santa Margarita and Santa Ana Rivers in Orange Co. o-SoCo; Stromberg & Patten (1992) report black cottonwood densities along streams in Inyo Co. TraSN.

Black oak series

Black oak sole, dominant, or important tree in canopy; California bay, California buckeye, canyon live oak, coast live oak, Douglas-fir, incense-cedar, Jeffrey pine, knobcone pine, madrone, Oregon white oak, ponderosa pine, valley oak, and/or white fir may be present. Trees < 40 m; canopy continuous, intermittent, or savanna-like. Shrubs infrequent or common. Ground layer sparse or grassy (Plate 21).

Uplands: various aspects and topographic settings. Soils moderately to excessively drained.

Distribution: NorCo, CenCo, l-KlaR, f-KlaR, m-KlaR, f-CasR, m-CasR, f-SN, m-SN, m-TraR, m-PenR.
Elevation: 60-2500 m.

NDDB/Holland type and status: Cismontane woodlands (71000), **Broadleaved upland forests** (81000), **Lower montane coniferous forests** (84000).
    Black oak woodland (71120) G3 S3.3.
    Black oak forest (81340) G4 S4.
    Sierran mixed coniferous forest (84320 *in part*) G4 S4.

Other types:
Barry type: G74 G7411121 CQUKE00.
Cheatham & Haller type: Northern oak woodland.
PSW-45 type: Black oak series.
Thorne type: Northern mixed evergreen forest.
WHR type: Montane hardwood.

General references: Black oak range in CA (Griffin & Critchfield 1972); silvics (McDonald 1990b). Allen *et al.* (1989, 1991), Allen-Diaz & Holtzman (1991), Barbour (1988), Gemmill (1980), Griffin (1977), MacDonald (1980a), Pavlik *et al.* (1991) Paysen *et al.* (1980).

Comments: Stands of this series are often described as forests or woodlands. This series is extensive at low and montane elevations in Cis-CA. Some stands appear self-maintaining, and others originate after disturbance.

Keeler-Wolf (1990e) qualitatively describes black oak-dominated stands at Hosselkus Limestone candidate RNA in Shasta Co. l-KlaR; Keeler-Wolf (1990d) at Indian Creek candidate RNA in Tehama Co. f-CasR.

Species mentioned in text	
Bigleaf maple	*Acer macrophyllum*
Black oak	*Quercus kelloggii*
Bracken	*Pteridium aquilinum*
California bay	*Umbellularia californica*
Canyon live oak	*Quercus chrysolepis*
Coast live oak	*Quercus agrifolia*
Deerbrush	*Ceanothus integerrimus*
Douglas-fir	*Pseudotsuga menziesii*
Grass nut	*Triteleia laxa*
Greenleaf manzanita	*Arctostaphylos patula*
Jeffrey pine	*Pinus jeffreyi*
Incense-cedar	*Calocedrus decurrens*
Knobcone pine	*Pinus attenuata*
Madrone	*Arbutus menziesii*
Ocean spray	*Holodiscus discolor*
Oregon white oak	*Quercus garryana*
Poison-oak	*Toxicodendron diversilobum*
Ponderosa pine	*Pinus ponderosa*
Storax	*Styrax officinalis*
Valley oak	*Quercus lobata*

Plot-based descriptions: Allen *et al.* (1989, 1991) define 11 black oak subseries based on tree and shrub data. They are called associations here. Thornburgh (1981), Keeler-Wolf (1989i) in Keeler-Wolf (1990e) describe stands at Ruth candidate RNA in Trinity Co., Newton (1987) in Keeler-Wolf (1990e) describes stands at Devil's Basin candidate RNA in Tehama Co. i-NorCo; Jimerson (1993) defines one association in Six Rivers National Forest, Wainwright & Barbour (1984) one association in Annadel State Park in Sonoma Co. o-NorCo; (Palmer 1981) in Keeler-Wolf (1990e) describes one stand at Sugar Pine Point RNA in Nevada Co., Keeler-Wolf (1987f) in Keeler-Wolf (1990e) defines one association at Grizzly Mountain RNA in Mariposa Co. m-SN; Keeler-Wolf (1986d, 1989m) in Keeler-Wolf (1990e) one association at Cahuilla Mountain RNA in Riverside Co. m-PenR.

Associations:
Allen *et al.* (1991):
>Black oak-canyon live oak/poison-oak association,
>Black oak-coast live oak-pine/ocean spray association,
>Black oak/deerbush association,
>Black oak/deerbush-poison-oak/bracken association,
>Black oak/grass association,
>Black oak/greenleaf manzanita association,
>Black oak-madrone-coast live oak association,
>Black-oak/poison-oak association,
>Black oak/poison-oak/grass association,
>Black oak/poison-oak storax/grass nut association,
>Black oak-valley oak/grass association,
>Canyon live oak-black oak association,
>Mixed oak-coast live oak/poison-oak association.

Keeler-Wolf (1986b, 1990b):
>Coulter pine-black oak association.

Keeler-Wolf (1987f):
>Black oak association.

Jimerson (1993):
>Black oak - Douglas-fir - bigleaf maple association.

Wainwright & Barbour (1984):
>Black oak - Douglas-fir association.

Black willow series

Black willow sole or dominant shrub or tree in canopy; California sycamore, coyote brush, Fremont cottonwood, Mexican elderberry, mulefat, white alder, and/or willows may be present. If shrubland, emergent trees may be present. Shrubs < 30 m; canopy continuous. Shrubs sparse under tree canopy. Ground layer variable.

Wetlands: habitats seasonally flooded, saturated. Water chemistry: fresh. Floodplains; low-gradient depositions along rivers, streams; meadow edges. Cowardin class: Palustrine forested or shrub-scrub wetland. The national inventory of wetland plants (Reed 1988) lists black willow as a OBL.

Distribution: i-NorCo, CenV, f-CasR, f-SN, SoCo, m-PenR, MojD. ColD.
Elevation: sea level-500 m.

NDDB/Holland type and status: Riparian forests (61000), **Riparian scrubs** (63000).
  Southern cottonwood-willow riparian forest (61330 *in part*) G3 S3.
  Great Valley cottonwood forest (61410 *in part*) G3 S2.1.
  Great Valley mixed riparian forest (61420 *in part*) G2 S2.1.
  Mojave riparian forest (61700 *in part*) G1 S1.1.
  Sonoran cottonwood-willow riparian forest (61810 *in part*) G2 S1.1.
  Central Coast riparian scrub (63200 *in part*) G3 S3.2.
  Southern willow scrub (63320 *in part*) G3 S3.2.

Other types:
Barry type: G7411121 BSALIX0.
Cheatham & Haller type: Bottomland woodlands and forest.
PSW-45 type: Willow series.
Thorne type: Riparian woodland.

WHR type: Desert riparian, Freshwater emergent wetland.

General references: Bowler (1989), Capelli & Stanley (1984), Faber *et al.* (1989), Holstein (1984), Paysen *et al.* (1980).

Comments: Willow stands may or may not be dominated by a single species. If no dominant willow is present at low elevations, then place the stands in the Mixed willow series. Montane and subalpine willow stands are placed in separate classes since other willow species are restricted to those elevations [*see* Montane wetland shrub habitat, Subalpine wetland shrub habitat].

Stands of Black willow have environmental conditions similar to alder, cottonwood, and other willow series [*see* Arroyo willow series, Black cottonwood series, Fremont cottonwood series, Hooker willow series, Pacific willow series, Red alder series, Red willow series, Sitka willow series, White alder series].

Species mentioned in text	
Arroyo willow	*Salix lasiolepis*
Black cottonwood	*Populus balsamifera*
Black willow	*Salix gooddingii*
California sycamore	*Platanus racemosa*
Coyote brush	*Baccharis pilularis*
Fremont cottonwood	*Populus fremontii*
Hooker willow	*Salix hookeriana*
Mexican elderberry	*Sambucus mexicana*
Mulefat	*Baccharis salicifolia*
Pacific willow	*Salix lucida* ssp.*lasiandra*
Red willow	*Salix laevigata*
Sitka willow	*Salix sitchensis*
White alder	*Alnus rhombifolia*
Willows	*Salix* species

Plot-based descriptions: not available.

Blue oak series

Blue oak sole, dominant, or important with California juniper, coast live oak, foothill pine, interior live oak, or valley oak in tree canopy; western juniper may be present. Trees < 18 m; canopy continuous, intermittent, or savanna-like, one or two-tiered. Shrubs infrequent or common. Ground layer grassy (Plate 20).

Uplands: valleys, slopes gentle to steep. Soils shallow, infertile, moderately to excessively-drained with extensive rock fragments. Surface may be covered with stones and rock outcrops.

Distribution: s. o-NorCo, i-NorCo, i-CenCo, f-KlaR, f-CasR, f-SN, SoCo, w. ModP. **Elevation:** 30-1700 m.

NDDB/Holland type and status: **Cismontane woodlands** (71000).
>   Blue oak woodland (71140) G3 S3.2.
>   Alvord oak woodland (71170) G2 S2.2.
>   Open digger pine woodland (71310 *in part*) G4 S4.
>   Digger pine-oak woodland (71410 *in part*) G4 S4.
>   Juniper-oak cismontane woodland (71430 *in part*) G4 S4.

Other types:
Barry type: G74 G7411121 CQUOD00.
Cheatham & Haller type: Blue oak woodland.
PSW-45 type: Blue oak series.
Rangeland type: SRM 201.
Thorne type: Foothill woodland.
WHR type: Blue oak woodland.

General references: Blue oak range in CA (Griffin & Critchfield 1972); silvics (McDonald 1990b). Allen *et al.* (1989, 1991), Allen-Diaz & Bartolome (1992), Allen-Diaz & Holtzman (1991), Baker *et al.* (1981), Barbour (1988), Borchert (1994), Borchert *et al.* (1989), Borchert *et al.* (1991), Gemmill (1980), Griffin (1977), Holland (1976), McClaran & Bartolome (1989,

1990), Mening (1990), Mensing (1990), Neal (1980), Paysen *et al.* (1980), Pavlik *et al.* (1991).

Species mentioned in text	
Alvord oak	*Quercus x alvordiana*
Bajada lupine	*Lupinus concinnus*
Birchleaf mountain-mahogany	*Cercocarpus betuloides*
Blue larkspur	*Delphinium parryi*
Blue oak	*Quercus douglasii*
Blue-eyed Mary	*Collinsia sparsiflora*
Bowlesia	*Bowlesia incana*
California phacelia	*Phacelia imbricata*
California plantain	*Plantago erecta*
Chile lotus	*Lotus subpinnatus*
Coast live oak	*Quercus agrifolia*
Common fiddleneck	*Amsinckia intermedia*
Foothill pine	*Pinus sabiniana*
Foxtail	*Hordeum leporinum*
Goldenback fern	*Pentagramma triangularis*
Hillside gooseberry	*Ribes californica*
Lemmon needlegrass	*Stipa lemmonii*
Linearleaf goldenbush	*Ericameria linearifolia*
Interior live oak	*Quercus wislizenii*
Johnny-jump-up	*Viola pedunculata*
Mission star	*Lithophragma cymbalaria*
Phloxleaf bedstraw	*Galium andrewsii*
Purple needlegrass	*Stipa pulchra*
Rigiopappus	*Rigiopappus leptocladus*
Ripgut brome	*Bromus diandrus*
Rusty popcorn flower	*Plagiobothrys nothofulvus*
Tree clover	*Trifolium ciliolatum*
Valley oak	*Quercus lobata*
Wand buckwheat	*Eriogonum elongatum*
Wart spurge	*Euphorbia spathulata*
Wedgeleaf ceanothus	*Ceanothus cuneatus*
Western juniper	*Juniperus occidentalis* ssp. *occidentalis*
Whiteleaf manzanita	*Arctostaphylos viscida*
Whitestem filaree	*Erodium moschatum*
Woodland star	*Lithophragma affine*

Comments: Stands of this series are commonly described as blue oak woodlands, but forest and savanna stands also qualify. Alvord oak stands are included in this series. Keeler-Wolf (1990e) qualitatively describes blue oak-dominated stands at Jawbone Ridge in Tuolumne Co. f-SN.

Plot-based descriptions: Allen *et al.* (1989, 1991) define 11 subseries in Cis-CA based on tree and shrub data. They are called associations here. Borchert *et al.* (1993a) define 13 associations in San Luis Obispo and Santa Barbara Co., Keeler-Wolf (1989a) describes stands at Wagon Caves candidate RNA in Monterey Co. CenCo.; Keeler-Wolf (1986b, 1990d) in Keeler-Wolf (1990e) stands at Indian Creek candidate RNA f-CasR; Newton (1987) in Keeler-Wolf (1990e) define one association at Devil's Basin candidate RNA in Tehama Co. i-NorCo.; McClaran (1987) describes stands in Tulare Co. f-SN.

Associations:
Allen, *et al.* (1991):
    Blue oak-coast live oak/grass association,
    Blue oak-foothill pine/grass association,
    Blue oak-foothill pine/wedgeleaf ceanothus - birchleaf mountain-mahogany association,
    Blue oak-foothill pine/whiteleaf manzanita/grass association,
    Blue oak/grass association,
    Blue oak-interior live oak/grass association,
    Blue oak-interior live oak/wedgeleaf ceanothus/ grass association,
    Blue oak/linearleaf goldenbush association,
    Blue oak-understory oak/grass association,
    Blue oak-valley oak-coast live oak/grass association,
    Blue oak-valley oak/grass association,
    Blue oak/wedgeleaf ceanothus/grass association,
    Interior live oak-blue oak-foothill pine/grass association.
Borchert *et al.* (1993a):
    Blue oak/bajada lupine-tree clover association,
    Blue oak/birchleaf mountain-mahogany/bowlesia-woodland star association,
    Blue oak/blue larkspur-California phacelia association,
    Blue oak/blue-eyed Mary-rigiopappus association,
    Blue oak/Chile lotus-purple needlegrass association,
    Blue oak/common fiddleneck-rusty popcorn flower association,
    Blue oak/foxtail-Johnny-jump-up association,
    Blue oak/hillside gooseberry/ripgut brome association,

Blue oak/phloxleaf bedstraw-bajada lupine association,
Blue oak/wand buckwheat/Chile lotus-California plantain association,
Blue oak/wart spurge-goldenback fern association,
Blue oak/whitestem filaree-foxtail association,
Blue oak-interior live oak/mission star association.
Newton (1987):
    Blue oak/Lemmon needlegrass association.

Blue palo verde-ironwood-smoke tree series

Blue palo verde, ironwood, or smoke tree sole, dominant, and/or important trees in canopy; desert-willow, black willow, honey mesquite, and/or screwbean may be present. Trees < 18 m tall. Tree canopy continuous, intermittent, or open. Shrubs common. Ground layer sparse; annuals seasonally present (Plate 22).

Wetlands: soils intermittently flooded, saturated. Water chemistry: fresh. Arroyo margins; seasonal watercourses. Cowardin class: Temporarily flooded palustrine forested wetland.

Distribution: ColD, Baja CA.
Elevation: 10 below sea level-500 m.

NDDB/Holland type and status: **Riparian woodlands** (62000).
> Desert dry wash woodland (62200) G3 S3.2.

Other types:
Barry type: G74111121 BCEFL20.
Bown Lowe Pase type: 153.141, 154.115 154.123, 154.127.
Cheatham & Haller type: Southern alluvial woodland.
PSW-45 type: Desert ironwood series, Palo verde series, Smoke tree series.
Rangeland type: SRM 507.
Thorne type: Desert microphyll woodland.
WHR type: Desert wash.

General references: Burk (1977), MacMahon (1988), Paysen *et al.* (1980), Thorne (1982), Turner & Brown (1982).

Comments: These trees form mixed or single species stands in ColD washes. The pure species stands of blue palo verde, desert-willow, ironwood, or smoke tree are included in this series at this time. Some extensive e. ColD washes are dominated by ironwood; w. ColD stands are dominated by smoke tree.

This series is similar in habitat to Catclaw acacia series which occurs at MojD elevations. Stands are transitional in character at intermediate elevations.

Species mentioned in text	
Black willow	*Salix gooddingii*
Blue palo verde	*Cercidium floridum*
Desert-willow	*Chilopsis linearis*
Honey mesquite	*Prosopis glandulosa*
Ironwood	*Olneya tesota*
Screwbean	*Prosopis pubescens*
Smoke tree	*Psorothamnus spinosa*

Plot-based descriptions: NDDB has plot data on file for on ironwood stands in Imperial Co., McHargue (1973) describes washes in Coachella Valley in Riverside Co., Spolsky (1979) defines five associations at Anza Borrego State Park in San Diego Co. ColD.

Associations:
Spolsky (1979):
> Blue palo verde wash woodland association,
> Desert-willow woodland association,
> Ironwood woodland association,
> Mixed wash woodland association,
> Smoketree wash woodland association.

Bristlecone pine series

Bristlecone pine sole or dominant tree in canopy; limber pine may be present. Trees < 18 m; canopy open. Shrubs infrequent or conspicuous. Ground layer sparse (Plate 28).

Uplands: all slopes, especially ridges and upper slopes below forestline. Soil dolomitic, limestone, or granitic-derived.

Distribution: su-WIS, su-DesR, inter-West.
Elevation: 2600-3600 m.

NDDB/Holland type and status: **Subalpine coniferous forests** (86000).
    Bristlecone pine forest (86400) G3 S2.3.

Other types:
Barry type: G7411112 BPILO00.
Cheatham & Haller type: Bristlecone pine forest.
PSW-45 type: Bristlecone pine series.
Thorne type: Bristlecone pine woodland.
WHR type: Subalpine conifer.

General references: Bristlecone pine range in CA (Griffin & Critchfield 1972). Beasley & Klemmedson (1980), Billings & Thompson (1957), Fritts (1969), Hawksworth & Bailey (1980), Lanner (1984), Lloyd & Mitchell (1973), Marchand (1973), Mooney (1973), Mooney *et al.* (1962), Paysen *et al.* (1980), Schulman (1954), Vasek & Thorne (1977), Wright & Mooney (1965).

Comments: Bristlecone pine may form single specific stands or mix with limber pine. The most famous stands are at Schulman Grove in the White Mountain RNA (Keeler-Wolf 1990e) and in the Ancient Bristlecone Pine Botanical Area in Inyo Co. WIS. Here the Bristlecone pine grows on dolomite.

Keeler-Wolf (1990e) qualitatively describes forests of Bristlecone pine and limber pine at Whippoorwill Flat RNA in Inyo Co. WIS. Bristlecone pine is rare (a CNPS List 4 plant) (Skinner & Pavlik 1994).

Species mentioned in text	
Bristlecone pine	*Pinus longaeva*
Limber pine	*Pinus flexilis*
Littleleaf mountain-mahogany	*Cercocarpus intricatus*

Plot-based descriptions: Taylor (1979) in Keeler-Wolf (1990e) defines two associations at the White Mountain RNA in Inyo Co. WIS.

Associations:
Taylor (1979):
    Bristlecone pine association,
    Bristlecone pine/littleleaf mountain-mahogany association.

California bay series

California bay sole or dominant tree in canopy; coast live oak, coast silktassel, Douglas-fir, interior live oak, madrone, redwood, and/or tanoak may be present. Trees < 25 m; canopy continuous. Shrubs infrequent. Ground layer sparse or absent (Plate 21).

Wetlands: soils intermittently flooded. Water chemistry: fresh. Streamsides. Cowardin class: Palustrine forested wetland. **Uplands:** all aspects. Soils sandstone, schist-derived. The national inventory of wetland plants (Reed 1988) lists California bay as a FAC+.

Distribution: o-NorCo, o-CenCo.
Elevation: sea level-500 m.

NDDB/Holland type and status: Broadleaved upland forests 981000).
    California bay forest (81200) G3 S3.2.
    Silktassel forest (81900) G3 S3.

Other types:
Barry type: G74 G741111 CUMCA00.
Cheatham & Haller type: California bay forest.
PSW-45 type: California bay series.
Thorne type: Northern evergreen forest.
WHR type: Coastal oak woodland.

General references: California bay range in CA (Griffin & Critchfield 1972); silvics (Stein 1990b). Barbour (1988), McBride (1974), Paysen *et al.* (1980), Unsiker (1974).

Comments: California bay is an important component of several o-NorCo and o-CenCo series. The species range is much more extensive than the series however. For a stand to be a member of this series, California bay must be the tree canopy dominant. In many cases, California bay is the only tree species present in old stands and few shrubs and herbs are present (McBride 1974). Most large stands have been cut and the land cleared (Stein 1990b).

Coast silktassel is a common component of this series in coastal stands. There are a few coast silktassel-dominated stands lacking California bay in o-NorCo [*ex.* Clam Beach cliffs, Humboldt Co.].

Species mentioned in text	
California bay	*Umbellularia californica*
Canyon live oak	*Quercus chrysolepis*
Coast live oak	*Quercus agrifolia*
Coast silktassel	*Garrya elliptica*
Interior live oak	*Quercus wislizenii*
Madrone	*Arbutus menziesii*
Redwood	*Sequoia sempervirens*
Tanoak	*Lithocarpus densiflora*

Plot-based descriptions: Fiedler & Leidy (1987) define one association at Ring Mountain, Parker (1990) one association at Mount Tamalpais in Marin Co. o-NorCo. McBride (1974) compares the species composition of what he calls California bay woodland stands and oak woodland stands in Contra Costa Co., Campbell (1980) defines one association in Santa Barbara Co. o-CenCo.

Associations:
Campbell (1980):
    California bay association
    [as *Umbellularia* community].
Fiedler & Leidy (1987):
    Tanoak-California bay association
    [as Mixed broadleaf evergreen forest].
Parker (1990):
    California bay-madrone association.

California buckeye series

California buckeye dominant or important tree in canopy; California bay, foothill ash, foothill pine, hollyleaf cherry, interior live oak, and/or toyon may be present. Trees < 10 m; canopy continuous or intermittent, one or two-tiered. Shrubs infrequent. Ground layer sparse (Plate 21).

Uplands: slopes north-facing, steep. Soils shallow, moderately to excessively drained.

Distribution: i-NorCo, i-CenCo, f-SN, m-TraR. Elevation: 100-1500 m.

NDDB/Holland type and status: **Broadleaved upland forests** (81000).
Mixed north slope forest (81500) G4 S4.
Mainland cherry forest (81820 *in part*) G1 S1.1.

Other types:
Barry type: G74 G7411121 BAECA00.
Cheatham & Haller type: Northern oak woodland.
PSW-45 type: California buckeye series.
Thorne type: Northern oak woodland.
WHR type: Montane hardwood forest.

General references: California buckeye range in CA (Griffin & Critchfield 1972). Griffin (1977), Paysen *et al.* (1980).

Comments: Griffin (1977) suggests plants of this series are tree-sized because they grow in locations of low fire frequency. California buckeye-dominated stands in f-SN lack hollyleaf cherry; otherwise they are similar to i-NorCo and i-CenCo stands which have hollyleaf cherry (Klyver 1931).

Species mentioned in text	
California bay	*Umbellularia californica*
California buckeye	*Aesculus californica*
Foothill ash	*Fraxinus dipetala*
Foothill pine	*Pinus sabiniana*
Hollyleaf cherry	*Prunus ilicifolia*
Interior live oak	*Quercus wislizenii*
Toyon	*Heteromeles arbutifolia*

Plot-based descriptions: not available.

California juniper series

California juniper sole or dominant tree in canopy; desert scrub oak, Joshua tree, Parry pinyon, and/or singleleaf pinyon may be present. Emergent trees may be present over a shrub canopy; big sagebrush, black bush, cliffrose, ephedra, Mojave yucca, and/or chaparral yucca may be present. Trees < 5 m. Trees scattered. Shrubs < 2m; intermittent or open. Ground layer sparse or grassy (Plate 10).

Uplands: ridges, slopes, valleys. Soils bedrock or alluvium-derived.

Distribution: i-CenCo, m-TraR, m-PenR, MojD, Baja CA.
Elevation: 1000-2450 m.

NDDB/Holland type and status: **Piñon and juniper woodlands** (72000).
    Mojavean juniper woodland and scrub (72220) G4 S4.
    Peninsular juniper woodland and scrub (72320 *in part*) G3 S3.
    Cismontane juniper woodland and scrub (72400) G2 S2.1.

Other types:
Barry type: G74 G7411112 BJUCA00.
Cheatham & Haller type: Mojavean pinyon-juniper woodland.
PSW-45 type: Juniper series.
Thorne type: Pinyon-juniper woodland.
WHR type: Juniper.

General references: Thorne (1982), Vasek & Thorne (1977).

Comments: Pinyon-juniper woodlands are better considered a collection of series. California juniper occurs in many stands and series. If a pinyon or Joshua tree occurs then the dominance rule applies to assign the stand [*see* Joshua tree series, Parry pinyon series, Singleleaf pinyon series]. California juniper is a component of Scalebroom stands in i-SoCo. Stands in w. MojD are restricted due to systematic removal for agriculture.

Keeler-Wolf (1990c) qualitatively describes stands on limestone at Long Canyon candidate RNA in Kern Co. m-SN.

Species mentioned in text	
Big sagebrush	*Artemisia tridentata*
Black bush	*Coleogyne ramosissima*
California juniper	*Juniperus californica*
Chaparral yucca	*Yucca whipplei*
Cliffrose	*Purshia mexicana*
Desert scrub oak	*Quercus turbinella*
Ephedra	*Ephedra* species
Joshua tree	*Yucca brevifolia*
Mojave yucca	*Yucca schidigera*
Parry pinyon	*Pinus quadrifolia*
Scalebroom	*Lepidospartum squamatum*
Singleleaf pinyon	*Pinus monophylla*

Plot-based descriptions: Boyd (1983) describes two stands in Gavilan Hills in Riverside Co. i. m-TraR; Phillips *et al.* (1980) two stands in Joshua Tree National Monument in San Bernardino Co. MojD; Spolsky (1979) defines a Desert transition chaparral association in Anza Borrego State Park in San Diego Co. ColD.

California sycamore series

California sycamore sole or dominant in canopy as widely spaced trees; arroyo willow, black willow, California bay, coast live oak, Fremont cottonwood, red willow, valley oak, white alder, and/or yellow willow may be present. Trees < 35 m; canopy open. Shrubs common or infrequent. Ground layer grassy (Plate 19).
Wetlands: soils permanently saturated at depth. Water chemistry: fresh. Riparian corridors; braided, depositional channels of intermittent streams; gullies; springs; seeps; stream and river banks, terraces adjacent to floodplains subject to high-intensity flooding. Soils alluvial, open cobbly, rocky. Cowardin class: Temporarily flooded palustrine forested wetland. **Uplands:** Slopes, commonly rocky. The national list of wetland plants (Reed 1988) lists California sycamore as a FACW.

Distribution: CenCo, sac-CenV, f-SN, SoCo, m-TraR, m-PenR, w. MojD, w. ColD, Baja CA. **Elevation:** sea level-2400 m.

NDDB/Holland type and status: Riparian forests (61000). **Riparian woodlands** (62000).
    Sycamore alluvial woodland (62100) G1 S1.1.
    Central Coast cottonwood-sycamore riparian forest (61210 *in part*) G3 S3.
    Southern sycamore-alder riparian woodland (62400 *in part*) G4 S4.
Other types:
Barry type: G7411121 BPLRA00.
Cheatham & Haller type: Northern riparian woodland, Southern riparian woodland.
PSW-45 type: Sycamore series.
Thorne type: Riparian woodlands.
WHR type: Valley foothill riparian.

General references: California sycamore range in CA (Griffin & Critchfield 1972). Bowler (1989), Capelli & Stanley (1984), Faber *et al.* (1989), Holstein (1984),

Keeler-Wolf *et al.* (1994), Paysen *et al.* (1980), Sands (1980), Thompson (1961).
Comments: Keeler-Wolf *et al.* (1994) summarizes natural history and condition of California sycamore woodland in CA. Grazing reduces tree regeneration (Shanfield 1984). Trees are affected by anthracnose in the spring.
Plot-based descriptions: Keeler-Wolf *et al.* (1994) define four associations through Cis-CA; Campbell (1980) one association in Santa Barbara Co. o-CenCo; Brothers (1985), Finn (1991) report sycamore density and cover along the San Gabriel River in Los Angeles Co., Zembal (1989) reports tree density and cover along Santa Margarita and Santa Ana Rivers in Orange Co. SoCo; White (1994a) defines one association at Cleghorn Canyon candidate RNA in San Bernardino Co. i. m-TraR.

Species mentioned in text	
Arroyo willow	*Salix lasiolepis*
Black willow	*Salix gooddingii*
California bay	*Umbellularia californica*
California sycamore	*Platanus racemosa*
Coast live oak	*Quercus agrifolia*
Fremont cottonwood	*Populus fremontii*
Mulefat	*Baccharis salicifolia*
Red willow	*Salix laevigata*
Slender wildoats	*Avena barbata*
Soft chess	*Bromus hordeaceus*
Valley oak	*Quercus lobata*
White alder	*Alnus rhombifolia*
Yellow willow	*Salix lutea*

Associations:
Campbell (1980):
    California sycamore association
    [as *Platanus* community].
Keeler-Wolf *et al.* (1994):
    California sycamore-coast live oak association
    [as Southern California group],
    California sycamore/mulefat association
    [as Foothill group, = White (1994a)],
    California sycamore/slender wildoats association
    [as Mid-coastal group],
    California sycamore/soft chess association
    [as Interior alluvial group].

California walnut series

California walnut sole or dominant tree in canopy; California bay, coast live oak, foothill ash, Mexican elderberry, and/or toyon may be present. Trees < 10 m tall; canopy closed or open. Shrubs common or infrequent. Ground layer sparse or grassy (Plate 20).

Wetlands: soils intermittently flooded, saturated. Water chemistry: fresh. Riparian corridors; floodplains; incised canyons, river and stream low-flow margins; seeps; stream and river banks, terraces. Cowardin class: Palustrine forested wetland. **Uplands:** northfacing slopes, rarely flooded, terraces, flats. Soils shale-derived, deep. The national list of wetland plants (Reed 1988) lists California walnut as a FAC.

Distribution: s. o-CenCo, SoCo.
Elevation: 150-900 m.

NDDB/Holland type and status: **Cismontane woodlands** (71000), **Broadleaved upland forests** (81000).
  California walnut woodland (71210) G2 S2.1.
  Walnut forest (81600) G1 S1.1.

Other types:
Barry type: G74 G7411121 BJUCA00.
Cheatham & Haller type: Southern oak woodland.
Thorne type: Southern oak woodland.
WHR type: Coastal oak woodland.

General references: California walnut range in CA (Griffin & Critchfield 1972). Barbour (1988), Bowler (1989), Holstein (1984), Keeley (1990), Mullally (1992, 1993), Quinn (1989), Swanson (1967).

Comments: Mullally (1992) describes woodlands dominated by California walnut and other woodlands where California walnut is a secondary component. The understories are composed of

coastal scrub, chaparral, and non-native grass species. California walnut is rare (a CNPS List 4 plant) (Skinner & Pavlik 1994).

Species mentioned in text	
California walnut	*Juglans californica* var. *californica*
California bay	*Umbellularia californica*
Coast live oak	*Quercus agrifolia*
Foothill ash	*Fraxinus dipetala*
Mexican elderberry	*Sambucus mexicana*

Plot-based descriptions: Keeley (1990) reports tree density and crown cover for three stands in SoCo.

Canyon live oak series

Canyon live oak sole, dominant, or important tree in canopy; bigleaf maple, black oak, California bay, bigcone Douglas-fir, Coulter pine, Douglas-fir, incense-cedar, madrone, ponderosa pine, sugar pine, tanoak, and/or white fir may be present. Trees < 30 m; canopy continuous, may be two-tiered. Shrubs infrequent. Ground layer sparse or absent.

Uplands: all aspects; raised, stream benches and terraces; may occur in canyon bottoms near streams. Soils shallow, well-drained. The national inventory of wetland plants (Reed 1988) does not list canyon live oak.

Distribution: o-NorCo, m-NorCo, m-CenCo, l-KlaR, m-KlaR, m-CasR, m-SN, m-TraR, m-PenR, Baja CA.
Elevation: 450-2000 m.

NDDB/Holland type and status: Riparian forests (61000), **Broadleaved upland forests** (81000).
  Canyon live oak ravine forest (61350) G3 S3.3.
  Canyon live oak forest (81320) G4 S4.

Other types:
Barry type: G74 G7411111 CQUCH20.
Cheatham & Haller type: Canyon live oak forest.
PSW-45 type: Canyon live oak series.
Thorne type: Northern mixed evergreen forest.
WHR type: Montane hardwood.

General references: Canyon live oak ranges in CA (Griffin & Critchfield 1972); silvics (Thornburgh 1990b). Barbour (1988), Borchert & Hibberd (1984), Cooper (1922), Mallory (1980), McDonald *et al.* (1983), Minnich (1976, 1980b, 1987), Myatt (1980), Paysen *et al.* (1980), Pavlik *et al.* (1991), Sawyer *et al.* (1977b), Shreve (1927).

Comments: Canyon live oak is common in several series, but it is more important in the tree canopy here

than in other series. The canopy can be two-tiered, with an upper tier of conifer trees above a lower tier of canyon live oak, but the conifer component cannot be important. If the conifer component is important, than the stand may be part of another series [*see* Douglas-fir - tanoak series, Coulter pine-canyon live oak series, Bigcone Douglas-fir - canyon live oak series]. If the stand is composed of shrubs of canyon live oak, then it is assigned to a shrub series [*see* Canyon live oak shrub series, Interior live oak-canyon live oak-shrub series].

Species mentioned in text

Bigcone Douglas-fir	*Pseudotsuga macrocarpa*
Bigleaf maple	*Acer macrophyllum*
Black oak	*Quercus kelloggii*
Blackberry	*Rubus* species
California bay	*Umbellularia californica*
Canyon live oak	*Quercus chrysolepis*
Coulter pine	*Pinus coulteri*
Deerbrush	*Ceanothus integerrimus*
Douglas-fir	*Pseudotsuga menziesii*
Goldenback fern	*Pentagramma triangularis*
Incense-cedar	*Calocedrus decurrens*
Indian manzanita	*Arctostaphylos mewukka*
Lemmon catchfly	*Silene lemmonii*
Madrone	*Arbutus menziesii*
Narrowleaf sword fern	*Polystichum imbricans*
Oregon white oak	*Quercus garryana*
Ponderosa pine	*Pinus ponderosa*
Sugar pine	*Pinus lambertiana*
White fir	*Abies concolor*

Keeler-Wolf (1990e) qualitatively describes canyon live oak-dominated forests at Devil's Basin candidate RNA in Mendocino Co. m-NorCo; at Bridge Creek candidate RNA in Siskiyou Co., at Hosselkus Limestone candidate RNA in Shasta Co., at Pearch Creek in Humboldt Co. l-KlaR, m-KlaR; Keeler-Wolf (1990d) at Indian Creek candidate RNA in Tehama Co. f-CasR; Keeler-Wolf (1992) at Graham Pinery candidate RNA in Butte Co. m-CasR.; Keeler-Wolf (1991a) at Big Pine Mountain proposed RNA in Santa Barbara Co. m-CenCo; Keeler-Wolf (1990e) at Millard Canyon RNA in Riverside Co., at King Creek RNA in San Diego Co., Minnich (1976) describes forest development from deerbrush shrublands in SoCo.

Plot-based descriptions: Gray (1978) describes stands at Snow Mountain in Lake Co. m-NorCo.; Gudmunds & Barbour (1987) stands in Middle Yuba River m-SN.

Mize (1973) defines one association in w. Marble Mountain Wilderness, Sawyer & Stillman (1978) in Keeler-Wolf (1990e) one association at Specimen Creek, Sawyer & Stillman (1977a) in Keeler-Wolf (1990e) two associations at Williams Point candidate RNA in Siskiyou Co., Taylor & Teare (1979a) in Keeler-Wolf (1990e) define one association at Manzanita Creek candidate RNA, Thornburgh (1987) in Keeler-Wolf 1990e) one association at Hennessey Ridge in Trinity Co. l-KlaR, m-KlaR; Talley (1974) describes stands in the Santa Lucia Mountains in Monterey Co. [Borchert's analysis of Talley's plot data, using TWINSPAN resulted in two associations in this series- *see* Santa Lucia fir-canyon live oak series, White alder series. This analysis is on file at NDDB]. Campbell (1980) defines one association in Santa Barbara Co. o-CenCo; Griffin (1976a), Borchert (1987) in Keeler-Wolf (1990e) one association at South Fork of Devil's Canyon RNA (now part of the Cone Peak Gradient RNA) in Monterey Co. m-CenCo; Taylor & Randall (1977a) in Keeler-Wolf (1990e) one association at Peavine Point RNA in El Dorado Co. m-SN; Meier (1979) in Keeler-Wolf (1990e) one association at Fern Canyon RNA in Riverside Co., Minnich (1980b) reports tree density for m-TraR stands, Myatt (1980) interprets moisture relations in m-SN trees, White & Sawyer (1995) define one association in San Bernardino Co., m-TraR; Keeler-Wolf (1986e, 1988e) in Keeler-Wolf (1990e) reports tree density at Hall Canyon RNA in Riverside Co., Gordon & White (1994) defines two associations m-PenR.

Associations:
Campbell (1980);
    Canyon live oak-madrone- tanoak association.
Griffin (1976a):
    Sugar pine-canyon live oak association
    [as Pine/oak woodland [= Borchert (1987)].
Gordon & White (1994):
    Canyon live oak association,
    Canyon live oak-deerbrush association
    [= Keeler-Wolf (1986e, 1988a), = White (1994)].

Meier (1979):
    Canyon live oak association
    [as Oak woodland type].
Mize (1973):
    Canyon live oak - Douglas-fir association.
NDDB:
    Canyon live oak/Lemmon catchfly association.
Sawyer & Stillman (1977a):
    Canyon live-Oregon white oak/goldenback fern association,
    Canyon live oak/narrowleaf sword fern association [= Sawyer & Stillman 1978b)].
Taylor & Randal (1977a):
    Canyon live oak/Indian manzanita association.
Taylor & Teare (1979a):
    Douglas-fir - canyon live oak association
    [= Thornburgh (1987)].

Coast live oak series

Coast live oak sole, dominant, or important tree in canopy; bigleaf maple, blue oak, box elder, California bay, Engelmann oak, laurel sumac, and/or madrone may be present. Trees < 30 m tall; canopy continuous, intermittent, or open. Shrubs occasional, or common. Ground layer grassy or absent (Plate 20).

Uplands: slopes often very steep; raised, stream banks and terraces. Soils mostly sandstone or shale-derived. The national inventory of wetland plants (Reed 1988) does not list coast live oak.

Distribution: s. o-NorCo, i-NorCo, o-CenCo, SoCo, m-TraR, m-PenR, Baja CA.
Elevation: sea level-1200 m.

NDDB/Holland type and status: Riparian forests (61000), **Cismontane woodlands** (71000), **Broadleaved upland forests** (81000).
    Central coast live oak riparian forest (61220) G3 S3.2.
    Southern coast live oak forest (61310) G4 S4.
    Coast live oak woodland (71160) G4 S4.
    Coast live oak forest (81310) G4 S4.

Other types:
Barry type: G74 G7411111 CQUAG00.
Cheatham & Haller type: Coast live oak forest.
PSW-45 type: Coast live oak series.
Rangeland type: SRM 202.
Thorne type: Foothill woodland.
WHR type: Coastal oak woodland.

General references: Coast live oak range in CA (Griffin & Critchfield 1972). Barbour (1988), Brown (1982b), Cooper (1922), Faber *et al.* (1989), Ferren (1989), Finch & McCleery (1980), Griffin (1977), Lathrop & Zuill (1984), Minnich (1987), Parker (1994), Parker & Muller (1982), Pavlik *et al.* (1991), Paysen *et al.* (1980), Plumb (1980), Shreve (1927), Snow (1973), Wells (1962), Zuill (1967).

Comments: Keeler-Wolf (1990e) qualitatively describes coast live oak-dominated stands at Limekiln Creek RNA (now part of the Cone Peak Gradient RNA) in Monterey Co. o-CenCo; at Fern Canyon RNA in Los Angeles Co., Gautier & Zedler (1980) at Guatay Mountain, Keeler-Wolf (1990e) at Organ Valley RNA San Diego Co. SoCo.

Species mentioned in text	
Bigleaf maple	*Acer macrophyllum*
Black oak	*Quercus kelloggii*
Black sage	*Salvia mellifera*
Blue oak	*Quercus douglasii*
Blackberry	*Rubus* species
Box elder	*Acer negundo*
Bracken	*Pteridium aquilinum*
California bay	*Umbellularia californica*
California coffeeberry	*Rhamnus californica*
California sagebrush	*Artemisia californica*
Chamise	*Adenostoma fasciculatum*
Coast live oak	*Quercus agrifolia*
Common snowberry	*Symphoricarpos albus*
Engelmann oak	*Quercus engelmannii*
Hairyleaf ceanothus	*Ceanothus oliganthus*
Hazel	*Corylus cornuta*
Laurel sumac	*Malosma laurina*
Madrone	*Arbutus menziesii*
Ocean spray	*Holodiscus discolor*
Poison-oak	*Toxicodendron diversilobum*
Scrub oak	*Quercus berberidifolia*
Toyon	*Heteromeles arbutifolia*

Plot-based descriptions: Allen *et al.* (1989, 1991) define 15 upland coast live oak subseries which are sorted by elevation and water relations in Nor-CA. Their subseries are called associations here. Parker (1990) defines one association at Mount Tamalpais in Marin Co., Wainwright & Barbour (1984) one association in Annadel State Park in Sonoma Co. o-NorCo; Cole (1980) reports cover for stands in the Purisima Hills, Campbell (1980) defines one association in Santa Barbara Co. o-CenCo; Keeler-Wolf (1990a) one association at King Creek RNA in San Diego Co., Zembal (1989) reports tree density and cover along Santa Margarita and Santa Ana Rivers in Orange Co. o-SoCo; Boyd (1983) describes stands in Gavilan Hills in Riverside Co. i. m-TraR; Gordon & White (1994) define one association in m-PenR.

Associations:
Allen *et al.* (1991):
   Blue oak-coast live oak/grass association,
   Coast live oak association,
   Coast live oak-bigleaf maple/California coffeeberry-ocean spray association,
   Coast live oak/blackberry/bracken association,
   Coast live oak-California bay/toyon-scrub oak association,
   Coast live oak/California coffeeberry-toyon association,
   Coast live oak/chamise-black sage association,
   Coast live oak/California sagebrush/grass association,
   Coast live oak/grass association,
   Coast live oak-madrone/hazel-blackberry association,
   Coast live oak/ocean spray-common snowberry association,
   Coast live oak/poison-oak/grass association,
   Coast live oak/poison-oak association,
   Coast live oak/toyon/grass association,
   Coast live oak/toyon-poison-oak association.
Campbell (1980):
   Coast live oak association.
Gordon & White (1994):
   Coast live oak/hairyleaf ceanothus association.
Keeler-Wolf (1990a):
   Coast live oak/snowberry association.
Parker (1990):
   Coast live oak association.
Wainwright & Barbour (1984):
   Coast live oak-black oak association.

Coulter pine series

Coulter pine sole or dominant tree in canopy; bigcone Douglas-fir, black oak, canyon live oak, coast live oak, interior live oak, foothill pine, and/or ponderosa pine may be present. Trees < 30 m; canopy continuous. Shrubs frequent or infrequent. Ground layer sparse (Plate 26).

Uplands: all aspects. Soils shallow, well-drained.

Distribution: m-CenCo, i-SoCo, m-TraR, m-PenR, Baja CA.
Elevation: 700-2000 m.

NDDB/Holland type and status: Lower montane coniferous forests (84000).
 Coulter pine forest (84140 *in part*) G3 S3.2.

Other types:
Barry type: G7411112 BPICO30.
Cheatham & Haller type: Coulter pine forest.
PSW-45 type: Coulter pine series.
Thorne type: Coulter pine forest.
WHR type: Montane hardwood conifer.

General references: Coulter pine range in CA (Griffin & Critchfield 1972). Barbour (1988), Borchert & Hibberd (1984), Minnich (1980b, 1987), Paysen *et al.* (1980), Sawyer *et al.* (1977b), Thorne (1977), Vale (1979).

Comments: Coulter pine dominates in this series; it is an important species in the Coulter pine-canyon live oak series where it shares dominance with canyon live oak. In the Coulter pine series the canopy is one-tiered. In the Coulter pine-canyon live oak series the canopy is two-tiered. In addition, Coulter pine may be present in the Canyon live oak series as a minor component.

Keeler-Wolf (1991a) qualitatively describes stands at Big Pine Mountain proposed RNA in Santa Barbara Co., Millard Canyon RNA in Riverside Co. i. m-TraR.

Species mentioned in text	
Bigcone Douglas-fir	*Pseudotsuga macrocarpa*
Black oak	*Quercus kelloggii*
Canyon live oak	*Quercus chrysolepis*
Coast live oak	*Quercus agrifolia*
Coulter pine	*Pinus coulteri*
Foothill pine	*Pinus sabiniana*
Interior live oak	*Quercus wislizenii*
Ponderosa pine	*Pinus ponderosa*

Plot-based descriptions: Burke (1992a) reports cover for a stand at Fishman's Camp candidate RNA in San Bernardino Co. i. m-TraR; Keeler-Wolf (1986d, 1989m) in Keeler-Wolf (1990e) describes a Coulter pine-black oak woodland at Cahuilla Mountain RNA in Riverside Co. m-PenR.

Coulter pine-canyon live oak series

Coulter pine and canyon live oak important in tree canopy; bigcone Douglas-fir, black oak, coast live oak, foothill pine, and/or ponderosa pine may be present. Trees < 30 m; canopy continuous, two-tiered in older stands. Shrubs infrequent. Ground layer sparse.

Uplands: all aspects. Soils shallow, sandstone or schist-derived.

Distribution: m-CenCo, m-TraR, m-PenR, Baja CA.
Elevation: 700-2000 m.

NDDB/Holland type and status: Lower montane coniferous forests (84000).
    Coulter pine forest (84140 *in part*) G3 S3.2.

Other types:
Cheatham & Haller type: Mixed evergreen forest.
Thorne type: Mixed evergreen forest.
WHR type: Montane hardwood conifer forest.

General references: Coulter pine and canyon live oak ranges in CA (Griffin & Critchfield 1972); silvics of canyon live oak (Thornburgh 1990b). Barbour (1988), Borchert & Hibberd (1984), Cooper (1922), Minnich (1980b, 1987), Pavlik *et al.* (1991), Sawyer *et al.* (1977b), Shreve (1927), Thorne (1977).

Comments: Coulter pine and canyon live oak are important in tree canopy in this series; Coulter pine dominates in the Coulter pine series, and canyon live oak dominates the Canyon live oak series. In this series the canopy is two-tiered, with an upper tier of conifer trees and a lower tier of canyon live oak.

Species mentioned in text	
Bigcone Douglas-fir	*Pseudotsuga macrocarpa*
Black oak	*Quercus kelloggii*
Canyon live oak	*Quercus chrysolepis*
Coast live oak	*Quercus agrifolia*
Coulter pine	*Pinus coulteri*
Foothill pine	*Pinus sabiniana*
Ponderosa pine	*Pinus ponderosa*

Plot-based descriptions: Keeler-Wolf (1985a) describes Coulter pine-canyon live oak stands at Big Pine Mountain proposed RNA in Santa Barbara Co. m-CenCo; Minnich (1980b) reports tree density for TraR stands. Borchert & Hibberd (1984) discuss fire relationships.

Curlleaf mountain-mahogany series

Curlleaf mountain-mahogany or littleleaf mountain-mahogany sole or dominant tree in canopy or as an emergent tree with foxtail pine, Jeffrey pine, lodgepole pine, mountain juniper, singleleaf pinyon, western juniper, and/or whitebark pine in shrub canopy; big sagebrush, bitterbrush, choke cherry, greenleaf manzanita, and/or service berry, may be present as scattered shrubs over a canopy of herbs and grass. Trees < 10 m; canopy continuous or scattered. Shrubs common or infrequent. Ground layer sparse or grassy (Plate 25).

Uplands: ridges, upper slopes. Soils sedimentary, ultramafic, volcanic-derived, shallow.

Distribution: m-NorCo, e. m-KlaR, su-KlaR, m-CasR, su-CasR, TraR, PenR, ModP, WarR, TraSN, WIS, DesR, inter-West.
Elevation: 1200-3000 m.

NDDB/Holland type and status: **Broadleaved upland forests** (81000).

Other types:
Barry type: G74 G7411214.
Cheatham & Haller type: High desert scrub.
PSW-45 type: Mountain mahogany series.
Rangeland types: SRM 415, SRM 417.
Thorne type: Desert rupicolous scrub, Mountain juniper woodland.
WHR type: Sagebrush.

General references: Davis (1994a, 1994b), Paysen *et al.* (1980), West (1988), Young *et al.* (1977).

Comments: Curlleaf mountain-mahogany has a wide range in CA. On rocky ridges and steep slopes with thin soil, this plant can be the sole tall shrub or small tree. Other trees may be present in these areas as well. Trees, if present, are from series of the region. The degree of canopy development

varies. Curlleaf mountain-mahogany can occur in other series as a secondary component as well. Young *et al.* (1977) qualitatively describe two types.

The Jepson Manual recognizes two varieties of curlleaf mountain-mahogany. *Cercocarpus ledifolius* var. *l.* is uncommon in comparison to *C. l.* var. *intermontanus*. Littleleaf mountain-mahogany may dominate on rock outcrops in WIS. Stands are included in this series at this time.

Keeler-Wolf (1990e) qualitatively describes curlleaf mountain-mahogany stands at Sugar Creek candidate RNA in Siskiyou Co. su-KlaR; Jenson (1989) Raider Basin candidate RNA in Modoc Co. WarR; Keeler-Wolf (1990e) littleleaf mountain-mahogany stands at Whippoorwill Flat RNA in Inyo Co. WIS.

Species mentioned in text	
Big sagebrush	*Artemisia tridentata*
Bitterbrush	*Purshia tridentata*
Choke cherry	*Prunus virginiana*
Curlleaf mountain-mahogany	*Cercocarpus ledifolius*
Foxtail pine	*Pinus balfouriana*
Greenleaf manzanita	*Arctostaphylos patula*
Jeffrey pine	*Pinus jeffreyi*
Littleleaf mountain-mahogany	*Cercocarpus intricatus*
Lodgepole pine	*Pinus contorta* ssp. *murrayana*
Mountain juniper	*Juniperus occidentalis* ssp. *australis*
Service berry	*Amelanchier alnifolia*
Singleleaf pinyon	*Pinus monophylla*
Western juniper	*Juniperus occidentalis* ssp. *occidentalis*
Whitebark pine	*Pinus albicaulis*

Plot-based descriptions: Keeler-Wolf (1987a) in Keeler-Wolf (1990e) describes stands at Crater Creek candidate RNA in Siskiyou Co. e. su-KlaR. The three structural forms of the series exist there. Jensen & Schierenbeck (1990) report tree density, frequency, and cover at Raider Basin candidate RNA in Modoc Co. WarR.

Douglas-fir series

Douglas-fir sole or dominant tree in canopy; canyon live oak, chinquapin, incense-cedar, ponderosa pine, sugar pine, and/or white fir may be present. Trees < 75 m; canopy continuous or intermittent. Shrubs infrequent or common. Ground layer sparse or abundant (Plate 23).

Uplands: all aspects; raised stream benches and terraces. Soils granite, sandstone, serpentine, schist, volcanic-derived.

Distribution: m-NorCo, m-CenCo, m-KlaR, m-CasR, m-SN, w ModP.
Elevation: 700-2700 m.

NDDB/Holland type and status: **North Coast coniferous forests** (82000), **Lower montane coniferous forests** (84000).
> Upland Douglas-fir forest (82420) G4 S2.1.
> Coast range mixed coniferous forest (84110 *in part*) G4 S4.

Other types:
Barry type: G7411112 BPSME00.
Cheatham & Haller type: Douglas-fir forest.
PSW-45 type: Douglas-fir forest.
Thorne type: Douglas-fir forest.
WHR type: Douglas-fir.

General references: Douglas-fir range in CA (Griffin & Critchfield 1972); silvics (Hermann & Lavender 1990). Barbour (1988), Bolsinger & Jaramillo (1990), Franklin (1988), Franklin *et al.* (1981), Keeler-Wolf (1988a), McKee (1990), Paysen *et al.* (1980), Rundel *et al.* (1977), Sawyer & Thornburgh (1977), Williamson (1980), Williamson & Twombly (1983), Zuckerman (1990).

Comments: *Pseudotsuga menziesii* has two varieties, but only *P. m.* var. *m.* occurs in CA; *P. m.* var. *glauca* grows in the Rocky Mountains. The CA variety is widespread, grows in many series, and is associated with numerous species. In addition, the

Species mentioned in text	
Bear-grass	*Xerophyllum tenax*
Bigleaf maple	*Acer macrophyllum*
Black oak	*Quercus kelloggii*
Bush tanoak	*Lithocarpus densiflora* var. *echinoides*
California bay	*Umbellularia californica*
California fescue	*Festuca californica*
Canyon live oak	*Quercus chrysolepis*
Chinquapin	*Chrysolepis chrysophylla*
Creeping snowberry	*Symphoricarpos mollis*
Douglas-fir	*Pseudotsuga menziesii*
Greenleaf manzanita	*Arctostaphylos patula*
Hazel	*Corylus cornuta*
Himalaya berry	*Rubus discolor*
Huckleberry oak	*Quercus vaccinifolia*
Incense-cedar	*Calocedrus decurrens*
Inside-out flower	*Vancouveria planipetala*
Jeffrey pine	*Pinus jeffreyi*
Leafless shinleaf	*Pyrola picta*
Little Oregon-grape	*Berberis nervosa*
Madrone	*Arbutus menziesii*
One-sided shinleaf	*Orthilia secunda*
Oregon white oak	*Quercus garryana*
Pacific dogwood	*Cornus nuttallii*
Pinemat manzanita	*Arctostaphylos nevadensis*
Poison-oak	*Toxicodendron diversilobum*
Ponderosa pine	*Pinus ponderosa*
Port Orford-cedar	*Chamaecyparis lawsoniana*
Prince's-pine	*Chimaphila umbellata*
Rattlesnake-plantain	*Goodyera oblongifolia*
Red fir	*Abies magnifica*
Redwood	*Sequoia sempervirens*
Rhododendron	*Rhododendron macrophyllum*
Sadler oak	*Quercus sadleriana*
Salal	*Gaultheria shallon*
Sugar pine	*Pinus lambertiana*
Sword fern	*Polystichum munitum*
Tanoak	*Lithocarpus densiflora*
Thimbleberry	*Rubus parviflorus*
Trail plant	*Adenocaulon bicolor*
Twinflower	*Linnaea borealis*
Vine maple	*Acer circinatum*
Vanilla leaf	*Achlys triphylla*
Western hemlock	*Tsuga heterophylla*
White fir	*Abies concolor*
White-veined shinleaf	*Pyrola picta*
Whiteflower hawkweed	*Hieracium albiflorum*
Wild rose	*Rosa gymnocarpa*
Yerba de selva	*Whipplea modesta*

literature applies the term Douglas-fir forest to stands of varying species composition.

The definition of this series best begins by listing stands that are not included: those with redwood [Redwood series], western hemlock [Western hemlock series], or white fir [White fir series] dominance, or those where Douglas-fir shares dominance with tanoak [Douglas-fir - Tanoak series] or other conifers [Douglas-fir - ponderosa pine series, White fir series].

The series is mainly a middle elevation one. White fir may be a component of the understory, and a secondary species of overstory canopy. For this reason, some authors place associations into other series [*ex.,* Jimerson (1994), Sawyer & Thornburgh (1977) placed their associations in the White fir series, Fites (1993) in the Mixed conifer series].

One confusion arises from not separating low elevation, Douglas-fir - tanoak series stands from lower montane, Douglas-fir series stands. Old-growth Douglas-fir - tanoak series stands have a two-tiered tree canopy (Bingham & Sawyer 1991, 1993). Douglas-fir series stands are, in contrast, single-tiered. Keeler-Wolf (1984a, 1988) studied chinquapin's status in transition stands between these series in w. m-KlaR.

Plot-based descriptions: Parker (1990) defines one association at Mount Tamalpais in Marin Co., Wainwright & Barbour (1984) one association at Annadel State Park in Sonoma Co. o-NorCo; Keeler-Wolf (1987b) in Keeler-Wolf (1990e) describes stands at Pearch Creek, Sawyer (1981a) defines one association at Adorni candidate RNA, Simpson (1980) four associations in s. Siskiyou Mountains in Humboldt Co., Keeler-Wolf (1985a, 1989c) in Keeler-Wolf (1990e) describes stands at Bridge Creek candidate RNA, Sawyer & Stillman (1977a) in Keeler-Wolf (1990e) define three associations at Specimen Creek, Stuart *et al.* (1992) four associations at Castle Crags State Park in Siskiyou Co. m-KlaR, Taylor & Teare (1979a) in Keeler-Wolf (1990e) one association at Manzanita Creek candidate RNA, Laidlaw-Holmes (1981) one association at South Fork Mountain, Sawyer (1978a) one association at Preacher Meadows candidate RNA, Taylor (1975a, 1975b) in Keeler-Wolf (1990e) one association at South Fork Mountain candidate RNA in Trinity Co. m-KlaR; Jimerson (1993) 45 associations in Six Rivers National Forest and Mendocino National Forest in m-NorCo, m-KlaR; Fites (1993) five associations in Lassen, Plumas, Tahoe, Eldorado National Forest m-CasR, m-SN. Atzet & Wheeler (1984) define sw. OR associations.

Associations:
Fites (1993):
  Douglas-fir - bigleaf maple/hazel association,
  Douglas-fir - bigleaf maple/trail plant association,
  Douglas-fir - Pacific dogwood/hazel association,
  Douglas-fir/hazel/trail plant association.
Jimerson (1993):
  Douglas-fir - California bay/poison-oak association,
  Douglas-fir - canyon live oak - madrone/poison-oak association,
  Douglas-fir - canyon live oak/rockpile association,
  Douglas-fir - canyon live oak - tanoak association,
  Douglas-fir - chinquapin/bear-grass association,
  Douglas-fir - chinquapin/rhododendron - little Oregon-grape association,
  Douglas-fir - chinquapin/rhododendron - Sadler oak/bear-grass association,
  Douglas-fir - chinquapin - tanoak association,
  Douglas-fir - chinquapin - tanoak/little Oregon-grape association,
  Douglas-fir/hazel association,
  Douglas-fir/huckleberry oak association,
  Douglas-fir/huckleberry oak - bush tanoak association,
  Douglas-fir - incense-cedar/California fescue association,
  Douglas-fir/yerba de selva association,
  Douglas-fir - Oregon white oak/grass association,
  Douglas-fir/vine maple - little Oregon-grape association,

Douglas-fir- bigleaf maple/sword fern association,
Port Orford-cedar - Douglas-fir/huckleberry oak association,
Port Orford-cedar - white fir/huckleberry oak association,
Port Orford-cedar - white-fir/Sadler oak association,
Red fir-white fir/pinemat manzanita association,
White fir - chinquapin association,
White fir - Douglas-fir/bear-grass association,
White fir - Douglas-fir - bigleaf maple association,
White fir - Douglas-fir - black oak association,
White fir - Douglas-fir - canyon live oak/ white-veined shinleaf association,
White fir - Douglas-fir - canyon live oak association,
White fir - Douglas-fir - canyon live oak/white-flower hawkweed-grass association,
White fir - Douglas-fir - canyon live oak/little Oregon-grape association,
White fir - Douglas-fir - chinquapin/little Oregon-grape/vanilla leaf association,
White fir - Douglas-fir/Sadler oak association,
White fir - Douglas-fir/Sadler oak - huckleberry oak association,
White fir - Douglas-fir/Sadler oak - pinemat manzanita association,
White fir - Douglas-fir/grass association,
White fir - Douglas-fir/hazel association,
White fir - Douglas-fir/huckleberry oak association,
White fir -Douglas-fir/yerba de selva association,
White fir - Douglas-fir/rhododendron - Sadler oak association,
White fir - Douglas-fir - tanoak/little Oregon-grape association,
White fir -Douglas-fir/thimbleberry association,
White fir - Douglas-fir/vine maple association,
White fir - Douglas-fir/wild rose/twinflower association,
White fir - incense-cedar/creeping snowberry association,
White fir - Jeffrey pine/California fescue association.

Keeler-Wolf (1985a, 1987a, 1989b):
Douglas-fir/vanilla leaf association [as White fir-Douglas-fir forest and Transitional mixed evergreen - white fir - Douglas-fir forest].
Laidlaw-Holmes (1981):
Douglas-fir/madrone association.
Sawyer (1981a):
Douglas-fir/rattlesnake plantain association.
Sawyer et al. (1978):
Douglas-fir/huckleberry oak association.
Sawyer & Stillman (1977):
Douglas-fir/greenleaf manzanita association,
Douglas-fir/prince's-pine association,
Douglas-fir/twinflower association.
Simpson (1980):
Douglas-fir/hazel association,
Douglas-fir/inside-out flower association,
Douglas-fir/rhododendron association,
Douglas-fir/salal association.
Stuart et al. (1992):
Black oak - Douglas-fir association,
Douglas-fir association,
Douglas-fir - Jeffrey pine - incense-cedar association,
Douglas-fir - white alder/Himalaya berry association.
Taylor (1975a,b):
Douglas-fir/little Oregon-grape association.
Taylor & Teare (1979a):
Douglas-fir/Pacific dogwood association.
Wainwright & Barbour (1984):
Douglas-fir - California bay association [as Group I, II].

Douglas-fir - ponderosa pine series

Douglas-fir and ponderosa pine important trees in canopy; black oak, canyon live oak, incense-cedar, Oregon white oak, and/or sugar pine may be present. Trees < 75 m; canopy continuous or intermittent. Shrubs infrequent or common. Ground layer sparse.

Uplands: all aspects; raised stream benches and terraces. Soils granite, schist, ultramafic-derived. The national inventory of wetland plants (Reed 1988) lists ponderosa pine as a FACU; Douglas-fir is not listed.

Distribution: m-NorCo, m-KlaR, m-CasR, n. m-SN.
Elevation: 600-1800 m.

NDDB/Holland type and status: Lower montane coniferous forests (84000).
  Coast Range mixed coniferous forest (84110 *in part*) G4 S4.
  Sierran mixed coniferous forest (84230 *in part*) G4 S4.

Other types:
Barry type: G7411112.
Cheatham & Haller type: Sierran mixed conifer forests.
Thorne type: Mixed conifer forest.
WHR type: Sierran mixed conifer.

General references: Douglas-fir and ponderosa pine ranges in CA (Griffin & Critchfield 1972); silvics (Hermann & Lavender 1990), (Oliver & Ryker 1990). Barbour (1988), Franklin (1988), Rundel *et al.* (1977).

Comments: The Holland description addresses only m-NorCo stands. The series is also extensive in central and eastern m-KlaR where it occurs at lower montane elevations. White fir is absent or infrequent there. Many definitions of mixed conifer would include these stands in that type. Mixed conifer forest is best thought of as a collection of series.

Plot-based descriptions: Jimerson (1993) defines one association in Six Rivers National forest m-NorCo, m-KlaR, Stuart *et al.* (1992) one association at Castle Crags State Park in Siskiyou Co., Taylor & Teare (1979b) in Keeler-Wolf (1990e) one association at Smoky Creek candidate RNA in Trinity Co. KlaR; Smith (1994) one association in Plumas National Forest m-CasR.; Gray (1978) reports importance values for stands in Sierra and Yuba Cos. m-SN.

Species mentioned in text	
Bigleaf maple	*Acer macrophyllum*
Black oak	*Quercus kelloggii*
Canyon live oak	*Quercus chrysolepis*
Douglas-fir	*Pseudotsuga menziesii*
Incense-cedar	*Calocedrus decurrens*
Jeffrey pine	*Pinus jeffreyi*
One-sided bluegrass	*Poa secunda*
Oregon white oak	*Quercus garryana*
Ponderosa pine	*Pinus ponderosa*
Sugar pine	*Pinus lambertiana*

Associations:
Jimerson (1993):
  Black oak-Douglas-fir - bigleaf maple association.
Stuart *et al.* (1992)
  Douglas-fir - ponderosa pine -incense-cedar association.
Smith (1994):
  Douglas-fir - ponderosa pine - Jeffrey pine/one-sided bluegrass association [as White fir-ponderosa pine - Jeffrey pine/one-sided bluegrass association].
Taylor & Teare (1979b):
  Douglas-fir-ponderosa pine association.

Douglas-fir - tanoak series

Douglas-fir and tanoak important in tree canopy; California bay, black oak, canyon live oak, madrone, ponderosa pine, and/or sugar pine may be present. Trees < 75 m; canopy continuous, two-tiered in older stands. Shrubs infrequent or common. Ground layer sparse or abundant (Plate 24).

Uplands: all aspects; raised stream benches and terraces. Soils mostly sandstone, schist-derived. The national inventory of wetland plants (Reed 1988) does not list Douglas-fir or tanoak.

Distribution: o-NorCo, l-KlaR, m-SN.
Elevation: 100-1500 m.

NDDB/Holland type and status: Broadleaved upland forests (81000), **North Coast coniferous forests** (82000).
> Mixed evergreen forest (81100 *in part*) G4 S4.
> Tanoak forest (81400 *in part*) G4 S4.
> Upland Douglas-fir forest (82420 *in part*) G4 S3.1.

Other types:
Barry type: G74 G7411114 BPSME00.
Cheatham & Haller type: Douglas-fir forest.
Thorne type: Northern mixed evergreen forest.
WHR type; Douglas-fir forest.

General references: Douglas-fir, tanoak ranges in CA (Griffin & Critchfield 1972); silvics (Hermann & Lavender 1990), (Tappeiner *et al.* 1990). Barbour (1988), Barrows (1984), Bingham & Sawyer (1991, 1993), Franklin *et al.* (1981), McDonald *et al.* (1983), McDonald & Tappeiner (1990), Sawyer & Thornburgh (1977), Sawyer (1980), Thornburgh (1982), Williamson (1980), Williamson & Twambly (1983), Zinke (1977).

Comments: Douglas-fir and tanoak are important components of several series. In addition, the literature applies the terms "Douglas-fir forest" and "mixed evergreen forest" to stands of varying species composition.

Species mentioned in text	
Beargrass	*Xerophyllum tenax*
Black huckleberry	*Vaccinium ovatum*
Black oak	*Quercus kelloggii*
Bracken	*Pteridium aquilinum*
California bay	*Umbellularia californica*
California fescue	*Festuca californica*
Canyon live oak	*Quercus chrysolepis*
Chinquapin	*Chrysolepis chrysophylla*
Douglas-fir	*Pseudotsuga menziesii*
Hairy honeysuckle	*Lonicera hispidula*
Hazel	*Corylus cornuta*
Little Oregon-grape	*Berberis nervosa*
Madrone	*Arbutus menziesii*
Oregon-grape	*Berberis aquifolium*
Pacific yew	*Taxus brevifolia*
Poison-oak	*Toxicodendron diversilobum*
Ponderosa pine	*Pinus ponderosa*
Port Orford-cedar	*Chamaecyparis lawsoniana*
Prince's-pine	*Chimaphila umbellata*
Red huckleberry	*Vaccinium parvifolium*
Rhododendron	*Rhododendron macrophyllum*
Salal	*Gaultheria shallon*
Starflower	*Trientalis latifolia*
Sugar pine	*Pinus lambertiana*
Sword fern	*Polystichum munitum*
Tanoak	*Lithocarpus densiflora*
Twinflower	*Linnaea borealis*
Vanilla leaf	*Achlys triphylla*
Vine maple	*Acer circinatum*
Western azalea	*Rhododendron occidentale*
White alder	*Alnus rhombifolia*
Wild rose	*Rosa gymnocarpa*

The definition of this series is best begun by explaining what stands are not included: those not dominated by Douglas-fir [Douglas-fir series] or tanoak [Tanoak series]. The largest problem is separating low montane, Douglas-fir series from low elevation Douglas-fir - tanoak series stands. Keeler-Wolf (1988) studied this transition in w. KlaR [*see* Douglas-fir series]. Old-growth Douglas-fir - tanoak series stands have a two-tiered tree canopy and are o-NorCo or l-KlaR in range (Bingham & Sawyer 1991, 1993). Old-growth Douglas-fir series stands are single-tiered and in m-KlaR.

Another source of confusion is the structural changes in Douglas-fir - tanoak series stands as they age (Oliver & Larson 1990). These recognize three stages: young, mature, old-growth. On the same site a mature, one-tiered Douglas-fir - dominated stand will develop two-tiered characteristics with age (Bingham & Sawyer 1991, 1993). Earlier ecologists (*ex.*, Whittaker 1960) considered these changes site-induced, and they distinguished them as different kinds of forests.

Plot-based descriptions: Mize (1973) defines three associations in w. Marble Mountain Wilderness, Keeler-Wolf (1985a, 1989c) in Keeler-Wolf (1990e) one association at Bridge Creek candidate RNA in Siskiyou Co., Keeler-Wolf (1987b) in Keeler-Wolf (1990e) describe stands at Pearch Creek, Simpson (1980) defines one association in s. Siskiyou Mountains in Humboldt Co., Thornburgh (1987) in Keeler-Wolf (1990e) defines two associations at Hennessy Ridge in Trinity Co., Jimerson (1993, 1994) defines 35 associations in l-KlaR. His names have been changed to include the important Douglas-fir component in his forests. In this way, Douglas-fir - tanoak stands will not be confused with Tanoak series associations lacking the important Douglas-fir component. Gudmunds & Barbour (1987) describe stands along Middle Yuba River m-SN. See Atzet & Wheeler (1979) for sw. OR associations.

Associations:
Keeler-Wolf (1985a, 1987b, 1989b):
   Douglas-fir - tanoak/prince's-pine association [as Mixed evergreen forest].
Jimerson (1993):
   Douglas-fir - tanoak - bigleaf maple/sword fern association,
   Douglas-fir - tanoak - black oak/wild rose association,
   Douglas-fir - tanoak - California bay/poison-oak association,
   Douglas-fir - tanoak - canyon live oak - black oak/poison-oak association,
   Douglas-fir - tanoak - canyon live oak/little Oregon-grape - salal association,
   Douglas-fir - tanoak - canyon live oak/little Oregon-grape association,
   Douglas-fir - tanoak - canyon live oak/black huckleberry association,
   Douglas-fir - tanoak - canyon live oak/black huckleberry - salal association,
   Douglas-fir - tanoak - canyon live oak/poison-oak association,
   Douglas-fir - tanoak - canyon live oak/rockpile association,
   Douglas-fir - tanoak - chinquapin/bracken association,
   Douglas-fir - tanoak - chinquapin/little Oregon-grape association,
   Douglas-fir - tanoak - chinquapin/rhododendron/beargrass association,
   Douglas-fir - tanoak - chinquapin/ rhododendron - salal association,
   Douglas-fir - tanoak - chinquapin/salal association,
   Douglas-fir - tanoak/little Oregon-grape association,
   Douglas-fir -tanoak/little Oregon-grape - salal association,
   Douglas-fir - tanoak/black huckleberry association,
   Douglas-fir - tanoak/black huckleberry - salal association,
   Douglas-fir - tanoak/hazel association,
   Douglas-fir - tanoak - incense-cedar/California fescue association,
   Douglas-fir - tanoak/poison-oak - hairy honeysuckle association,
   Douglas-fir - tanoak -Port Orford-cedar - California bay/black huckleberry association,
   Douglas-fir - tanoak - Port Orford-cedar/little Oregon-grape/twinflower association,
   Douglas-fir - tanoak - Port Orford-cedar/black huckleberry association,
   Douglas-fir - tanoak - Port Orford-cedar/black huckleberry -western azalea association,
   Douglas-fir - tanoak - Port Orford-cedar/red huckleberry association,
   Douglas-fir - tanoak - Port Orford-cedar/salal association,
   Douglas-fir - tanoak - Port Orford-cedar/vine maple association,
   Douglas-fir - tanoak- Port Orford-cedar - white alder/riparian association,

Douglas-fir - tanoak/rhododendron-black
huckleberry association,
Douglas-fir-tanoak/rhododendron-huckleberry
oak association,
Douglas-fir - tanoak/rhododendron-salal
association,
Douglas-fir - tanoak/salal association,
Douglas-fir - tanoak/vine maple association,
Douglas-fir - tanoak/vine maple-salal
association.
Mize (1973):
Douglas-fir - tanoak/Pacific yew association,
Douglas-fir - tanoak - sugar pine association,
Douglas-fir - tanoak/vanilla leaf association.
Simpson (1980):
Douglas-fir-tanoak/black huckleberry
association.
Thornburgh (1987):
Douglas-fir - tanoak/poison-oak association,
Douglas-fir - tanoak/little Oregon-grape
association.

Engelmann oak series

Engelmann oak sole, dominant, or important with coast live oak in tree canopy; black oak, and/or California walnut may be present. Trees < 18 m tall; canopy closed or open. Shrubs common or infrequent. Ground grassy or sparse.

Uplands: gentle slopes; valley bottoms; raised stream terraces along stream corridors in canyons. Soils deep. The national inventory of wetland plants (Reed 1988) does not list Engelmann oak.

Distribution: w. m-PenR, SoCo, ChaI, Baja CA.
Elevation: 50-1220 m.

NDDB/Holland type and status: Cismontane woodlands (71000).
  Open Engelmann oak woodland (71181) G2 S2.2.
  Dense Engelmann oak woodland (71182) G2 S2.1.

Other types:
Barry type: G74 G7411121 CQUEN00.
Cheatham & Haller type: Southern oak woodland.
PSW-45 type: Engelmann oak series.
Thorne type: Southern oak woodland.
WHR type: Coastal oak woodland.

General references: Engelmann oak range in CA (Griffin & Critchfield 1972). Barbour (1988), Bowler (1989), Brown (1982b), Griffin (1977), Holstein (1984), Lathrop & Zuill (1984), Pavlik *et al.* (1991), Scott (1991), Snow (1973), Zuill (1967).

Comments: Both open and dense woodlands are included in this series as long as Engelmann oak is important or dominant. If coast live oak is more important than Engelmann oak, then the stand is a member of the Coast live oak series.

Engelmann oak hybridizes with scrub oak. Oaks intermediate between the two species are common on margins between scrub oak stands and Engelmann oak stands.

Keeler-Wolf (1990e) qualitatively describes riparian forests with Engelmann oak at King Creek RNA, at Organ Valley RNA in San Diego Co. SoCo.

Engelmann oak is rare (a CNPS List 4 plant) (Skinner & Pavlik 1994).

Species mentioned in text	
Black oak	*Quercus kelloggii*
California walnut	*Juglans californica* var. *californica*
Coast live oak	*Quercus agrifolia*
Engelmann oak	*Quercus engelmannii*
Scrub oak	*Quercus berberidifolia*

Plot-based descriptions: Griffin (1977), Griggs (1987), Lathrop & Arct (1987) describe stands at the Santa Rosa Plateau in Riverside Co., Griffin (1977) stands Ramona-Santa Ysabel area in San Diego Co. SoCo.

Engelmann spruce series

Engelmann spruce sole or dominant tree in canopy; Brewer spruce, Douglas-fir, incense-cedar, lodgepole pine, mountain hemlock, Pacific yew, ponderosa pine, Shasta fir, sugar pine, western white pine, and/or white fir may be present. Trees < 55 m; canopy continuous. Shrubs common or infrequent. Ground layer sparse or abundant (Plate 28).

Wetlands: soils intermittently flooded. Water chemistry: fresh. Bottom and streamsides. Cowardin class: Palustrine forested wetland. The national inventory of wetland plants (Reed 1988) lists Engelmann spruce as a FAC.

Distribution: m-KlaR, su-KlaR, OR, WA, Rocky Mountains.
Elevation: 1200-2100 m.

NDDB/Holland type and status: **Upper montane coniferous forests** (85000).
  Salmon-Scott enriched conifer forests (85420 *in part*) G1 S1.2.

Other types:
Barry type: G7411112 BPIENOO.
Cheatham & Haller type: Klamath enriched conifer forest.
Thorne type: Enriched conifer forest.
WHR type: Klamath enriched mixed conifer.

General references: Engelmann spruce range in CA (Griffin & Critchfield 1972). Sawyer & Thornburgh (1970).

Comments: Engelmann spruce forms a series in OR, WA and the Rocky Mountains where it is extensive. In CA it occurs in six isolated stands. The southernmost CA stand is along Clark Creek in Shasta Co.; the remaining stands are in the Salmon Mountains near Russian Peak [Blakes Fork of the South Russian Creek, Duck Lake Creek, Horse Range Creek, Music Creek, Sugar Creek]. Only the Sugar Creek drainage population is in the Sugar Creek candidate RNA Keeler-Wolf (1990e). Engelmann spruce is rare (CNPS List 2 plant) in CA (Skinner & Pavlik 1994).

Species mentioned in text	
Brewer spruce	*Picea breweriana*
Douglas-fir	*Pseudotsuga menziesii*
Engelmann spruce	*Picea engelmannii*
Incense-cedar	*Calocedrus decurrens*
Lodgepole pine	*Pinus contorta* ssp. *murrayana*
Mountain hemlock	*Tsuga mertensiana*
Pacific yew	*Taxus brevifolia*
Ponderosa pine	*Pinus ponderosa*
Shasta fir	*Abies magnifica* var. *shastensis*
Sugar pine	*Pinus lambertiana*
Western white pine	*Pinus monticola*
White fir	*Abies concolor*

Plot-based descriptions: Keeler-Wolf (1984b, 1989f), Sawyer & Thornburgh (1971) in Keeler-Wolf (1990e) summarize three surveys in Salmon Mountains including the Sugar Creek candidate RNA in Siskiyou Co. m-KlaR. In this drainage Keeler-Wolf maps stands dominated by Engelmann spruce.

Eucalyptus series

Eucalyptus sole or dominant tree in the canopy; few other species present. Trees < 50 m; canopy continuous. Shrubs infrequent. Ground layer sparse.

Uplands: all slopes.

Distribution: o-NorCo, o-CenCo, CenV, So-Co, m-TraR, m-Pen, ChaI.
Elevation: sea level-300 m.

NDDB/Holland type and status: none.

Other types:
Barry type: G7411111 CEUGO00.
PSW-45 type: Eucalyptus series.

General references: Boyd (1985), Bulman (1988), Paysen *et al.* (1980), Skolmen (1990), Skolmen & Ledig (1990).

Comments: Bulman (1988) discusses the introduction history of eucalyptus into CA. *The Jepson Manual* lists nine *Eucalyptus* species naturalized in the state [Forest red gum, lemon-scented gum, manna gum, red iron bark, red gum, silver dollar gum, silverleaf gum, and sugar gum]. Some, like blue gum, are common, while others are infrequently encountered. The inclusion of this series recognizes the importance of these introduced species in CA vegetation.

Species mentioned in text	
Blue gum	Eucalyptus globulus
Forest red gum	Eucalyptus tereticornis
Lemon-scented gum	Eucalyptus citriodora
Manna gum	Eucalyptus viminalis
Red gum	Eucalyptus camaldulensis
Red iron bark	Eucalyptus sideroxylon
Silver dollar gum	Eucalyptus polyanthemos
Silverleaf gum	Eucalyptus pulverulenta
Sugar gum	Eucalyptus cladocalyx

Plot-based descriptions: not available.

Fan palm series

Fan palm sole or dominant tree in canopy; arroyo willow, black willow, California sycamore, canyon live oak, Fremont cottonwood, narrowleaf willow, and/or velvet ash may be present. Trees < 30 m; canopy continuous or open. Shrubs occasional or absent. Ground layer sparse (Plate 18).

Wetlands: soils intermittently flooded, saturated. Water chemistry: fresh mixosaline. Canyon waterways; along fault lines. Cowardin class: Temporarily flooded palustrine forested wetland. The national inventory of wetland plants (Reed 1988) lists fan palm as a FACW.

Distribution: ColD, AZ, NV, Baja CA.
Elevation: sea level-900 m.

NDDB/Holland type and status: **Riparian woodlands** (62000).
      Desert fan palm oasis woodland (62300) G3 S3.

Other types:
Barry type: G7411111 CWAFI00.
Cheatham & Haller type: Palm oasis woodland.
PSW-45 type: Palm series.
Thorne type: Desert oasis woodland.
WHR type: Palm oasis.

Comments: Vogl & McHargue (1966) report great floristic variation among oases, however, canyon waterway oases differ consistently from those along fault lines in importance of common plants other than the fan palm.

General references: Burk (1977), Cornett (1987), MacMahon (1988), McClintock (1978), Minckley & Brown (1982) Paysen *et al.* (1980).

Species mentioned in text	
Arroyo willow	*Salix lasiolepis*
Black willow	*Salix gooddingii*
California sycamore	*Platanus racemosa*
Canyon live oak	*Quercus chrysolepis*
Fan palm	*Washingtonia filifera*
Fremont cottonwood	*Populus fremontii*
Honey mesquite	*Prosopis glandulosa*
Narrowleaf willow	*Salix exigua*
Velvet ash	*Fraxinus velutina*

Plot-based descriptions: Vogl & McHargue (1966) report on 24 oases in ColD. Spolsky (1979) defines a Fan palm-sycamore association at Anza Borrego State Park in San Diego Co. ColD.

Foothill pine series

Foothill pine sole or dominant in tree canopy or an emergent tree over a shrub canopy; birchleaf mountain-mahogany, coffeeberries, chamise, leather oak, manzanitas, scrub oak, toyon, and/or wedgeleaf ceanothus may be present. Trees < 20 m; canopy open. Shrubs common or infrequent. Ground layer sparse or grassy.

Wetlands: soils intermittently flooded. Water chemistry: fresh. Terraces. Cowardin class: Palustrine forested wetland. **Uplands:** valleys, slopes gentle to steep. Soils shallow, infertile, moderately to excessively drained. Surface may be covered with stones and rock outcrops.

Distribution: NorCo, CenCo, sac-CenV, l-KlaR, f-KlaR, f-CasR, f-SN, m-TraR, ModP, w. MojD.
Elevation: 300-2100 m.

NDDB/Holland type and status: Cismontane woodlands (71000).
> Serpentine digger pine chaparral (71321) G3 S3.
> Non-serpentine digger pine chaparral (71322) G4 S4.

Other types:
Barry type: G7411112 BPISA20.
Cheatham & Haller type: Blue oak woodland, Mixed chaparral.
PSW-45 type: Digger pine series.
Thorne type: Foothill woodland.
WHR type: Blue oak woodland.

General references: Foothill pine range in CA (Griffin & Critchfield 1972); silvics (Powers 1990). Graves (1932), Griffin (1965), Hanes (1977), Kruckeberg (1984), Klyver (1931), Paysen *et al.* (1980), Yeaton *et al.* (1980).

Comments: Several series have foothill pine as a component. It is especially common in Blue oak series stands. In Foothill pine series stands, however, blue oak is infrequent or absent. Foothill pine commonly emerges from a shrub canopy in foothill locations. The tree forms stands over grass in valleys. Other trees (black oak, California buckeye, coast live oak, Coulter pine, interior live oak, valley oak, western juniper) may mix with foothill pine locally. Wetland stands of foothill pine are rare due to systematic removal of the species to increase pasture according to Powers (1990).

Species mentioned in text	
Birchleaf mountain-mahogany	*Cercocarpus betuloides*
Black oak	*Quercus kelloggii*
Blue oak	*Quercus douglasii*
California buckeye	*Aesculus californica*
Chamise	*Adenostoma fasciculatum*
Coast live oak	*Quercus agrifolia*
Coffeeberries	*Rhamnus* species
Coulter pine	*Pinus coulteri*
Foothill pine	*Pinus sabiniana*
Interior live oak	*Quercus wislizenii*
Leather oak	*Quercus durata*
Manzanitas	*Arctostaphylos* species
Scrub oak	*Quercus berberidifolia*
Toyon	*Heteromeles arbutifolia*
Valley oak	*Quercus lobata*
Western juniper	*Juniperus occidentalis*
Wedgeleaf ceanothus	*Ceanothus cuneatus*

Plot-based descriptions: Keeler-Wolf (1983) in Keeler-Wolf (1990e) describes stands at Frenzel Creek RNA in Colusa Co. i-NorCo. Allen *et al.* (1989, 1991) define vegetation types with foothill pine. All have blue oak as an important tree with this pine. They are included in the Blue oak series.

Foxtail pine series

Foxtail pine sole or dominant tree in canopy; lodgepole pine, limber pine, mountain hemlock, Shasta fir, western white pine, and/or whitebark pine may be present. Trees < 20 m; canopy intermittent or open. Shrubs usually infrequent. Ground layer sparse (Plate 28).

Uplands: all slopes, especially ridges, upper slopes below forestline. Soils granitic, metamorphic, or ultramafic-derived.

Distribution: m-NorCo, su-KlaR, s. su-SN.
Elevation: 2100-3500 m.

NDDB/Holland type and status: Subalpine coniferous forests (86000).
Foxtail pine forest (86300) G3 S3.3.

Other types:
Barry type: G7411112 BPIBA00.
Cheatham & Haller type: Foxtail pine forest.
Thorne type: Subalpine coniferous forest.
WHR type: Subalpine conifer.

General references: Foxtail pine range in CA (Griffin & Critchfield 1972). Barbour (1988), Ryerson (1983), Sawyer & Thornburgh (1977).

Comments: Separate populations of foxtail pine represent varieties of *Pinus balfouriana* [*P. b.* var. *balfouriana* in NorCo and KlaR and *P. b.* var. *aus-trina* in s. SN]. In m-NorCo the series occurs on Franciscan greywacke sandstone, metavolcanic, and schist-derived soils. In su-KlaR, foxtail pine is mostly found on ultramafic-derived soils. In SN it occurs on granitic-derived soils. The two varieties are included in this series.

Species mentioned in text	
Drummond windflower	*Anemone drummondii*
Foxtail pine	*Pinus balfouriana*
Limber pine	*Pinus flexilis*
Lodgepole pine	*Pinus contorta* var. *murrayana*
Mountain hemlock	*Tsuga mertensiana*
Shasta fir	*Abies magnifica* var. *shastensis*
Western white pine	*Pinus monticola*
Whitebark pine	*Pinus albicaulis*

Plot-based descriptions: Keeler-Wolf & Keeler-Wolf (1984a) describe stands at South Yolla Bolly Mountain in Tehama Co. su-NorCo; Keeler-Wolf (1987a) in Keeler-Wolf (1990e) at the Crater Creek candidate RNA, Whipple & Cope (1979) in Keeler-Wolf (1990e) define one association at Mount Eddy candidate RNA, Keeler-Wolf (1984b, 1989f), Sawyer & Thornburgh (1971) in Keeler-Wolf (1990e) describe stands at Sugar Creek candidate RNA in Siskiyou Co. su-KlaR; Ball (1976) in Keeler-Wolf (1990e) defines one association at Last Chance RNA in Inyo Co. su-SN.

Associations:
Ball (1976):
Sierra foxtail pine forest.
Whipple & Cope (1979):
Foxtail pine forest/Drummond windflower association.

Fremont cottonwood series

Fremont cottonwood sole, dominant, or important tree in canopy; black willow, box elder, California sycamore, narrowleaf willow, Oregon ash, Pacific willow, red willow, walnuts, and/or yellow willow may be present. Trees < 25 m; canopy continuous or open. Shrubs and grape lianas infrequent to common. Ground layer variable (Plate 19).

Wetlands: soils intermittently or seasonally flooded, saturated. Water chemistry: fresh. Riparian corridors; floodplains subject to high-intensity flooding; floodplains; low-gradient depositions along rivers, streams; seeps; stream and river banks, terraces. Cowardin class: Palustrine forested wetland. The national inventory of wetland plants (Reed 1988) lists Fremont cottonwood as a FACW.

Distribution: i-NorCo, n. CenCo, CenV, f-KlaR, f-CasR, f-SN, SoCo, m-TraR, m-PenR, GB, MojD, ColD, Baja CA.

Elevation: sea level-2400 m.

NDDB/Holland type and status: **Riparian forests** (61000).

    Central Coast cottonwood-sycamore riparian forest (61210 *in part*) G3 S3.3.

    Southern cottonwood-willow riparian forest (61330 *in part*) G3 S3.

    Great Valley cottonwood riparian forest (61410) G2 S2.1.

    Great Valley mixed riparian forest (61420) G2 S2.1.

    Modoc-Great Basin cottonwood-willow riparian forest (61610) G3 S2.1.

    Mojave riparian forest (61700) G1 S1.1.

    Sonoran cottonwood-willow riparian forest (61810) G2 S1.1.

Other types:

Barry type: G7411121 BPPFR20.

Cheatham & Haller type: Central Valley bottomland woodland, Northern riparian woodlands, Southern riparian woodland, Southern alluvial woodland.

PSW-45 type: Cottonwood series.

Rangeland type: SRM 203.

Thorne type: Riparian woodlands.

WHR type: Valley foothill riparian, Desert riparian.

General references: Range in CA (Griffin & Critchfield 1972). Bowler (1989), Capelli & Stanley (1984), Conard & Robichaux (1980), Faber *et al.* (1989), Holstein (1984), MacMahon (1988), McBride (1994), Minckley & Brown (1982), Paysen *et al.* (1980), Sands (1980).

Comments: In CenV Fremont cottonwood may dominate or mix with other trees. Most cottonwood-dominated stands are altered by grazing. McCarten & Patterson (1989) mapped stands along the main streams of the Sacramento River.

Species mentioned in text

Arroyo willow	*Salix lasiolepis*
Black willow	*Salix gooddingii*
Box elder	*Acer negundo*
California wild grape	*Vitis californica*
California sycamore	*Platanus racemosa*
Fremont cottonwood	*Populus fremontii*
Narrowleaf willow	*Salix exigua*
Oregon ash	*Fraxinus latifolia*
Pacific willow	*Salix lucida* ssp. *lasiandra*
Red willow	*Salix laevigata*
Walnuts	*Juglans californica* var. *californica* in So-CA; *J. c.* var. *hindsii, J. nigra, J. c.* var. *hindsii x nigra* in Nor-CA.
Yellow willow	*Salix lutea*

Plot-based descriptions: Gray & Greaves (1984) describe stands in Santa Barbara Co. CenCo; Conard & Robichaux (1980) mixed stands in Glenn Co., Bahre *et al.* (1982) on dredge fields CenV; Hanes (1984) along Santa Ana River, Zembal (1989) reports tree density and cover along Santa Margarita and Santa Ana Rivers in Orange Co. o-SoCo; Brothers (1984) along Owens River in Inyo Co. TraSN; NDDB has plot data for stands near Victorville and Palmdale in San Bernardino Co. MojD; Spolsky (1979) defines a Pacific willow-Fremont cottonwood association at Anza Borrego State Park in San Diego Co. ColD.

Giant sequoia series

Giant sequoia sole, dominant, or important tree in canopy; black oak, Douglas-fir, incense-cedar, Jeffrey pine, lodgepole pine, ponderosa pine, red fir, sugar pine, and/or white fir may be present. Trees < 95 m; canopy continuous or intermittent. Shrubs infrequent or common. Ground layer sparse or abundant (Plate 27).

Uplands: Soils commonly granitic-derived, well-drained.

Distribution: m-SN.
Elevation: 1400-2600 m.

NDDB/Holland type and status: **Lower montane coniferous forests** (84000).
    Big tree forest (84250) G3 S3.

Other types:
Barry type: G7411112 BSEGI00.
Cheatham & Haller type: Sierra big tree forest.
Thorne type: Mixed conifer forest.
WHR type: Sierran mixed conifer.

General references: Giant sequoia range in CA (Griffin & Critchfield 1972); silvics (Weatherspoon 1990). Barbour (1988), Barbour *et al.* (1981b), Biswell *et al.* (1966), Bonnicksen (1982), Kilgore & Taylor (1979), Nichols (1989), Rundel (1971, 1972a, 1972b), Rundel *et al.* (1977), Willard (1994).

Comments: There are approximately 75 groves in m-SN. Most are concentrated south of the Kings River. Groves range from 1 to 1620 ha in extent. Species associated with giant sequoia are those commonly considered part of mixed conifer forest [*see* Douglas-fir series, Douglas-fir - ponderosa pine series, Mixed conifer series, Ponderosa pine series, Red fir series, White fir series].

Species mentioned in text	
Black oak	*Quercus kelloggii*
Douglas-fir	*Pseudotsuga menziesii*
Giant sequoia	*Sequoiadendron giganteum*
Incense-cedar	*Calocedrus decurrens*
Jeffrey pine	*Pinus jeffreyi*
Lodgepole pine	*Pinus contorta* ssp. *murrayana*
Ponderosa pine	*Pinus ponderosa*
Red fir	*Abies magnifica* var. *magnifica*
Sugar pine	*Pinus lambertiana*
White fir	*Abies concolor*

Plot-based descriptions: Anderson (1990) describes Lost Grove and Tuolumne Grove in Tuolumne Co. In both cases white fir is very important. Keeler-Wolf (1989b) describes two stands at Moses Mountain RNA in Tulare Co. m-SN. Red fir is important in one stand.

Grand fir series

Grand fir sole or dominant tree in canopy; bishop pine, red alder, redwood, Sitka spruce, and/or western hemlock may be present. Trees < 70 m tall. Tree canopy continuous. Shrubs infrequent or common. Ground layer infrequent or abundant (Plate 23).

Uplands: slopes, maritime terraces. Soils sandstone-derived, shallow.

Distribution: o-NorCo, OR, WA, ID.
Elevation: sea level-50 m.

NDDB/Holland type and status: North Coast coniferous forests (82000).
   Sitka spruce-grand fir forest (82100 *in part*) G4 S1.1.

Other types:
Barry type: G7411112 BABGR00.
Cheatham & Haller type: Sitka spruce-grand fir forest.
Thorne type: North coastal coniferous forest.
WHR type: Douglas-fir.

General references: Grand fir range in CA (Griffin & Critchfield 1972); silvics (Foiles *et al* 1990). Boyd (1980a), Franklin (1980b, 1988), Franklin *et al.* (1983), Minore (1980, 1990), Zinke (1977).

Comments: Grand fir is a common series outside California. In CA the species mixes as a secondary species in Sitka spruce, redwood, and Douglas-fir - tanoak series forests in Del Norte and Humboldt Co. Stands dominated by grand fir occur in Humboldt, Mendocino, and Sonoma Cos.

Grand fir x white fir hybrids occur at low and montane elevations in ne. NorCo and nw. KlaR. These populations are considered part of Douglas-fir - tanoak, Douglas-fir, and White fir series. In CA

grand fir is coastal and occurs at low elevations in NorCo.

Species mentioned in text	
Bishop pine	*Pinus muricata*
Douglas-fir	*Pseudotsuga menziesii*
Grand fir	*Abies grandis*
Red alder	*Alnus rubra*
Redwood	*Sequoia sempervirens*
Sitka spruce	*Picea sitchensis*
Tanoak	*Lithocarpus densiflora*
Western hemlock	*Tsuga heterophylla*
White fir	*Abies concolor*

Plot-based descriptions: NDDB has plot data on file for stands in Mendocino Co., for stands along the Mad River in Humboldt Co. o-NorCo.

Hooker willow series

Hooker willow sole or dominant shrub or tree in canopy; black cottonwood, coyote brush, red alder, wax-myrtle, and/or willows may be present. If shrubland, emergent trees may be present. Shrubs < 8 m; canopy continuous. Shrubs sparse under tree canopy. Ground layer variable (Plate 18).

Wetlands: habitats seasonally flooded, saturated. Water chemistry: fresh. Along streams; floodplains; depositions; sand dunes. Cowardin class: Palustrine shrub-scrub wetland. The national inventory of wetland plants (Reed 1988) lists Hooker willow as a FACW.

Distribution: o-NorCo to AK.
Elevation: sea level-100 m.

NDDB/Holland type and status: **Marshes and swamps** (52000), **Riparian forests** (61000), **Riparian scrubs** (63000).
    Freshwater swamp (52600 *in part*) G1 S1.1.
    Red alder riparian forest (61130 *in part*) G3 S2.2.
    North Coast riparian scrub (63100 *in part*) G3 S3.2.

Other types:
Barry type: G7411221 BSAHO00.
Cheatham & Haller type: Bottomland woodlands and forest.
PSW-45 type: Willow series.
Thorne type: Riparian woodland.
WHR type: Freshwater emergent wetland.

General references: Brayshaw (1976), Paysen *et al.* (1980).

Comments: Willow stands may or may not be dominated by a single species. If no dominant willow is present at low elevations, then place the stands in the Mixed willow series. Montane and subalpine willow stands are placed in separate classes since different willow species are restricted to those elevations [*see* Montane wetland shrub habitat, Subalpine wetland shrub habitat].

Stands of Hooker willow have environment conditions similar to alder, cottonwood and other willow series [*see* Black cottonwood series, Black willow series, Fremont cottonwood series, Pacific willow series, Red alder series, Red willow series, Sitka willow series, White alder series].

Species mentioned in text	
Black cottonwood	*Populus balsamifera*
Black willow	*Salix gooddingii*
California blackberry	*Rubus ursinus*
Coyote brush	*Baccharis pilularis*
Fremont cottonwood	*Populus fremontii*
Hooker willow	*Salix hookeriana*
Pacific willow	*Salix lucida* ssp. *lasiandra*
Red alder	*Alnus rubra*
Red willow	*Salix laevigata*
Sitka willow	*Salix sitchensis*
Wax-myrtle	*Myrica californica*
White alder	*Alnus rhombifolia*
Willows	*Salix* species

Plot-based descriptions: Duebendorfer (1989) defines a Hooker willow association [as Woody hollows] on coastal dunes in Humboldt Bay area, Imper & Sawyer (1987) a Hooker willow/ California blackberry association [as Willow scrub] at Table Bluff Ecological Reserve, NDDB has data for Humboldt Co. o-NorCo.

Incense-cedar series

Incense-cedar sole or dominant tree in canopy; canyon live oak, Douglas-fir, lodgepole pine, ponderosa pine, sugar pine, and/or white fir may be present. Trees < 65 m tall; canopy continuous, intermit-tent or open. Shrubs infrequent or common. Ground layer variable.

Uplands: slopes, ridges; raised stream benches and terraces. Soils around wet meadows, upper slopes, ridges. The national inventory of wetland plants (Reed 1988) does not list incense-cedar.

Distribution: m-NorCo, m-CenCo, m-KlaR, m-CasR, m-SN, m-TraR, m-PenR, Baja CA. **Elevation:** 500-2100 m.

NDDB/Holland type and status: Lower montane coniferous forests (84000).
> Sierran mixed coniferous forest (84230 *in part*) G4 S4.
> White fir forest (84240 *in part*) G4 S4.

Other types:
Barry type: G7411112 BCADE00.
Cheatham & Haller type: Sierra mixed conifer forest.
Thorne type: Mixed conifer forest.
WHR type: Sierra mixed conifer.

General references: Incense-cedar range in CA (Griffin & Critchfield 1972); silvics (Powers & Oliver 1990). Barbour (1988), Kruckeberg (1984), Minnich (1987).

Comments: Incense-cedar dominated stands have commonly been included in definitions of mixed conifer forest. It is best to consider the mixed conifer forest as a collection of series.

Plot-based descriptions: Muldavin (1982) defines one association in Humboldt Co., Keeler-Wolf (1986c, 1988b) in Keeler-Wolf (1990e) describes a gully forest at Doll Basin candidate RNA in Mendocino Co. m-NorCo; Stuart *et al.* (1992) define one association at Castle Crags State Park in Siskiyou Co. l-KlaR; Keeler-Wolf (1986e, 1988e) in Keeler-Wolf (1990e) describes a canyon bottom forest at Hall Canyon RNA in Riverside Co. m-PenR.

Species mentioned in text	
Canyon live oak	*Quercus chrysolepis*
Douglas-fir	*Pseudotsuga menziesii*
Incense-cedar	*Calocedrus decurrens*
Lodgepole pine	*Pinus contorta* var. *murrayana*
Ponderosa pine	*Pinus ponderosa*
Sugar pine	*Pinus lambertiana*
Twayblane	*Listera convallarioides*
White fir	*Abies concolor*

Associations:
Muldavin (1982):
> Incense-cedar/twayblane association
> [as *Listera convallarioides-Athyrium filix-femina* association].

Stuart *et al.* (1992):
> Incense-cedar - Douglas-fir association.

Interior live oak series

Interior live oak sole or dominant tree in canopy; black oak, blue oak, foothill pine, madrone, and/or tanoak may be present. Trees < 15 m; canopy continuous, intermittent, or savanna-like. Shrubs infrequent or common. Ground layer sparse.

Uplands: slopes; valleys; raised stream benches and terraces. Soils shallow, moderately to excessively drained. The national inventory of wetland plants (Reed 1988) does not list interior live oak.

Distribution: s. o-NorCo, i-NorCo, CenCo, f-KlaR, f-CasR, f-SN, m-TraR, m-PenR.
Elevation: 500-4500 m.

NDDB/Holland type and status: **Cismontane woodlands** (71000), **Broadleaved upland forests** (81000).
    Interior live oak woodland (71150) G3 S3.2.
    Digger pine-oak woodland (71410 *in part*) G4 S4.
    Interior live oak forest (81330) G4 S4.

Other types:
Barry type: G74 G7411111 CQUW100.
Cheatham & Haller type: Northern oak woodland.
PSW-45 type: Interior live oak series.
Thorne type: Northern oak woodland forest.
WHR type: Montane hardwood.

General references: Interior live oak range in CA (Griffin & Critchfield 1972). Allen *et al.* (1989, 1991), Allen-Diaz & Holtzman (1991), Barbour (1988), Cooper (1922), Griffin (1977), Paysen *et al.* (1980), Pavlik *et al.* (1991), Shreve (1927).

Comments: Interior live oak is common in several series. Stands where it dominates may be forests or woodlands. If the stand is composed of shrubs, it is a member of the Interior live oak shrub series. Justification for recognizing two series is presented in White & Sawyer (1995). Allen *et al.* (1991) mention probable wetland associations. Keeler-Wolf (1990d)

qualitatively describes interior live oak-dominated stands at Indian Creek candidate RNA in Tehama Co. f-CasR.

Species mentioned in text	
Black oak	*Quercus kelloggii*
Blue oak	*Quercus douglasii*
Common manzanita	*Arctostaphylos manzanita*
Foothill pine	*Pinus sabiniana*
Interior live oak	*Quercus wislizenii*
Madrone	*Arbutus menziesii*
Poison-oak	*Toxicodendron diversilobum*
Tanoak	*Lithocarpus densiflora*
Whiteleaf manzanita	*Arctostaphylos viscida*
Yerba santa	*Eriodictyon californicum*

Plot-based descriptions: Allen *et al.* (1989, 1991) define six subseries in Cis-CA, based on tree and shrub data only. They are called associations here. Keeler-Wolf (1988d) in Keeler-Wolf (1990e) describes stands at Millard Canyon RNA in Riverside Co. TraR.

Associations:
Allen *et al.* (1991):
    Interior live oak-blue oak-foothill pine association,
    Interior live oak-foothill pine/common manzanita association,
    Interior live oak-madrone/poison-oak association,
    Interior live oak/whiteleaf manzanita association,
    Interior live oak/yerba santa/grass association.

Island oak series

Island oak sole or dominant in tree canopy; Lyon cherry or MacDonald oak may be present. Trees < 15 m; canopy continuous. Shrubs absent. Ground layer sparse.

Uplands: slopes in sub-riparian settings. Soil sandstone-derived.

Distribution: ChaI.
Elevation: sea level-100 m.

NDDB/Holland type and status: Cismontane woodlands (71000).
    Island oak woodland (71190) G2 S2.1.

Other types:
Barry type: G74 G7411214.
Cheatham & Haller type: Island woodland.
Thorne type: Island woodland.
WHR type: Coastal oak woodland.

General references: Island oak range in CA (Griffin & Critchfield 1972). Minnich (1980a), Muller (1967), Pavlik *et al.* (1991), Philbrick & Haller (1977).

Comments: The mix of other oaks with island oak varies; stands merge with chaparral types as well.

Species mentioned in text	
Island oak	*Quercus tomentella*
Lyon cherry	*Prunus ilicifolia* ssp. *lyonii*
MacDonald oak	*Quercus* x *macdonaldii*

Plot-based descriptions: NDDB has plot data on file for Santa Catalina Island in Los Angeles Co. ChaI.

Jeffrey pine series

Jeffrey pine sole or dominant tree in canopy; black oak, canyon live oak, foxtail pine, incense-cedar, interior live oak, knobcone pine, lodgepole pine, ponderosa pine, Port Orford-cedar, red fir, Shasta fir, Washoe pine, western juniper, western white pine, and/or white fir may be present. Trees < 60 m; canopy intermittent or open. Shrubs infrequent or common. Ground layer sparse, abundant, or grassy (Plate 26).

Uplands: in NorCo, CenCo, KlaR most commonly found on ultramafic-derived soils; in other regions on productive or harsh soils regardless of parent material. Soils well-drained.

Distribution: m-NorCo, m-CenCo, l-KlaR, m-KlaR, su-KlaR, m-CasR, su-CasR, m-SN, su-SN, m-TraR, su-TraR, m-PenR, su-PenR, GB, Baja CA.
Elevation: 60-2900 m.

NDDB/Holland type and status: **Lower montane coniferous forests (84000), Upper montane coniferous forests (85000).**
>   Northern ultramafic Jeffrey pine forest (84171) G3 S3.
>   Ultramafic mixed coniferous forest (84180) G4 S4.
>   Southern ultramafic Jeffrey pine (84180) G2 S2.1.
>   Jeffrey pine forest (85100) G4 S4.
>   Jeffrey pine-fir forest (85210) G4 S4.

Other types:
Barry type: G7411112 BPIJE00.
Cheatham & Haller type: Jeffrey pine forest.
PSW-45 type: Ponderosa/Jeffrey pine series.
Thorne type: Yellow pine forest.
WHR type: Jeffrey pine, Eastside pine.

General references: Jeffrey pine range in CA (Griffin & Critchfield 1972); silvics (Jenkinson 1990). Jenkinson (1980), Kruckeberg (1984), Minnich (1987), Paysen *et al.* (1980), Rundel *et al.* (1977), Sawyer *et al.* (1977), Sawyer & Thornburgh (1977), Thorne (1977), Vasek & Thorne (1977), Zinke (1977), Zobel (1952).

Species mentioned in text

Antelope bitterbrush	*Purshia tridentata* var.*tridentata*
Basket bush	*Rhus trilobata*
Bear-grass	*Xerophyllum tenax*
Black oak	*Quercus kelloggii*
Bush chinquapin	*Chrysolepis sempervirens*
California fescue	*Festuca californica*
Canyon live oak	*Quercus chrysolepis*
Curlleaf mountain-mahogany	*Cercocarpus ledifolius*
Del Norte iris	*Iris innominata*
Desert snowberry	*Symphoricarpos longiflorus*
Douglas-fir	*Pseudotsuga menziesii*
Foxtail pine	*Pinus balfouriana*
Huckleberry oak	*Quercus vaccinifolia*
Idaho fescue	*Festuca idahoensis*
Incense-cedar	*Calocedrus decurrens*
Interior live oak	*Quercus wislizenii*
Jeffrey pine	*Pinus jeffreyi*
Knobcone pine	*Pinus attenuata*
Lodgepole pine	*Pinus contorta* ssp. *murrayana*
Mountain big sagebrush	*Artemisia tridentata* var. *vaseyana*
Mountain whitethorn	*Ceanothus cordulatus*
Mule's ears	*Wyethia mollis*
One-sided bluegrass	*Poa secunda*
Pinemat manzanita	*Arctostaphylos nevadensis*
Ponderosa pine	*Pinus ponderosa*
Port Orford-cedar	*Chamaecyparis lawsoniana*
Red fir	*Abies magnifica* var. *magnifica*
Sadler oak	*Quercus sadleriana*
Serpentine-haplopappus	*Ericameria ophitidis*
Shasta fir	*Abies magnifica* var. *shastensis*
Tailed lupine	*Lupinus caudatus*
Tufted reedgrass	*Calamagrostis koelerioides*
Western needlegrass	*Stipa occidentalis*
Western white pine	*Pinus monticola*
Wheeler bluegrass	*Poa wheeleri*
White fir	*Abies concolor*
Yellow pine	*Pinus jeffreyi* and *P. ponderosa*

Comments: Jeffrey pine dominates in self-replacing stands on shallow, stony, strongly serpentinized peridotite at elevations as low as 200 m in Del Norte Co. w. l-KlaR (Duebendorfer 1987). This

---

vegetation was first described by Whittaker (1960) as a woodland, with Jeffrey pine in a patchy mosaic with western white pine shrublands [*see* Western white pine series].

Keeler-Wolf (1990e) qualitatively defines a Jeffrey pine woodland at the Bell Meadow RNA in Calaveras Co. m-SN; at the Grass Lake RNA in El Dorado Co. su-SN.

Plot-based descriptions: Keeler-Wolf (1991a) reports tree density, frequency for stands at Big Pine Mountain proposed RNA in Santa Barbara Co. m-CenCo; Keeler-Wolf (1986a) in Keeler-Wolf (1990e) describes stands at the L. E. Horton candidate RNA. Duebendorfer (1987) defines one association at nearby at Pine Flat Mountain in Del Norte Co. w. l-KlaR; Taylor & Teare (1979b) in Keeler-Wolf (1990e two associations at Smoky Creek candidate RNA in Trinity Co. m-KlaR, Jimerson (1993) six associations in Klamath National Forest KlaR, Waddell (1982) one association in Yolla Bolly Mountains in Tehama and Trinity Cos. NorCo, KlaR; Potter (1994) four associations in c. m-SN, su-SN; Parikh (1993b) reports tree density and basal area for stands at Sawmill Mountain candidate RNA in Ventura Co. o. m-TraR; Riegel *et al.* (1990) define one associa-tion in South Warner Wilderness WaR, Smith (1994) eight associations in Klamath, Lassen, Modoc, Shasta-Trinity, Tahoe National Forests m-CasR, ModP, Vasek (1978) importance values for stands in Mono and Placer Cos. n. TraSN; Taylor (1980) in Keeler-Wolf (1990e) defines one association at Indiana Summit RNA, Talley (1978) in Keeler-Wolf (1990e) one association at Sentinel Meadow RNA in Mono Co. TraSN.

Associations:
Jimerson (1993):
  Jeffrey pine/Sadler oak/bear-grass association,
  Jeffrey pine - Douglas-fir/huckleberry oak/ California fescue association,
  Jeffrey pine/Idaho fescue association,
  Jeffrey pine - incense-cedar/huckleberry oak association,
  Jeffrey pine - incense-cedar/huckleberry oak/ bear-grass association,

  Jeffrey pine- western white pine/Del Norte iris association.
Duebendorfer (1987):
  Jeffrey pine/Idaho fescue association [as Jeffrey pine woodland].
Potter (1994):
  Jeffrey pine association,
  Jeffrey pine/huckleberry oak association,
  Jeffrey pine/greenleaf manzanita association,
  Jeffrey pine/mountain whitethorn.
Riegel *et al.* (1990):
  Jeffrey pine/tailed lupine association [as *Pinus jeffreyi* phase of *Abies concolor/ Lupinus caudatus* association].
Smith (1994):
  Jeffrey pine/antelope bitterbrush/mule's ears association,
  Jeffrey pine/antelope bitterbrush-curlleaf mountain-mahogany/western needlegrass association,
  Jeffrey pine/antelope bitterbrush-desert snow-berry/Wheeler bluegrass association,
  Jeffrey pine-black oak/one-sided bluegrass association [as Yellow pine-black oak/one-sided bluegrass association],
  Jeffrey pine-black oak/basket bush association,
  Jeffrey pine/curlleaf mountain-mahogany association,
  Jeffrey pine/desert snowberry/Wheeler blue-grass association,
  Jeffrey pine/mountain big sagebrush/Idaho fescue association.
Talley (1978):
  Jeffrey pine/bush chinquapin association [as Jeffrey pine forest].
Taylor (1980):
  Jeffrey pine/antelope bitterbrush association.
Taylor & Teare(1979):
  Jeffrey pine/serpentine-haplopappus association,
  Jeffrey pine/tufted reedgrass association.
Waddell (1982):
  Jeffrey pine/pinemat manzanita association.

Jeffrey pine-ponderosa pine series

Jeffrey pine and ponderosa pine important trees in canopy; black oak, incense-cedar, western juniper, and/or white fir may be present. Trees < 60 m; canopy intermittent or open. Shrubs infrequent or common. Ground layer sparse or abundant.

Uplands: slopes, all aspects; raised, benches and terraces. Soils well-drained. The national inventory of wetland plants (Reed 1988) lists ponderosa pine as a FACU. Jeffrey pine is not listed.

Distribution: m-CasR, ModP, m-SN.
Elevation: 1000-2400 m.

NDDB/Holland type and status: **Lower montane coniferous forests (84000), Upper montane coniferous forests** (85000).
Jeffrey pine forest (85100 *in part*) G4 S4.
Jeffrey pine-white fir forest (85210 *in part)* G4 S4.
Eastside ponderosa pine forest (84220 *in part*) G4 S2.1.

Other types:
Barry type: G7411112.
Cheatham & Haller type: Eastside ponderosa pine forest.
PSW-45 type: Ponderosa/Jeffrey pine series.
Thorne type: Yellow pine forest.
WHR type: Eastside pine.

General references: Jeffrey pine and ponderosa pine ranges in CA (Griffin & Critchfield 1972); silvics (Jenkinson 1990), (Oliver & Ryker 1990). Paysen *et al.* (1980), Rundel (1977), Sawyer *et al.* (1977), Sawyer & Thornburgh (1977), Thorne (1977), Vasek & Thorne (1977).

Comments: Stands of this series differ from the Jeffrey pine and Ponderosa pine series, where one species dominates. Here both are equally important [*see* Jeffrey pine series, Ponderosa pine series].

Stands of all three series occur in Cis-CA and Trans-CA in ne. Nor-CA.

Plot-based descriptions: Smith (1994) describes seven associations [as yellow pine series] in Lassen, Plumas, Tahoe National Forests m-CasR, ModP, m-SN.

Species mentioned in text	
Antelope bitterbrush	*Purshia tridentata*
	var. *tridentata*
Arrowleaf balsam root	*Balsamorhiza sagittata*
Black oak	*Quercus kelloggii*
Columbia needlegrass	*Stipa nelsonii*
Creeping snowberry	*Symphoricarpos mollis*
Huckleberry oak	*Quercus vaccinifolia*
Idaho fescue	*Festuca idahoensis*
Incense-cedar	*Calocedrus decurrens*
Jeffrey pine	*Pinus jeffreyi*
Modoc coffeeberry	*Rhamnus rubra*
Mule's ears	*Wyethia mollis*
One-sided bluegrass	*Poa secunda*
Oregon ash	*Fraxinus latifolia*
Ponderosa pine	*Pinus ponderosa*
Western juniper	*Juniperus occidentalis*
	ssp. *occidentalis*
White fir	*Abies concolor*
Yellow pine	*Pinus jeffreyi* and *P. ponderosa*

Associations:
Smith (1994):
Jeffrey pine-ponderosa pine/arrowleaf balsam root association,
Jeffrey pine-ponderosa pine/antelope bitterbrush/Idaho fescue association,
Jeffrey pine-ponderosa pine/antelope bitterbrush/Idaho fescue/granite association,
Jeffrey pine-ponderosa pine/Modoc coffeeberry/one-sided bluegrass association,
Jeffrey pine-ponderosa pine/huckleberry oak association,
Jeffrey pine-ponderosa pine/Columbia needlegrass/Oregon ash association,
Jeffrey pine-ponderosa pine/creeping snowberry/mule's ears association.

Knobcone pine series

Knobcone pine sole or dominant tree in canopy; canyon live oak, Coulter pine, foothill pine, interior live oak, lodgepole pine, Monterey pine, and/or western white pine may be present. Trees < 25 m; canopy continuous, intermittent, or open, may be two-tiered. Shrubs infrequent or continuous. Ground layer sparse (Plate 23).

Uplands: ridges, upper slopes. Noted in literature for growing on ultramafic parent materials, but common on others as well. Soils infertile, rocky, dry.

Distribution: i-NorCo, m-NorCo, CenCo, l-KlaR, f-KlaR, m-KlaR, f-CasR, m-CasR, m-SN, m-TraR, w. m-PenR, ModP, Baja CA, OR. **Elevation:** 180-2000 m.

NDDB/Holland type and status: **Closed-cone coniferous forests** (83000).
 Knobcone pine forest (83210) G4 S4.

Other types:
Barry type: G741112 BPIAT00.
Cheatham & Haller type: Knobcone forest.
PSW-45 type: Knobcone pine series.
Thorne type: Inland closed-cone coniferous woodland.
WHR type: Closed cone pine-cypress.

General references: Knobcone pine range in CA (Griffin & Critchfield 1972). Colwell (1980), Keeley & Keeley (1978), Kruckeberg (1984), Paysen *et al.* (1980), Vogl (1973), Vogl *et al.* (1977).

Comments: Knobcone pine is present, often an important species in other series throughout its range, but not dominant as in this series. In some stands the trees are dense with few other species present; others are open with a sparse or abundant shrub layer. Often the pine is found as individual trees emerging though a dense, continuous shrub canopy [*see* Key to series dominated by shrubs].

Many stands are even-aged and related to fire; others are all-aged. The pine is known for its persistent, closed-cone habit, but the percentage of trees in a stand with persistent, closed cones varies greatly among stands.

Keeler-Wolf (1990e) qualitatively describes stands at Frenzel Creek RNA in Colusa Co. i-NorCo; Bridge Creek candidate RNA in Siskiyou Co., Pearch Creek in Humboldt Co. m-KlaR.

Species mentioned in text	
Canyon live oak	*Quercus chrysolepis*
Coulter pine	*Pinus coulteri*
Foothill pine	*Pinus sabiniana*
Greenleaf manzanita	*Arctostaphylos patula*
Hairy manzanita	*Arctostaphylos columbiana*
Huckleberry oak	*Quercus vaccinifolia*
Interior live oak	*Quercus wislizenii*
Knobcone pine	*Pinus attenuata*
Lodgepole pine	*Pinus contorta*
Monterey pine	*Pinus radiata*
Tanoak	*Lithocarpus densiflora*
Western white pine	*Pinus monticola*

Plot-based descriptions: Keeler-Wolf (1987e) in Keeler-Wolf (1990c) report tree density and basal area at Hale Ridge candidate RNA Lake Co. o-NorCo; Imper (1991a) defines one association at Craigs Creek candidate RNA in Del Norte Co., Taylor & Teare (1979b) in Keeler-Wolf (1990e) one association at Manzanita Creek candidate RNA in Trinity Co. m-KlaR; Imper (1991b) one association at Mayfield candidate RNA in Shasta Co. m-CasR; Vogl *et al.* (1977) describes stands in the Santa Ana Mountains in Orange Co. o-SoCo.
Associations:
Imper (1991a):
 Knobcone pine/hairy manzanita association [including Knobcone pine forest and Knobcone pine-tanoak forest].
Imper (1991b):
 Knobcone pine/greenleaf manzanita association [as Knobcone pine forest].
Taylor & Teare (1979a):
 Knobcone pine/huckleberry oak association.

Limber pine series

Limber pine sole or dominant tree in canopy; bristlecone pine, foxtail pine, Jeffrey pine, lodgepole pine, white fir, and/or whitebark pine may be present. Trees < 18 m; canopy open. Shrubs infrequent or common. Ground layer sparse.

Uplands: all slopes, especially ridges and upper slopes below forestline. Soils commonly granitic-derived.

Distribution: s. su-SN, su-TraR, su-PenR, su-WIS, su-DesR.
Elevation: 2200-3350 m.

NDDB/Holland type and status: **Subalpine coniferous forests** (86000).
  Limber pine forest (86700) G4 S2.3.

Other types:
Barry type: G7411112 BPIFL20.
Cheatham & Haller type: Southern California subalpine forest.
PSW-45 type: Limber pine series.
Thorne type: Limber pine forest.
WHR type: Subalpine conifer.

General references: Limber pine range in CA (Griffin & Critchfield 1972), silvics (Steele 1990). Barney (1980), Paysen *et al.* (1980), Thorne (1977) Vasek & Thorne (1977).

Comments: Limber pine may form monospecific stands or mix with other conifers. Thorne (1982) found limber pine with white fir (Sweetwater Mountains) and bristlecone pine in su-WIS. In the su-SN, limber pine grows with foxtail pine and whitebark pine. It occurs with bristlecone pine and lodgepole pine in su-DesR (Panamint Mountains), and with lodgepole pine, Jeffrey pine, and white fir in su-TraR and su-PenR.

Species mentioned in text	
Bristlecone pine	*Pinus longaeva*
Curlleaf mountain-mahogany	*Cercocarpus ledifolius*
Foxtail pine	*Pinus balfouriana*
Jeffrey pine	*Pinus jeffreyi*
Limber pine	*Pinus flexilis*
Lodgepole pine	*Pinus contorta* ssp. *murrayana*
White fir	*Abies concolor*
Whitebark pine	*Pinus albicaulis*

Plot-based descriptions: Keeler-Wolf & Keeler-Wolf (1976), Keeler-Wolf (1989h) in Keeler-Wolf (1990e) describe stands at Whippoorwill Flat RNA, Taylor (1979) in Keeler-Wolf (1990e) defines one association in Inyo Co. su-WIS; Talley (1978) in Keeler-Wolf (1990e) describes stands at Sentinel Meadow RNA in Inyo Co. TraSN; Ball (1976) in Keeler-Wolf (1990e) stands at Last Chance RNA in Inyo Co. su-SN.

Associations:
Taylor (1979):
  Limber pine/curlleaf mountain-mahogany association.

Lodgepole pine series

Lodgepole pine sole or dominant tree in canopy; foxtail pine, limber pine, mountain hemlock, red fir, Shasta fir, western white pine, white fir, and/or whitebark pine may be present. Trees < 40 m; canopy continuous, intermittent, or open. Shrubs usually infrequent. Ground layer sparse or abundant (Plate 26).

Wetlands: soils seasonally flooded, saturated. Water chemistry: fresh. Cowardin class: Palustrine forested wetland. **Uplands:** all slopes to forestline, best developed in subalpine elevations. Soils well-drained. The national inventory of wetland plants (Reed 1988) lists lodgepole as a FAC.

Distribution: l-KlaR, m-KlaR, su-KlaR, m-CasR, su-CasR, m-SN, su-SN, su-TraR, su-PenR, ModP, WarR, TraSN, su-WIS, Baja CA. **Elevation:** 1500-3400 m.

NDDB/Holland type and status: **Subalpine coniferous forests** (86000).
>   Lodgepole pine forest (86100) G4 S4.
>   Whitebark pine-lodgepole pine forest (86220 *in part*) G4 S4.

Other types:
Barry type: G7411112 BPICOMO.
Cheatham & Haller type: Lodgepole pine forest.
PSW-45 type: Lodgepole pine series.
Thorne type: Lodgepole pine forest.
WHR type: Lodgepole pine.

General references: Lodgepole pine range in CA (Griffin & Critchfield 1972), silvics (Lotan & Critchfield 1990). Alexander *et al.* (1983), Barbour (1988), Minnich (1976), Paysen *et al.* (1980), Pfister & McDonald (1980), Sawyer & Thornburgh (1977).

Comments: *Pinus contorta* includes four ecologically distinct subspecies (Lotan & Critchfield 1990). The Rocky Mountain race (*P. c.* ssp. *latifolia*) grows outside CA. The coastal race (*P. c.* ssp. *contorta*) is the dominant tree of the Beach pine series. The Mendocino white plains race (*P. c.* ssp. *bolanderi*) is a component of the Pygmy cypress series. The Sierra-Cascade race (*P. c.* ssp. *murrayana*) dominates this series. Since the varieties have very distinct ecology, each is treated in different series [*see* Beach pine series, Knobcone pine series, Pygmy cypress series, Western white pine series].

The Cascade-Sierra race's cones tend to open at maturity and then quickly fall. Stands are self-replacing and not closely associated with fire. Outbreaks of lodgepole pine needle miner may result in stand replacement. Stands in KlaR, CasR, ModP are associated with depressions which are seasonally saturated and cold; m-SN stands are associated with meadow margins; su-SN, WI, So-CA lodgepole pine stands are most commonly of upland settings among subalpine meadows. The species may be the only canopy tree or mix with other conifers.

Species mentioned in text	
Fendler meadow-rue	*Thalictrum fendleri*
Foxtail pine	*Pinus balfouriana*
Gray lovage	*Ligusticum grayi*
Labrador-tea	*Ledum glandulosum*
Limber pine	*Pinus flexilis*
Lodgepole pine	*Pinus contorta* ssp. *murrayana*
Mountain hemlock	*Tsuga mertensiana*
Pussypaws	*Calyptridium monosperma*
Red fir	*Abies magnifica* var. *magnifica*
Ross sedge	*Carex rossii*
Shasta fir	*Abies magnifica* var. *shastensis*
Western white pine	*Pinus monticola*
White fir	*Abies concolor*
Whitebark pine	*Pinus albicaulis*

Plot-based descriptions: Keeler-Wolf (1985c, 1989l) in Keeler-Wolf (1990e) describes stands at Bell Meadow RNA in Tuolumne Co., Talley (1976) in Keeler-Wolf (1990e) stands at W. B. Critchfield (formerly Bourland Meadow) RNA Tuolumne Co., Beguin & Major (1975), Burke (1987) in Keeler-Wolf (1990e) stands at Grass Lake RNA in El Dorado Co., Potter (1994) defines four associations in c. SN, Taylor (1984) in Keeler-Wolf (1990e) three associations at Harvey Monroe Hall RNA in Mono Co., Ball (1976) in Keeler-Wolf (1990e) describes stands at Last Chance RNA in Inyo Co., Keeler-Wolf & Keeler-Wolf (1981), Keeler-Wolf (1989k) in Keeler-Wolf (1990e) stands at Mt. Pleasant RNA in Plumas Co., Keeler-Wolf (1985b, 1988c) in Keeler-Wolf (1990e) stands at Mud Lake RNA in Plumas Co., Griffin (1975) in Keeler-Wolf (1990e) stands at Teakettle Creek Experimental Forest in Fresno Co. m-SN, su-SN; Taylor (1980) in Keeler-Wolf (1990e) defines one association at Indiana Summit RNA in Mono Co., Talley (1978) in Keeler-Wolf (1990e) describes stands at Sentinel Meadow RNA in Mono Co. TraSN.

Associations:
Potter (1994):
    Lodgepole pine association,
    Lodgepole pine/big sagebrush association,
    Lodgepole pine/Gray lovage association,
    Lodgepole pine/open association.
Taylor (1980):
    Lodgepole pine/pussypaws association.
Taylor (1984):
    Lodgepole pine/Ross sedge association,
    Lodgepole pine/Fendler meadow-rue association,
    Lodgepole pine/Labrador-tea association.

McNab cypress series

McNab cypress sole or dominant tree or shrub in canopy; blue oak, foothill pine, interior live oak, knobcone pine, and/or Sargent cypress may be present. Trees < 10 m; canopy continuous, intermittent, or open. Shrubs common or infrequent. Ground layer sparse.

Uplands: slopes, ridges. Soils basalt, conglomerate, gabbro, greenstone, ultramafic-derived, sterile.

Distribution: o-NorCo, i-NorCo, f KlaR, f-CasR, m-CasR, f-SN.
Elevation: 300-800 m.

NDDB/Holland type and status: Closed-cone coniferous forests (83000).
  Northern interior cypress forest (83220 *in part*) G2 S2.2.

Other types:
Barry type: 11112 BCUMA20.
Cheatham & Haller type: Northern interior cypress forest.
PSW-45 type: Cypress series.
Thorne type: Inland closed-cone coniferous woodland.
WHR type: Closed-cone pine-cypress.

General references: McNab cypress range in CA (Griffin & Critchfield 1972), Kruckeberg (1984), Paysen *et al.* (1980), Vogl *et al.* (1977).

Comments: This is the most abundant and widespread cypress in the state. There are several large stands, and many smaller ones scattered in Nor-CA. It can grow with Sargent cypress, but McNab cypress tends to grow on upper slopes while Sargent cypress grows on lower slopes and ravines [*see* Sargent cypress series]. Both can be found in the same stand.

Species mentioned in text	
Blue oak	*Quercus douglasii*
Foothill pine	*Pinus sabiniana*
Interior live oak	*Quercus wislizenii*
Knobcone pine	*Pinus attenuata*
McNab cypress	*Cupressus macnabiana*
Sargent cypress	*Cupressus sargentii*

Plot-based descriptions: Keeler-Wolf (1983) in Keeler-Wolf (1990e) reports density and basal area at Frenzel Creek RNA in Colusa Co. i-NorCo. This survey considered the cypress part of his serpentine chaparral. Sargent cypress also grows in this RNA.

Mesquite series

Honey mesquite or screw bean sole canopy tree. Trees < 10 m; canopy continuous or open. Shrubs as saltbushes occasional. Ground layer sparse or grassy (Plate 16).

Wetlands: soils intermittently flooded, saturated. Water chemistry: fresh. floodplains; fringes of lake beds; sand dunes; streambanks; surrounding alkali sinks; washes. **Uplands:** Rarely flooded margins of arroyos and washes. The national inventory of wetland plants (Reed 1988) lists honey mesquite as a FACU and screw bean as a FAC.

Distribution: s. i-CenCo, s. sj-CenV, MojD, ColD, Baja CA.
Elevation: sea level-1200 m.

NDDB/Holland type and status: Riparian forests (61000), **Riparian scrubs** (63000).
　　Mesquite bosque (61820) G3 S1.1.
　　Great Valley mesquite scrub (63420) G1 S1.1.

Other types:
Barry type: G7411121.
Brown Lowe Pase type: 143.112 143.152 144.331 153.131 154.114 154.173.
Cheatham & Haller type: Desert dry wash woodland, Southern alluvial woodland.
PSW-45 type: Mesquite series.
Thorne type: Desert riparian woodland.
WHR type: Desert riparian.

General references: Brown (1982e), Burkart (1976), Hilu *et al.* (1982), MacMahon (1988), Martin (1980a), Paysen *et al.* (1980), Sharf *et al.* (1982), Vasek & Barbour (1977).

Comments: This series includes the tall, closed-canopied mesquite stands called Mesquite bosque of the MojD and ColD and the short, open-canopied honey mesquite scrub in the sj-CenV. The sj-CenV stands were probably established 120 years ago (Holland 1987, 1988). There are two introduced *Prosopis* species [*P. strombulifera, P. velutina*] in mesquite in ColD which are included in this series.

Species mentioned in text	
Honey mesquite	*Prosopis glandulosa*
Saltbushes	*Atriplex* species
Screw bean	*Prosopis pubescens*

Plot-based descriptions: Bradley (1970) describes stands in Death Valley National Monument in Inyo Co. MojD; Spolsky (1979) defines three associations at Anza Borrego State Park in San Diego Co. ColD.

Spolsky (1979):
　　Mesquite dune scrub association,
　　Acacia-mesquite thickets association,
　　Mesquite dry lake association.

Mixed conifer series

Three or more equally important coniferous trees in canopy; black oak, Douglas-fir, incense-cedar, Jeffrey pine, ponderosa pine, sugar pine, and/or white fir may be present. Trees < 70 m; canopy intermittent. Shrubs infrequent or common. Ground layer sparse or abundant (Plate 27).

Uplands: all aspects. Soils shallow, well-drained.

Distribution: m-NorCo, m-KlaR, m-CasR, m-SN, o. m-TraR, m-PenR, Baja CA.
Elevation: 900-2200 m.

NDDB/Holland type and status: Lower coniferous forests (84000).
> North Range mixed coniferous forest (84110 *in part*) G4 S4.
> Sierran mixed coniferous forest (84230 *in part*) G4 S4.

Other types:
Barry type: G74 G7411112.
Cheatham & Haller type: Sierran mixed conifer forest.
PSW-45 type: Mixed conifer series.
Thorne type: Mixed conifer forest.
WHR type: Sierran mixed conifer forest.

General references: Conifers range in CA (Griffin & Critchfield 1972). Barbour (1988), Kinloch & Scheuner (1990), Laacke & Fiske (1983b), Minnich (1987), Pase (1982c), Minore & Kingsley (1983), Paysen *et al.* (1980), Rundel *et al.* (1977), Sawyer & Thornburgh (1977), Stone & Sumida (1983), Tappeiner (1980), Vankat (1970, 1978, 1982).

Comments: The term mixed conifer forest or mixed coniferous forest has been used in many ways. In CA the term generally refers to m-SN forests with up to five conifers [including Douglas-fir, incense-cedar, ponderosa pine, sugar pine, or white fir].Stands having this mix, or several of these conifers, can be found in NorCo, KlaR, CasR. In

TraR, PenR, bigcone Douglas-fir may be present. Stands with this makeup that are dominated by one species or have two important trees in canopy have been called mixed conifer as well.

Species mentioned in text	
Bigcone Douglas-fir	*Pseudotsuga macrocarpa*
Bitter cherry	*Prunus emarginata*
Black oak	*Quercus kelloggii*
Bolander bedstraw	*Galium bolanderi*
Canyon live oak	*Quercus chrysolepis*
Creeping snowberry	*Symphoricarpos mollis*
Douglas-fir	*Pseudotsuga menziesii*
Hazel	*Corylus cornuta*
Huckleberry oak	*Quercus vaccinifolia*
Incense-cedar	*Calocedrus decurrens*
Irises	*Iris* species
Jeffrey pine	*Pinus jeffreyi*
Little Oregon-grape	*Berberis nervosa*
Mahala carpet	*Ceanothus prostratus*
Milkwort	*Polygala cornuta*
Mountain misery	*Chamaebatia foliolosa*
Naked-stemmed buckwheat	*Eriogonum latifolium*
Ponderosa pine	*Pinus ponderosa*
Rattlesnake-plantain	*Goodyera oblongifolia*
Rosy everlasting	*Antennaria rosea*
Service berry	*Amelanchier alnifolia*
Starflower	*Trientalis latifolia*
Sugar pine	*Pinus lambertiana*
Sword fern	*Polystichum munitum*
Tanoak	*Lithocarpus densiflora*
White fir	*Abies concolor*

The Mixed conifer series is more narrowly defined here. Mixed conifer series refers only to stands where three or more conifer species similarly share importance. If one species dominates the tree canopy, the stand belongs to the Bigcone Douglas-fir series, Douglas-fir series, Incense-cedar series, Jeffrey pine series, Ponderosa pine series, or White fir series even though other conifer species are present. If Douglas-fir and ponderosa pine are important trees in canopy, and other species are less conspicuous, then the stand belongs to the Douglas-fir - ponderosa pine series even though other conifer species are present. If Douglas-fir shares dominance with tanoak, it belongs to Douglas-fir - tanoak series even though other conifer species are present.

Plot-based descriptions: Keeler-Wolf (1986c, 1988b) in Keeler-Wolf (1990e) describes stands at Doll Basin candidate RNA in Mendocino Co., Thornburgh (1981), Keeler-Wolf (1989i) in Keeler-Wolf (1990e) at Ruth candidate RNA in Humboldt Co. m-NorCo; Keeler-Wolf (1991a) reports tree density, basal area, frequency for stands at Big Pine Mountain proposed RNA in Santa Barbara Co. m-CenCo; Keeler-Wolf (1987b) in Keeler-Wolf (1990e) describes stands at Pearch Creek in Humboldt Co., Keeler-Wolf (1987c) in Keeler-Wolf (1990e) at Rock Creek Butte candidate RNA in Siskiyou Co., Sawyer & Thornburgh (1977) define two associations m-KlaR; Taylor & Randall (1978) in Keeler-Wolf (1990e) at Cub Creek RNA in Butte Co. m-CasR; Palmer (1981) in Keeler-Wolf (1990e) at Sugar Pine Point RNA in Placer Co., Fites (1993) defines 12 associations in Lassen, Plumas, Eldorado National Forests m-SN; Sproul (1981) in Keeler-Wolf (1990e) describes stands at Falls Canyon RNA in Los Angeles Co. i. m-TraR; Keeler-Wolf (1986e, 1988e) in Keeler-Wolf (1990e) at Hall Canyon RNA in Riverside Co. m-PenR.

Associations:
Fites (1993):
    Mixed conifer/Bolander bedstraw-milkwort association,
    Mixed conifer/huckleberry oak association,
    Mixed conifer/rosy everlasting- naked-stemmed buckwheat association,
    Mixed conifer/service berry association,
    Mixed conifer/canyon live oak-huckleberry oak association,
    Mixed conifer/starflower association,
    Mixed conifer-canyon live oak/hazel association,
    Mixed conifer-canyon live oak/Bolander bedstraw association,
    Mixed conifer-canyon live oak/mountain misery association,
    Mixed conifer-canyon live oak/sword fern association,
    Mixed conifer-tanoak/mountain misery association,
    Mixed conifer-tanoak/creeping snowberry/iris association.
Sawyer & Thornburgh (1977):
    Mixed conifer/little Oregon-grape association [as White fir/little Oregon-grape association],
    Mixed conifer/mahala carpet association [as White fir/mahala carpet association].

Mixed oak series

Black oak, blue oak, coast live oak, interior live oak, and/or valley oak important trees in canopy; California bay, California buckeye, Douglas-fir, foothill pine, madrone and/or ponderosa pine may be present. Trees < 30 m; canopy continuous, may be two-tiered. Shrubs infrequent or common. Ground layer sparse or abundant, may be grassy (Plate 20).

Uplands: valleys, gentle to steep slopes. Soils moderately deep. The national inventory of wetland plants (Reed 1988) lists valley oak as a FAC*.

Distribution: o-NorCo, i-NorCo, o-CenCo, i-CenCo, f-SN.
Elevation: 250-2000 m.

NDDB/Holland type and status:: **Cismontane woodlands** (71000), **Broadleaved upland forests** (81000).
　　Digger pine-oak woodland (71410 *in part*) G4 S4.

Other types:
Barry type: G74 G7411121.
Cheatham & Haller type: Northern oak woodland.
Thorne type: Oak woodland.
WHR type: Coastal oak woodland.

General references: Oak species ranges in CA (Griffin & Critchfield 1972). Barbour (1988), Pavlik *et al.* (1991), Sawyer *et al.* (1977).

Comments: For a stand to be classified as a member of the mixed oak series, no single oak can dominate. Stands dominated by a single oak species are assigned to other series [*see* Black oak series, Blue oak series, Interior live oak series, Valley oak series].
Plot-based descriptions: Allen *et al.* (1989, 1991), define 11 Mixed oak subseries, based on tree and shrub data in Cis-CA. These subseries are called

associations here. Jimerson (1993) defines two associations in o-NorCo.

Species mentioned in text	
Black oak	*Quercus kelloggii*
Blue oak	*Quercus douglasii*
California bay	*Umbellularia californica*
California buckeye	*Aesculus californica*
California coffeeberry	*Rhamnus californica*
Coast live oak	*Quercus agrifolia*
Coyote brush	*Baccharis pilularis*
Douglas-fir	*Pseudotsuga menziesii*
Firecracker flower	*Dichelostemma ida-maia*
Foothill pine	*Pinus sabiniana*
Interior live oak	*Quercus wislizenii*
Madrone	*Arbutus menziesii*
Oregon white oak	*Quercus garryana*
Poison-oak	*Toxicodendron diversilobum*
Ponderosa pine	*Pinus ponderosa*
Tall-oatgrass	*Arrhenatherum elatius*
Toyon	*Heteromeles arbutifolia*
Valley oak	*Quercus lobata*

Associations:
Allen *et al.* (1991):
　　Black oak-valley oak/grass association,
　　Blue oak-valley oak-coast live oak/grass association,
　　Interior live oak/toyon association,
　　Mixed oak-black oak/grass association,
　　Mixed oak-California buckeye/grass association,
　　Mixed oak-coast live oak/poison-oak association,
　　Mixed oak-foothill pine/grass association,
　　Mixed oak/grass association,
　　Mixed oak-interior live oak-foothill pine association,
　　Mixed oak/poison-oak - coyote brush association,
　　Mixed oak-valley oak/poison-oak - California coffeeberry association.
Jimerson (1993):
　　Oregon white oak-black oak/firecracker flower association,
　　Oregon white oak-black oak/tall-oatgrass association.

Mixed subalpine forest series

Foxtail pine, lodgepole pine, mountain hemlock, red fir, Shasta fir, western white pine, and/or whitebark pine may be present in similar importance. Trees < 40 m. Tree canopy intermittent. Shrubs infrequent or common. Ground layer sparse or abundant (Plate 28).

Uplands: all slopes, but most extensive on crests, ridges, summits, upper slopes to forestline; concave slopes; raised stream benches and terraces. Soils granitic, ultramafic, volcanic-derived, well-drained. The national inventory of wetland plants (Reed 1988) lists lodgepole pine as a FAC, mountain hemlock as a FACU+, red fir as a FACU, western white pine as a FACU.

Distribution: su-KlaR, su-CasR, su-SN, su-TraR, su-PenR.
Elevation: 1500-3660 m.

NDDB/Holland type and status: **Subalpine coniferous forests** (86000).
>   Whitebark pine-mountain hemlock forest (86210) G4 S4.
>   Whitebark pine-lodgepole pine forest (86220) G4 S4.
>   Southern California subalpine subslpine forest (86500) G4 S4.

Other types:
Barry type: G74 G74112.
Cheatham & Haller type: Sierran mixed subalpine coniferous forest, Southern California subalpine forest.
Thorne type: Subalpine coniferous forest.
WHR type: Subalpine conifer forest.

General references: Species ranges in CA (Griffin & Critchfield 1972). Arno (1980), Barbour (1988), Franklin (1980a), Parsons (1976, 1980), Parsons & Stohlgrem (1989), Pase (1982b), Rundel *et al.* (1977), Sawyer & Thornburgh (1977), Waring (1969).

Comments: The composition of these stands varies regionally. Foxtail pine mixes with Shasta fir, in some cases with whitebark pine, in su-KlaR. Whitebark pine and lodgepole pine mix in su-CasR; mountain hemlock and whitebark pine in su-SN.

These stands differ from other subalpine series in that no tree species dominates stands of this series. Instead two or three trees are similarly important [*see* Foxtail pine series, Lodgepole pine series, Red fir series, Western white pine series, Whitebark pine series].

Species mentioned in text	
Foxtail pine	*Pinus balfouriana*
Lodgepole pine	*Pinus contorta*
Mountain hemlock	*Tsuga mertensiana*
Red fir	*Abies magnifica* var. *magnifica*
Shasta fir	*Abies magnifica* var. *shastensis*
Western white pine	*Pinus monticola*
Whitebark pine	*Pinus albicaulis*

Plot-based descriptions: Keeler-Wolf (1987a) in Keeler-Wolf (1990e) describes foxtail pine, mountain hemlock, whitebark pine stands at Crater Creek candidate RNA, Keeler-Wolf (1984b, 1989f), Sawyer & Thornburgh (1971) in Keeler-Wolf (1990e) at Sugar Creek candidate RNA in Siskiyou Co. su-KlaR.

Mixed willow series

More than one willow species important shrub or tree in canopy; arroyo willow, big-leaf maple, black cottonwood, black willow, California sycamore, Fremont cottonwood, Hooker willow, narrowleaf willow, Pacific willow, red alder, red willow, Sitka willow, and/or white alder may be present. If shrub-land, emergent trees may be present. Trees < 10 m; canopy continuous. Shrubs sparse under tree canopy. Ground layer sparse (Plate 19).

Wetlands: habitats seasonally flooded, saturated. Water chemistry: fresh. Flood-plains; low gradient depositions along rivers, streams. Cowardin class: Palustrine forested or shrub-scrub wetland. The national inventory of wetland plants (Reed 1988) lists willows as a OBL-FACW.

Distribution: Cis-CA, Trans-CA.
Elevation: sea level-1800 m.

NDDB/Holland type and status: Marshes and swamps (52000), **Riparian forests** (61000), **Riparian scrubs** (63000).
   Freshwater swamp (52600 *in part*) G1 S1.1.
   Red alder riparian forest (61130 *in part*) G3 S2.1.
   Central Coast cottonwood-sycamore riparian forest (61210 *in part*) G3 S3.
   Central Coast arroyo willow riparian forest (61230) G3 S3.
   Southern arroyo willow riparian forest (61320) G2 S2.1.
   Southern cottonwood-willow riparian forest (61330 *in part*) G3 S3.
   Great Valley cottonwood riparian forest (61410 *in part*) G2 S2.1.
   Great Valley mixed riparian forest (61420 *in part*) G2 S2.1.
   Modoc-Great Basin cottonwood-willow riparian forest (61610 *in part*) G2 S2.1.
   Mojave riparian forest (61700 *in part*) G1 S1.1.

Sonoran cottonwood-willow riparian forest (61810 *in part*) G2 S1.1.
North Coast riparian forest (63100 *in part*) G3 S3.
Central Coast riparian scrub (63200 *in part*) G3 S3.
Southern willow scrub (63320 *in part*) G3 S2.1.
Great Valley willow scrub (63410 *in part*) G3 S3.

Other types:
Barry type: G7411121.
Cheatham & Haller type: Bottomland woodlands and forest.
PSW-45 type: Willow series.
Thorne type: Riparian woodland.
WHR type: Desert riparian, Freshwater emergent wetland.
General references: Bowler (1989), Capelli & Stanley (1984), Faber *et al.* (1989), Holstein (1984), Paysen *et al.* (1980).

Species mentioned in text	
Arroyo willow	*Salix lasiolepis*
Bigleaf maple	*Acer macrophyllum*
Black cottonwood	*Populus balsamifera*
Black willow	*Salix gooddingii*
California sycamore	*Platanus racemosa*
Fremont cottonwood	*Populus fremontii*
Hooker willow	*Salix hookeriana*
Narrowleaf willow	*Salix exigua*
Pacific willow	*Salix lucida* ssp. *lasiandra*
Red alder	*Alnus rubra*
Red willow	*Salix laevigata*
Sitka willow	*Salix sitchensis*
White alder	*Alnus rhombifolia*

Comments: Willow stands may or may not be do-minated by a single species. If no dominant willow is present at low elevations, then place the stands in this series. Montane and subalpine willow stands are placed in separate classes since different willow species are restricted to those elevations [*see* Montane wetland shrub habitat, Subalpine wetland shrub habitat]. Stands of the Mixed willow series have environ-ment conditions similar to alder, cottonwood, and other willow series.

Plot-based descriptions: not available.

Monterey pine series

Monterey pine sole or dominant tree in canopy, coast live oak may be important; Bishop pine, Douglas-fir, knobcone pine, madrone, and/or redwood may be present. Trees < 30 m; canopy continuous or intermittent. Shrubs absent, infrequent, or common. Ground layer sparse or abundant (Plate 23).

Uplands: maritime terraces, headlands. Soils excessively drained.

Distribution: o-CenCo, ChaI, Baja CA.
Elevation: sea level-300 m.

NDDB/Holland type and status: **Closed-cone coniferous forests** (83000).
  Monterey pine forest (83130) G1 S1.1.

Other types:
Barry type: G7411112 BPIRA00.
Cheatham & Haller type: Monterey pine forest.
PSW-45 type: Monterey pine series.
Thorne type: Coastal closed-cone coniferous forest.
WHR type: Closed cone pine-cypress.

General references: Monterey pine range in CA (Griffin & Critchfield 1972). Axelrod (1980a, 1980b), Cylinder (1995) Deghi *et al.* (1995), Jones & Stokes (1994a,b), MacDonald & Laacke (1990), Mathews & Nedeff (1995), Paysen *et al.* (1980), Vogl et al. (1977).

Comments: Plantations exist in the state and worldwide, but only three natural areas exist in California near Año Nuevo in Santa Cruz and San Mateo Cos., Cambria in San Luis Obispo Co., and on the Monterey peninsula in Monterey Co.

Qualitative descriptions (Jones & Stokes 1994) point out the high level of variation in species composition among the three areas where the pine grows. At the Año Nuevo area, Monterey pine associates with coast live oak, Douglas-fir, knobcone

pine, madrone, ponderosa pine, and redwood. In the other areas, the pine grows with Bishop pine and coast live oak.

Cylinder (1995) describes the link between marine terrace conditions and Monterey pine success. His eight proposed vegetation types suggest that Monterey pine dominates in stands of distinct species composition on different terraces, and that it is also a secondary species in other series [*see* Bishop pine series, Coyote brush series, Knobcone pine series].

Monterey pine is rare (a CNPS List 1B plant) (Skinner & Pavlik 1994).

Species mentioned in text	
Bishop pine	*Pinus muricata*
Coast live oak	*Quercus agrifolia*
Douglas-fir	*Pseudotsuga menziesii*
Knobcone pine	*Pinus attenuata*
Madrone	*Arbutus menziesii*
Monterey pine	*Pinus radiata*
Ponderosa pine	*Pinus ponderosa*
Redwood	*Sequoia sempervirens*

Plot-based descriptions: Vogl *et al.* (1977) describe three stands. All were affected by differing degrees of cutting and grazing.

Mountain hemlock series

Mountain hemlock sole or dominant tree in canopy; lodgepole pine, red fir, Shasta fir, western white pine, white fir, and/or whitebark pine may be present. Trees < 40 m; canopy continuous or intermittent. Shrubs infrequent or common. Ground layer sparse to abundant (Plate 28).

Uplands: all slopes, but most extensive on north-facing aspects with late-lasting snow; concave slopes; raised stream benches and terraces. Soils commonly granitic-derived. The national inventory of wetland plants (Reed 1988) lists mountain hemlock as a FACU+.

Distribution: su-KlaR, su-CasR, n. su-SN, c. su-SN.
Elevation: 1500-3050 m.

NDDB/Holland type and status: Subalpine coniferous forests (86000).
> Whitebark pine-mountain hemlock forest (86210 *in part*) G4 S4.

Other types:
Barry type: G7411112 BTSME00.
Cheatham & Haller type: Whitebark pine-mountain hemlock forest.
Thorne type: Whitebark pine-mountain hemlock forest.
WHR type: Subalpine conifer.

General references: Mountain hemlock range in CA (Griffin & Critchfield 1972); silvics (Means 1990). Arno (1980), Barbour (1988), Franklin (1980a, 1988), Rundel *et al.* (1977), Sawyer & Thornburgh (1977).

Comments: Stands included in this series occur north of Sequoia National Park (Parsons 1972), Rundel *et al.* 1977). At its southern range limit in SN, the species forms local populations mixed with other subalpine conifers, hence mountain hemlock

can be important in other subalpine series [*see* Lodgepole pine series, Mixed subalpine series, Red fir series, Western white pine series, Whitebark pine series].

Species mentioned in text

Cascade heather	*Phyllodoce empetriformis*
Dwarf bilberry	*Vaccinium caespitosum*
Heartleaf arnica	*Arnica cordifolia*
Huckleberry oak	*Quercus vaccinifolia*
Lodgepole pine	*Pinus contorta*
Mountain hemlock	*Tsuga mertensiana*
Parry rush	*Juncus parryi*
Red fir	*Abies magnifica* var. *magnifica*
Sadler oak	*Quercus sadleriana*
Shasta fir	*Abies magnifica* var. *shastensis*
Western white pine	*Pinus monticola*
White fir	*Abies concolor*
White-veined shinleaf	*Pyrola picta*
Whitebark pine	*Pinus albicaulis*

Plot-based descriptions: Jimerson (1993) defines one association in Six River National Forest, Keeler-Wolf (1982, 1989d) in Keeler-Wolf (1990e) describes stands at Cedar Basin candidate RNA, Keeler-Wolf (1987a) in Keeler-Wolf (1990e) stands at Crater Creek candidate RNA, and Keeler-Wolf (1984b, 1989f), Sawyer & Thornburgh (1971) in Keeler-Wolf (1990e) stands at Sugar Creek candidate RNA in Siskiyou Co., Sawyer & Thornburgh (1977) define three regional associations, Sawyer & Thornburgh (1969) report tree age for stands at Diamond Lake in Siskiyou Co. su-KlaR; Fiedler (1987), Keeler-Wolf (1989g) in Keeler-Wolf (1990e) describe stands at Antelope Creek Lakes candidate RNA, Imper (1988a) in Keeler-Wolf (1990e) defines two associations at Shasta Fir candidate RNA (on Mount Shasta) in Siskiyou Co. CasR; Nachlinger (1988) reports tree density and size for two stands at Lyon Peak/Needle Peak RNA in Placer Co., Potter (1994) defines two regional associations in c. su-SN, Taylor (1984) one association at Harvey Monroe Hall RNA in Mono Co. su-SN.

Associations:
Imper (1988a):
    Mountain hemlock/Cascade heather
    association,
    Mountain hemlock/Parry rush association.
Jimerson (1993):
    Mountain hemlock/Sadler oak association.
Potter (1994):
    Mountain hemlock association.
    Mountain hemlock/steep association.
Sawyer & Thornburgh (1977):
    Mountain hemlock/dwarf bilberry association
    [as *Tsuga mertensiana/Phyllodoce empetriformis*
    association],
    Mountain hemlock/huckleberry oak
    association,
    Mountain hemlock/white-veined shinleaf
    association.
Taylor (1984):
    Mountain hemlock/heartleaf arnica association.

Mountain juniper series

Mountain juniper sole or dominant tree in canopy; black oak, canyon live oak, Jeffrey pine, singleleaf pinyon, and/or white fir, or as an emergent tree over a shrub canopy of big sagebrush, antelope bitterbrush, choke cherry, and/or curlleaf mountain-mahogany may be present. Trees < 20 m. Tree canopy intermittent or open, or trees scattered. Shrubs common or infrequent. Ground layer sparse or grassy.

Uplands: gentle slopes; sloping alluvial fans; canyon slopes; steep, rocky escarpments. Aspects vary. Soils bedrock, eolian, colluvial, or alluvial-derived, commonly shallow.

Distribution: m-SN, su-SN, m-TraR, ModP, TraSN.
Elevation: 900-3000 m.

NDDB/Holland type and status: Pinyon and juniper woodlands (72000).
    Great Basin juniper woodland and scrub (72123) G4 S4.

Other types:
Barry type: G7411112 BJUOCA0.
Cheatham & Haller type: Upper montane coniferous forest.
PSW-45 type: Western juniper series.
Thorne type: Mountain juniper woodland.
WHR type: Juniper.

General references: Mountain juniper range in CA (Griffin & Critchfield 1972); silvics (Dealy 1990). Barbour (1988), Paysen *et al.* (1980), Thorne (1982), Vasek & Thorne (1977).

Comments: There are taxonomic and ecological challenges involved with *Juniperus occidentalis*. The species has two subspecies, western juniper (*J. o.* ssp. *occidentalis*) and mountain juniper (*J. o.* ssp. *australis*). The two subspecies have generally differ-

ent ranges and ecology, so the two are segregated into two series [*see* Western juniper series]. There is some overlap in s. ModP and n. TraSN. Vasek & Thorne (1977) report that mountain juniper tends to occupy ridges and escarp-ments, while western juniper occupies the valleys.

Species mentioned in text	
Antelope bitterbrush	*Purshia tridentata*
Big sagebrush	*Artemisia tridentata*
Black oak	*Quercus kelloggii*
Canyon live oak	*Quercus chrysolepis*
Choke cherry	*Prunus virginiana*
Curlleaf mountain-mahogany	*Cercocarpus ledifolius*
Jeffrey pine	*Pinus jeffreyi*
Lodgepole pine	*Pinus contorta* ssp. *murrayana*
Mountain juniper	*Juniperus occidentalis* ssp. *australis*
Singleleaf pinyon	*Pinus monophylla*
White fir	*Abies concolor*

Another problem arises in separating stands of the Mountain juniper series from those of the Singleleaf pinyon-western juniper series. If only one species is present, the appropriate series is obvious. If both occur, then the dominance rule applies. In addition, Mountain juniper series includes two physiognomic types: one where mountain juniper dominates the tree canopy in m-SN and su-SN, and a second where mountain juniper trees are emergent over a shrub canopy of big sagebrush and other species in TraSN and s. ModP.

Stands dominated by Mountain juniper may occur as small enclaves within more extensive series [*see* Jeffrey pine series, Lodgepole pine series, Mountain hemlock series, Red fir series, Singleleaf pinyon series, White fir series]. On occasion, mountain juniper may grow in stands of other series [*see* White fir-ponderosa pine series]. Trees of mountain juniper grow outside the range of this series [in m-NorCo, m-PenR, ModP, WIS, DesR] as well.

Keeler-Wolf (1990e) qualitatively describes stands at Onion Creek in Placer Co. m-SN.

Plot-based descriptions: Keeler-Wolf (1974) describes stands for Soldier Ridge in Yolla Bolly Mountains m-NorCo; Potter (1994) defines two regional associations in m-SN, su-SN; Vasek & Thorne (1977) report density and cover for a stand in Lassen Co. n. TraSN.

Associations:
Potter (1994):
    Mountain juniper association
    [as Western juniper association],
    Mountain juniper/big sagebrush association
    [as Western juniper/big sagebrush association].

Oregon white oak series

Oregon white oak sole or dominant tree in canopy; California bay, black oak, canyon live oak, Douglas-fir, foothill pine, incense-cedar, Jeffrey pine, madrone, ponderosa pine, and/or western juniper may be present. Trees < 30 m; canopy continuous, intermittent, or savanna-like, may be two-tiered. Shrubs infrequent or common. Ground layer sparse or grassy (Plate 21).

Uplands: dry slopes, ridges; raised stream benches and terraces. The national inventory of wetland plants (Reed 1988) does not list Oregon white oak.

Distribution: NorCo, l-KlaR, f-KlaR, m-KlaR, m-CasR, ModP.
Elevation: 60-2500 m.

NDDB/Holland type and status: **Cismontane woodlands** (71000).
Oregon oak woodland (71110) G3 S3.
Mixed north cismontane woodland (71420 *in part*) G3 S3.2.

Other types:
Barry type: G74 G7411121 CQUGA40.
Cheatham & Haller type: Northern oak woodland.
Thorne type: Northern oak woodland.
WHR type: Montane hardwood.

General references: Oregon white oak range in CA (Griffin & Critchfield 1972); silvics (Stein 1990a). Barbour (1988), Griffin (1977) Pavlik *et al.* (1991), Riegel *et al.* (1991), Stein (1980b).

Comments: Stands of this series are forests or woodlands. This series includes associations of the tree form [*Quercus garryana* var. *g.*] of the species. The shrub form [*Quercus garryana* var. *breweri*] is treated in a shrub series. This tree is intolerant of shade, so is replaced over time by more tolerant conifers and hardwoods on drained, moist soils. We

include the bald hills woodlands (Griffin 1977) of o-NorCo in this series.

Keeler-Wolf (1990e) qualitatively describes Oregon white oak stands at Twin Rocks in Mendocino Co. o-NorCo.

Species mentioned in text	
Black oak	*Quercus kelloggii*
Brewer oak	*Quercus garryana* var. *breweri*
California brome	*Bromus californica*
California fescue	*Festuca californica*
Canyon live oak	*Quercus chrysolepis*
Common snowberry	*Symphoricarpos albus*
Dogtail	*Cynosurus cristatus*
Douglas-fir	*Pseudotsuga menziesii*
Incense-cedar	*Calocedrus decurrens*
Jeffrey pine	*Pinus jeffreyi*
Klamath gooseberry	*Ribes roezlii*
Madrone	*Arbutus menziesii*
Mock-orange	*Philadelphus lewisii*
Orchid grass	*Dactylis glomerata*
Oregon white oak	*Quercus garryana* var. *garryana*
Poison larkspur	*Delphinium trolliifolium*
Poison-oak	*Toxicodendron diversilobum*.
Ponderosa pine	*Pinus ponderosa*
Tall-oatgrass	*Arrhenatherum elatius*
Western juniper	*Juniperus occidentalis* ssp. *occidentalis*

Plot-based descriptions: Saenz (1983), Saenz & Sawyer (1986) describe understories in Oregon white oak woodlands correlated with two grazing regimes in Humboldt Co., Anderson & Pasquinelli (1984) report tree age, basal area, and density for one stand in Sonoma Co. o-NorCo, Sugihara & Reed (1987) define six associations in Redwood National Park, Leitner & Leitner (1988) in Keeler-Wolf (1990e) two associations at Soldier candidate RNA in Humboldt Co., Jokerst (1987) in Keeler-Wolf (1990e) reports tree density and cover at Twin Rocks in Mendocino Co., Jimerson (1993) defines three associations in Six Rivers National Forest in o-NorCo; Taylor & Teare (1979a) in Keeler-Wolf (1990e) one association at Manzanita Creek candidate RNA in Trinity Co. m-KlaR.

Associations:
Jimerson (1993):

> Oregon white oak - black oak/tall-oatgrass association,
> Oregon white oak-Brewer oak/California fescue association,
> Oregon white oak - Douglas-fir/California fescue association.

Leitner & Leitner (1988):

> Oregon white oak/poison-oak association [as Oregon white oak dense stands].

Sugihara *et al.* (1987):

> Oregon white oak/common snowberry association
> [as *Quercus/Symphoricarpos* vegetation type],
> Oregon white oak/dogtail association
> [as *Quercus/Cynosurus* vegetation type],
> Oregon white oak/Klamath gooseberry association [as *Ribes/Phacelia* vegetation type],
> Oregon white oak/mock-orange association
> [as *Philadelphus/Cystopteris* vegetation type],
> Oregon white oak/poison larkspur association
> [as *Quercus/Delphinium* vegetation type],
> Oregon white oak/orchid grass association
> [as *Quercus/Dactylis* vegetation type].

Taylor & Teare (1979a):

> Oregon white oak/California brome association [as Oregon white oak association].

Pacific willow series

Pacific willow sole or dominant shrub or tree in canopy; bigleaf maple, black cottonwood, California sycamore, Fremont cottonwood, Mexican elderberry, red osier, white alder, and/or willows may be present. If shrubland, emergent trees may be present. Shrubs < 15 m; canopy continuous. Shrubs sparse under tree canopy. Ground layer variable.

Wetlands: habitats seasonally flooded, saturated. Water chemistry: fresh. Floodplains; low-gradient depositions along rivers, streams. Cowardin class: Palustrine shrub-scrub wetland. The national inventory of wetland plants (Reed 1988) lists Pacific willow as a OBL.

Distribution: Cis-CA.
Elevation: sea level-2700 m.

NDDB/Holland type and status: **Marshes and swamps** (520000, **Riparian forests** (61000), **Riparian scrubs** (63000).

Freshwater swamp (52600 *in part*) G1 S1.1.
Red alder riparian forest (61130 *in part*) G3 S2.2.
Central Coast cottonwood-sycamore riparian forest (61210 *in part*) G3 S3.3.
Southern cottonwood-willow riparian forest (61330 *in part*) G3 S3.
Great Valley mixed riparian forest (61420 *in part*) G2 S2.1.
Central Coast riparian scrub (63200 *in part*) G3 S3.2.
Southern willow scrub (63320 *in part*) G3 S3.2.
Great Valley willow scrub (634100 *in part*) G3 S3.2.

Other types:
Barry type: G7411221.
Cheatham & Haller type: Bottomland woodlands and forest.
PSW-45 type: Willow series.
Thorne type: Riparian woodland.
WHR type: Freshwater emergent wetland.

General references: Bowler (1989), Brayshaw (1976), Capelli & Stanley (1984), Faber *et al.* (1989), Holstein (1984), Paysen *et al.* (1980), Shanfield (1984).

Comments: Willow stands may or may not be dominated by a single species. If no dominant willow is present at low elevations, then place the stands in the Mixed willow series. Montane and subalpine willow stands are placed in separate classes since different willow species are restricted to those elevations [*see* Montane wetland shrub habitat, Subalpine wetland shrub habitat].

Stands of Pacific willow have environment conditions similar to alder, cottonwood, and other willow series [*see* Black cottonwood series, Black willow series, Fremont cottonwood series, Hooker willow series, Red alder series, Red willow series, Sitka willow series, White alder series].

Species mentioned in text	
Bigleaf maple	*Acer macrophyllum*
Black cottonwood	*Populus balsamifera*
Black willow	*Salix gooddingii*
California sycamore	*Platanus racemosa*
Fremont cottonwood	*Populus fremontii*
Hooker willow	*Salix hookeriana*
Mexican elderberry	*Sambucus mexicana*
Pacific willow	*Salix lucida* ssp. *lasiandra*
Red alder	*Alnus rubra*
Red osier	*Cornus sericea*
Red willow	*Salix laevigata*
Sitka willow	*Salix sitchensis*
White alder	*Alnus rhombifolia*
Willows	*Salix* species

Plot-based descriptions: Zembal (1989) reports tree density and cover along Santa Margarita and Santa Ana Rivers in Orange Co. o-SoCo.

Parry pinyon series

Parry pinyon sole or dominant tree in canopy; California juniper, singleleaf pinyon, and/or Jeffrey pine may be present. Trees < 15 m; canopy intermittent or open. Shrubs common. Ground layer sparse.

Uplands: slopes, ridges; aspect north-facing. Soils commonly well-drained.

Distribution: m-PenR, Baja CA.
Elevation: 900-1800 m.

NDDB/Holland type and status: **Piñon and juniper woodlands** (72000).
> Peninsular pinyon woodland (72310 *in part*) G3 S3.
> Peninsular juniper woodland and scrub (72320 *in part*) G3 S3.

Other types:
Barry type: G74 G7411112.
Cheatham & Haller type: Baja California pinyon-juniper woodland.
PSW-45 type: Pinyon pine series.
Thorne type: Pinyon-juniper woodland.
WHR type: Pine-juniper.

General references: Parry pinyon range in CA (Griffin & Critchfield 1972). Minnich (1987), Paysen *et al.* (1980), Thorne (1982), Vasek & Thorne (1977).

Comments: Pinyon-juniper woodlands are better considered a collection of series. This series occurs on north-facing, cismontane slopes in Laguna Mountains in San Diego Co. Parry pinyon occasionally grows some stands dominated by singleleaf pinyon on transmontane PenR slopes. Parry pinyon may be emergent trees over a shrub canopy of chamise, Eastwood manzanita, Muller oak, red shank [*see* Chamise series, Eastwood manzanita series, Cupleaf ceanothus-fremontia-oak series, Red shank series].

Species mentioned in text	
Chamise	*Adenostoma fasciculatum*
Eastwood manzanita	*Arctostaphylos glandulosa*
Jeffrey pine	*Pinus jeffreyi*
Muller oak	*Quercus cornelius-mulleri*
Parry pinyon	*Pinus quadrifolia*
Red shank	*Adenostoma sparsifolium*
Singleleaf pinyon	*Pinus monophylla*

Plot-based descriptions: not available.

Ponderosa pine series

Ponderosa pine sole, dominant, or important tree with black oak or incense-cedar in canopy; canyon live oak, Coulter pine, Douglas-fir, interior live oak, Jeffrey pine, sugar pine, and/or white fir may be present. Trees < 70 m; canopy intermittent or open. Shrubs infrequent or common. Ground layer sparse, abundant, grassy (Plate 26).

Uplands: slopes, all aspects; raised stream benches and terraces. Soils well-drained. The national inventory of wetland plants (Reed 1988) lists ponderosa pine as a FACU.

Distribution: NorCo, CenCo, KlaR, CasR, SN, m-TraR, m-PenR, GB.
Elevation: 300-2100 m.

NDDB/Holland type and status: **Lower montane coniferous forests** (84000).
    Upland Coast Range ponderosa pine forest (84131) G3 S3.2.
    Maritime Coast Range ponderosa pine forest (84132) G1 S1.1.
    Westside ponderosa pine forest (84210) G3 S2.1.
    Eastside ponderosa pine forest (84220) G4 S2.1.
    Ponderosa dune forest (84221) G1 S1.1.
    Sierran mixed coniferous forest (84230 *in part*) G4 S4.

Other types:
Barry type: G7411112 BPIPO00.
Cheatham & Haller type: Coast Range ponderosa pine forest, Westside ponderosa pine forest, Eastside ponderosa pine forest.
PSW-45 type: Ponderosa/Jeffrey pine series.
Thorne type: Yellow pine forest.
WHR type: Ponderosa pine, Eastside pine.

General references: Ponderosa pine range in CA (Griffin & Critchfield 1972); silvics (Oliver & Ryker 1990). Barbour (1988), Barrett *et al.* (1980, 1983), Griffin (1985), McDonald (1980b), Paysen *et al.* (1980), Rundel *et al.* (1977), Solinas *et al.* (1985), Talley & Griffin (1980), Thorne (1977, 1982), Yeaton *et al.* (1980).

Species mentioned in text

Antelope bitterbrush	*Purshia tridentata* var. *tridentata*
Arrowleaf balsam root	*Balsamorhiza sagittata*
Big sagebrush	*Artemisia tridentata*
Black oak	*Quercus kelloggii*
Blue wheatgrass	*Pseudoroegneria spicata*
Bolander bedstraw	*Galium bolanderi*
California brome	*Bromus carinatus*
Canyon live oak	*Quercus chrysolepis*
Choke cherry	*Prunus virginiana*
Columbia needlegrass	*Stipa nelsonii*
Coulter pine	*Pinus coulteri*
Creeping Oregon-grape	*Berberis repens*
Curlleaf mountain-mahogany	*Cercocarpus ledifolius*
Desert snowberry	*Symphoricarpos longiflorus*
Douglas-fir	*Pseudotsuga menziesii*
Greenleaf manzanita	*Arctostaphylos patula*
Heartleaf arnica	*Arnica cordifolia*
Idaho fescue	*Festuca idahoensis*
Incense-cedar	*Calocedrus decurrens*
Jeffrey pine	*Pinus jeffreyi*
Lodgepole pine	*Pinus contorta* ssp. *murrayana*
Mahala carpet	*Ceanothus prostratus*
Mountain big sagebrush	*Artemisia tridentata* var. *vaseyana*
Mountain misery	*Chamaebatia foliolosa*
Mule's ears	*Wyethia mollis*
Orcutt brome	*Bromus orcuttii*
Ponderosa pine	*Pinus ponderosa*
Service berry	*Amelanchier alnifolia*
Shrubby bedstraw	*Galium angustifolium*
Sugar pine	*Pinus lambertiana*
Tobacco brush	*Ceanothus velutinus*
Tower butterweed	*Senecio integerrimus*
Wax currant	*Ribes cereum*
Wedgeleaf ceanothus	*Ceanothus cuneatus*
White fir	*Abies concolor*
Yellow pine	*Pinus jeffreyi* and *P. ponderosa*

Comments: *Pinus ponderosa* has three varieties, one of which grows in CA [*P. p.* var. *ponderosa*]. This variety has an extensive range and it grows in many habitats in the State. It is present in many series, even important in some with other yellow pines [*see* Jeffrey pine series, Jeffrey pine-ponderosa pine series, Washoe pine series]. Yellow pine genetics is reviewed by Critchfield (1984), especially in terms of hybridization.

Fites (1993) and Smith (1994) place many associations dominated by ponderosa pine in their White fir series based on the proposed replacement of ponderosa pine by white fir in time. These associations are placed here based on canopy dominance of ponderosa pine.

Plot-based descriptions: Sawyer & Thornburgh (1977) define one association in m-KlaR, Simpson (1980) one association in s. Siskiyou Mountains in Humboldt Co., Waddell (1982) one association in Yolla Bolly Mountains in Tehama and Trinity Cos. NorCo, KlaR.; Keeler-Wolf (1992) describes stands at Graham Pinery candidate RNA in Butte Co., Keeler-Wolf (1984c) in Keeler-Wolf (1990e) defines three associations at Shasta mudflow RNA in Siskiyou Co., Conard & Robichaux (1980) in Keeler-Wolf (1990e) describe stands at Soda Ridge candidate RNA in Butte Co. m-CasR; Fites (1993) defines three associations in Lassen, Plumas, Tahoe, Eldorado National Forests m-CasR, m-SN; Vora (1988) two associations at Black's Mountain Experimental Forest in Lassen Co. m-CasR, Smith (1994) 24 associations in Klamath, Lassen, Modoc, Shasta-Trinity, Tahoe National Forests m-CasR, ModP, n. TraSN; Gray (1978) reports importance values for stands in Lake and Sierra Cos. m-NorCo, m-SN, Taylor & Randall (1977a) in Keeler-Wolf (1990e) define one association at Peavine Point RNA in El Dorado Co., Keeler-Wolf (1987f) in Keeler-Wolf (1990e) describes stands at Grizzly Mountain RNA, Talley (1981) in Keeler-Wolf (1990e) stands at South Fork Merced River candidate RNA in Mariposa Co. m-SN; Meier (1979) in Keeler-Wolf (1990e) stands at Fern Canyon RNA in San Bernardino Co. i. m-TraR; Keeler-Wolf (1986e, 1988e) in Keeler-Wolf (1990e) defines one association at Hall Canyon RNA in Riverside Co. m-PenR.

Associations:
Fites (1993):
   Ponderosa pine/antelope bitterbrush/ Bolander bedstraw association [as Mixed conifer/
   antelope bitterbrush/Bolander bedstraw association],
   Ponderosa pine/mountain misery [as Mixed conifer/mountain misery association],
   Ponderosa pine/greenleaf manzanita-mountain misery association [as Mixed conifer/greenleaf manzanita-mountain misery association].
Keeler-Wolf (1984c):
   Ponderosa pine/big sagebrush association [as subtype 1],
   Ponderosa pine/antelope bitterbrush association [as subtype 2],
   Ponderosa pine/California brome association [as subtypes 3,4].
Keeler-Wolf (1986e, 1988e):
   Ponderosa pine/shrubby bedstraw association.
Sawyer & Thornburgh (1977):
   Ponderosa pine/mahala carpet association [= *Pinus ponderosa/Ceanothus prostratus* association (Muldavin 1982)].
Simpson (1980):
   Ponderosa pine/wedgeleaf ceanothus association.
Smith (1994):
   Ponderosa pine/antelope bush/arrowleaf balsam root association,
   Ponderosa pine/antelope bitterbrush-choke cherry/Orcutt brome association,
   Ponderosa pine/antelope bitterbrush/Columbia needlegrass/pumice association,
   Ponderosa pine/antelope bitterbrush-greenleaf manzanita/Columbia needlegrass association,
   Ponderosa pine/antelope bitterbrush-tobacco brush association,
   Ponderosa pine/antelope bitterbrush/tower butterweed/granite association,
   Ponderosa pine/antelope bitterbrush-wax currant/Orcutt brome association,
   Ponderosa pine-black oak/curlleaf mountain-mahogany association,
   Ponderosa pine/Columbia needlegrass association,
   Ponderosa pine/curlleaf mountain-mahogany-antelope bitterbrush/Idaho fescue association,
   Ponderosa pine/curlleaf mountain-mahogany/ blue wheatgrass association,
   Ponderosa pine - Douglas-fir/antelope bitterbrush/mule's ears association,

Ponderosa pine-interior live oak association,
Ponderosa pine-lodgepole pine/service berry associaton,
Ponderosa pine/mountain big sagebrush/Idaho fescue association,
Ponderosa pine/mountain big sagebrush-antelope bitterbrush association,
Ponderosa pine/service berry-choke cherry association,
Ponderosa pine/service berry-creeping Oregon-grape/heartleaf arnica association,
Ponderosa pine/tobacco bush/Columbia needlegrass association.

Taylor & Randall (1977a):
Ponderosa pine/mountain misery association.

Vora (1988):
Ponderosa pine/antelope bitterbrush association,
Ponderosa pine/desert snowberry association.

Waddell (1982):
Ponderosa pine-canyon live oak association.

Port Orford-cedar series

Port Orford-cedar sole, dominant, or important tree in canopy; Sitka spruce, Douglas-fir, Jeffrey pine, ponderosa pine, sugar pine, Shasta fir, tanoak, western hemlock, white fir, and/or western white pine may be present. Trees < 60 m tall; canopy continuous, intermittent or open. Shrubs infrequent or common. Ground layer sparse or grassy (Plate 24).

Uplands: concave slopes; raised stream benches and terraces. Soils commonly ultramafic, most commonly peridotite-derived. The national inventory of wetland plants (Reed 1988) lists Port Orford-cedar as a FACU.

Distribution: m-KlaR.
Elevation: sea level-3000 m.

NDDB/Holland type and status: **North Coast coniferous forests** (82000).
　　Port Orford-cedar forest (82500) G3 S2.1.

Other types:
Barry type: G7411112 BCHLA00.
Cheatham & Haller type: North Coast coniferous forest.
Thorne type: North coastal coniferous forest.
WHR type: Douglas-fir.

General references: Port Orford-cedar range in CA (Griffin & Critchfield 1972); silvics (Zobel 1990). Franklin (1988), Kruckeberg (1984), Stein (1980a), Zobel *et al.* (1985), Zobel & Hawk (1980).

Comments: Port Orford-cedar is present in several forest series besides this one where it dominates. Root rot infection now exists in the Smith River drainage (Zobel 1990), Del Norte Co. l-KlaR, m-KlaR. A substantial decline in population size is conceivable for this KlaR endemic.

Species mentioned in text	
Black huckleberry	*Vaccinium ovatum*
California bay	*Umbellularia californica*
Darlingtonia	*Darlingtonia californica*
Douglas-fir	*Pseudotsuga menziesii*
Huckleberry oak	*Quercus vaccinifolia*
Jeffrey pine	*Pinus jeffreyi*
Little Oregon-grape	*Berberis nervosa*
Ponderosa pine	*Pinus ponderosa*
Port Orford-cedar	*Chamaecyparis lawsoniana*
Rhododendron	*Rhododendron macrophyllum*
Sadler oak	*Quercus sadleriana*
Salal	*Gaultheria shallon*
Shasta fir	*Abies magnifica var. shastensis*
Sitka spruce	*Picea sitchensis*
Spikenard	*Aruncus dioicus*
Sugar pine	*Pinus lambertiana*
Tanoak	*Lithocarpus densiflora*
Thinleaf huckleberry	*Vaccinium membranaceum*
Twinflower	*Linnaea borealis*
Vine maple	*Acer circinatum*
Western azalea	*Rhododendron occidentale*
Western white pine	*Pinus monticola*
White fir	*Abies concolor*

Plot-based descriptions: Keeler-Wolf (1986a) in Keeler-Wolf (1990e) describes a bog forest dominated by Port Orford-cedar with darlingtonia at L.E. Horton candidate RNA, Keeler-Wolf (1987d) in Keeler-Wolf (1990e) a Port Orford-cedar - Douglas-fir - western hemlock forest at Upper Goose Creek candidate RNA in Del Norte Co. l-KlaR; Simpson (1980) defines one association in s. Siskiyou Mountains in Humboldt Co., Sawyer & Thornburgh (1969) describe stands at Youngs Valley area in Siskiyou Mountains m-KlaR; Keeler-Wolf (1982, 1989d) in Keeler-Wolf (1990e) inland, high elevation stands at Cedar Basin candidate RNA in Siskiyou Co. su-KlaR; Jimerson (1994) defines 17 associations in Six Rivers National Forest KlaR. Atzet & Wheeler (1982) define associations in sw. OR.

Associations:
Jimerson (1994):
> Port Orford-cedar/western azalea association,
> Port Orford-cedar - Douglas-fir/huckleberry oak association,
> Port Orford-cedar - Shasta fir/Sadler oak - thinleaf huckleberry association,
> Port Orford-cedar/rhododendron - salal association,
> Port Orford-cedar/salal association,
> Port Orford-cedar - white fir/azalea association,
> Port Orford-cedar - white fir/Sadler oak association,
> Port Orford-cedar - white fir/herb association,
> Port Orford-cedar - white fir/huckleberry oak association,
> Port Orford-cedar - western white pine/ huckleberry oak association,
> Port Orford-cedar - western white pine/ huckleberry oak association,
> Tanoak -Port Orford-cedar - California bay association,
> Tanoak - Port Orford-cedar/little Oregon-grape/twinflower association,
> Tanoak -Port Orford-cedar/black huckleberry association,
> Tanoak - Port Orford-cedar/black huckleberry-azalea association,
> Tanoak - Port Orford-cedar/spikenard association,
> Tanoak - Port Orford-cedar/vine maple association.

Simpson (1980):
> Pine/rhododendron association.

Pygmy cypress series

Pygmy cypress dominant or important tree in canopy; bishop pine and Bolander pine may be present. Trees < 18 m; canopy intermittent or open, taller or the same height as shrubs. Shrubs common. Ground layer sparse (Plate 25).

Uplands: maritime terraces. Soils sandstone-derived, acid, sterile, poorly-drained with iron hardpan.

Distribution: o-NorCo.
Elevation: 100-300 m.

NDDB/Holland type and status: **Closed-cone coniferous forests** (83000).
    Mendocino pygmy cypress forest (83161) G2 S2.1.

Other types:
Barry type: G7411112 BCUPY00.
Cheatham & Haller type: Pygmy cypress forest.
PSW-45 type: Cypress series.
Thorne type: Pygmy-forest coniferous woodland.
WHR type: Closed-cone pine-cypress.

General references: Pygmy cypress range in CA (Griffin & Critchfield 1972). Howard (1992), Jenny *et al.* (1969), Paysen *et al.* (1980), Sholars (1979, 1982, 1984), Sposito (1992), Vogl *et al.* (1977), Wolf & Wagener (1948).

Comments: Pygmy cypress [*Cupressus goveniana* ssp. *pygmaea*] is one of two dwarfed cypresses in the state. The other [*Cupressus goveniana* ssp. *goveniana*] occurs in Monterey Co. [*see* Gowen cypress stands].

The two taxa have also been treated as species in some manuals. Both occur in similar habitats; this subspecies has a wider range. The species has been studied ecologically, and stands differ from surrounding vegetation (Vogl *et al.* 1977).

This is the only series in which Bolander pine is present. It is closely related to beach pine, which occurs on coastal dunes [*see* Beach pine series].

Pygmy cypress is rare (a CNPS List 1B plant) (Skinner & Pavlik 1994).

Species mentioned in text	
Beach pine	*Pinus contorta* ssp. *contorta*
Bishop pine	*Pinus muricata*
Bolander pine	*Pinus contorta* ssp. *bolanderi*
Gowen cypress	*Cupressus goveniana* ssp. *goveniana*
Pygmy cypress	*Cupressus goveniana* ssp. *pygmaea*
Redwood	*Sequoia sempervirens*

Plot-based descriptions: Westman & Whittaker (1975) defines four associations as part of vegetation gradients from pygmy cypress to redwood-dominated stands in Mendocino Co. o-NorCo. Lichen species and tree height separate associations. Sholars (1979) reports tree sizes.

Associations:
Westman & Whittaker (1975):
    Pygmy cypress/*Cladonia bellidiflora* association [as Short hydric pygmy cypress type],
    Pygmy cypress/*Cladina impexa* association [as Mesophytic pygmy cypress type],
    Pygmy cypress/*Ramalina tharusta* association [as Tall hydric pygmy cypress type],
    Pygmy cypress/*Usnea subfloridana* association [as Extreme pygmy cypress type].

Red alder series

Red alder sole or dominant tree in canopy; arroyo willow, black cottonwood, Hooker willow, and/or Sitka spruce may be present. Trees < 40 m; canopy continuous. Shrubs common or infrequent. Ground layer continuous, especially with ferns (Plate 18).

Wetlands: soils seasonally flooded, seasonally saturated, permanently saturated. Water chemistry: fresh. Stream and river backwaters, banks, bottoms, floodplains, mouths, terraces. Cowardin class: Palustrine forested wetland. **Uplands:** all aspects. Soils sandstone, schist-derived. The national inventory of wetland plants (Reed 1988) lists red alder as a FACW.

Distribution: o-NorCo, n. o-CenCo.
Elevation: sea level-750 m.

NDDB/Holland type and status: **Riparian forests** (61000), **Riparian scrubs** (63000), **Broadleaved upland forests** (81000).
>   Red alder riparian forest (61130) G3 S3.2.
>   North Coast riparian scrub (63100) G3 S2.2.
>   Woodwardia thicket (63110 *in part*) G3 S3.2
>   Red alder forest (81A00) G4 S3.2.

Other types:
Barry type: G74
Cheatham & Haller type: Red alder groves.
PSW-45 type: Alder series.
Rangeland type: SRM 203.
Thorne type: Riparian woodlands.
WHR type: Montane riparian.

General references: Red alder range in CA (Griffin & Critchfield 1972); silvics (Harrington 1990). Antos & Allen (1990), DeBell & Turpin (1983), Holstein (1984), Miller (1980), McBride (1994), Paysen *et al.* (1980).

Comments: Self-perpetuating stands of this series occur on alluvial sites. Temporary stands dominated by red alder establish after logging on upland settings. Douglas-fir, grand fir, redwood, Sitka spruce, and/or western hemlock replace red alder in time. Chain fern thickets are common in NorCo and l-KlaR stands.

Species mentioned in text	
Arroyo willow	*Salix lasiolepis*
Black cottonwood	*Populus balsamifera*
Candyflower	*Claytonia sibirica*
Chain fern	*Woodwardia fimbriata*
Douglas-fir	*Pseudotsuga menziesii*
Grand fir	*Abies grandis*
Hooker willow	*Salix hookeriana*
Red alder	*Alnus rubra*
Redwood	*Sequoia sempervirens*
Salal	*Gaultheria shallon*
Sitka spruce	*Picea sitchensis*
Vine maple	*Acer circinatum*
Western hemlock	*Tsuga heterophylla*

Plot-based descriptions: Taylor (1982) in Keeler-Wolf (1990e) describes stands at Yurok RNA in Del Norte Co., Stuart *et al.* (1986) define one association at Stagecoach Hill Azalea Preserve, Newton (1989) describes riparian stands in n. Humboldt Co. o-NorCo; Jimerson (1993) defines one association in Six Rivers National Forest l-KlaR.

Associations:
Jimerson (1993):
>   Douglas-fir-red alder/vine maple/candyflower association.

Stuart *et al.* (1986):
>   Red alder/salal association.

Red fir series

Noble fir, red fir, or Shasta fir sole or dominant tree in canopy; lodgepole pine, Jeffrey pine, mountain hemlock, sugar pine, western white pine, and/or white fir may be present. Trees < 60 m; canopy continuous or intermittent. Shrubs infrequent or common. Ground layer sparse or abundant (Plate 27).

Uplands: slopes, all aspects; raised stream benches and terraces. Soils shallow. The national inventory of wetland plants (Reed 1988) lists red fir as a FACU.

Distribution: m-NorCo, m-KlaR, m-CasR, m-SN.
Elevation: 1400-2700 m.

NDDB/Holland type and status: **Upper montane coniferous forests** (85000).
    Red fir forest (85310) G4 S4.

Other types:
Barry type: G7411112 BABMA00.
Cheatham & Haller type: Red fir forest.
Thorne type: Red fir forest.
WHR type: Red fir.

General references: Noble fir, Red fir ranges in CA (Griffin & Critchfield 1972); silvics (Laacke 1990b). Barbour (1984, 1988), Franklin (1990), Gordon (1980b), Heckart & Hickman (1984), Laacke & Fiske (1983a), Oosting & Billings (1943), Rundel *et al.* (1977), Sawyer & Thornburgh (1977).

Comments: *Abies magnifica* has two varieties [*A. m.* var. *magnifica* (red fir) and *A. m.* var. *shastensis* (Shasta fir)]. Red fir grows in NorCo, SN, CasR; Shasta fir in NorCo, KlaR, CasR, s. SN. The closely related *A. procera* (noble fir) grows in w. m-KlaR. The nature and extent of noble fir in CA is unresolved. Griffin & Critchfield (1972) map it north of the Klamath River in the Siskiyou Mountains, and report populations in w. m-KlaR and m-NorCo. Parker (1991) argues that it is

separate from Shasta fir where they grow in KlaR and CasR, but others [Franklin (1990), Laacke (1990b)] consider Nor-CA and s. OR a hybrid zone between *Abies magnifica* and *A. procera*. Liu (1972) goes as far as to not recognize *A. m.* var. *shastensis,* since he considers it an *A. magnifica* x *procera* hybrid.

Species mentioned in text	
Black-laurel	*Leucothoe davisiae*
Bracken	*Pteridium aquilinum*
Brewer spruce	*Picea breweriana*
Bush chinquapin	*Chrysolepis sempervirens*
Creeping snowberry	*Symphoricarpos mollis*
Heartleaf arnica	*Arnica cordifolia*
Huckleberry oak	*Quercus vaccinifolia*
Incense-cedar	*Calocedrus decurrens*
Jeffrey pine	*Pinus jeffreyi*
Lodgepole pine	*Pinus contorta* ssp. *murrayana*
Mountain hemlock	*Tsuga mertensiana*
Noble fir	*Abies procera*
One-sided shinleaf	*Orthilia secunda*
Pinemat manzanita	*Arctostaphylos nevadensis*
Port Orford-cedar	*Chamaecyparis lawsoniana*
Prince's-pine	*Chimaphila umbellata*
Red fir	*Abies magnifica* var. *magnifica*
Rhododendron	*Rhododendron macrophyllum*
Sadler oak	*Quercus sadleriana*
Shasta fir	*Abies magnifica* var. *shastensis*
Silver lupine	*Lupinus albifrons*
Slender penstemon	*Penstemon gracilentus*
Sugar pine	*Pinus lambertiana*
Thinleaf huckleberry	*Vaccinium membranaceum*
Trail penstemon	*Penstemon gracilis*
Twinflower	*Linnaea borealis*
Vanilla leaf	*Achlys triphylla*
Western white pine	*Pinus monticola*
Wild rose	*Rosa gymnocarpa*
White fir	*Abies concolor*
White-veined shinleaf	*Pyrola picta*
Whitebark pine	*Pinus albicaulis*

We adopt Forest Service judgment (Atzet *et al.* 1992) and include stands with noble fir at Pearch Creek (Keeler-Wolf 1990e) and Rock Creek Butte candidate RNA (Keeler-Wolf 1990e) in Humboldt Co. w. m-KlaR in the Red fir series.

Plot-based descriptions: Barbour & Woodward (1985) report tree density, basal area, cover) for 86 red fir and Shasta fir-dominated stands throughout the species range; composition was consistent throughout its range. Gray (1978) reports importance values for stands in Lake and Sierra Cos.

Simpson (1980) defines two associations in s. Siskiyou Mountains in Humboldt Co., Keeler-Wolf (1982, 1989d) in Keeler-Wolf (1990e) describes stands at Cedar Basin candidate RNA, Imper (1988a) in Keeler-Wolf (1990e) defines three associations at Haypress Meadows candidate RNA, Keeler-Wolf (1984b, 1989f), Sawyer & Thornburgh (1971) in Keeler-Wolf (1990e) describe stands at Sugar Creek candidate RNA in Siskiyou Co. m-KlaR, su-KlaR, Waddell (1982) defines two associations in Yolla Bolly Mountains in Tehama and Trinity Cos. NorCo, KlaR; Jimerson (1993) defines 17 associations in Mendocino and Six Rivers National Forests m-NorCo, m-KlaR, su-KlaR; Imper (1988b) in (Keeler-Wolf 1990) three associations at Shasta Fir candidate RNA (on Mt. Shasta), Fiedler (1987), Keeler-Wolf (1989g) in Keeler-Wolf (1990e) describe stands at Antelope Creek Lakes candidate RNA in Siskiyou Co., Taylor & Randall (1978) in Keeler-Wolf (1990e) stands at Cub Creek RNA in Tehama Co. m-CasR; Talley (1977), Keeler-Wolf (1989j) in Keeler-Wolf (1990e) stands at Babbitt Peak RNA in Sierra Co., Keeler-Wolf & Keeler-Wolf (1981), Keeler-Wolf (1989k) in Keeler-Wolf (1990e) stands at Mt. Pleasant RNA in Plumas Co., Keeler-Wolf (1991b) stands at Mountaineer Creek candidate RNA in Tulare Co., Keeler-Wolf (1985b, 1988c) in Keeler-Wolf (1990e) stands at Mud Lake RNA in Plumas Co., Keeler-Wolf (1985c, 1989l) in Keeler-Wolf (1990e) stands at Bell Meadow RNA, Talley (1976a) in Keeler-Wolf (1990e) stands at W. B. Critchfield (formerly Bourland Meadow) RNA in Calaveras Co., Griffin (1975) in Keeler-Wolf (1990e) stands at Teakettle Creek Experimental Forest in Fresno Co., Jensen (1992) reports tree density, basal area, frequency at Home Camp candidate RNA in Fresno Co. m-SN, Potter (1994) defines 12 regional associations in c. SN.

Associations:
Imper (1988a):
    Shasta fir/Sadler oak association,
    Shasta fir/thinleaf huckleberry association,
    Shasta fir/vanilla leaf association.
Imper (1988b):
    Shasta fir/pinemat manzanita association,
    Shasta fir/prince's-pine association,
    Shasta fir/slender penstemon association,
Jimerson (1993):
    Port Orford-cedar - red fir/Sadler oak - thinleaf huckleberry association,
    Red fir/Sadler oak association,
    Red fir/Sadler oak-pinemat manzanita association,
    Red fir -incense-cedar association,
    Red fir/one-sided shinleaf association,
    Red fir/rhododendron association,
    Red fir/white-veined shinleaf association,
    Red fir-Brewer spruce/Sadler oak-thinleaf huckleberry association,
    Red fir-mountain hemlock/one-sided shinleaf association,
    Red fir-white fir/bracken association,
    Red fir-white fir/heartleaf arnica association,
    Red fir-white fir/creeping snowberry/white-veined shinleaf association,
    Red fir-white fir/creeping snowberry-wild rose association,
    Red fir-white fir/Sadler oak association,
    Red fir-white fir/pinemat manzanita association,
    Red fir-white fir/vanilla leaf association.
Potter (1994):
    Red fir association,
    Red fir/lodgepole pine/whiteflower hawkweed association,
    Red fir/lodgepole pine association,
    Red fir/mule's ears association,
    Red fir/pinemat manzanita association,
    Red fir-western white pine/pinemat manzanita association,
    Red fir-western white pine-lodgepole pine association,
    Red fir-western white pine association,
    Red fir/western white pine/bush chinquapin association,

Red fir-white fir association,
Red fir-white fir-Jeffrey pine association,
White fir-sugar pine-red fir association,
Sawyer & Thornburgh (1977):
Shasta fir/black-laurel association,
Shasta fir/huckleberry oak association,
Shasta fir/prince's-pine association
[= Imper (1988b)],
Shasta fir/twinflower association,
Shasta fir/white-veined shinleaf association.
Simpson (1980):
Shasta fir/pinemat manzanita association,
Shasta fir/Sadler association.
Waddell (1982):
Red fir/silver lupine association,
Red fir/white-veined shinleaf association.

Red willow series

Red willow sole or dominant shrub or tree in canopy; California sycamore, coyote brush, Fremont cottonwood, Mexican elderberry, mulefat, white alder, and/or willows may be present. If shrubland, emergent trees may be present. Shrubs < 15 m; canopy continuous. Shrubs sparse under tree canopy. Ground layer variable.

Wetlands: habitats seasonally flooded, saturated. Water chemistry: fresh. Ditches; floodplains; lake edges; low-gradient depositions along rivers, streams. Cowardin class: Palustrine forested or shrub-scrub wetland. The national inventory of wetland plants (Reed 1988) does not list red willow.

Distribution: Cis-CA, Trans-CA, inter-West.
Elevation: sea level-1700 m.

NDDB/Holland type and status: Riparian forests (61000), **Riparian scrubs** (63000).
    Central Coast cottonwood-sycamore riparian forest (61210 *in part*) G3 S3.
    Great Valley mixed riparian forest (61420 *in part*) G2 S2.1.
    Modoc-Great Basin cottonwood-willow riparian forest (61610 *in part*) G3 S2.1.
    Mojave riparian forest (61700 *in part*) G1 S1.1.
    Central Coast riparian scrub (63200 *in part*) G3 S3.2.
    Southern willow scrub (63320 *in part*) G3 S2.1.

Other types:
Barry type: G7411121 BSALA60.
Cheatham & Haller type: Bottomland woodlands and forest.
PSW-45 type: Willow series.
Thorne type: Riparian woodland.
WHR type: Freshwater emergent wetland.

General references: Bowler (1989), Capelli & Stanley (1984), Faber *et al.* (1989), Holstein (1984), Paysen *et al.* (1980), Read & Sprackling (1980).

Comments: Willow stands may or may not be dominated by a single species. If no dominant willow is present at low elevations, then place the stands in the Mixed willow series. Montane and subalpine willow stands are placed in separate classes since different willow species are restricted to those elevations [*see* Montane wetland shrub habitat, Subalpine wetland shrub habitat].

Stands of the Red willow series have environmental conditions similar to alder, cottonwood, and other willow series [*see* Black cottonwood series, Black willow series, Fremont cottonwood series, Hooker willow series, Pacific willow series, Sitka willow series, White alder series].

Species mentioned in text

California sycamore	*Platanus racemosa*
Coyote brush	*Baccharis pilularis*
Fremont cottonwood	*Populus fremontii*
Mexican elderberry	*Sambucus mexicana*
Mulefat	*Baccharis salicifolia*
Red willow	*Salix laevigata*
White alder	*Alnus rhombifolia*
Willows	*Salix* species

Plot-based descriptions: not available.

Redwood series

Redwood sole, dominant, or important tree in canopy; California bay, Douglas-fir, madrone, tanoak, and/or western hemlock may be present. Trees < 120 m tall; canopy continuous or intermittent, may be two-tiered. Shrubs infrequent or common. Ground layer absent or abundant (Plate 24).

Uplands: slopes, all aspects; reased stream benches and terraces. Soils mostly sandstone or schist-derived. The national inventory of wetland plants (Reed 1988) does not list redwood.

Distribution: o-NorCo, n. o-CenCo, OR.
Elevation: 10-600 m.

NDDB/Holland type and status: Riparian forests (61000), **North Coast coniferous forests** (82000).
  North coast alluvial redwood forest (61120) G2 S2.2.
  Alluvial redwood forest (82310) G2 S2.2.
  Upland redwood forest (82320) G4 S2.3.

Other types:
Barry type: G7411112 BSESE30.
Cheatham & Haller type: Redwood forest.
PSW-45 type: Coast redwood series.
Thorne type: Redwood forest.
WHR type: Redwood.

General references: Redwood range in CA (Griffin & Critchfield 1972); silvics (Olson *et al.* 1990). Franklin (1988), Olson (1983), Olson *et al.* (1990), Paysen *et al.* (1980), Pillers & Stuart (1993), Roy (1980), Veirs (1982), Zinke (1977).

Comments: Ecologists differentiate forests on alluvial streamside terraces, where redwood is the only canopy tree, from those on upland settings, where redwood shares the canopy with other conifer and broadleaf tree species. These differences are emphasized at the association level. Keeler-Wolf & Keeler-Wolf (1977), Borchert (1987) in Keeler-

Wolf (1993e) qualitatively describe stands at Limekiln RNA (now part of the Cone Gradient RNA) in Monterey Co. o-CenCo.

Species mentioned in text	
Bigleaf maple	*Acer macrophyllum*
Black huckleberry	*Vaccinium ovatum*
Bracken	*Pteridium aquilinum*
California bay	*Umbellularia californica*
California polypody	*Polypodium californicum*
Chain fern	*Woodwardia fimbriata*
Common vetch	*Vicia angustifolia*
Deer fern	*Blechnum spicant*
Douglas iris	*Iris douglasiana*
Douglas-fir	*Pseudotsuga menziesii*
Grand fir	*Abies grandis*
Little Oregon-grape	*Berberis nervosa*
Madrone	*Arbutus menziesii*
Man root	*Marah fabaceus*
Redwood	*Sequoia sempervirens*
Redwood oxalis	*Oxalis oregana*
Round-fruited sedge	*Carex globosa*
Salal	*Gaultheria shallon*
Sword fern	*Polystichum munitum*
Tanoak	*Lithocarpus densiflora*
Trillium	*Trillium ovatum*
Western hemlock	*Tsuga heterophylla*

Plot-based descriptions: Borchert *et al.* (1988) define seven associations in Monterey Co. o-CenCo; Matthew (1986a, 1986b) four associations at Humboldt Redwood State Park in s. Humboldt Co.; Combs (1984) describes stand structure in Little Lost Man Creek RNA, Lenihan (1990) defines three associations at Redwood National Park in n. Humboldt Co., Taylor (1982) in Keeler-Wolf (1990e) one association in Del Norte Co. o-NorCo.

Associations:
Borchert *et al.* (1988):
  Redwood-bigleaf maple/California polypody/Gamboa association,
  Redwood/bracken-chain fern/streamsides association,
  Redwood/bracken-trillium/Gamboa-Sur association,

Redwood/Gamboa-Sur association,
Redwood/man root-common vetch/Gamboa-
Sur association,
Redwood-tanoak/round-fruited sedge - Douglas
iris/Gamboa association.
Lenihan (1990):
Redwood/deer fern association,
Redwood/little Oregon-grape association
[= Redwood/sword fern association (Taylor
1982)],
Redwood/madrone association.
Matthews (1986a, 1986b):
Redwood-Douglas-fir/madrone association,
Redwood-Douglas-fir/salal association,
Redwood-Douglas-fir/black huckleberry
association,
Redwood/redwood oxalis association.

Santa Lucia fir series

Santa Lucia fir sole, dominant, or important with canyon live oak in the tree canopy; Coulter pine, madrone, ponderosa pine, sugar pine, and/or tanoak may be present. Trees < 50 m; canopy continuous, may be two-tiered. Shrubs infrequent. Ground layer sparse (Plate 21).

Uplands: steep slopes, ridges; aspects north or easterly; canyon bottoms, raised stream benches and terraces. Soils sandstone, shale-derived. The national inventory of wetland plants (Reed 1988) does not list Santa Lucia fir.

Distribution: m-CenCo.
Elevation: 250-1400 m.

NDDB/Holland type and status: Lower montane coniferous forests (84000).
    Santa Lucia fir forest (84120) G3 S3.

Other types:
Barry type: G7411112 BABBR00.
Cheatham & Haller type: Santa Lucia fir forest.
PSW-45 type: Santa Lucia series.
Thorne type: Northern mixed evergreen forest.
WHR type: Montane hardwood conifer.

General references: Santa Lucia fir range in CA (Griffin & Critchfield 1972). Axelrod (1976), Barbour (1988), Paysen *et al.* (1980), Talley (1974), Sawyer *et al.* (1977).

Comments: This series occurs in a complex vegetation mosaic with other series in rugged topography. Santa Lucia fir stands are best developed on fire-protected sites (Talley 1974). Keeler-Wolf (1990e) qualitatively describes stands at Limekiln RNA and adjacent South Fork of Devil's Canyon RNA (now part of the Cone Peak Gradient RNA) in Monterey Co. m-CenCo.

Santa Lucia fir is rare (a CNPS List 4 plant) (Skinner & Pavlik 1994).

Plot-based descriptions: Talley (1974) describes stands in the Santa Lucia Mountains in Monterey Co. m-CenCo. Borchert's analysis of Talley's plot data using TWINSPAN yielded two associations for this series [*see* also Canyon live oak series, White alder series]. This analysis is on file at NDDB.

Species mentioned in text

Canyon live oak	*Quercus chrysolepis*
Coulter pine	*Pinus coulteri*
Madrone	*Arbutus menziesii*
Ponderosa pine	*Pinus ponderosa*
Santa Lucia bedstraw	*Galium clementis*
Santa Lucia fir	*Abies bracteata*
Sugar pine	*Pinus lambertiana*
Sword fern	*Polystichum munitum*
Tanoak	*Lithocarpus densiflora*

Associations:
NDDB:
    Santa Lucia fir/Santa Lucia bedstraw association,
    Santa Lucia fir/sword fern association.

Sargent cypress series

Sargent cypress sole, dominant, or important tree in canopy; California bay, foothill pine, interior live oak, knobcone pine, McNab cypress may be present. Trees < 15 m; canopy intermittent or open. Shrubs common or infrequent. Ground layer sparse (Plate 25).

Uplands: slopes, ridges; raised stream benches and terraces. Soils ultramafic-derived; sterile. The national inventory of wetland plants (Reed 1988) does not list Sargent cypress.

Distribution: o-NorCo, i-NorCo, i-CenCo.
Elevation: 200-900 m.

NDDB/Holland type and status: Closed-cone coniferous forests (83000).
Northern interior cypress forest (83220 *in part*) G2 S2.2.

Other types:
Barry type: G7411112 BCUSA00.
Cheatham & Haller type: Northern interior cypress forest.
PSW-45 type: Cypress series.
Thorne type: Inland closed-cone coniferous woodland.
WHR type: Closed-cone pine-cypress.

General references: Sargent cypress range in CA (Griffin & Critchfield 1972), Kruckeberg (1984), Mendocino National Forest (1987), Paysen *et al.* (1980), Vogl *et al.* (1977), Wolf & Wagener (1948).

Comments: This is an abundant and widespread cypress. There are several large stands and many smaller ones scattered along the interior Coast Ranges. It can grow with McNab cypress in NorCo, but it tends to grow on upper slopes whereas Sargent cypress grows on lower slopes and in ravines [*see* McNab cypress series]. Stands where both cypresses are equally important are placed in this series.

Species mentioned in text	
California bay	*Umbellularia californica*
Foothill pine	*Pinus sabiniana*
Interior live oak	*Quercus wislizenii*
Knobcone pine	*Pinus attenuata*
McNab cypress	*Cupressus macnabiana*
Sargent cypress	*Cupressus sargentii*

Plot-based descriptions: Keeler-Wolf (1983) in Keeler-Wolf (1990e) reports tree density and basal area at Frenzel Creek RNA in riparian and valley bottom settings in Colusa Co., Parker (1990) describes stands at Mount Tamalpais in Marin Co. i-NorCo.

Singleleaf pinyon series

Singleleaf pinyon sole or dominant tree in canopy; California juniper, canyon live oak, Jeffrey pine, mountain juniper, Tucker oak, and/or Utah juniper may be present. Emergent trees may be present over a shrub canopy; bitterbrush, big sagebrush, black sagebrush, green ephedra, and/or low sagebrush may be present. Trees < 15 m; canopy intermittent or open. Shrubs common. Ground layer absent, sparse, or grassy (Plate 22).

Uplands: alluvial fans, pediments, slopes, ridges. Soils commonly well-drained.

Distribution: m-TraR, m-PenR, TraSN, WIS, MojD, Baja CA.
Elevation: 1000-2800 m.

NDDB/Holland type and status: Piñon and juniper woodlands (72000).
    Great Basin pinyon woodland (72122) G3 S3.2.
    Mojavean pinyon woodland (72210) G4 S4.

Other types:
Barry type: G74 G7411112 BPIMO10.
Cheatham & Haller type: Pinyon-juniper woodland.
PSW-45 type: Pinyon series.
Rangeland type: SRM 412.
Thorne type: Pinyon-juniper woodland.
WHR type: Pine-juniper.

General references: Singleleaf pinyon range in CA (Griffin & Critchfield 1972); silvics (Meeuwig *et al.* 1990). Brown (1982a), Meeuwig & Bassett (1983), Minnich (1987), Paysen *et al.* (1980), Thorne (1982), Tueller *et al.* (1979), Vasek & Thorne (1977), West (1994).

Comments: Pinyon-juniper woodlands are better considered a collection of series. If singleleaf pinyon dominates the stand, it is a member of this series. In the case of the Singleleaf pinyon-Utah juniper

series, the two conifers are equally important [*see also* California juniper series, Mountain juniper series, Utah juniper series].

Species mentioned in text	
Big sagebrush	*Artemisia tridentata*
Bitterbrush	*Purshia tridentata*
Black sagebrush	*Artemisia nova*
California juniper	*Juniperus californica*
Canyon live oak	*Quercus chrysolepis*
Green ephedra	*Ephedra viridis*
Jeffrey pine	*Pinus jeffreyi*
Low sagebrush	*Artemisia arbuscula*
Mountain juniper	*Juniperus occidentalis*
	ssp. *australis*
Ponderosa pine	*Pinus ponderosa*
Singleleaf pinyon	*Pinus monophylla*
Tucker oak	*Quercus john-tuckeri*
Utah juniper	*Juniperus osteosperma*

Plot-based descriptions: Keeler-Wolf (1990c) describes stands at Long Canyon candidate RNA in Kern Co. m-SN; Parikh (1993a) reports tree density and basal area at San Emigdio Mesa in Ventura Co., Vasek & Thorne (1977) tree density and cover at Burns Pinyon Ridge in San Bernardino Co. and Pine Mountain Summit in Ventura Co. o. m-TraR; Keeler-Wolf & Keeler-Wolf (1976), Keeler-Wolf (1989h) in Keeler-Wolf (1990e) describe stands at Whippoorwill Flat RNA in Inyo Co. WIS.

Singleleaf pinyon-Utah juniper series

Singleleaf pinyon and Utah juniper important trees in canopy or as emergent trees over a shrub canopy; bitterbrush, big sagebrush, cliffrose, ephedra, rabbitbrush, and/or hop-sage may be present. Trees < 15 m; canopy intermittent or open. Shrubs < 1m; continuous or intermittent. Ground layer sparse or grassy (Plate 22).

Uplands: pediments, slopes, ridges. Aspects vary. Soils commonly well-drained.

Distribution: TraSN, WIS, MojD, m-DesR, NV. **Elevation:** 1000-2800 m.

NDDB/Holland type and status: Piñon and juniper woodlands (72000).
　　Great Basin pinyon juniper woodland (72121) G4 S4.

Other types:
Barry type: G74 G7411112.
Cheatham & Haller type: Nevadean pinyon-juniper woodland.
Rangeland type: SRM 412.
Thorne type: Pinyon-juniper woodland.
WHR type: Pine-juniper.

General references: Utah juniper ranges in CA (Griffin & Critchfield 1972), silvics (Meeuwig *et al.* (1990). Brown (1982a), Meeuwig & Bassett (1983), Tueller *et al.* (1979), Taylor (1976b), Vasek & Thorne (1977), West (1994).

Comments: Pinyon-juniper woodlands are better considered a collection of series. A problem arises in separation of stands belonging to the Singleleaf pinyon-Utah juniper series from others where only one of these species is dominant or incidental. Here the two species are equally important. If one species is noticeably more abundant, the stand is classed as a member of another series. If either tree is incidental, see the alternative series.

Species mentioned in text	
Bitterbrush	*Purshia tridentata*
Big sagebrush	*Artemisia tridentata*
Black bush	*Coleogyne ramosissima*
Cliffrose	*Purshia mexicana*
Ephedra	*Ephedra* species
Hop-sage	*Grayia spinosa*
Rabbitbrush	*Chrysothamnus* species
Singleleaf pinyon	*Pinus monophylla*
Utah juniper	*Juniperus osteosperma*

Plot-based descriptions: Vasek & Thorne (1977) report tree density and cover at Clark Mountain and New York Mountains in San Bernardino Co. m-DesR; Keeler-Wolf (1990e) describes stands at Whippoorwill Flat RNA in Inyo Co. WIS. Spolsky (1979) defines a Colorado desert pinyon-juniper woodland association in Anza Borrego State Park in San Diego Co. ColD.

Sitka spruce series

Sitka spruce sole or dominant tree in canopy; grand fir, red alder, redwood, and/or western hemlock may be present. Trees < 75 m; canopy continuous or intermittent. Shrubs infrequent or common. Ground layer abundant, especially with ferns (Plate 24).

Wetlands: soils seasonally flooded, seasonally saturated, permanently saturated. Water chemistry: fresh. Stream and river backwaters, bottoms, floodplains; raised maritime terraces with perched watertables. Cowardin class: Palustrine forested wetland. **Uplands:** steep, seaward slopes near ocean. Soils sandstone, schist-derived. The national inventory of wetland plants (Reed 1988) lists Sitka spruce as a FAC.

Distribution: c. & n. o-NorCo, OR, WA, Canada, Alaska.
Elevation: sea level-20 m.

NDDB/Holland type and status: **Marsh and swamp** (52000), **North Coast coniferous forests** (82000).
Freshwater swamp (52600) G1 S1.1.
Sitka spruce-grand fir forest (82100) G4 S1.1.

Other types:
Barry type: G7411112 BPISI00.
Cheatham & Haller type: North Coast coniferous forest.
Thorne type: Douglas-fir forest.
WHR type: Douglas-fir.

General references: Sitka spruce range in CA (Griffin & Critchfield 1972); silvics (Harris 1990b). Franklin (1988), Harris (1980a, 1990b), Harris & Johnson (1983), Minore (1980), Zinke (1977).

Comments: Old-growth stands of Sitka spruce series forests are rare in NorCo. Most stands are in shrub or young forest stages. Stands of cascara, hazel, Oregon crab apple, western azalea, and other shrubs are quickly invaded by Sitka spruce. Azalea

Reserve and Stagecoach Hill in Lagoons State Park are managed to maintain azalea. Young forests of these settings support stands of the Red alder series as well.

Species mentioned in text	
Cascara	*Rhamnus purshiana*
False lily-of-the-valley	*Maianthemum dilatatum*
Grand fir	*Abies grandis*
Hazel	*Corylus cornuta*
Oregon crab apple	*Malus fusca*
Red alder	*Alnus rubra*
Redwood	*Sequoia sempervirens*
Salmonberry	*Rubus spectabilis*
Sitka spruce	*Picea sitchensis*
Sword fern	*Polystichum munitum*
Western azalea	*Rhododendron occidentale*
Western hemlock	*Tsuga heterophylla*

Plot-based descriptions: Imper & Sawyer (1987) defines three associations at Table Bluff Ecological Reserve, NDDB has descriptions at Prairie Creek Redwood State Park and Redwood National Park in Humboldt Co., Westman & Whittaker (1975) de-fine one association in Mendocino Co. o-NorCo.

Associations:
Imper & Sawyer (1987):
Sitka spruce/false lily-of-the-valley association,
Sitka spruce/salmonberry association,
Sitka spruce/sword fern association.
Westman & Whittaker (1975):
Sitka spruce-western hemlock association
[as Spruce-hemlock type].

Sitka willow series

Sitka willow sole or dominant shrub or tree in canopy; bigleaf maple, black cottonwood, California sycamore, Fremont cottonwood, Mexican elderberry, red osier, white alder, and/or willows may be present. If shrubland, emergent trees may be present. Shrubs < 7 m; canopy continuous. Shrubs sparse under tree canopy. Ground layer variable.

Wetlands: habitats seasonally flooded, saturated. Water chemistry: fresh. Floodplains; low-gradient depositions along rivers, streams. Cowardin class: Palustrine forested or shrub-scrub wetland. The national inventory of wetland plants (Reed 1988) lists Sitka willow as a FAC+.

Distribution: o-NorCo, o-CenCo to Alaska.
Elevation: sea level-400 m.

NDDB/Holland type and status: **Marshes and swamps** (52000), **Riparian forests** (61000), **Riparian scrubs** (63000).
    Freshwater swamp (52600 *in part*) G1 S1.1.
    North Coast riparian forest (63100 *in part*) G3 S3.2.
    Central Coast riparian scrub (63200 *in part*) G3 S3.2.

Other types:
Barry type: G7411221 BALSI30.
Cheatham & Haller type: Bottomland woodlands and forest.
PSW-45 type: Willow series.
Thorne type: Riparian woodland.
WHR type: Freshwater emergent wetland.

General references: Brayshaw (1976), Paysen *et al.* (1980), Shanfield (1984).

Comments: Willow stands may or may not be dominated by a single species. If no dominant willow is present at low elevations, then place the stands in the Mixed willow series. Montane and subalpine willow stands are placed in separate classes since different willow species are restricted to those elevations [*see* Montane wetland shrub habitat, Subalpine wetland shrub habitat].

Stands of this series have environmental conditions similar to alder, cottonwood, and other willow series [*see* Black cottonwood series, Black willow series, Fremont cottonwood series, Hooker willow series, Pacific willow series, Red alder series, Red willow series, White alder series].

Species mentioned in text	
Bigleaf maple	*Acer macrophyllum*
Black cottonwood	*Populus balsamifera*
Black willow	*Salix gooddingii*
California sycamore	*Platanus racemosa*
Fremont cottonwood	*Populus fremontii*
Hooker willow	*Salix hookeriana*
Mexican elderberry	*Sambucus mexicana*
Pacific willow	*Salix lucida* ssp. *lasiandra*
Red alder	*Alnus rubra*
Red osier	*Cornus sericea*
Red willow	*Salix laevigata*
Sitka willow	*Salix sitchensis*
White alder	*Alnus rhombifolia*
Willows	*Salix* species

Plot-based descriptions: not available.

Subalpine fir series

Subalpine fir sole or dominant tree in canopy; Brewer spruce, Engelmann spruce, lodgepole pine, mountain hemlock, Pacific yew, Shasta fir, western white pine, and/or white fir may be present. Trees < 50 m; canopy continuous. Shrubs common or infrequent. Ground layer sparse or abundant (Plate 7).

Wetlands: soils intermittently flooded. Water chemistry: fresh. Streamsides. Cowardin class: Palustrine forested wetland. **Uplands:** raised terraces, morainal rocks and boulders, ridges. Soils granitic, metasediment-derived. The national inventory of wetland plants (Reed 1988) lists subalpine fir as a FACW*.

Distribution: m-KlaR, su-KlaR, OR, WA, Rocky Mountains.
Elevation: 1700-2200 m.

NDDB/Holland type and status: **Upper montane coniferous forests** (85000).
    Salmon-Scott enriched coniferous forest (85420) G1 S1.2.

Other types:
Barry type: G7411112 BABLA00.
Cheatham & Haller type: Klamath enriched conifer forest.
Thorne type: Enriched conifer forest.
WHR type: Klamath enriched mixed conifer.

General references: Subalpine fir range in CA (Griffin & Critchfield 1972); silvics (Alexander *et al.* 1990). Sawyer *et al.* (1970), Sawyer & Cope (1982), Sawyer (1987).

Comments: Subalpine fir forms a series in OR, WA, and Rocky Mountains, but its presence in CA was not verified until 1970 (Sawyer *et al.* 1970). Earlier reports in the manuals were based on specimens of Pacific silver fir collected at English Peak in the Marble Mountains (Haddock 1961). Since 1970 four additional stands have been identified; three in the Marble Mountains (Sawyer & Cope 1982). The newest and most southern stand occurs between Virginia Lake and Marvis Lake in the Scott Mountains (Sawyer 1987). Subalpine fir is rare (a CNPS List 2 plant) in CA (Skinner & Pavlik 1994).

Species mentioned in text	
Brewer spruce	*Picea breweriana*
Engelmann spruce	*Picea engelmannii*
Lodgepole pine	*Pinus contorta* spp. *murrayana*
Mountain hemlock	*Tsuga mertensiana*
Pacific silver fir	*Abies amabilis*
Pacific yew	*Taxus brevifolia*
Shasta fir	*Abies magnifica* var. *shastensis*
Subalpine fir	*Abies lasiocarpa*
Western white pine	*Pinus monticola*
White fir	*Abies concolor*

Plot-based descriptions: Keeler-Wolf (1984b, 1989f), Sawyer & Thornburgh (1971) in Keeler-Wolf (1990e) summarize surveys in Salmon Mountains including the Sugar Creek candidate RNA. Keeler-Wolf (1984b) maps stands dominated by subalpine fir in Sugar Creek. Sawyer & Thornburgh (1969) report density and size, and De Jager (1991) age patterns for stands in the Duck Lake Creek drainage in Siskiyou Co. m-KlaR, su-KlaR.

Tanoak series

Tanoak sole or dominant tree in canopy; black oak, California bay, canyon live oak, coast live oak, Coulter pine, madrone, and/ or sugar pine may be present. Trees < 75 m; canopy continuous, may be two-tiered. Shrubs infrequent or common. Ground layer sparse to abundant.

Uplands: all aspects. Soils mostly sandstone, schist-derived, well-drained.

Distribution: o-NorCo, i-CenCo, m-CenCo, l-KlaR, m-KlaR, m-SN.
Elevation: 100-1500 m.

NDDB/Holland type and status:s: **Broadleaved upland forests** (81000).
> Mixed evergreen forest (81100 *in part*) G4 S4.
> Tanoak forest (81400) G4 S4.

Other types:
Barry type: G74 G7411111 CLIDE00.
Cheatham & Haller type: Northern oak woodland.
PSW-45 type: Tanoak series.
Thorne type: Northern mixed evergreen forest.
WHR type: Montane hardwood forest.

General references: Tanoak ranges in CA (Griffin & Critchfield 1972); silvics (Tappeiner *et al.* 1990). Barbour (1988), Cooper (1922), McDonald *et al.* (1983), McDonald & Tappeiner (1990), Paysen *et al.* (1980), Sawyer & Thornburgh (1977b), Shreve (1927), Zuckerman (1990).

Comments: Tanoak is an important component in several series. The term "mixed evergreen forest" is most commonly associated with tanoak, but stands called mixed evergreen forest vary in composition (Barbour 1988b, Sawyer *et al.* 1977). The original concept of mixed evergreen forests was used to describe hardwood stands in the Santa Lucia Mountains where tanoak, madrone, live oaks and California bay grow. This was Cooper's (1922) broad sclerophyll formation.

Munz (1959) and Whittaker (1960) expanded the term mixed evergreen to include Douglas-fir - tanoak series stands. Azet *et al.* (1992) argue that tanoak is the primary naturally regenerating species in sw. OR, whereas Douglas-fir importance is a function of fire history (Azet & Wheeler 1982). Jimerson (1993) comes to the same conclusion for l-KlaR. The low elevation associations are called Tanoak series even though Douglas-fir is an important component.

Sawyer *et al.* (1977c) called stands in Nor-CA of Douglas-fir, tanoak, and madrone *Pseudotsuga/ hardwood* forests. Other authors (Keeler-Wolf 1990, Bingham & Sawyer 1991, 1993, Stuart 1993) refer to Douglas-fir/hardwood forests in Nor-CA. [The slash indicates the two tiered character of the stands].

Species mentioned in text	
Bigcone Douglas-fir	*Pseudotsuga macrocarpa*
Black oak	*Quercus kelloggii*
California bay	*Umbellularia californica*
California coffeeberry	*Rhamnus californica*
Canyon live oak	*Quercus chrysolepis*
Coast live oak	*Quercus agrifolia*
Coulter pine	*Pinus coulteri*
Douglas-fir	*Pseudotsuga menziesii*
Madrone	*Arbutus menziesii*
Poison-oak	*Toxicodendron diversilobum*
Sugar pine	*Pinus lambertiana*
Tanoak	*Lithocarpus densiflora*

The reason for not following Forest Service terminology (Jimerson 1993) here is to be able to recognize associations with tanoak dominance [Tanoak series] as different from stands where tanoak shares the canopy with Douglas-fir [Douglas-fir - tanoak series]. There is parallel classification of canyon live oak stands in CenCo and So-CA. There, canyon live oak sometimes mixes with Coulter pine or bigcone Douglas-fir [*see* Coulter pine-canyon live oak series, Bigcone Douglas-fir - canyon live oak series] and sometimes

is a sole dominant [*see* Canyon live oak series].
Differences in interpretation can be accommodated
by remembering that an association can be placed
in different categories by following the conventions
of each classification.

Plot-based descriptions: Griffin (1976), Borchert
1987) in Keeler-Wolf (1990e) define a Sugar pine -
tanoak/poison-oak association at South Fork of
Devil's Canyon (part of Cone Peak Gradient RNA)
in Monterey Co. m-CenCo; Sawyer (1981a) a
Tanoak/California coffeeberry association at Adorni
candidate RNA in Humboldt Co. l-KlaR; Gudmunds
& Barbour (1987) describe stands along Middle
Yuba River m-SN.

Utah juniper series

Utah juniper sole or dominant tree or as emergent trees over shrub canopy; singleleaf pinyon may be present. Bitterbrush, big sagebrush, ephedra, rabbitbrush, and/or hop-sage may be present. Trees < 15 m; canopy intermittent or open. Shrubs <1 m; continuous or intermittent. Ground layer sparse or grassy.

Uplands: pediments, slopes, ridges. Soils commonly well-drained.

Distribution: m-TraR, TraSN, WIS, MojD, m-DesR, NV.
Elevation: 1000-2800 m.

NDDB/Holland type and status: **Piñon and juniper woodlands** (72000).
> Great Basin pinyon juniper woodland (72121 *in part*) G4 S4.

Other types:
Barry type: G74 G7411112 BJUOS00.
Cheatham & Haller type: Nevadean pinyon-juniper woodland.
Rangeland type: SRM 414.
Thorne type: Pinyon-juniper woodland.
WHR type: Pine-juniper.

General references: Utah juniper ranges in CA (Griffin & Critchfield 1972). Meeuwig & Bassett (1983), Tueller *et al.* (1979), Vasek & Thorne (1977a).

Comments: Pinyon-juniper woodlands are better considered a collection of series. A problem arises in separation of stands belonging to the Utah juniper series from others where the species is only important or incidental. Pine and juniper share importance in the Singleleaf pinyon-Utah juniper series, and Utah juniper is incidental in the Singleleaf pinyon series.

Species mentioned in text	
Big sagebrush	*Artemisia tridentata*
Bitterbrush	*Purshia tridentata*
Black bush	*Coleogyne ramosissima*
Ephedra	*Ephedra* species
Hop-sage	*Grayia spinosa*
Rabbitbrush	*Chrysothamnus* species
Singleleaf pinyon	*Pinus monophylla*
Utah juniper	*Juniperus osteosperma*

Plot-based descriptions: not available.

Valley oak series

Valley oak sole or dominant tree in canopy; black oak, blue oak, California sycamore, coast live oak, and/or Oregon ash may be present. Trees < 30 m tall; canopy continuous, intermittent or open. Shrubs occasional and lianas common. Ground layer grassy (Plate 20).

Wetlands: soils intermittently flooded, seasonally saturated. Water chemistry: fresh. Floodplains. Cowardin class: Palustrine forested wetland. **Uplands:** valley bottoms; gentle slopes; summit valleys. Soils alluvial or residual. The national inventory of wetland plants (Reed 1988) lists valley oak as a FAC*.

Distribution: NorCo; CenCo; CenV; f-KlaR; f-CasR; f-SN; SoCo.
Elevation: sea level-775 m.

NDDB/Holland type and status: Riparian forests (61000), **Cismontane woodlands** (71000).
>   Great Valley valley oak riparian forest (61430) G1 S1.1.
>   Valley oak woodland (71130) G2 S2.1.

Other types:
Barry type: G74 G7411121 CQULO00.
Cheatham & Haller type: Valley bottomland woodland.
PSW-45 type: Valley oak series.
Thorne type: Foothill woodland.
WHR type: Valley oak woodland.

General references: Valley oak range in CA (Griffin & Critchfield 1972). Allen *et al.* (1989, 1991a), Allen-Diaz & Holtzman (1991), Danielsen (1990), Barbour (1988), Griffin (1976b, 1977), Holstein (1984), Pavlik *et al.* (1991), Paysen *et al.* (1980), Thompson (1961), Warner & Hendrix (1984), White (1966a).

Comments: CNPS is very interested in biology and conservation of oaks, especially valley oak (CNPS Oak Hardwood Committee 1989, Faber 1990, Plumb & Pillsbury 1987).

Species mentioned in text	
Black oak	*Quercus kelloggii*
Blue oak	*Quercus douglasii*
California coffeeberry	*Rhamnus californica*
California sycamore	*Platanus racemosa*
California wild grape	*Vitis californica*
Coast live oak	*Quercus agrifolia*
Oregon ash	*Fraxinus latifolia*
Poison-oak	*Toxicodendron diversilobum*
Valley oak	*Quercus lobata*

Plot-based descriptions: Griffin (1977) describes upland stands in the Santa Lucia Mountains, Keeler-Wolf (1989a) reports tree density, basal area, and frequency for different aged stands at Wagon Caves candidate RNA in Monterey Co. i-CenCo; Conard & Robichaux (1980) describe a riparian valley oak forest stand along the Cosumnes River, Sacramento Co., Knudsen (1987) describes stands at Boblaine Audubon Sanctuary in Sutter Co. CenV; Thomas (1987) describes stands in Santa Monica Mountains National Recreation Area in Los Angeles Co. SoCo; Allen *et al.* (1989, 1991a) define seven upland associations in Cis-CA.

Associations:
Allen *et al.* (1991a):
>   Black oak-valley oak/grass association,
>   Blue oak-valley oak/grass association,
>   Coast live oak-valley oak/poison-oak association,
>   Mixed oak-valley oak/poison-oak - California coffeeberry association,
>   Blue oak-valley oak/grass association,
>   Valley oak/grass association,
>   Valley oak-coast live oak/grass association.

Washoe pine series

Washoe pine sole or dominant tree in canopy; lodgepole pine, Jeffrey pine, ponderosa pine, red fir, western white pine, and/or white fir may be present. Trees < 25 m; canopy intermittent or open. Shrubs sparse. Ground layer grassy (Plate 26).

Uplands: all slopes, but most extensive on upper slopes. Soils granitic or volcanic-derived.

Distribution: m-CasR, su-CasR, m-SN, su-SN, m-WarR, su-WarR.
Elevation: 2100-2850 m.

NDDB/Holland type and status: Upper montane coniferous forests (85000).
Washoe pine-fir forest (85220) G1 S1.2.

Other types:
Barry type: G7411112 BPITO00.
Cheatham & Haller type: Washoe pine-fir forest
Thorne type: Washoe pine forest.
WHR type: Eastside pine.

General references: Washoe pine range in CA (Griffin & Critchfield 1972). Rundel *et al.* (1977).

Comments: It is unclear whether this species establishes a separate series, is a rare to important component in other series, or is found in distinct stands. The problem of hybridization between ponderosa pine, Jeffrey pine, and Washoe pine creates additional confusion [*see* Riegel *et al.* (1990) for a discussion of taxonomic problems in s. WarR].

Plot-based descriptions: Talley (1977), Keeler-Wolf (1989j) in Keeler-Wolf (1990e) describe stands at Babbitt Peak RNA in Sierra Co. m-SN; Riegel *et al.* (1990) define one association, Smith (1994) two associations in Modoc National Forest in Modoc Co. WarR.

Species mentioned in text

Desert snowberry	*Symphoricarpos longiflorus*
Jeffrey pine	*Pinus jeffreyi*
Lodgepole pine	*Pinus contorta* ssp. *murrayana*
Pinemat manzanita	*Arctostaphylos nevadensis*
Ponderosa pine	*Pinus ponderosa*
Red fir	*Abies magnifica* var. *magnifica*
Sticky starwort	*Pseudostellaria jamesiana*
Tailed lupine	*Lupinus caudatus*
Washoe pine	*Pinus washoensis*
Western white pine	*Pinus monticola*
White fir	*Abies concolor*

Associations:
Riegel *et al.* (1990):
  Washoe pine/tailed lupine association [as a Washoe pine phase of *Abies concolor/Lupinus caudatus* association].
Smith (1994):
  Washoe pine/desert snowberry/sticky starwort association [as White fir-Washoe pine/desert snowberry/starwort association],
  Washoe pine/pinemat manzanita association.

Water birch series

Water birch sole or dominant shrub or tree in canopy; arroyo willow, black cottonwood, Fremont cottonwood, narrowleaf willow, red willow, and/or shinning willow may be present. Trees < 30 m; canopy continuous. Shrubs sparse under canopy. Ground layer sparse (Plate 7).

Wetlands: habitats seasonally flooded, saturated. Water chemistry: fresh. Depositions along streams. Cowardin class: Palustrine forested or shrub-scrub wetland. The national inventory of wetland plants (Reed 1988) lists water birch as a FACW.

Distribution: e. m-KlaR, m-CasR, GB, DesR.
Elevation: 600-2500 m.

NDDB/Holland type and status: **Riparian forests** (61000), **Riparian scrubs** (63000).
    Modoc-Great Basin cottonwood-willow riparian forest (61610 *in part*) G2 S2.1.

Other types:
Thorne type: Riparian woodland.
WHR type: Freshwater emergent wetland.

General references: Holstein (1984).

Comments: Water birch produces distinct riparian corridors along streams in Trans-CA, but regional descriptions include it in general riparian descriptions. Stands of water birch have environmental conditions similar to cottonwood and willow series [*see* Arroyo willow series, Black cottonwood series, Fremont cottonwood series, Narrowleaf willow series, Red willow series].

Species mentioned in text	
Arroyo willow	*Salix lasiolepis*
Black cottonwood	*Populus balsamifera*
Fremont cottonwood	*Populus fremontii*
Narrowleaf willow	*Salix exigua*
Red willow	*Salix laevigata*
Shinning willow	*Salix lucida* ssp. *caudata*
Water birch	*Betula occidentalis*

Plot-based descriptions: not available.

Western hemlock series

Western hemlock sole or dominant tree in canopy; California bay, Douglas-fir, madrone, Port Orford-cedar, redwood, and/or tanoak may be present. Trees < 60 m tall; canopy continuous. Shrubs common or infrequent. Ground layer abundant or sparse (Plate 24).

Uplands: slopes; raised stream benches and terraces. Soils sandstone, schist-derived. The national inventory of wetland plants (Reed 1988) lists western hemlock as a FACU.

Distribution: o-NorCo, l-KlaR.
Elevation: sea level-400 m.

NDDB/Holland type and status: North Coast coniferous forests (82000).
   Western hemlock forest (82200) G4 S3.
   Douglas-fir - western hemlock forest (82410) G4 S2.1.

Other types:
Barry type: G7411112 BTSHE00.
Cheatham & Haller type: Western hemlock forest.
Thorne type: North coastal coniferous forest.
WHR type: Douglas-fir.

General references: Western hemlock range in CA (Griffin & Critchfield 1972); silvics (Packee 1990). Franklin (1988), Franklin & Dyrness (1973), Franklin *et al.* (1983), Harris (1980b), Olson *et al.* (1990), Roy, (1980), Zinke (1977).

Comments: This series is restricted in extent in Nor-CA, but is extensive in OR and WA. In CA stands of the series are mainly represented by stands logged two or more times. This species is also present, even important, in redwood and Sitka spruce stands as well [*see* Douglas-fir -tanoak series, Redwood series, Sitka spruce series].

Species mentioned in text	
California bay	*Umbellularia californica*
Douglas-fir	*Pseudotsuga menziesii*
Madrone	*Arbutus menziesii*
Port Orford-cedar	*Chamaecyparis lawsoniana*
Redwood	*Sequoia sempervirens*
Sitka spruce	*Picea sitchensis*
Tanoak	*Lithocarpus densiflora*
Western hemlock	*Tsuga heterophylla*

Plot-based descriptions: Westman & Whittaker (1975) define a coastal ravine Spruce-hemlock type in Mendocino Co., Keeler-Wolf (1987d) in Keeler-Wolf (1990e) a Western hemlock subtype of Port-Orford-cedar - Douglas-fir - western hemlock forests at Upper Goose Creek candidate RNA in Del Norte Co., NDDB has plot data on a Mad River stand in Humboldt Co. o-NorCo.

Western juniper series

Western juniper sole or dominant tree canopy; black oak, Jeffrey pine, Oregon white oak, ponderosa pine, Washoe pine, and/or white fir may be present. Emergent trees may be present over a shrub canopy; antelope bitterbrush, big sagebrush, black sagebrush, curlleaf mountain-mahogany, and/or rabbitbrush may be present. Trees < 20 m. Trees scattered. Shrubs common. Ground layer grassy (Plate 25).

Uplands: gentle slopes; valleys. Soils bedrock basalt, aeolian, colluvial, or alluvial-derived. Indurated layers, rock fragments may be may be present.

Distribution: e. m-KlaR, m-CasR, ModP, m-WarR, TraSN, ID, OR, WA.
Elevation: 900-1700 m.

NDDB/Holland type and status: **Piñon and juniper woodlands** (72000).
    Northern juniper woodland (72110) G4 S4.
    Great Basb juniper woodland (72123) G4 S4.

Other types:
Barry type: G74 G7411112 BJUOC00.
Cheatham & Haller type: Northern juniper woodland.
PSW-45 type: Western juniper series.
Rangeland type: SRM 107.
Thorne type: Northern juniper woodland.
WHR type: Juniper.

General references: Western juniper range in CA (Griffin & Critchfield 1972), silvics (Dealy 1990). Adams (1975), Dealy *et al.* (1977), Driscoll (1964), Eddleman (1987), Hall (1977), Martin (1980a), Paysen *et al.* (1980), Vasek (1966), Vasek & Thorne (1977a), Volland (1982).

Comments: There are taxonomic and ecological challenges involved with *Juniperus occidentalis*. The species has two subspecies, western juniper (*J. o.*

ssp. *occidentalis*) and mountain juniper (*J. o.* ssp. *australis*). The two subspecies have generally different ranges and ecology, so they represent two different series [*see* Mountain juniper series]. There is some overlap in s. ModP and n. TraSN, where Vasek & Thorne (1977a) report that mountain juniper tends to occur on ridges and escarpments, while western juniper occurs in valleys. Occasional western juniper trees may grow in stands of other series [*see* Jeffrey pine series, Oregon white oak series, Ponderosa pine series, Washoe pine series].

Species mentioned in text	
Antelope bitterbrush	*Purshia tridentata* var. *tridentata*
Big sagebrush	*Artemisia tridentata*
Black oak	*Quercus kelloggii*
Black sagebrush	*Artemisia nova*
Curlleaf mountain-mahogany	*Cercocarpus ledifolius*
Jeffrey pine	*Pinus jeffreyi*
Low sagebrush	*Artemisia arbuscula*
Mountain juniper	*Juniperus occidentalis* ssp. *australis*
Oregon white oak	*Quercus garryana*
Ponderosa pine	*Pinus ponderosa*
Rabbitbrush	*Chrysothamnus nauseosus*
Washoe pine	*Pinus washoensis*
Western juniper	*Juniperus occidentalis* ssp. *occidentalis*
White fir	*Abies concolor*

Plot-based descriptions: Vasek & Thorne (1977a) report density for a stand in Siskiyou Co., Keeler-Wolf (1984d) in Keeler-Wolf (1990e) describes a stand at Devil's Garden RNA in Modoc Co. ModP.

Western white pine series

Western white pine sole or dominant tree in canopy; Douglas-fir, foxtail pine, Jeffrey pine, knobcone pine, lodgepole pine, red fir, Shasta fir, and/or white fir may be present. Trees < 70 m; canopy intermittent or open. Shrubs common. Ground layer sparse or abundant (Plate 27).

Uplands: all slopes, but most extensive on plateaus and upper slopes; raised stream benches and terraces. Soils granitic or ultramafic-derived. The national inventory of wetland plants (Reed 1988) lists western white pine as a FACU.

Distribution: w. l-KlaR, m-KlaR, su-KlaR, su-CasR, su-SN.
Elevation: 200-2300 m.

NDDB/Holland type and status: Lower montane coniferous forests (84000), **Upper montane coniferous forests** (85000).
Ultramafic western white pine forest (84160) G3 S3.

Other types:
Barry type: G7411112 BPIMO00.
Cheatham & Haller type: Klamath enriched conifer forest.
Thorne type: Enriched conifer forest.
WHR type: Klamath mixed conifer.

General references: Western white pine range in CA (Griffin & Critchfield 1972); silvics (Graham 1990). Boyd (1980b), Kruckeberg (1984), Sawyer & Thornburgh (1977b).

Comments: Western white pine grows in a wide range of mountain settings. It is also a common secondary species in several series. At elevations as low as 200 m in Del Norte Co. w. l-KlaR, western white pine dominates self-replacing stands on deep, partially serpentinized peridotite (Duebendorfer 1987). This vegetation was first described by Whittaker (1960) as the shrubland half of a patchy mosaic that includes Jeffrey pine woodland [*see* Jeffrey pine series]. Duebendorfer (1987) concluded that the mosaic is caused by soil differences. This shrubland grows on western white pine forest sites disturbed by fire and white pine rust. On relatively undisturbed sites, western white pine shares the canopy with Douglas-fir, knobcone pine, and lodgepole pine. In many areas the "Del Norte race" of *Pinus contorta* is common (Griffin & Critchfield 1972). These stands are being affected by white pine blister rust during the 1990s drought.

Western white pine may be an important component of the Foxtail pine series and may dominate on ultramafic or granitic-derived soils (Sawyer & Thornburgh 1977b) in su-KlaR, Keeler-Wolf (1991b) at Mountaineer Creek candidate RNA in Tulare Co. su-SN.

Species mentioned in text

Angelica	*Angelica arguta*
Bear-grass	*Xerophyllum tenax*
Bush tanoak	*Lithocarpus densiflora* var. *echinoides*
Douglas-fir	*Pseudotsuga menziesii*
Dwarf silktassel	*Garrya buxifolia*
Foxtail pine	*Pinus balfouriana*
Greenleaf manzanita	*Arctostaphylos patula*
Jeffrey pine	*Pinus jeffreyi*
Knobcone pine	*Pinus attenuata*
Lodgepole pine	*Pinus contorta* ssp. *murrayana*
Ocean spray	*Holodiscus discolor*
Red fir	*Abies magnifica* var. *magnifica*
Shasta fir	*Abies magnifica* var. *shastensis*
Western white pine	*Pinus monticola*
Western coffeeberry	*Rhamnus californica* var. *occidentalis*
White fir	*Abies concolor*

Plot-based descriptions: Keeler-Wolf (1986a) in Keeler-Wolf (1990e) defines one association at L. E. Horton candidate RNA, Duebendorfer (1987) two associations at nearby Pine Flat Mountain in Del Norte Co., Simpson (1980) one association in s. Siskiyou Mountains in Humboldt Co., Sawyer & Thornburgh (1977b) one association, Whipple & Cope (1979) in

Keeler-Wolf (1990e) one association at Mount Eddy candidate RNA in Siskiyou Co. KlaR; Talley (1977) in Keeler-Wolf (1990e) describes stands at Babbitt Peak RNA in Sierra Co. su-SN.

Associations:
Duebendorfer (1987):
    Western white pine/bush tanoak association
    [as Western white pine shrubland, = Dwarf forest
    (Keeler-Wolf 1986a)].
Sawyer & Thornburgh (1977b):
    Western white pine/ocean spray association.
Simpson (1980):
    Pine/bear-grass association.
Whipple & Cope (1979);
    Western white pine/angelica association.

White alder series

White alder sole or dominant tree in canopy; bigleaf maple, California sycamore, and/or Oregon ash may be present. Trees < 35 m; canopy intermittent or open. Shrubs common or infrequent. Ground layer variable (Plate 18).

Wetlands: soils intermittently flooded, saturated. Water chemistry: fresh. Riparian corridors; floodplains subject to high-intensity flooding; incised canyons, river and stream low-flow margins; seeps; stream and river banks, terraces. Cowardin class: Palustrine forested wetland. The national inventory of wetland plants (Reed 1988) lists red alder as a FACW.

Distribution: i-NorCo, m-NorCo, CenCo, l-KlaR, f-KlaR, f-CasR, m-CasR, f-SN, m-SN, SoCo, m-TraR, m-PenR.
Elevation: sea level-2500 m.

NDDB/Holland type and status: Riparian forests (61000) Riparian woodlands (62000).
> White alder riparian forest (61510) G3 S3.
> Southern sycamore-alder riparian woodland (62400 *in part*) G4 S4.

Other types:
Barry type: G7411121 BALRH00.
Cheatham & Haller type: Northern riparian woodland, Southern riparian woodland.
PSW-45 type: Alder series.
Rangeland type: SRM 203.
Thorne type: Riparian woodlands.
WHR type: Valley foothill riparian.

General references: White alder range in CA (Griffin & Critchfield 1972). Bowler (1989), Capelli & Stanley (1984), Faber *et al.* (1989), Ferren (1989), Holstein (1984), McBride (1994), Paysen *et al.* (1980).

Comments: Keeler-Wolf (1990e) qualitatively describes stands at Doll Basin candidate RNA in Mendocino Co. m-NorCo; Keeler-Wolf (1990d) at Indian Creek candidate RNA in Tehama Co. m-CasR.

Species mentioned in text	
Bigleaf maple	*Acer macrophyllum*
California polypody	*Polypodium californicum*
California sycamore	*Platanus racemosa*
Douglas-fir	*Pseudotsuga menziesii*
Fragrant bedstraw	*Galium trifolium*
Himalaya berry	*Rubus discolor*
Indian rhubarb	*Darmera peltata*
Miner dogwood	*Cornus sessilis*
Mulefat	*Baccharis salicifolia*
Oregon ash	*Fraxinus latifolia*
Port Orford-cedar	*Chamaecyparis lawsoniana*
Red osier	*Cornus sericea*
Spikenard	*Aruncus dioicus*
Tanoak	*Lithocarpus densiflora*
White alder	*Alnus rhombifolia*

Plot-based descriptions: Borchert *et al.* (1988) define one association in Los Padres National Forest CenCo; Jimerson (1993) one association in Six Rivers National Forest, Taylor & Teare (1979a) in Keeler-Wolf (1990e) two associations at Manzanita Creek candidate RNA, Taylor & Teare (19-79b) in Keeler-Wolf (1990e) one association at Smoky Creek candidate RNA, Taylor (1975a,b) in Keeler-Wolf (1990e) one association at South Fork Mountain in Trinity Co., Stuart *et al.* (1992) two associations at Castle Crags State Park in Siskiyou Co. KlaR; Taylor & Randall (1977a) in Keeler-Wolf (1990e) one association at Peavine Point RNA in El Dorado Co. m-SN; Brothers (1985) reports densities along the San Gabriel River in Los Angeles Co. SoCo; White (1994a) defines one association at Cleghorn Canyon candidate RNA in San Bernardino Co. i. m-TraR.

Associations:
Borchert *et al.* (1988):
     White alder/California polypody association.
Jimerson (1993):
     White alder/spikenard association [as Tanoak-
     Port Orford-cedar/spikenard association].
Stuart *et al.* (1992):
     White alder-bigleaf maple association,
     Douglas-fir -white alder/Himalaya berry
     association.
Taylor (1975a,b):
     White alder/Indian rhubarb association
     [=*Carex senta/Darmera peltata* association,
     (Taylor & Randall 1977a)].
Taylor & Teare (1979a):
     White alder/miner dogwood association,
     White alder/red osier association.
Taylor & Teare (1979b):
     White alder/fragrant bedstraw association.
White (1994a):
     White alder/mulefat association.

White fir series

White fir sole or dominant tree in canopy; black oak, Douglas-fir, incense-cedar, Jeffrey pine, lodgepole pine, ponderosa pine, red fir, Shasta fir, singleleaf pinyon, and/or sugar pine may be present. Trees < 60 m; canopy continuous or intermittent. Shrubs infrequent or common. Ground layer sparse or abundant (Plate 27).

Uplands: slopes, all aspects; raised stream benches and terraces. Soils well-drained. The national inventory of wetland plants (Reed 1988) does not list white fir.

Distribution: m-NorCo, m-KlaR, m-CasR, m-SN, m-TraR, m-PenR, m-WarR, TraSN, m-DesR, Baja CA.
Elevation: 1400-2700 m.

NDDB/Holland type and status: **Lower montane coniferous forests** (84000), **Upper montane coniferous forests** (85000).
    Sierran mixed coniferous forest (84230 *in part*) G4 S4.
    Sierran white fir forest (84240) G4 S4.
    Southern California white fir forest (85320) G4 S4.
    Desert mountain white fir forest (85330) G4 S1.2.

Other types:
Barry type: G7411112 BABCO00.
Cheatham & Haller type: Sierran white fir forest, Southern California white fir forest.
PSW-45 type: White fir series.
Thorne type: Mixed conifer forest, White fir forest, White fir-sugar pine forest.
WHR type: White fir.

General references: White fir range in CA (Griffin & Critchfield 1972); silvics (Laacke 1990a). Barbour (1988), Franklin (1988), Gordon (1980b), Hendrickson & Prigge (1975), Laacke & Fiske (1983a), Minnich (1987), Paysen *et al.* (1980), Rundel *et al.* (1977), Sawyer & Thornburgh (1977b), Stone & Sumida

(1983), Thorne (1982), Vasek (1985).

Species mentioned in text	
Alaska onion grass	*Melica subulata*
American vetch	*Vicia americana*
Bear-grass	*Xerophyllum tenax*
Bigleaf maple	*Acer macrophyllum*
Bitter cherry	*Prunus emarginata*
Black oak	*Quercus kelloggii*
Bracken	*Pteridium aquilinum*
Brewer spruce	*Picea breweriana*
Bush chinquapin	*Chrysolepis sempervirens*
Canyon live oak	*Quercus chrysolepis*
Chinquapin	*Chrysolepis chrysophylla*
Creeping snowberry	*Symphoricarpos mollis*
Douglas-fir	*Pseudotsuga menziesii*
Grand fir	*Abies grandis*
Hazel	*Corylus cornuta*
Heartleaf arnica	*Arnica cordifolia*
Hooker fairybells	*Disporum hookeri*
Huckleberry oak	*Quercus vaccinifolia*
Incense-cedar	*Calocedrus decurrens*
Jeffrey pine	*Pinus jeffreyi*
Kelloggia	*Kelloggia galioides*
Leafless shinleaf	*Pyrola picta*
Little Oregon-grape	*Berberis nervosa*
Little prince's-pine	*Chimaphila menziesii*
Lodgepole pine	*Pinus contorta* ssp. *murrayana*
Madrone	*Arbutus menziesii*
Mahala carpet	*Ceanothus prostratus*
Mountain maple	*Acer glabrum*
Pacific dogwood	*Cornus nuttallii*
Pennyroyal	*Monardella odoratissima*
Pinemat manzanita	*Arctostaphylos nevadensis*
Ponderosa pine	*Pinus ponderosa*
Port Orford-cedar	*Chamaecyparis lawsoniana*
Prince's-pine	*Chimaphila umbellata*

Continued on next page

Comments: *Abies concolor* has two varieties [*A. c.* var. *concolor* (Rocky Mountain white fir or race) and *A. c.* var. *lowiana* (Sierran white fir or race (Hamrick & Libby 1972)]. Rocky Mountain white fir grows in the m-TraR, m-PenR, m-DesR,; Sierran white fir in NorCo, KlaR, CasR, SN, GB. In the Flora of North America the races are awarded species rank, and populations in KlaR are considered *A. lowiana* (Flora of America

Committee 1993). The closely related *A. grandis* (grand fir) grows in o-NorCo, and transition populations between Sierra white fir and grand fir occur in w. m-NorCo, w. m-KlaR (Griffin & Critchfield 1972). Since the two varieties have a similar ecology (Barbour 1988a, Laacke 1990b), they are included in this series. Varietal differences can be handled at the association level.

Species mentioned in text	
Rattlesnake-plantain	*Goodyera oblongifolia*
Red fir	*Abies magnifica* var. *magnifica*
Rhododendron	*Rhododendron macrophyllum*
Ross sedge	*Carex rossii*
Sadler oak	*Quercus sadleriana*
Serviceberry	*Amelanchier alnifolia*
Shasta fir	*Abies magnifica* var. *shastensis*
Singleleaf pinyon	*Pinus monophylla*
Solomon's seal	*Maianthemum racemosum*
Sticky starwort	*Pseudostellaria jamesiana*
Sugar pine	*Pinus lambertiana*
Thimbleberry	*Rubus parviflorus*
Thinleaf huckleberry	*Vaccinium membranaceum*
Threeleaf anemone	*Anemone deltoidea*
Trail penstemon	*Penstemon anguineus*
Trail plant	*Adenocaulon bicolor*
Trillium	*Trillium ovatum*
Vanilla leaf	*Achlys triphylla*
Vine maple	*Acer circinatum*
White-veined shinleaf	*Pyrola picta*
Wild rose	*Rosa gymnocarpa*
White fir	*Abies concolor*
Begins on previous page	

Plot-based descriptions: Keeler-Wolf (1986c, 1988b) in Keeler-Wolf (1990e) describes stands at Doll Basin candidate RNA, Muldavin (1982) one association at Big Butte area in Mendocino Co., Thornburgh (1981), Keeler-Wolf (1989i) in Keeler-Wolf (1990e) stands at Ruth candidate RNA in Humboldt Co. m-NorCo.; Laidlaw-Holmes (1981) defines one association at South Fork Mountain in Trinity Co., Waddell (1982) two associations in Yolla Bolly Mountains in Tehama and Trinity Cos. in NorCo & KlaR; Imper (1988a) in Keeler-Wolf (1990e) three associations at Haypress Meadows candidate RNA in Siskiyou Co., Sawyer (1981b) in Keeler-Wolf (1990e) two associations at North Trinity Mountain candidate RNA in Humboldt

Co., Taylor & Teare (1979a) in Keeler-Wolf (1990e) one association inManzanita Creek candidate RNA in Trinity Co., Jimerson (1993) associations in Six Rivers National Forest, Sawyer & Thornburgh (1977b) five associations in m-KlaR generally; Conard & Robichaux (1980) in Keeler-Wolf (1990e) describe stands at Soda Ridge candidate RNA in Tehama Co. m-CasR; Conard & Radosevich (1982) report tree density and basal area for stands in Sierra Co., Jensen (1992) tree density, basal area, and frequency for stands and understory frequency at Home Camp candidate RNA in Fresno Co., Keeler-Wolf (1991b) tree density, basal area, and frequency for white fir-sugar pine stands at Mountaineer Creek candidate RNA in Tulare Co., Fites (1993) define 9 associations [as Mixed conifer associations] in Lassen, Plumas, Tahoe, Eldorado National Forests m-CasR, m-SN, Talley (1977), Keeler-Wolf (1989j) in Keeler-Wolf (1990e) describe stands at Babbitt Peak RNA in Sierra Co., Keeler-Wolf (1985b, 1988c) in Keeler-Wolf (1990e) stands at Mud Lake RNA in Plumas Co., Talley (1976b) in Keeler-Wolf (1990e) stands at Onion Creek in Placer Co., Taylor & Randall (1977b) in Keeler-Wolf (1990e) stands at Station Creek RNA in El Dorado Co., Keeler-Wolf (1985c, 1989l) in Keeler-Wolf (1990e) stands at Bell Meadow RNA in Calaveras Co. m-SN; Burke (1992b) reports cover for stands at Horse Meadows candidate RNA in San Bernardino Co. i. m-TraR; Jensen & Schierenback (1990), Schierenback & Jensen (1994) tree density, basal area, and frequency at Raider Basin candidate RNA in Modoc Co. WarR; Taylor (1980) in Keeler-Wolf (1990e) defines two associations at Indiana Summit RNA in Mono Co. TraSN.

Associations:
Fites (1993):
    White fir/bush chinquapin association,
    White fir/creeping snowberry/kelloggia association,
    White fir-Pacific dogwood/bush chinquapin association,
    White fir-Pacific dogwood/hazel association,
    White fir-Pacific dogwood/trail plant association,
    White fir/Ross sedge association,
    White fir/Solomon's seal-Hooker fairybells association,
    White fir/trail plant association,
    White fir/vine maple-bush chinquapin association.

Imper (1988a):
> White fir/pinemat manzanita association,
> White fir/prince's-pine association,
> White fir/vanilla leaf association.

Jimerson (1993):
> Port Orford-cedar - Shasta fir/Sadler oak-thinleaf huckleberry association,
> Shasta fir-white fir/mountain maple association,
> Shasta fir-white fir/pinemat manzanita association,
> Shasta fir-white fir/trail penstemon-pennyroyal association,
> White fir/heartleaf arnica association,
> White fir/creeping snowberry association,
> White fir - Douglas-fir/beargrass association,
> White fir -Douglas-fir/bigleaf maple association,
> White fir - Douglas-fir/heartleaf arnica association,
> White fir -Douglas-fir - canyon live oak association,
> White fir -Douglas-fir/Sadler oak association,
> White fir - Douglas-fir/Sadler oak-pinemat manzanita association,
> White fir - Douglas-fir/mountain maple association,
> White fir - Douglas-fir/rhododendron-Sadler oak association,
> White fir - Douglas-fir/vanilla leaf association,
> White fir - Douglas-fir/wild rose-creeping snowberry association,
> White fir - incense-cedar - black oak association,
> White fir - incense-cedar/creeping snowberry association,
> White fir - incense-cedar/white-veined shinleaf association,
> White fir/little prince's-pine - white-veined shinleaf association,
> White fir/serviceberry association,
> White fir/wild rose association,
> White fir/wild rose-creeping snowberry association,
> White fir-Brewer spruce/Sadler oak-thinleaf huckleberry association,
> White fir-chinquapin association,
> White fir-chinquapin-sugar pine/prince's-pine association,
> White fir-Douglas-fir/Alaska oniongrass association,
> White fir-Douglas-fir/bear-grass association,
> White fir-Douglas-fir/bigleaf maple association,
> White fir-Douglas-fir/heartleaf arnica association,
> White fir-Douglas-fir/Sadler oak association,
> White fir-Douglas-fir/Sadler oak-huckleberry oak association,
> White fir-Douglas-fir/Sadler oak-pinemat manzanita association,
> White fir-Douglas-fir/Sadler oak-rhododendron association,
> White fir-Douglas-fir/hazel association,
> White fir-Douglas-fir/huckleberry oak association,
> White fir-Douglas-fir/mountain maple association,
> White fir-Douglas-fir/thimbleberry association,
> White fir-Douglas-fir/vanilla leaf association,
> White fir-Douglas-fir/vine maple association,
> White fir-Douglas-fir/wild rose-twinflower-creeping snowberry association,
> White fir-Douglas-fir/wild rose-twinflower association,
> White fir-incense-cedar/creeping snowberry association,
> White fir-incense-cedar/white-veined shinleaf association,
> White fir-Shasta fir/Sadler oak association,
> White fir-Shasta fir/white-veined shinleaf association,
> White fir-Shasta fir/threeleaf anemone association.

Laidlaw-Holmes (1981):
> White fir/huckleberry oak association [including Chinquapin phase, Madrone phase].

Sawyer (1981b):
> White fir/bracken association,
> White fir/Sadler oak association.

Sawyer & Thornburgh (1977b):
> White fir/American vetch association,
> White fir/little Oregon-grape association,
> White fir/prince's-pine association,
> White fir/mahala carpet association,
> White fir/trillium association.

Taylor & Randall (1977b):
> White fir/bitter cherry association,
> White fir/rattlesnake-plantain association.

Taylor & Teare (1979b):
> White fir/mountain maple association [including White fir association].

Waddell (1982):
> White fir/sticky starwort association
> White fir/white-veined shinleaf association.

Whitebark pine series

Whitebark pine sole or dominant tree in canopy; lodgepole pine, mountain hemlock, red fir, Shasta fir, and/or western white pine may be present. Trees < 30 m; canopy intermittent or open. Shrubs infrequent or common. Ground layer sparse or abundant (Plate 25).

Uplands: all slopes, but most extensive on crests, ridges, summits, upper slopes to forestline and above. Soils loose, well-drained, may be rocky.

Distribution: su-KlaR, su-CasR, su-SN, su-WarR.

Elevation: 2220-3660 m.

NDDB/Holland type and status: **Subalpine coniferous forests** (86000).
> Whitebark pine-mountain hemlock forest (86210 *in part)* G4 S4.
> Whitebark pine-lodgepole pine forest (86220 *in part)* G4 S4.
> Whitebark pine forest (86600) G4 S4.

Other types:
Barry type: G7411112 BPIAL00.
Cheatham & Haller type: Whitebark pine-mountain hemlock, Whitebark pine-lodgepole pine forest.
Thorne type: Whitebark pine-mountain hemlock forest.
WHR type: Subalpine conifer.

General references: Whitebark pine range in CA (Griffin & Critchfield 1972); silvics (Arno & Hoff 1990). Arno (1980), Barbour (1988), Franklin (1980a), Rundel *et al.* (1977), Sawyer & Thornburgh (1977b).
Comments: Whitebark pine grows at timberline in many CA mountains. "Timberline" is a vague term used to describe the elevation of the last stand of trees on the mountain (forestline), or the last erect tree (treeline) or the last shrub even if it is < 1 dm tall (krummholzline) (Price 1981). Forestline is preferred by many ecologists to differentiate subalpine forests from the herb-dominated vegetation of the alpine zone with scattered trees or shrubs. But in SN, forests are open at their upper elevations making even forestline difficult to interpret. This Whitebark pine series defini-tion includes both forests and woodlands. Individual trees or krummholz occur in the alpine zone.

Species mentioned in text	
California needlegrass	*Stipa californica*
Davidson penstemon	*Penstemon davidsonii*
Foxtail pine	*Pinus balfouriana*
Lodgepole pine	*Pinus contorta* var. *murrayana*
Mountain hemlock	*Tsuga mertensiana*
Ocean spray	*Holodiscus discolor*
Red fir	*Abies magnifica* var. *magnifica*
Shasta fir	*Abies magnifica* var. *shastensis*
Slender penstemon	*Penstemon gracilentus*
Western white pine	*Pinus monticola*
Wheeler bluegrass	*Poa wheeleri*
Whitebark pine	*Pinus albicaulis*
Woody sandwort	*Arenaria pumicola*

Plot-based descriptions: Keeler-Wolf (1987a) in Keeler-Wolf (1990e) describes stands at Crater Creek candidate RNA, Keeler-Wolf (1984b, 1989f), Sawyer & Thornburgh (1971) in Keeler-Wolf (1990e) at Sugar Creek candidate RNA, Sawyer & Thornburgh (1977b) define one association in Siskiyou Co. su-KlaR; Feidler (1987), Keeler-Wolf (1989g) in Keeler-Wolf (1990e) describe stands at Antelope Creek candidate RNA in Siskiyou Co. su-CasR; Tomback (1986) reports post-fire tree regeneration in Yosemite National Park, Taylor (1984) in Keeler-Wolf (1990e) defines two associations at Harvey Monroe Hall RNA in Mono Co. su-SN; Talley (1978) in Keeler-Wolf (1990e) describes stands at Sentinel Meadow RNA in Inyo Co. TraSN; Riegel *et al.* (1990) define three associations in the South Warner Mountains in Modoc Co. s. su-WarR.

Associations:
Riegel *et al.* (1990):
> Whitebark pine/California needlegrass association,
> Whitebark pine/slender penstemon association,
> Whitebark pine/woody sandwort association
> [as Whitebark pine/*Arenaria aculeata* association].

Sawyer & Thornburgh (1977b):
> Whitebark pine/ocean spray association.

Taylor (1984):
> Whitebark pine/Wheeler bluegrass association
> [as *Pinus albicaulis*/*Poa nervosa* association],
> Whitebark pine/Davidson penstemon association.

Unique Stands

The reader encounters the phrase "See the UNIQUE STANDS section" in the series keys. These stands are listed and described here using the same format as used for series.

Reasons for separating unique stands from series have their foundations in our philosophy of vegetation science. Any stand [an actual piece of vegetation in which plant composition and structure are uniform] is unique in some way from all other stands. What allows ecologists to recognize vegetation types is the fact that many stands are quite similar in species composition and structure.

Stands of a single vegetation type have what ecologists call "redundancy" (Gauch 1982). This redundancy has two aspects; stands are 1) consistently similar among themselves for one vegetation type, and 2) constantly distinct from stands of another vegetation type. In this way the ecologist can distinguish and define different series and associations.

Redundancy should be a feature not only of common or extensive vegetation types, but of rare ones as well. Following this principle, the ecologist can distinguish between a rare vegetation type and a rare species which does not demonstate redundancy.

Based on this outlook, we have divided this section into three categories: 1) stands that are really unique in species composition or *one of a kind* locales, 2) stands defined by the presence of a CNPS-listed plant, and 3) stands that are structurally distinctive and rare.

One of a kind stands:
Enriched stands in the Klamath Mountains,
Stands on San Benito Mountain.

CNPS-listed plant stands:
Some descriptions of NDDB vegetation types (Holland 1986) appear to use *the presence of a rare plant* as the rule for their definition. If the presence of a rare species were the measure for defining vegetation types, a fast scan of the California Native Plant Society's *Inventory of rare and endangered vascular plants of California* is all that would be needed; each rare plant would get its own community. Such an approach is not worthwhile.

CNPS-listed species may be restricted to rare vegetation types or may be a part of common ones. We can decide whether the following proposed forests, woodlands, and dune vegetation uphold as vegetation types as data are collected by CNPS members and others.

CYPRESS FORESTS
Baker cypress stands,
Cuyamaca cypress stands,
Gowen cypress stands,
Monterey cypress stands,
Piute cypress stands,
Santa Cruz cypress stands,
Tecate cypress stands.
DESERT WOODLANDS
All-thorn stands,
Crucifixion-thorn stands,
Elephant tree stands.

DUNES
Stands at Antioch dunes.
FORESTS
Alaska yellow-cedar stands,
Catalina ironwood stands,
Hinds walnut stands,
Pacific silver fir stands,
Torrey pine stands,
Twoleaf pinyon stands.

Distinctive stands:

As the state's native vegetation is altered, certain facets of species and vegetation types may become rare. In the case of species, stands of old or large specimens may be recognized to aid in their conservation. In the case of vegetation, the old-growth stage, for example, may be rare even though the vegetation type is not. At this time we have distinguished only one stand dominated by tree-sized specimens of a typical shrub species as a member of this group.

DISTINCTIVE STANDS
Hollyleaf cherry stands

Alaska yellow-cedar stands

Alaska yellow-cedar sole or dominant tree in canopy; Brewer spruce, Douglas-fir, ground juniper, lodgepole pine, mountain hemlock, noble fir, Port Orford-cedar, Pacific yew, western white pine, and/or white fir may be present. Trees < 30 m; canopy continuous. Shrubs absent. Ground layer sparse.

Wetlands: soils intermittently flooded. Water chemistry: fresh. Bottomlands and streamsides. Cowardin class: Palustrine forested wetland. **Uplands:** steep, north-facing slopes, ridges. Soils granitic or ultramafic-derived. The national inventory of wetland plants (Reed 1988) lists Alaska yellow-cedar as a FAC.

Distribution: w. m-KlaR, su-KlaR, OR, WA, AK, Canada.
Elevation: 1500-2150 m.

NDDB/Holland type and status: Upper montane coniferous forests (85000).
 Siskiyou enriched coniferous forest (85410) G1 S1.2.

Other types:
Barry type: G7411112 BCHN000.
Cheatham & Haller type: Klamath enriched conifer forest.
Thorne type: Enriched conifer forest.
WHR type: Klamath enriched mixed conifer.

General references: Alaska yellow-cedar range in CA (Griffin & Critchfield 1972); silvics (Harris 1990a). Franklin (1988), Sawyer & Thornburgh (1977b).

Comments: An Alaska yellow-cedar series is recognized in OR, WA, BC and AK. In CA, Alaska yellow-cedar associates with local series rather than forming one.

Griffin & Critchfield (1972) map seven stands in the Siskiyou Mountains. The species is probably more common; the very rugged, unroaded, and untrailed Siskiyou Mountains need more exploration. Much of the probable range is in the Siskiyou Wilderness. In addition, Alaska yellow-cedar is easily confused with Port Orford-cedar, which grows in the area. Alaska yellow-cedar is rare (a CNPS List 4 plant) in CA (Skinner & Pavlik 1994).

Species mentioned in text	
Alaska yellow-cedar	*Chamaecyparis nootkatensis*
Brewer spruce	*Picea breweriana*
Douglas fir	*Pseudotsuga menziesii*
Ground juniper	*Juniperus communis*
Lodgepole pine	*Pinus contorta* spp. *murrayana*
Mountain hemlock	*Tsuga mertensiana*
Noble fir	*Abies procera*
Pacific yew	*Taxus brevifolia*
Port Orford-cedar	*Chamaecyparis lawsoniana*
Western white pine	*Pinus monticola*
White fir	*Abies concolor*

Plot-based descriptions: not available.

All-thorn stands

All-thorn sole or dominant tree in canopy; brittlebush, creosote bush, and/or white bursage may be present. Trees < 5 m. Trees scattered. Shrubs common. Ground layer herbs seasonally present.

Uplands: sandy areas. Active, partially stabilized dunes; partially stabilized sand fields.

Distribution: ColD, AZ, Mexico.
Elevation: sea level-650 m.

NDDB/Holland type and status: **Sonoran thorn woodlands** (75000).
    All-thorn woodland (75300) G3 S1.1.

Other types:
Cheatham & Haller type: Low desert scrub.
Thorne type: Creosote bush scrub.
WHR type: Desert scrub.

General references: MacMahon (1988).

Comments: Five populations are known from the Chocolate Mountains in Imperial Co. ColD. Descriptions suggest that all-thorn associates with local series rather than forming one. All-thorn is rare (a CNPS List 2 plant) in CA (Skinner & Pavlik 1994).

Species mentioned in text	
All-thorn	*Koeberlinia spinosa*
Brittlebush	*Encelia farinosa*
Creosote bush	*Larrea tridentata*
White bursage	*Ambrosia dumosa*

Plot-based descriptions: not available.

Stands at Antioch dunes

Scattered forbs and grasses form a ground canopy; Antioch dunes evening-primrose, California croton, California matchweed, Contra Costa wallflower, devil's-lettuce, lessingia, nude buckwheat, and/or telegraph weed may be present. Individual emergent shrubs or coast live oak trees may be present over a ground canopy. Ground layer open; annuals seasonally present.

Uplands: sandy areas. Stabilized or partially stabilized dunes.

Distribution: d-CenV.
Elevation: 20 m.

NDDB/Holland type and status: **Interior dunes** (23000).
    Stabilized interior dunes (23100) G1 S1.1.

Other types: none.

General references: Holland (1986).

Comments: This unique area supports populations of the federal and state endangered Antioch dunes evening-primrose (a CNPS List 1B plant) and Contra Costa wallflower (a CNPS List 1B plant) (Skinner & Pavlik 1994) in Contra Costa Co. d-CenV. Vegetation descriptions suggest that these plants associate with local series rather than forming a series.

Species mentioned in text	
Antioch dunes evening-primrose	*Oenothera deltoides* ssp. *howellii*
California croton	*Croton californicus*
California matchweed	*Gutierrezia californica*
Coast live oak	*Quercus agrifolia*
Contra Costa wallflower	*Erysimum capitatum* ssp. *angustatum*
Devil's-lettuce	*Amsinckia tessellata*
Lessingia	*Lessingia glandulifera*
Nude buckwheat	*Eriogonum nudum* var. *auriculatum*
Telegraph weed	*Heterotheca grandiflora*

Plot-based descriptions: not available.

Baker cypress stands

Baker cypress sole or dominant tree in canopy; Douglas-fir, incense-cedar, Jeffrey pine, knobcone pine, ponderosa pine, red fir, sugar pine, western juniper, and/or white fir may be present. Trees < 30 m; canopy continuous, intermittent, or open. Shrubs common or infrequent. Ground layer sparse (Plate 29).

Uplands: Soils basalt, ultramafic-derived, sterile.

Distribution: m-KlaR, m-CasR, m-SN, OR.
Elevation: 1150-2100 m.

NDDB/Holland type and status: Closed-cone coniferous forests (83000).
> Northern interior cypress forest (83220 *in part*) G2 S2.2.

Other types:
Barry type: G7411112 BCUBA00.
Cheatham & Haller type: Northern interior cypress forest.
PSW-45 type: Cypress series.
Thorne type: Inland closed-cone coniferous woodland.
WHR type: Closed cone pine-cypress.

General references: Baker cypress range in CA (Griffin & Critchfield 1972). Kruckeberg (1984), Murray (1988), Paysen *et al.* (1980), Vogl *et al.* (1977), Wolf & Wagener (1948).

Comments: Nine Baker cypress populations exist. They occur on local parent materials over a wide elevation range. Two subspecies, Baker cypress (*Cupressus bakeri* ssp. *bakeri*) and Matthew cypress (*C. b.* ssp. *matthewsii*) are recognized; the latter growing in m-KlaR and at Goosenest in m-CasR, whereas Baker cypress grows in m-CasR both north and south of the Goosenest population, and in m-SN.

The Jepson Manual does not recognize the subspecies. Both subspecies are included here.

This cypress associates with local series, since other species growing with Baker cypress are drastically different among populations. Baker cypress is rare (a CNPS List 4 plant) in CA (Skinner & Pavlik 1994) and endangered in OR.

Species mentioned in text

Baker cypress	*Cupressus bakeri*
Douglas-fir	*Pseudotsuga menziesii*
Incense-cedar	*Calocedrus decurrens*
Jeffrey pine	*Pinus jeffreyi*
Knobcone pine	*Pinus attenuata*
Ponderosa pine	*Pinus ponderosa*
Red fir	*Abies magnifica* var. *magnifica*
Sugar pine	*Pinus lambertiana*
Western juniper	*Juniperus occidentalis* ssp. *occidentalis*
White fir	*Abies concolor*

Plot-based descriptions: Keeler-Wolf (1985b, 1988c) in Keeler-Wolf (1990e) describes the Mud Lake and Wheeler Peak populations included in the Mud Lake RNA in Plumas Co. m-SN; Keeler-Wolf (1990b) stands at Timbered Crater candidate RNA in Siskiyou Co. ModP.

Catalina ironwood stands

Catalina ironwood sole or dominant tree in canopy; scrub oak and Catalina manzanita may be present as trees. Trees < 15 m; canopy continuous. Shrubs infrequent or absent. Ground layer sparse (Plate 30).

Uplands: slopes in sub-riparian settings. The trees are associated with calcareous soils.

Distribution: ChaI.
Elevation: sea level-300 m.

NDDB/Holland type and status: Broadleaved upland forests (81000).
 Island ironwood forest (81700) G2 S2.1.

Other types:
Cheatham & Haller type: Island woodland.
PSW-45 type: Island ironwood series.
Thorne type: Island woodland.
WHR type: Coastal oak woodland.

General references: Catalina ironwood range in CA (Griffin & Critchfield 1972). Minnich (1980a), Philbrick & Haller (1977).

Comments: This tree occurs as groves of 50-100 stems on four southern ChaI. Some 150-300 groves exist on each island. Each grove may represent a clone (Mark Dunn personal communication).

Santa Catalina Island ironwood [*Lyonothamnus floribundus* ssp. *floribundus*] grows on Santa Catalina Island. Santa Cruz Island ironwood, with pinnate leaves [*L. f.* ssp. *asplenifolius*], grows on Santa Cruz, Santa Rosa, and San Clemente Islands. Santa Catalina Island ironwood and Santa Cruz Island ironwood are both rare (CNPS List 1B plants) (Skinner & Pavlik 1994).

Species mentioned in text	
Catalina ironwood	*Lyonothamnus floribundus*
Catalina manzanita	*Arctostaphylos catalinae*
Scrub oak	*Quercus berberidifolia*

Plot-based descriptions: NDDB has plot data on file for Santa Catalina Island in Los Angeles Co. ChaI.

Crucifixion-thorn stands

Crucifixion-thorn emergent over a shrub canopy; brittlebush, creosote bush, and/or white bursage may be present. Shrubs < 2m; canopy intermittent or open. Ground layer seasonally present (Plate 30).

Uplands: pediments; fans; sand fields. Soils alluvial, bedrock, or aeolian-derived.

Distribution: MojD, ColD, AZ, Mexico.
Elevation: 60-700 m.

NDDB/Holland type and status: **Sonoran thorn woodlands** (75000).
Crucifixion thorn woodland (75200) G2 S1.1.

Other types:
Cheatham & Haller type: Low desert scrub.
Thorne type: Creosote bush scrub.
WHR type: Desert scrub.

General references: MacMahon (1988).

Comments: Eighteen populations are known from Imperial, Riverside, and San Bernardino Cos. Descriptions suggest that Crucifixion-thorn associates with local series rather than forming one.

Crucifixion-thorn is rare (a CNPS List 2 plant) in California (Skinner & Pavlik 1994). The Crucifixion-thorn Natural Area (BLM) occurs along SR 98 in Imperial Co.

Species mentioned in text	
Brittlebush	*Encelia farinosa*
Creosote bush	*Larrea tridentata*
Crucifixion-thorn	*Castela emoryi*
White bursage	*Ambrosia dumosa*

Plot-based descriptions: NDDB has plot data on file for stands in San Bernardino Co. MojD and Imperial Co. ColD.

Cuyamaca cypress stands

Cuyamaca cypress present with Coulter pine emerging above a shrub canopy. Trees < 16 m; canopy intermittent or open. Shrubs < 2 m; canopy dominated by chamise, continuous. Ground layer sparse (Plate 29).

Uplands: Soils gabbro-derived, deep.

Distribution: SoCo, m-PenR, AZ, Baja CA.
Elevation: 300-1500 m.

NDDB/Holland type and status: **Closed-cone coniferous forest** (83000).
    Southern interior cypress forest (83330 *in part*) G2 S2.1.

Other types:
Barry type: G7411112 BCUST00.
Cheatham & Haller type: Southern interior cypress forest.
PSW-45 type: Cypress series.
Thorne type: Inland closed-cone coniferous woodland.
WHR type: Closed-cone pine-cypress.

General references: Cuyamaca cypress range in CA (Griffin & Critchfield 1972). Armstrong (1974), Kurmes & Wommanack (1980), Paysen *et al.* (1980), Reveal (1978), Minnich (1987), Vogl *et al.* (1977), Wolf & Wagener (1948).

Comments: Two occurrences occur near Cuyamaca Peak in San Diego Co. A 1950 fire extirpated the cypress over part of its range. Descriptions suggest that Cuyamaca cypress associates with local series rather than forming one.

In most CA floras this taxon is called *Cupressus stephensonii* or *C. arizonica* var. *s. The Jepson Manual* refers to these populations as *C. arizonica.* var. *arizonica.* The species is rare (a CNPS List 1B plant) and worthy of special treatment (Skinner & Pavlik 1994).

Keeler-Wolf (1990e) presents a qualitative description of stands at the King Creek RNA in San Diego Co. SoCo. Dunn (1987) considers that recent high fire frequency threatens this cypress' existence.

Species mentioned in text	
Cuyamaca cypress	*Cupressus stephensonii*
Chamise	*Adenostoma fasciculatum*
Coulter pine	*Pinus coulteri*

Plot-based descriptions: Keeler-Wolf (1990 reviews the data from Scheid & Zedler (1989)

Elephant tree stands

Elephant trees emergent over a shrub canopy; brittlebush, creosote bush, and/or white bursage may be present. Trees < 3 m. Trees scattered. Shrubs <2 m; canopy intermittent. Ground layer herbs seasonally present (Plate 30).

Uplands: slopes; fans. Soils bedrock, alluvial-derived.

Distribution: w. ColD.
Elevation: sea level-500 m.

NDDB/Holland type and status: **Sonoran thorn woodlands** (75000).
    Elephant tree woodland (75100) G3 S1.2.

Other types:
Cheatham & Haller type: Enriched desert scrub.
Thorne type: Semi-succulent scrub.
WHR type: Desert succulent shrub.

General references: Burk (1977), MacMahon (1988), Thorne (1982).

Comments: Fewer than 20 populations are known in CA. Descriptions suggest that elephant tree associates with local series rather than forming one.

This species is rare (a CNPS List 2 plant) in CA with most of its range in Mexico. In AZ it is state- listed as highly safeguarded (Skinner & Pavlik 1994).

Species mentioned in text	
Brittlebush	*Encelia farinosa*
Creosote bush	*Larrea tridentata*
Elephant tree	*Bursera microphylla*
White bursage	*Ambrosia dumosa*

Plot-based descriptions: NDDB has plot data on file for four stands at Anza Borrego Desert State Park in San Diego Co. ColD.

Enriched stands in the Klamath Mountains

Alaska yellow-cedar, Brewer spruce, Douglas-fir, Engelmann spruce, foxtail pine, ground juniper, incense-cedar, Jeffrey pine, knobcone pine, lodgepole pine, mountain hemlock, noble fir, Pacific yew, ponderosa pine, Port Orford-cedar, Shasta fir, subalpine fir, sugar pine, western white pine, white fir, and/or whitebark pine may be present in stands. Trees < 60 m; canopy continuous or intermittent. Shrubs common or infrequent. Ground layer sparse or abundant (Plate 30).

Wetlands: soil intermittently flooded. Water chemistry: fresh. Streamsides. Cowardin class: Palustrine forested wetland. **Uplands:** ridges, slopes, morainal rocks and boulders; raised terraces. Soils diorite, metasedimentary, ultramafic-derived.

Distribution: m-KlaR, su-KlaR.
Elevation: 1500-2200 m.

NDDB/Holland type and status: **Upper montane coniferous forests** (85000).
    Salmon Scott enriched coniferous forest (85420) G1 S1.2.
    Siskiyou enriched conifer forest (85410) G1 S1.2.

Other types:
Cheatham & Haller type: Klamath enriched coniferous forest.
Thorne type: Enriched conifer forest.
WHR type: Klamath mixed conifer forest.

General references: Species ranges in CA (Griffin & Critchfield 1972). Kruckeberg (1984), Sawyer (1986a, 1987), Sawyer & Thornburgh (1969, 1970, 1971, 1977), Thornburgh (1990a), Waring (1969).

Comments: Local stands with different mixes of species occur in the Klamath Mountains. Descriptions suggest these stands are better assigned to local series rather than forming one.

Salmon Mountains: Horse Range and Sugar Creek drainages in the Russian Wilderness support 17 conifer species in many mixtures. Other drainages in the Salmon Mountains have different mixtures of species, but lack such high diversity. For this reason these drainages are considered vegetationally unique, and not part of any series (Sawyer 1987).

Species mentioned in text	
Alaska yellow-cedar	*Chamaecyparis nootkatensis*
Brewer spruce	*Picea breweriana*
Douglas-fir	*Pseudotsuga menziesii*
Engelmann spruce	*Picea engelmannii*
Foxtail pine	*Pinus balfouriana*
Ground juniper	*Juniperus communis*
Incense-cedar	*Calocedrus decurrens*
Jeffrey pine	*Pinus jeffreyi*
Knobcone pine	*Pinus attenuata*
Lodgepole pine	*Pinus contorta* ssp. *murrayana*
Mountain hemlock	*Tsuga mertensiana*
Noble fir	*Abies procera*
Pacific yew	*Taxus brevifolia*
Ponderosa pine	*Pinus ponderosa*
Port Orford-cedar	*Chamaecyparis lawsoniana*
Red fir	*Abies magnifica*
Shasta fir	*Abies magnifica* var. *shastensis*
Subalpine fir	*Abies lasiocarpa*
Sugar pine	*Pinus lambertiana*
Western white pine	*Pinus monticola*
White fir	*Abies concolor*
Whitebark pine	*Pinus albicaulis*

Siskiyou Mountains: Bear Basin Butte Botanical Area on the Six Rivers National Forest was the original location of high diversity stands. Here 16 conifer species grow in an area of 500 ha with seven series. Enriched stands have been described at Rock Creek Butte as well. Other stands are known in the western Marble Mountains.

The nature of the enrichment in the w. m-KlaR differs from enrichment in the Salmon Mountains. There individual species mix in one stand, whereas here different species grow in distinct stands in a small area. Each stand is dominated by a species, so it can be assigned to a series.

Trinity Mountains: Cedar Basin area is similar in character to the Siskiyou Mountain case. Numerous conifer species grow in stands of local series in a small area. This mix is different in detail from those of the Siskiyou or Salmon Mountains.

Plot-based descriptions: Keeler-Wolf (1984a, 1989f) in (Keeler-Wolf 1990e), Sawyer *et al.* 1970), Sawyer & Thornburgh (1969, 1971) in Keeler-Wolf (1990e) summarize three surveys of the Salmon Mountains including the Sugar Creek candidate RNA. In this drainage Keeler-Wolf recognizes 15 plant communities in addition to the enriched stands. Sawyer & Thornburgh (1969) report tree density and basal area at Cyclone Gap and Indian Creek in the Siskiyou Mountains, Keeler-Wolf (1987c) in Keeler-Wolf (1990e) describes enriched stands of Douglas-fir series, Port Orford-cedar series, and red fir series stands in Rock Creek Butte candidate RNA, Keeler-Wolf (1982, 1989b) in Keeler-Wolf (1990e) describes seven plant communities at Cedar Basin candidate RNA in five series in 400 ha in Siskiyou Co. This area is a good representation of the disjunct, high elevation part of Port Orford-cedar's range.

Gowen cypress stands

Gowen cypress sole or dominant tree in canopy; bishop pine or Monterey pine may be present. Trees < 7 m; canopy open. Shrubs common. Ground layer sparse.

Uplands: maritime terraces. Soils sandstone-derived, acid, sterile, poorly drained.

Distribution: o-CenCo.
Elevation: 300 m.

NDDB/Holland type and status: **Closed-cone coniferous forests** (83000).
   Monterey pygmy cypress forest (83162) G1 S1.1.

Other types:
Barry type: G7411112 BCUGO00.
Cheatham & Haller type: Pygmy cypress forest.
PSW-45 type: Cypress series.
Thorne type: Coastal closed-cone coniferous woodland.
WHR type: Close-cone pine-cypress.

General references: Gowen cypress range in CA (Griffin & Critchfield 1972). Paysen *et al.* (1980), Vogl *et al.* (1977), Wolf & Wagener (1948).

Comments: Two Gowen cypress populations exist, often called the Huckleberry Hill and Gibson Creek stands (Vogl *et al.* 1977). Descriptions suggest that Gowen cypress associates with regional series rather than forming its own series.

Gowen cypress is one of two pygmy cypresses in the state. The other [*Cupressus goveniana* ssp. *pygmaea*] occurs in Mendocino Co. Some botanists consider these taxa best represented as separate species; *The Jepson Manual* treats them as subspecies. Both occur in similar habitats, but the northern subspecies has a larger range. Its stands have been shown to differ from surrounding series [*see* Pygmy cypress series]. Gowen cypress is rare (a CNPS List 1B plant) (Skinner & Pavlik 1994).

Species mentioned in text	
Bishop pine	*Pinus muricata*
Gowen cypress	*Cupressus goveniana* ssp. *goveniana*
Monterey pine	*Pinus radiata*

Plot-based descriptions: not available.

Hinds walnut stands

Hinds walnut present in tree canopy. Trees < 25 m; canopy continuous or intermittent. Shrubs infrequent. Ground layer sparse.

Wetlands: soils intermittently flooded, saturated. Water chemistry: fresh. Riparian corridors; floodplains, river and stream low-flow margins; stream and river banks, terraces. Cowardin class: Palustrine forested wetland. **Uplands:** valley bottoms. Soil alluvial. The national inventory of wetland plants (Reed 1988) lists Hinds walnut as a FAC.

Distribution: d-CenV.
Elevation: sea level-300 m.

NDDB/Holland type and status: **Cismontane woodlands** (71000).
  Hinds walnut woodland (71220) G1 S1.1.

Other types: none.

General references: Hinds walnut range in CA (Griffin & Critchfield 1972). Holstein (1984).

Comments: The situation with Hinds walnut is taxonomically and ecologically confused. It is closely related to California walnut, and in *The Jepson Manual* it is considered a variety (*Juglans californica* var. *hindsii*). The natural range of the walnut is assumed to be restricted, but it has been planted in CenV and used for rootstock. Native Americans may have planted or not planted it in prehistoric times (Thompson 1961).

In addition, Hinds walnut hybridizes with black walnut (*J. nigra*), which, along with its hybrids, have naturalized in CenV.

Griffin & Critchfield (1972) map three stands in Contra Costa Co., one in Solano Co., and 13 questionable stands in seven counties. Descriptions suggest that Hinds walnut associates with local series rather than forming one. Skinner & Pavlik (1994) consider three stands native, and two are extant. Hinds walnut is rare (a CNPS List 1B plant) (Skinner & Pavlik 1994).

Species mentioned in text

California walnut	*Juglans californica* var. *californica*
Hinds walnut	*Juglans californica* var. *hindsii*

Plot-based descriptions: none available.

Hollyleaf cherry stands

Lyon cherry or hollyleaf cherry sole tree in canopy. Trees < 15 m; canopy continuous. Shrubs absent. Ground layer sparse.

Uplands: slopes steep, dry, north-facing. Soils sandstone-derived, rocky.

Distribution: o-SoCo, m-TraR, ChaI.
Elevation: sea level-100 m.

NDDB/Holland type and status: **Broadleaved upland forests** (81000).
>   Island cherry forest (81810) G2 S2.1.
>   Mainland cherry forest (81820) G1 S1.1.

Other types:
Cheatham & Haller type: Northern oak woodland, Southern oak woodland.
PSW-45 type: Prunus series
Thorne type: Island woodland.
WHR type: Coastal oak woodland.

General references: Hollyleaf cherry in CA (Griffin & Critchfield 1972). Paysen *et al.* (1980), Philbrick & Haller (1977).

Comments: Populations of *Prunus ilicifolia* once occurred on five of ChaI as the tree form [Lyon cherry]; two island populations survive. The often shrubby form of the species [hollyleaf cherry] has been reported from the islands, but identification is questioned (Griffin & Critchfield 1972). The typical form also grows on mainland So-CA where it can reach tree size, but is more often seen as shrubs. Stands with large trees are exceptional. Descriptions suggest that large trees of hollyleaf cherry associate with local series rather than forming one.

Species mentioned in text	
Lyon cherry	*Prunus ilicifolia* ssp. *lyonii*
Hollyleaf cherry	*Prunus ilicifolia* ssp. *ilicifolia*

Plot-based descriptions: NDDB has plot data on file for Santa Catalina Island in Los Angeles Co. ChaI.

Monterey cypress stands

Monterey cypress sole tree in canopy. Trees < 25 m; canopy open. Shrubs infrequent. Ground layer sparse.

Uplands: headlands. Soils granitic-derived.

Distribution: o-CenCo.
Elevation: sea level-20 m.

NDDB/Holland type and status: **Closed-cone coniferous forests** (83000).
    Monterey cypress forest (83150) G1 S1.1.

Other types:
Barry type: G7411112 BABAM00.
Cheatham & Haller type: Monterey cypress forest.
PSW-45 type: Cypress series.
Thorne type: Coastal closed-cone coniferous woodland.
WHR type: Closed-cone pine-cypress.

General references: Monterey cypress range in CA (Griffin & Critchfield 1972). Paysen *et al.* (1980), Vogl *et al.* (1977), Wolf & Wagener (1948).

Comments: Although plantations exist in the state and worldwide, only two natural populations occur in Monterey Co. Green (1929) reported that 10,000 trees grew in these populations at that time; three-fourths in the Cypress Point grove, and the rest near Seventeen Mile Drive at Pebble Beach. A description (Vogl *et al.* (1977) suggests that Monterey cypress associates with series rather than forming one.

Monterey cypress is rare (a CNPS List 1B plant) (Skinner & Pavlik 1994).

Species mentioned in text	
Monterey cypress	*Cupressus macrocarpa*

Plot-based descriptions: not available.

Pacific silver fir stands

Pacific silver fir sole or dominant tree in canopy; Brewer spruce, lodgepole pine, mountain hemlock, noble fir, Shasta fir, and/or western white pine may be present. Trees < 75 m; canopy continuous. Shrubs infrequent. Ground layer sparse.

Uplands: morainal rocks and boulders, north-facing slopes. Soils granitic-derived, well-drained. The national inventory of wetland plants (Reed 1988) lists Pacific silver fir as a FACU.

Distribution: su-KlaR.
Elevation: 1700-2100 m.

NDDB/Holland type and status: **Upper montane coniferous forests** (86000).

Other types:
Barry type: G7411112 BABAM00.
Cheatham & Haller type: Klamath enriched conifer forest.
Thorne type: Enriched conifer forest.
WHR type: Klamath enriched mixed conifer.

General references: Pacific silver fir range in CA (Griffin & Critchfield 1972). Langer (1988), Parker (1988).

Comments: Only two California populations have been confirmed, stands at English Peak in the Marble Mountains and near Joe Creek in the Siskiyou Mountains. The Ukonom Lake stand (Haddock 1961) is not confirmed (Griffin & Critchfield 1972).

Local stand descriptions suggest that Pacific silver fir mingles with local series rather than being part of the Pacific silver fir series of OR and WA. Pacific silver fir is rare (a CNPS List 2 plant) in CA (Skinner & Pavlik 1994).

Species mentioned in text	
Brewer spruce	*Picea breweriana*
Lodgepole pine	*Pinus contorta* ssp. *murrayana*
Mountain hemlock	*Tsuga mertensiana*
Noble fir	*Abies procera*
Pacific silver fir	*Abies amabilis*
Shasta fir	*Abies magnifica* var. *shastensis*
Western white pine	*Pinus monticola*

Plot-based descriptions: Hunt (1976) describes species, size, and age composition for stands at English Peak area su-KlaR, NDDB has stand density by size class for the Joe Creek stand in Siskiyou Co. m-KlaR.

Piute cypress stands

Piute cypress sole or dominant tree in canopy; blue oak, California juniper, foothill pine, and/or singleleaf pinyon may be present. Trees < 10 m; canopy intermittent or open. Shrubs common or infrequent. Ground layer sparse (Plate 29).

Uplands: slopes and ridges. Soils shallow.

Distribution: s. f-SN, m-SN.
Elevation: 1200-1850 m.

NDDB/Holland type and status: **Closed-cone coniferous forests** (83000).
    Southern interior cypress forest (83330 *in part*)
    G2 S2.1.

Other types:
Barry type: G7411112 BCUNE00.
Cheatham & Haller type: Northern interior cypress forest.
PSW-45 type: Cypress series.
Thorne type: Inland closed-cone coniferous woodland.
WHR type: Closed-cone pine-cypress.

General references: Piute cypress range in CA (Griffin & Critchfield 1972). Bartel (1980), Brown (1980c), Paysen *et al.* (1980), Twisselmann (1967), Vogl *et al.* (1977).

Comments: Several Piute cypress populations exist. Bartel (1980) describes nine stands, and at least three additional ones have been discovered (Jim Shevock, personal communication). The largest stand is in the Bodfish Grove Botanical Area on the Sequoia National Forest. The most recent find is the Bartolas Creek grove in Domeland Wilderness.

Descriptions suggest that Piute cypress associates with local series rather than forming one. Piute cypress is rare (a CNPS List 1B plant) (Skinner & Pavlik 1994).

Species mentioned in text	
Blue oak	*Quercus douglasii*
California juniper	*Juniperus californica*
Foothill pine	*Pinus sabiniana*
Piute cypress	*Cupressus nevadensis*
Singleleaf pinyon	*Pinus monophylla*

Plot-based descriptions: Bartel (1980) describes the distribution and ecology of Piute cypress stands. Two of the larger stands are surveyed for basal area, density, and frequency for woody species. Keeler-Wolf (1990c) reports basal area, density, and frequency for two stands, one recently burned, at Long Canyon candidate RNA in Kern Co. m-SN.

Stands on San Benito Mountain

Coulter pine, foothill pine, incense-cedar, and/or Jeffrey pine may be present in tree canopy; bigberry manzanita, chamise, Congdon silktassel, Jepson barberry, leather oak, pointleaf manzanita, and/or wedgeleaf ceanothus may be present. Trees < 20 m; canopy continuous or intermittent. Shrubs common or infrequent. Ground layer sparse (Plate 29).

Uplands: ridges, slopes. Soils ultramafic-derived.

Distribution: m-CenCo.
Elevation: 500-1500 m.

NDDB/Holland type and status: Lower montane coniferous forests (84000)
>   Southern ultramafic Jeffrey pine forest (84172) G2 S2.1.
>   Southern ultramafic mixed coniferous forest (84180) G4 S4.

Other types:
Barry type: G74
Cheatham & Haller type: Coast Range coniferous forests.
Thorne type: Lower montane coniferous forest.
WHR type: Mixed conifer.

General references: Species ranges in CA (Griffin & Critchfield 1972). Griffin (1974), Kruckeberg (1984).

Comments: The area of Clear Creek and San Benito Mountain in San Benito Co. is a complex patchwork of forests, woodlands, shrublands, and barren talus slopes. This area has the only CenCo populations of Jeffrey pine and the largest populations of incense-cedar in the region. Coulter pine and Jeffrey pine, and possibly foothill pine, hybridize here.

Descriptions suggest that this area is a mosaic of series rather than forming its own, and the mosaic is the point of conservation in this case. The BLM San Benito Mountain RNA was established in 1971.

The Federal-listed San Benito evening-primrose (a CNPS List 1B plant), and state-listed rayless layia (a CNPS List 1B plant), talus fritillary (a CNPS List 1B plant) and San Benito fritillary (a CNPS List 4 plant) occur here (Skinner & Pavlik 1994).

Species mentioned in text	
Bigberry manzanita	*Arctostaphylos glauca*
Chamise	*Adenostoma fasciculatum*
Congdon silktassel	*Garrya congdonii*
Coulter pine	*Pinus coulteri*
Foothill pine	*Pinus sabiniana*
Incense-cedar	*Calocedrus decurrens*
Jeffrey pine	*Pinus jeffreyi*
Jepson barberry	*Berberis aquifolium* var. *dictyota*
Leather oak	*Quercus durata*
Pointleaf manzanita	*Arctostaphylos pungens*
Rayless layia	*Layia discoidea*
San Benito fritillary	*Fritillaria viridea*
San Benito evening-primrose	*Camissonia benitensis*
Talus fritillary	*Fritillaria falcata*
Wedgeleaf ceanothus	*Ceanothus cuneatus*

Plot-based descriptions: not available.

Santa Cruz cypress stands

Santa Cruz cypress sole or dominant tree in canopy; canyon live oak, knobcone pine, and/or ponderosa pine may be present. Trees < 10 m; canopy open. Shrubs common or infrequent. Ground layer sparse (Plate 29).

Uplands: Soils granitic and sandstone-derived, shallow or deep, sterile.

Distribution: o-CenCo.
Elevation: 300-760 m.

NDDB/Holland type and status: Closed-cone coniferous forests (83000).
　　Northern interior cypress forest (83220 *in part*) G2 S2.2.

Other types:
Barry type: G7411112 BCUAB00.
Cheatham & Haller type: Northern interior cypress forest.
PSW-45 type: Cypress series.
Thorne type: Coastal closed-cone coniferous woodland.
WHR type: Closed-cone pine-cypress.

General references: Santa Cruz cypress range in CA (Griffin & Critchfield 1972). Paysen *et al.* (1980), Vogl *et al.* (1977), Wolf & Wagener (1948).

Comments: Five Santa Cruz cypress populations exist (Vogl *et al.* 1977). The wide range of associated species among stands suggests that this cypress is a member of local series. This cypress is rare (a CNPS List 1B plant) (Skinner & Pavlik 1994).

Keeler-Wolf (personal communication) reports that at stands near Smith Grade on Ben Lomond Mountain Santa Cruz cypress grows in Knobcone pine and Woollyleaf manzanita series.

Species mentioned in text	
Canyon live oak	*Quercus chrysolepis*
Knobcone pine	*Pinus attenuata*
Ponderosa pine	*Pinus ponderosa*
Santa Cruz cypress	*Cupressus abramsiana*
Woollyleaf manzanita	*Arctostaphylos tomentosa*

Plot-based descriptions: none available.

Tecate cypress stands

Tecate cypress emergent above a shrub canopy; chamise, ceanothus, and/or scrub oak may be present. Trees < 8 m. Shrub < 2m; canopy continuous. Ground layer sparse.

Uplands: Soils gabbro-derived, deep.

Distribution: m-PenR, Baja CA.
Elevation: 300-1500 m.

NDDB/Holland type and status: **Closed-cone coniferous forests** (83000).
   Southern interior cypress forest (83330 *in part*) G2 S2.1.

Other types:
Barry type: G7411112 BCUFO00.
Cheatham & Haller type: Southern interior cypress forest.
PSW-45 type: Cypress series.
Thorne type: Inland closed-cone coniferous woodland.
WHR type: Closed-cone pine-cypress.

General references: Tecate cypress range in CA (Griffin & Critchfield 1972). Armstrong (1974), Brown (1982c), Dunn (1985), Keeler-Wolf (1990e), Minnich (1987), Paysen *et al.* (1980), Spring (1985), Vogl *et al.* (1977), Wolf & Wagener (1948), Zedler (1977, 1981).

Comments: Four Tecate cypress populations exist in CA. Descriptions of Tecate cypress stands suggest that Tecate cypress associates with local series rather than forming one. Many authors suggest that recent high fire frequencies threaten the cypress' existence. Tecate cypress is rare (a CNPS List 1B plant) in CA (Skinner & Pavlik 1994).

Species mentioned in text	
Ceanothus	*Ceanothus* species
Chamise	*Adenostoma fasciculatum*
Scrub oak	*Quercus berberidifolia*
Tecate cypress	*Cupressus forbesii*

Plot-based descriptions: Zedler (1981) reports tree age and density.

Torrey pine stands

Torrey pine sole tree in canopy or emergent trees over a shrub canopy. Trees < 15 m; canopy intermittent or open. Shrubs < 2m; absent, infrequent, or common. Ground layer sparse or abundant (Plate 30).

Uplands: eroded coastal bluffs, maritime terraces, slopes. Soils sandstone or diatomaceous-derived.

Distribution: s. o-SoCo, ChaI.
Elevation: sea level-175 m.

NDDB/Holland type and status: **Closed-cone coniferous forests** (83000)
    Torrey pine forest (83140) G1 S1.1.

Other types:
Barry type: G7411112 BPITO00.
Cheatham & Haller type: Torrey pine forest.
PSW-45 type: Torrey pine series.
Thorne type: Torrey pine woodland.
WHR type: Closed cone pine-cypress.

General references: Torrey pine range in CA (Griffin & Critchfield 1972). Ledig (1984), Paysen *et al.* (1980), Philbrick & Haller (1977), Vogl *et al.* (1977).

Comments: Three populations exist [about 9000 trees]. The mainland population occurs at Torrey Pines State Reserve in San Diego Co. o-SoCo. This is the typical subspecies (*Pinus torreyana* ssp. *torreyana*), and it occurs with common and local coastal scrub and chaparral species. Local endemics include Encinitas baccharis (a CNPS List 1B plant), Nuttall scrub oak (a CNPS List 1B plant), sea-dahlia (a CNPS List 2 plant), and white coast ceanothus (a CNPS List 2 plant) (Skinner & Pavlik 1994).

The two Santa Rosa Island Torrey pine populations (*P. t.* ssp. *insularis*) occur in Santa Barbara Co. ChaI. The two subspecies differ in needle and cone characters (Haller 1967). Both subspecies are rare (CNPS List 1B plants) (Skinner & Pavlik 1994).

Descriptions suggest that Torrey pine associates with local series rather than forming one. Mainland and island populations have few species in common (Philbrick & Haller 1977, Vogl *et al.* 1977).

Species mentioned in text	
Encinitas baccharis	*Baccharis vanessae*
Nuttall scrub oak	*Quercus dumosa*
Sea-dahlia	*Coreopsis maritima*
Torrey pine	*Pinus torreyana*
White coast ceanothus	*Ceanothus verrucosus*

Plot-based descriptions: NDDB has plot data on file for Torrey Pines State Park in San Diego Co. o-SoCo.

Twoleaf pinyon stands

Twoleaf pinyon sole or dominant tree in canopy; desert scrub oak, and/or canyon live oak may be present. Emergent trees may be present over a shrub canopy; black sagebrush, cupleaf ceanothus, desert snowberry, pale silktassel, flat sagebrush, green ephedra, and/or pointleaf manzanita may be present. Trees < 15 m; canopy intermittent or open. Shrubs < 1 m; continuous or intermittent. Ground layer sparse.

Uplands: slopes. Soils limestone-derived.

Distribution: m-MojD.
Elevation: 1800-2300 m.

NDDB/Holland type and status: none.

Other types:
Barry type: G74 G7411112 BPIED00.
Cheatham & Haller type: Mojavean pinyon-juniper woodland.
Thorne type: Pinyon-juniper woodland.
WHR type: Pine-juniper.

General references: Twoleaf pinyon range in CA (Griffin & Critchfield 1972), silvics (Ronco 1990), des Lauriers & Ikeda (1986), Trombulak & Cody (1980), Vasek & Thorne (1977).

Comments: There are four populations of this tree in the New York Mountains and Mid Hills in e. San Bernardino Co. Descriptions suggest that twoleaf pinyon associates with local series rather than forming its own.

Another issue is the taxonomic treatment of these trees. Some botanists consider them to be *Pinus edulis* (Trombulak & Cody 1980), others two-leaved populations of *P. monophylla* (Vasek & Thorne 1977). A third interpretation argues that they are two-leaved populations of *P. californianum* of PenR (Bailey 1987). *The Jepson Manual* considers Bailey's new taxon to be singleleaf pinyon. Twoleaf pinyon [as *P. edulis*] is a rare (a CNPS List 2 plant) in CA (Skinner & Pavlik 1994).

Species mentioned in text	
Black sagebrush	*Artemisia nova*
Canyon live oak	*Quercus chrysolepis*
Cupleaf ceanothus	*Ceanothus greggii*
Desert scrub oak	*Quercus turbinella*
Desert snowberry	*Symphoricarpos longiflorus*
Flat sagebrush	*Artemisia bigelovii*
Green ephedra	*Ephedra viridis*
Pale silktassel	*Garrya flavescens*
Pointleaf manzanita	*Arctostaphylos pungens*
Twoleaf pinyon	*Pinus edulis*

Plot-based descriptions: not available.

Habitats

Some California vegetation types cannot be classified using series rules at this time. This problem arises from two sources. In some cases there is little information, so definition is difficult. In other cases plot-based data exist at a local, fine [association] scale, but it is unclear as to the best categorization at the series level. For now these associations are grouped into habitats [associations that share relatively homogenous environmental conditions] and described using the series format.

Alpine habitat: extensive areas above forestline in a-CasR, a-SN, a-WIS, and local areas in su-KlaR, a-TraR, a-PenR, su-WarR. Seven series are recognized at alpine elevations, but do not include 23 additional associations listed in this habitat.

Fen habitat: local depressions with accumulations of organic matter (peat) associated with springs, seeps, and streams. California's areas do not fit the conventional definition of a bog (Gore 1983), even though the term is commonly used in California literature. Peat accumulations occur at various elevations in NorCo and KlaR, and at m-CasR, m-SN, su-CasR, su-SN. Species composition among isolated fens will not be known until extensive sampling suggests series designation.

Montane meadow habitat: openings exist in m-NorCo, m-CenCo, m-KlaR, m-CasR, m-SN, m-TraR, m-PenR, m-WaR, m-WIS forests. Seven meadow series are recognized at montane elevations, but do not include 11 additional associations listed in this habitat.

Montane wetland shrub habitat: six willow species dominate wetlands in m-NorCo, m-CenCo, m-KlaR, m-CasR, m-SN, m-TraR, m-PenR, m-WaR, m-WIS. The character of willow thickets will not be known until extensive sampling suggests series designation. Mountain alder series and Sitka alder series are associated with this habitat.

Subalpine wetland shrub habitat: ten willow species and mountain spiraea dominate local wetlands in su-KlaR, su-CasR, su-SN, su-TraR, su-PenR, su-WarR, su-WIS. Associations characterized by grayleaf willow, Sierra willow, and tealeaf willow have been defined in su-SN. The character of the willow thickets will not be known until extensive sampling suggests series designation. Mountain alder series and Sitka alder series are associated with this habitat.

Subalpine meadow habitat: extensive meadows mix with forests and woodlands in su-KlaR, su-CasR, su-SN, su-TraR, su-PenR, su-WarR, su-WIS. Seven meadow series are recognized, but do not include seven additional associations listed in this habitat.

Subalpine upland shrub habitat: local areas of shrubs are common in su-KlaR, su-CasR, su-SN, su-TraR, su-PenR, su-WarR, su-WIS. Granite-phlox, red elderberry, shrub cinquefoil, and wax currant are locally common at these elevations, for example. Four shrub series are recognized at these elevations, but they are not dominated by these species.

Alpine habitat

Perennial broad-leaved herbs, grasses, sedges important in ground canopy. In some areas emergent shrubs may be present. Plants < 1 m. Cover continuous to sparse (Plate 31).

Wetlands: habitat seasonally flooded, semi-permanently flooded, permanently saturated, seasonally saturated. Water chemistry: fresh. Margins of channels, lakes, ponds, overflow areas, streams; wet meadows. Cowardin classes: Lacustrine littoral unconsolidated shore wetland, Palustrine nonpersistent emergent freshwater wetland, Palustrine persistent emergent freshwater wetland, Palustrine unconsolidated freshwater shore wetland.

Uplands: habitat above forest line; includes moist sods; steppes; patches of plants; individual plants, including shrubs in rock crevices, talus. Mountain range, parent material, snow ecology are important in explaining pattern.

Distribution: a-CasR, a-SN, a-WIS, a-PenR.
Elevation: 2500-4400 m.

NDDB/Holland type and status: **Meadows and seeps** (45000), **Alpine boulder and rock fields** (91000).
　　Wet subalpine or alpine meadow (45210) G3 S3.
　　Dry subalpine or alpine meadow (45220) G3 S3.
　　Klamath Cascade fell field (91110) G4 S4.
　　Sierra Nevada fell field (91120) G4 S4.
　　Southern California fell field (91130) G2 S2.2.
　　White Mountains fell field (91140) G2 S2.2.
　　Alpine glacier (91200) G5 S2.3.
　　Wet alpine talus and scree slope (91210) G5 S4.
　　Dry alpine talus and scree slope (91220) G5 S4.
　　Alpine snowfield (93100) G5 S4.

Other types:
Cheatham & Haller type: Alpine habitats.
Thorne type: Alpine fell-field cushion, Mountain meadow.

Species mentioned in text	
Alpine alumroot	Heuchera rubescens
Alpine goldenbush	Ericameria discoidea
Alpine ipomopsis	Ipomopsis congesta
Alpine phoenicaulis	Anelsonia eurycarpa
Alpine pussypaws	Antennaria alpina
Alpine pyrrocoma	Haplopappus apargioides
Alpine saxifrage	Saxifraga tolmiei
Alpine sedum	Sedum roseum
Alpine smartweed	Polygonum minimum
Alpine sorrel	Oxyria digyna
Alpine timothy	Phleum alpinum
Arctic willow	Salix arctica
Arrowhead butterweed	Senecio triangularis
Baltic rush	Juncus balticus
Bilberry	Vaccinium caespitosum
Bulrush	Scirpus species
Compact phlox	Phlox pulvinata
Cordilleran arnica	Arnica mollis
Coville phlox	Phlox covillei
Coville ragwort	Senecio scorzonella
Davidson arabis	Arabis davidsonii
Davidson penstemon	Penstemon davidsonii
Dense draba	Draba densiflora
Fragile fern	Cystopteris fragilis
Gordon ivesia	Ivesia gordonii
Hairy draba	Draba oligosperma
Heretic penstemon	Penstemon heterodoxus
Holodiscus	Holodiscus species
Inyo buckwheat	Eriogonum gracilipes
Inyo rock-cress	Arabis inyoensis
King ricegrass	Oryzopsis kingii
King sandwort	Arenaria kingii
Lemmon draba	Draba lemmonii
Mason sky pilot	Polemonium chartaceum
Merten rush	Juncus mertensianus
Moss saxifrage	Saxifraga bryophora
Mountain heather	Phyllodoce species
Mountain muhly	Muhlenbergia montana
Mountain sedum	Sedum obtusatum
Muir ivesia	Ivesia muirii
Needle-and-thread	Stipa comata
Nested saxifrage	Saxifraga nidifica

See next box

Rangeland type: SRM 213.
WHR type: Wet meadow.

General references: Barbour & Billings (1988), Ives & Barry (1976), Jackson & Bliss (1982), Klikoff (1965), LaMarche & Mooney (1967), Major (1994), Major & Taylor (1977), Taylor (1976a), Thorne (1982).

Species mentioned in text

Netted willow	*Salix nivalis*
Nevada claytonia	*Claytonia nevadensis*
Nude buckwheat	*Eriogonum nudum*
Nuttall sandwort	*Minuartia nuttallii*
Old man's whiskers	*Geum canescens*
One-spike oatgrass	*Danthonia unispicata*
One-sided bluegrass	*Poa secunda*
Parry rush	*Juncus parryi*
Podistera	*Podistera nevadensis*
Pygmy daisy	*Erigeron pygmaeus*
Round-leaved buckwheat	*Eriogonum ovalifolium*
Sibbaldia	*Sibbaldia procumbens*
Sedge	*Carex* species
Shorthair reedgrass	*Calamagrostis breweri*
Shorthair sedge	*Carex fifilolia*
Showy fescue	*Festuca minutiflora*
Sierra primrose	*Primula suffrutescens*
Silky raillardella	*Raillardella argentea*
Small haplopappus	*Ericameria suffruticosa*
Spikerush	*Eleocharis* species
Stemless haplopappus	*Stenotus acaulis*
Suksdorf monkeyflower	*Mimulus suksdorfii*
Sweetwater lupine	*Lupinus montigenus*
Sweetwater mountain milkvetch	*Astragalus kentrophyta* var. *danaus*
Tawny buckwheat	*Eriogonum incanum*
Tiling monkeyflower	*Mimulus tilingii*
Timberline phacelia	*Phacelia frigida*
Tufted hairgrass	*Deschampsia caespitosa*
Vagus fleabane	*Erigeron vagus*
Vagus buckwheat	*Eriogonum incanum*
Watson spikemoss	*Selaginella watsonii*
Western needlegrass	*Stipa occidentalis*
White corn-lily	*Veratrum californicum*
Woodrush	*Luzula divaricata*

See previous box

Comments: The fine-scale nature of the environment makes it difficult to place associations in dominance types. Stands of Bulrush series, Needle and thread series, Sedge series, Shorthair series, Shorthair sedge series, Spikerush series and Tufted hairgrass series occur at alpine elevations.

Alpine communities can occur at subalpine elevations [*see* Subalpine meadows habitat]. Some stands above and below tree line have an important component of shrubs taller than the herbs [*see* Holodiscus series, Mountain heather-bilberry series, Subalpine wetland shrub habitat, Subalpine upland shrub habitat].

Plants typical in alpine communities of the western mountains are part of subalpine communities in KlaR, TraR, PenR. Floristic descriptions exist for Cascade and So-CA peaks (Major & Taylor 1977, Thorne 1982).

Plot-based descriptions: Major & Taylor (1977) define 12 associations in a-SN and WIS; Taylor (1984) in Keeler-Wolf (1990e) 22 associations at Harvey Monroe Hall RNA in Mono Co., Burke (1982) two associations at Rae Lakes in King's Canyon National Park in Fresno Co. a-SN.

Associations:
Burke (1982):
> Sierra primrose association [containing *Primula suffrutescens-Silene sargentii* association (Major & Taylor 1977b)].

Major & Taylor (1977):
> Alpine pussypaws-heretic penstemon association,
> Alpine pyrrocoma association,
> Dense draba-Sweetwater mountain milkvetch association,
> Hairy draba-Inyo rock-cress association,
> Muir ivesia association,
> Netted willow association,
> Podistera-King sandwort association,
> Podistera-pygmy daisy association,
> Sibbaldia-Merten rush association,
> Silky raillardella-tawny buckwheat association,
> Stemless haplopappus-old man's whiskers association,
> Watson spikemoss - Round-leaved buckwheat association,
> Western needlegrass-nude buckwheat association in su-SN, a-SN.

Major & Taylor (1977):
> Mason sky pilot-Vagus fleabane association,
> Coville phlox-Inyo buckwheat association

[containing *Phlox covillei-Lilium lewisii* association] in White Mountains WIS.

Major & Taylor (1977):

Compact phlox-Gordon ivesia association,
Compact phlox-alpine phoenicaulis association,
Compact phlox-Small haplopappus-alpine ipomopsis association,
Compact phlox-Sweetwater lupine association [containing *Phlox pulvinata-Festuca brachyphylla* association, *Lupinus montigenus* association] in Sweetwater Mountains WIS.

Taylor (1984):

Alpine alumroot-fragile fern association,
Alpine goldenbush-timberline phacelia association,
Alpine saxifrage-woodrush association,
Alpine sedum-Waston spikemoss association,
Alpine smartweed association,
Alpine timothy - one-spike oatgrass association,
Baltic rush association,
Arctic willow association,

Cordilleran arnica-Davidson arabis association,
Coville phlox-vagus buckwheat association,
Dense draba-Sweetwater mountain milkvetch association,
King ricegrass-Coville ragwort association,
Lemmon draba-alpine sorrel association,
Merten rush association,
Moss saxifrage association,
Mountain sedum-mountain muhly association,
Nested saxifrage-Suksdorf monkeyflower association,
Nevada claytonia association,
Tiling monkeyflower - one-sided bluegrass association,
Nuttall sandwort association,
Parry rush-vagus buckwheat association,
Podistera-pygmy daisy association,
Showy fescue-Davidson penstemon association,
Vagus buckwheat-silky raillardella association,
White corn-lily - arrowhead butterweed association.

Fen habitat

Moss, perennial herbs, ericaceous shrubs important plants in ground and shrub canopy; bog bilberry, bog-bean, cotton grass, deer fern, Labrador-tea, rynchosposa moss, salal, sedges, sphagnum moss, spikerushes, spiraea, sun dew, and/or tinker's penney may be present. Plants < 1 m; canopy continuous (Plate 32).

Wetlands: habitat saturated. Water chemistry: fresh. Depressions with accumulations of organic matter (peat) associated with springs, seeps, streams. Cowardin class: Palustrine moss-lichen wetland.

Distribution: NorCo, KlaR, m-CasR, su-CasR, m-SN, su-SN.
Elevation: sea level-2750 m.

NDDB/Holland type and status: Bogs and fens (51000), **Marshes and swamps** (52000).
    Sphagnum bog (51110) G3 S1.1.
    Fen (51120) G2 S1.2.
    Ledum swamp (5251A) G2 S2.1.

Other types:
Cheatham & Haller type: Meadows and swamps.
Rangeland: 216.
Thorne type: Freshwater aquatic.
WHR type: Wet meadows.

General references: Bartolome *et al.* (1990), Gore (1983).

Comments: Local wet areas with peat supporting genera and species associated with peatlands are called bogs in Canada and Europe. In California, local wet areas are called bogs, even though they do not satisfy the standard definition (Sawyer 1986a).

Bogs are ombrogenous (*i.e.* receiving nutrients from precipitation only). Fens are minerotropic (*i.e.* receiving nutrients from rocks, including via streams). The term "mire" is used to include bogs and fens in one category, since the distinction is not always easily made. California's mires better fit the concept of a fen (Gore 1983). Species composition of fens varies at a fine scale. Bulrush series, Sedge series and Spikerush series occur in this habitat.

Baker (1972) qualitatively describes the sea-level Ingelnook fen in Mendocino Co. o-NorCo; Rae (1970) describes high elevation fens at the Sagehen Field Station in Nevada Co. m-SN, su-SN.

Species mentioned in text	
Bog bilberry	*Vaccinium uliginosum*
Bog-bean	*Menyanthes trifoliata*
Bulrushes	*Scirpus* species
Cotton-grass	*Eriophorum gracile*
Deer fern	*Blechnum spicant*
Labrador-tea	*Ledum glandulosum*
Primrose monkeyflower	*Mimulus primuloides*
Rynchosposa moss	*Rynchosposa* species
Salal	*Gaultheria shallon*
Sedges	*Carex* species
Shore sedge	*Carex limosa*
Sphagnum	*Sphagnum* species
Spikerushes	*Eleocharis* species
Spiraea	*Spiraea douglasii*
Sun dew	*Drosera rotundifolia*
Tinker's penney	*Hypericum anagalloides*

Plot-based descriptions: Beguin & Major (1975) define a Primrose monkeyflower-shore sedge association [as Mimulo-Cericetum limosae, = Category A (Rae 1970)] at Grass Lake RNA in El Dorado Co., Ratliff (1982, 1985) a tinker's penney association [as Tinker's penney vegetative series] in m-SN. Shore sedge is rare (a CNPS List 2 plant) (Skinner & Pavlik 1994).

Montane meadow habitat

A mixture of perennial herbs, grasses and sedges important herbs in ground canopy. Plants < 2 m; canopy continuous or intermittent (Plate 32).

Wetlands: habitat seasonally flooded, semipermanently flooded, permanently saturated, seasonally saturated. Water chemistry: fresh. Margins of channels, lakes, ponds, overflow areas, streams; wet meadows. Cowardin classes: Lacustrine littoral unconsolidated shore wetland, Palustrine nonpersistent emergent freshwater wetland, Palustrine persistent emergent freshwater wetland, Palustrine unconsolidated freshwater shore wetland. **Uplands:** habitat seasonally dry to seasonally saturated soils, openings in montane forests on slopes or concave settings near or surrounding creeks.

Distribution: m-NorCo, m-KlaR, m-CasR, m-SN, m-TraR, m-PenR, m-WarR, m-WIS, m-DesR.
Elevation: 1200-2900 m.

NDDB/Holland type and status: Meadows and seeps (45000).
> Wet montane meadow (45110) G3 S3.
> Dry montane meadow (45120) G3 S3.

Other Types:
Cheatham & Haller type: Meadows and swamps.
Rangeland: 216.
Thorne type: Freshwater aquatic.
WHR type: Wet meadow.

General references: Benedict (1982), Benedict & Major (1982), Halpern (1985), Major & Taylor (1977), Ratliff (1979, 1982, 1985).

Comments: The fine-grain nature of the environment makes it difficult to place local associations into dominance types. Bulrush series, Kentucky bluegrass series, One-sided bluegrass series, Sedge series, Shorthair series, Spikerush series and Tufted hairgrass series occur at montane elevations. Thorne (1982) treats the So-CA meadows floristically.

Species mentioned in text	
Alpine aster	*Aster alpigenus*
Angelica	*Angelica tomentosa*
Blue wildrye	*Elymus glaucus*
Bluejoint reedgrass	*Calamagrostis canadensis*
Bracken	*Pteridium aquilinum*
Bulrushes	*Scirpus* species
Carpet clover	*Trifolium monanthum*
Cow-parsnip	*Heracleum lanatum*
Diego bentgrass	*Agrostis diegoensis*
Gentian	*Gentiana newberryi*
Indian paintbrush	*Castilleja minuata*
Kentucky bluegrass	*Poa pratensis*
Longstalk clover	*Trifolium longipes*
One-sided bluegrass	*Poa secunda*
Pale hedge-nettle	*Stachys rigida*
Ribbed sedge	*Carex multicostata*
Rough bentgrass	*Agrostis scabra*
Sedges	*Carex* species
Shorthair reedgrass	*Calamagrostis breweri*
Shorthair sedge	*Carex filifolia*
Small-fruited bulrush	*Scirpus microphyllus*
Spikerushes	*Eleocharis* species
Stream deervetch	*Lotus longifolius*
Tall mannagrass	*Glyceria elata*
Tufted hairgrass	*Deschampsia cespitosa*

Plot-based descriptions: Halpern (1986) defines five associations in Sequoia National Park in Tulare Co., Ratliff (1982, 1985) three associations in central Sierra, especially Yosemite National Park in m-SN; Palmer (1979) one association in Trinity Alps in Trinity Co. Murray (1991) re-sampled meadows in the Marble Mountains first sampled in 1977 by Stillman (1980). Only series were present in both studies in Siskiyou Co. m-KlaR.

Associations:
Halpern (1986):
> Blue wildrye-cow-parsnip association,
> Bluejoint reedgrass - small-fruited bulrush

association [including in Bluejoint reedgrass vegetative series (Ratliff 1985)],

Rough bentgrass association, [included in Bentgrass vegetative series (Ratliff 1985)],

Tall mannagrass - small-fruited bulrush association,

Tall mannagrass-stream deervetch association.

Palmer (1979):

Bracken - pale hedge-nettle association.

Ratliff (1982, 1985):

Gentian-alpine aster association,

Longstalk clover association,

Carpet clover association [as vegetative series].

Stillman (1980):

Angelica-Indian paintbrush association [containing *Hordeum brachyantherum-Senecio triangularis* type (Stillman 1980), *Agastache urticifolia-Veratrum californicum, Polygonum phytolaccaefolium-Hackelia micrantha* types (Murray 1991)],

Diego bentgrass-ribbed sedge association [containing *Ligusticum californicum-Erigeron aliceae, Calyptridium umbellatum-Castilleja arachnoidea* types (Murray 1991)].

Montane wetland shrub habitat

Willows sole or dominant shrubs in canopy (dusky willow, Jepson willow, Lemmon willow, MacKenzie willow, Scouler willow, and/or strapleaf willow); arroyo willow, mountain alder, red osier, shinning willow, and/or Sitka alder may be present. Emergent trees may be present. Shrubs < 10 m; canopy continuous. Ground layer variable (Plate 31).

Wetlands: habitats seasonally flooded, saturated. Water chemistry: fresh. Floodplains; low-gradient depositions along streams. Cowardin class: Palustrine shrub/scrub wetland.

Distribution: m-NorCo, m-CenCo, m-KlaR, m-CasR, m-SN, m-TraR, m-PenR, m-WarR, m-WIS.
Elevation: 1200-2800 m.

NDDB/Holland type and status: **Riparian scrubs** (63000).
    Montane riparian scrub (63500 *in part*) G4 S4.

Other types:
Cheatham & Haller type: Willow thickets.
Thorne type: Mountain meadow.
WHR type: Montane riparian.

General references: Brayshaw (1976).

Comments: Species composition of montane willow thickets varies, and descriptions are lacking. Mountain alder series occur in similar habitat [*see* Mountain alder series, Sitka alder series].

Keeler-Wolf (1990e) qualitatively describes strapleaf willow stands at Bell Meadow RNA in Calaveras Co., mixed willow stands at Mud Lake RNA in Plumas Co., and grayleaf Sierra willow and Jepson willow stands at Mt. Pleasant RNA in Plumas Co. m-SN.

Species mentioned in text	
Arroyo willow	*Salix lasiolepis*
Dusky willow	*Salix melanopsis*
Grayleaf Sierra willow	*Salix orestera*
Jepson willow	*Salix jepsonii*
Lemmon willow	*Salix lemmonii*
Mackenzie willow	*Salix prolixa*
Mountain alder	*Alnus incana*
Red osier	*Cornus sericea*
Scouler willow	*Salix scouleriana*
Shinning willow	*Salix lucida* ssp. *caudata*
Sitka alder	*Alnus viridis*
Strapleaf willow	*Salix ligulifolia*

Plot-based descriptions: not available.

Subalpine meadow habitat

Perennial broad-leaved herbs, grasses, sedges important herbs in canopy. Emergent shrubs may be present. Herbs < .5 m; canopy continuous or intermittent (Plate 31).

Wetlands: habitat seasonally flooded, semi-permanently flooded, permanently saturated, seasonally saturated. Water chemistry: fresh. Margins of channels, lakes, ponds, overflow areas, streams; wet meadows. Cowardin classes: Lacustrine littoral unconsolidated shore wetland, Palustrine nonpersistent emergent freshwater wetland, Palustrine persistent emergent freshwater wetland, Palustrine unconsolidated freshwater shore wetland.
Uplands. habitat below cover forest line; including moist and dry meadows.

Distribution: su-KlaR, su-CasR, su-SN, su-TraR, su-PR, su-WarR, su-WIS.
Elevation: 1800-3330 m.

NDDB/Holland type and status: Meadows and seeps (45000).
> Wet subalpine or alpine meadow (45110) G3 S3.
> Dry subalpine or alpine meadow (45220) G3 S3.

Other types:
Cheatham & Haller type: Meadows and swamps.
Thorne type: Freshwater aquatic.
WHR type: Wet meadow.
General references: Major & Taylor (1977), Thorne (1982).

Comments: Species composition of subalpine meadows varies. Meadows are extensive, and mixed with forests and woodlands. Stands with a conspicuous shrub component occur at subalpine and alpine elevations [*see* Alpine habitat, Sedge series, Spikerush series]. Thorne (1982) describes meadows in su-So-CA floristically.

Plot-based descriptions: Beguin & Major (1975),

Burke (1987) in Keeler-Wolf (1990e) define one association at Grass Lake RNA in El Dorado Co., Major & Taylor (1977) two associations at Carson Pass area in Alpine Co., Benedict (1983) three associations at Rock Creek, Sequoia National Park in Tulare Co. su-SN; Murray (1991) re-sampled subalpine meadows in the Marble Mountains first sampled in 1977 by Stillman (1980) in Siskiyou Co., Palmer (1979) defines one association in Trinity Co. su-KlaR.

Species mentioned in text

Davis knotweed	*Polygonum davisiae*
Heretic penstemon	*Penstemon heterodoxus*
Jeffrey shooting star	*Dodecatheon jeffreyi*
Many-nerved sedge	*Carex multicostata*
Merten rush	*Juncus mertensianus*
Pussypaws	*Calyptridium umbellatum*
Sedges	*Carex* species
Skyline bluegrass	*Poa cusickii* ssp. *epilis*
Smooth-beaked sedge	*Carex integra*
Spikerushes	*Eleocharis* species
Tawny buckwheat	*Eriogonum incanum*
Woolly mountain-parsley	*Oreonana vestita*
Yarrow	*Achillea lanulosa*

Associations:
Beguin & Major (1975):
> Skyline bluegrass - smooth-beaked sedge association [as the Poo-Caricetum integrae].

Benedict (1983):
> Heretic penstemon-yarrow association,
> Many-nerved sedge - yarrow association,
> Tawny buckwheat - woolly mountain-parsley association, [included in Buckwheat vegetative series (Ratliff 1985)]

Major & Taylor (1977):
> Davis knotweed-tawny buckwheat association,
> Pussypaws-heretic penstemon association [including Heretic penstemon vegetative series (Ratliff 1985)].

Palmer (1979):
> Jeffrey shooting star-Mertens rush association.

Subalpine upland shrub habitat

Shrubs emergent over a ground layer canopy; alpine ipomopsis, alpine goldenbush, granite-gilia, shrub cinquefoil, and/or wax currant may be present. Shrubs < 1 m; canopy scattered. Ground layer open (Plate 31).

Uplands: dry slopes, summits, talus. Soil skeletal.

Distribution: su-CasR, su-SN, su-PenR, su-WarR, su-WIS.
Elevation: 1800-3600 m.

NDDB/Holland type and status: Montane dwarf scrub (38000), **Meadow and seep** (45000), **Alpine boulder and rock field** (91000).
> Montane dwarf scrub (38000) G3 S3.2.
> Dry subalpine or alpine meadow (45220 *in part*) G3 S3.2.
> White Mountain fell-field (91140 *in part*) G2 S2.2.

Other types:
Cheatham & Haller type: Meadows and seeps, Alpine boulder and rock fields.
Thorne type: Alpine fell-field cushion.
WHR type: Alpine dwarf scrub.

General references: Major & Taylor (1977).

Comments: Stands of several shrub series occur at subalpine and alpine elevations. This habitat type does not include high elevation stands dominated by rock-spiraea or sagebrush [*see* Holodiscus series, Low sagebrush series, Rothrock sagebrush series] or wetland scrubs [*see* Mountain heather - bilberry series]. Other shrub species may be common at high elevations. Stands that do not fit those series are placed here. Thorne (1982) describes subalpine upland shrubs floristically in su-TraR, su-PenR.

Plot-based descriptions: Major & Taylor (1977) define six regional associations, Burke (1982) two

associations at Rae Lakes area, King's Canyon National Park in Fresno Co., Taylor (1984) three associations at Harvey Monroe Hall RNA in Mono Co. su-SN, a-SN.

Species mentioned in text	
Alpine goldenbush	*Ericameria discoidea*
Alpine ipomopsis	*Ipomopsis congesta*
Bilberry	*Vaccinium caespitosum*
Compact phlox	*Phlox pulvinata*
Congdon sedge	*Carex congdonii*
Granite-gilia	*Leptodactylon pungens*
King ricegrass	*Oryzopsis kingii*
Low sagebrush	*Artemisia arbuscula*
One-seeded oatgrass	*Danthonia unispicata*
Podistera	*Podistera nevadensis*
Purple reedgrass	*Calamagrostis purpurascens*
Pygmy fleabane	*Eriogonum pygmaeus*
Rock-spiraea	*Holodiscus microphyllus*
Red elderberry	*Sambucus racemosa*
Rothrock sagebrush	*Artemisia rothrockii*
Shrub cinquefoil	*Potentilla fruticosa*
Stemless haplopappus	*Stenotus acaulis*
Wax currant	*Ribes cereum*

Associations:
Burke (1982):
> Shrub cinquefoil association [= *Potentilla fruticosa-Potentilla breweri* association (Major & Taylor 1977c)].

Major & Taylor (1977):
> Compact phlox-stemless haplopappus-alpine ipomopsis association,
> Granite-gilia/King ricegrass association,
> Granite-gilia/alpine goldenbush association [= *Festuca brachyphylla-Eriogonum ovalifolium* association (Burke 1982)],
> Podistera-pygmy fleabane association,
> Wax currant/purple reedgrass association.

Taylor (1984):
> Red elderberry-Congdon sedge association,
> Purple reedgrass - granite-gilia association,
> Shrub cinquefoil - one-seeded oatgrass association.

Subalpine wetland shrub habitat

Willows or mountain spiraea sole or dominant shrubs in canopy; Arctic willow, Booth willow, grayleaf willow, Jepson willow, Lemmon willow, netted willow, Sierra willow, small-fruited willow, tealeaf willow, and/or yellow willow may be present. Emergent trees may be present. Shrubs < 5 m; canopy continuous. Ground layer variable (Plate 31).

Wetlands: habitat seasonally flooded, saturated. Water chemistry: fresh. Floodplains; low-gradient depositions along streams. Cowardin class: Palustrine shrub-scrub wetland.

Distribution: su-KlaR, su-CasR, su-SN, su-PenR, su-WarR, su-WIS.
Elevation: 1800-3600 m.

NDDB/Holland type and status: **Riparian scrubs** (63000).
　　Montane riparian scrub (63500 *in part*) G4 S4.

Other types:
Cheatham & Haller type: Willow thickets.
Thorne type: Mountain meadow.
WHR type: Montane riparian.

General references: Brayshaw (1976), Major & Taylor (1977), Thorne (1982).

Comments: Species composition of subalpine willow thickets varies. At this time series definitions are unclear. As stands are sampled independent of location, series and associations will be defined. The Mountain heather-bilberry series is another wetland, subalpine series [*see* Subalpine upland preliminary shrub series].

Plot-based descriptions: Burke (1982) defines one association at the Rae Lakes area, King Canyon

National Park in Fresno Co., Major & Taylor (1977) one association at Carson Pass area in Alpine Co., Taylor (1984) three associations at Harvey Monroe Hall RNA in Mono Co. su-SN.

Species mentioned in text	
Arctic willow	*Salix arctica*
Arrowhead butterweed	*Senecio triangularis*
Bilberry	*Vaccinium caespitosum*
Booth willow	*Salix boothii*
Coville ragwort	*Senecio scorzonella*
Grayleaf willow	*Salix orestera*
Jepson willow	*Salix jepsonii*
Lemmon willow	*Salix lemmonii*
Meadow onion	*Allium validum*
Mountain heather	*Phyllodoce* species
Mountain spiraea	*Spiraea densiflora*
Netted willow	*Salix nivalis*
Shorthair reedgrass	*Calamagrostis breweri*
Showy sedge	*Carex spectabilis*
Sierra willow	*Salix eastwoodiae*
Small-fruited willow	*Salix brachycarpa*
Tealeaf willow	*Salix planifolia*
Yellow willow	*Salix lutea*

Associations:
Burke (1982):
　　Coville ragwort-showy sedge association.
Major & Taylor (1977):
　　Sierra willow/arrowhead butterweed association.
Taylor (1984):
　　Grayleaf willow-meadow onion association,
　　Grayleaf willow-shorthair association,
　　Mountain spiraea association,
　　Tealeaf willow-showy sedge association.

Vernal Pools

Vernal pools are ephemeral wetlands forming in shallow depressions underlain by a substrate near the surface that restricts the percolation of water. They are characterized by a barrier to overland flow that causes water to collect and pond. These depressions fill with rainwater and runoff from adjacent areas during the winter and may remain inundated until spring or early summer, sometimes filling and emptying several times during the wet season (CNPS vernal pool monitoring guidelines 1994).

Vernal pools are a exceptional facet of the state's landscape, but ecologists disagree as to whether these marvels are better treated from an ecosystem or vegetation viewpoint. We take the ecosystem approach here.

Vernal pools are divided into Nor-CA and So-CA sections.

Nor-CA pools:

Five vernal pool groups are recognized in Nor-CA: Northern basalt flow vernal pools, Northern claypan vernal pools, Northern hardpan vernal pools, Northern volcanic ash-flow vernal pools, Northern volcanic mud-flow vernal pools.

The focus of interest has been in Nor-CA pools themselves, their residing plant species, especially in terms of obligate species (Jain (1977), and island biogeography (Holland & Jain 1981, Stebbins 1976) rather than on defining vegetation types.

Even more research has centered on soil environment, especially the nature of the soil water restricting layers. Holland & Dains

(1990) show that regional floras vary in consort with soil variation.

Taylor *et al.* (1990) studied 21 pool complexes in Nor-CA, analyzing an extensive data set using Twinspan and Coenos programs. A pool was the sampling unit. Resulting pool types are defined by characteristic plant species which may not be of high cover.

Schlising & Sanders (1982) defined vegetation types for sac-CenV pools. Plots within pools were the sampling units, not the pools themselves. Some of the associations were assigned to the California annual grassland series.

Olson (1992) describes northern Santa Barbara Co. pools in o-CenCo, but data are insufficient for definition at this time.

So-CA pools:

Three vernal pool groups are recognized in So-CA: San Diego mesa vernal pools, San Jacinto Valley vernal pools, and Santa Rosa Plateau vernal pools.

Zedler (1987) surveyed the environment, history, plant and animal ecology associated with So-CA pools. Soil type along with climatic variables are probably the most important factors influencing the frequency and duration of ponding. Bauder (1987) and Zedler (1987) suggest that frequency and ponding duration best explain the variation in pool composition and structure.

Most research has been done on pools in San Diego Co. and on the Santa Rosa Plateau in

Riverside Co. White (1994) described pools in the San Jacinto Valley in Riverside Co.

Kepecko & Lathrop (1975) define four vegetation zones at Santa Rosa Plateau. Plots within pools were the sampling units, not the pools themselves. Some categories are assigned to existing series as the California annual grassland series and Spikerush series.

Northern basalt flow vernal pools

Bacigalupi downingia and/or Mathias button-celery may be present with silver sagebrush, other herbs or/and grasses. Species composition varies among pools. Herbs < 0.25 m; canopy intermittent or open (Plate 32).

Wetlands: habitat seasonally flooded, seasonally saturated. Water chemistry: mixosaline, fresh. Pools form after winter rains in settings of impeded water over rock-bound depressions. Cowardin class: Palustrine nonpersistent emergent wetland.

Distribution: NorCo, CenV, f-CasR, m-CasR, ModP, TraSN.
Elevation: 200-1900 m.

NDDB/Holland type and status: **Vernal pool** (44000). Northern basalt flow vernal pool (44131) G1 S2.1.

Other types:
Cheatham & Haller type: Vernal pools.
Thorne type: Vernal pool ephemeral.
WHR habitat type: Annual grassland.

General references: Keeler-Wolf (1990), Taylor *et al.* (1992).

Comments: These pools occur over a greater elevation range than other vernal pool types. The soils are variable as well. Small pools form directly over bedrock and larger pools form over clay-rich soils, as the Supan soil series, and other heavily weathered flows.

Pools and intermittent streams in ModP are dominated by silver sagebrush and probably represent a series. The subspecies, *Artemisia cana* ssp. *bolanderi*, grows in TraSN.

The following rare taxa may occur in these vernal pools: Ahart paronychia (a CNPS List 1B plant),

Ahart dwarf rush (a CNPS List 1B plant), alkali milkvetch (a CNPS List 1B plant), Baker navarretia (a CNPS List 1B plant), Boggs Lake hedge-hyssop (a CNPS List 1B plant), Bolander horkelia (a CNPS

Species mentioned in text	
Ahart paronychia	*Paronychia ahartii*
Ahart dwarf rush	*Juncus leiospermus* var. *ahartii*
Alkali milkvetch	*Astragalus tener* var. *tener*
Bacigalupi downingia	*Downingia bacigalupii*
Baker navarretia	*Navarretia leucocephala* ssp. *bakeri*
Boggs Lake hedge-hyssop	*Gratiola heterosepala*
Bolander horkelia	*Horkelia bolanderi*
Burke goldfields	*Lasthenia burkei*
Butte County meadowfoam	*Limnanthes floccosa* ssp. *californica*
Calistoga popcorn flower	*Plagiobothrys strictus*
Contra Costa goldfields	*Lasthenia conjugens*
Douglas pogogyne	*Pogogyne douglasii* ssp. *parviflora*
Dwarf downingia	*Downingia pusilla*
Egg Lake monkeyflower	*Mimulus pygmaeus*
Few-flowered navarretia	*Navarretia leucocephala* ssp. *pauciflora*
Greene tuctoria	*Tuctoria greenei*
Hairy Orcutt grass	*Orcuttia pilosa*
Lake County stonecrop	*Parvisedum leiocarpum*
Legenere	*Legenere limosa*
Loch Lomond button-celery	*Eryngium constancei*
Mathias button-celery	*Eryngium mathiasiae*
Modoc County knotweed	*Polygonum polygaloides* ssp. *esotericum*
Pincushion navarretia	*Navarretia meyersi*
Profuse-flowered pogogyne	*Pogogyne floribunda*
Red Bluff dwarf rush	*Juncus leiospermus* var. *leiospermus*
Sebastopol meadowfoam	*Limnanthes vinculans*
Silver sagebrush	*Artemisia cana*
Slender Orcutt grass	*Orcuttia tenuis*

List 1B plant), Burke goldfields (a CNPS List 1B plant), Butte County meadowfoam (a CNPS List 1B plant), Calistoga popcorn flower (a CNPS List 1B plant), Contra Costa goldfields (a CNPS List 1B plant), Douglas pogogyne (a CNPS List 3 plant), dwarf downingia (a CNPS List 2 plant), Egg Lake monkeyflower (a CNPS List 1B plant), few-flowered navarretia (a CNPS List 1B plant), Greene tuctoria (a CNPS List 1B plant), hairy Orcutt grass (a CNPS List 1B plant), Lake County stonecrop (a CNPS List 1B plant), legenere (a CNPS List 1B plant), Loch Lomond button-celery (a CNPS List 1B plant), Modoc County knotweed (a CNPS List 1B plant), profuse-flowered pogogyne (a CNPS List 1B plant), Red Bluff dwarf rush (a CNPS List 1B plant), Sebastopol meadowfoam (a CNPS List 1B plant), slender Orcutt grass (a CNPS List 1B plant) (Skinner & Pavlik 1994).

Plot-based descriptions: Taylor *et al.* (1990) describe a Mathias button-celery - Bacigalupi downingia pool type for the Fall River pools in Shasta Co. ModP.

Northern claypan vernal pools

Alkali heath, California goldfields, coyote-thistle, dwarf blennosperma, Fremont goldfields, Fremont tidytips, and/or saltgrass may be present with other herbs and grasses. Species composition varies among pools. Herbs < 0.25 m; canopy intermittent or open (Plate 32).

Wetlands: habitat seasonally flooded, seasonally saturated. Water chemistry: mixosaline, fresh. Pools form after winter rains in settings of impeded water by claypans. Cowardin class: Palustrine nonpersistent emergent wetland.

Distribution: i-CenCo, CenV.
Elevation: sea level-100 m.

NDDB/Holland type and status: Vernal pool (44000). Northern claypan vernal pool (44120) G1 S1.1.

Other types:
Cheatham & Haller type: Great Valley vernal pools.
Thorne type: Vernal pool ephemeral.
WHR habitat type: Annual grassland.

General references: Barry (1981), Holland (1978a,b), Holland & Jain (1977, 1981), Ikeda & Schlising (1990), Jain (1976), Jain & Moyle (1981), Taylor *et al.* (1992).

Comments: These pools occur on neutral to alkaline, silica-cemented hardpan soils which are often saline. These pools are more widespread in s. sj-CenV, but range north in the sac-CenV.

The following rare taxa occur in these vernal pools: Ahart dwarf rush (a CNPS 1B plant), alkali milkvetch (a CNPS 1B plant), Boggs Lake hedge-hyssop (a CNPS 1B plant), Colusa grass (a CNPS 1B plant), Contra Costa goldfields (a CNPS 1B plant), Coulter goldfields (a CNPS 1B plant), dwarf downingia (a CNPS 2 plant), Nelson pepperwort (a CNPS 2

Species mentioned in text	
Ahart dwarf rush	*Juncus leiospermus* var. *ahartii*
Alkali heath	*Frankenia salina*
Alkali milkvetch	*Astragalus tener* var. *tener*
Bearded popcorn flower	*Plagiobothrys hystriculus*
Boggs Lake hedge-hyssop	*Gratiola heterosepala*
California goldfields	*Lasthenia californica*
Colusa grass	*Neostapfia colusana*
Contra Costa goldfields	*Lasthenia conjugens*
Coulter goldfields	*Lasthenia glabrata* ssp. *coulteri*
Coyote-thistle	*Eryngium castrense*
Crampton tuctoria	*Tuctoria mucronata*
Dwarf downingia	*Downingia pusilla*
Dwarf blennosperma	*Blennosperma nana*
Fragrant fritillary	*Fritillaria liliacea*
Fremont tidytips	*Layia fremontia*
Fremont goldfields	*Lasthenia fremontii*
Greene tuctoria	*Tuctoria greenei*
Hairy Orcutt grass	*Orcuttia pilosa*
Henderson bentgrass	*Agrostis hendersonii*
Hoover spurge	*Chamaesyce hooveri*
Hoover button-celery	*Eryngium aristulatum* var. *hooveri*
Legenere	*Legenere limosa*
Little mousetail	*Myosurus minimus* ssp. *apus*
Lost Hills crownscale	*Atriplex vallicola*
Nelson pepperwort	*Marsilea oligosperma*
Pincushion navarretia	*Navarretia myersii*
Red Bluff dwarf rush	*Juncus leiospermus* var. *leiospermus*
Sacramento Orcutt grass	*Orcuttia viscida*
Saltgrass	*Distichlis spicata*
San Joaquin Valley Orcutt grass	*Orcuttia inaequalis*
Sanford arrowhead	*Sagittaria sanfordii*
Shinning navarretia	*Navarretia nigelliformis* ssp. *radians*
Slender Orcutt grass	*Orcuttia tenuis*
Spiny-sepaled button-celery	*Eryngium spinosepalum*
Succulent owl's-clover	*Castilleja campestris* ssp. *succulenta*
Vernal barley	*Hordeum intercedens*

CNPS 2 plant), Greene tuctoria (a CNPS 1B plant), hairy Orcutt grass (a CNPS 1B plant), Henderson bentgrass (a CNPS 3 plant), Hoover button-celery (a CNPS 4 plant), Hoover spurge (a CNPS 1B plant), legenere (a CNPS 1B plant), little mousetail (a CNPS 3 plant), Lost Hills crownscale (a CNPS 1B plant), pincushion navarretia (a CNPS 1B plant), Red Bluff dwarf rush (a CNPS 1B plant), Sacramento Orcutt grass (a CNPS 1B plant), San Joaquin Valley Orcutt grass (a CNPS 1B plant), Sanford arrowhead (a CNPS 1B plant), shinning navarretia (a CNPS 1B plant), slender Orcutt grass (a CNPS 1B plant), spiny-sepaled button-celery (a CNPS 1B plant), succulent owl's-clover (a CNPS 1B plant), vernal barley (a CNPS 3 plant) (Skinner & Pavlik 1994).

Plot-based descriptions: Taylor *et al.* (1990) describe two vernal pool types at Jepson Prairie in Solano Co. in sac-CenV.

Pool types:
Taylor *et al.* (1990):
    Coyote-thistle - alkali heath pools,
    Fremont goldfields-saltgrass pools.

Northern hardpan vernal pools

Bladder clover, coyote-thistle, Fremont goldfields, popcorn flower, rose meadowfoam, and water pygmy may be present with other herbs and grasses. Species composition varies among pools. Herbs < 0.25 m; canopy intermittent or open.

Wetlands: habitat seasonally flooded, seasonally saturated. Water chemistry: mixosaline, fresh. Pools form after winter rains in settings of impeded water over areas with hardpans. Cowardin class: Palustrine nonpersistent emergent wetland.

Distribution: s. o-NorCo, CenV.
Elevation: sea level-100 m.

NDDB/Holland type and status: **Vernal pool** (44000). Northern hardpan vernal pool (44110) G3 S3.

Other types:
Cheatham & Haller type: Great Valley vernal pools.
Thorne type: Vernal pool ephemeral.
WHR habitat type: Annual grassland.

General references: Holland (1976, 1978), Holland & Jain (1977), Ikeda & Schlising (1990), Jain (1976), Jain & Moyle (1981), Taylor *et al.* (1992).

Comments: These pools occur on old, acidic, iron-silica cemented soils including Corning, Redding, and San Joaquin soil series. Topography is typified by hogwallows and mima mounds which occur on aggregations most commonly on old alluvial fans ringing CenV.

The following rare taxa may occur in these vernal pools: Ahart dwarf rush (a CNPS List 1B plant), alkali milkvetch (a CNPS List 1B plant), Baker navarretia (a CNPS List 1B plant), Boggs Lake hedge-hyssop (a CNPS List 1B plant), Burke goldfields (a CNPS List 1B plant), Colusa grass (a CNPS List 1B plant), Coulter goldfields (a CNPS List

1B plant), dwarf downingia (a CNPS List 2 plant), Greene tuctoria (a CNPS List 1B plant), hairy Orcutt grass (a CNPS List 1B plant), Henderson

Species mentioned in text	
Ahart dwarf rush	*Juncus leiospermus* var. *ahartii*
Alkali milkvetch	*Astragalus tener* var. *tener*
Baker navarretia	*Navarretia leucocephala* ssp. *bakeri*
Bladder clover	*Trifolium depauperatum*
Boggs Lake hedge-hyssop	*Gratiola heterosepala*
Burke goldfields	*Lasthenia burkei*
Colusa grass	*Neostapfia colusana*
Coulter goldfields	*Lasthenia glabrata* ssp. *coulteri*
Coyote-thistle	*Eryngium castrense*
Dwarf downingia	*Downingia pusilla*
Dwarf brodiaea	*Brodiaea minor*
Fitch tarweed	*Hemizonia fitchii*
Fremont goldfields	*Lasthenia fremontii*
Greene tuctoria	*Tuctoria greenei*
Hairy Orcutt grass	*Orcuttia pilosa*
Henderson bentgrass	*Agrostis hendersonii*
Hoover spurge	*Chamaesyce hooveri*
Legenere	*Legenere limosa*
Little mousetail	*Myosurus minimus* ssp. *apus*
Lost Hills crownscale	*Atriplex vallicola*
Many-flowered navarretia	*Navarretia leucocephala* ssp. *plieantha*
Nelson pepperwort	*Marsilea oligosperma*
North Coast semaphore grass	*Pleuropogon hooverianus*
Paintbrush	*Castilleja campestris*
Perennial ryegrass	*Lolium perenne*
see next box	

bentgrass (a CNPS List 1B plant), Hoover spurge (a CNPS List 1B plant), legenere (a CNPS List 1B plant), little mousetail (a CNPS List 3 plant), Lost Hills crownscale (a CNPS List 1B plant), many-flowered navarretia (a CNPS List 1B plant), Nelson pepperwort (a CNPS List 2 plant), North Coast semaphore grass (a CNPS List 1B plant), pincushion navarretia (a CNPS List 1B plant), Red Bluff dwarf rush (a CNPS List 1B plant), Sacramento Orcutt grass (a CNPS List 1B plant), San Joaquin Valley

Orcutt grass (a CNPS List 1B plant), Sanford arrowhead (a CNPS List

---

Species mentioned in text

Pincushion navarretia	*Navarretia myersii*
Popcorn flower	*Plagiobothrys stipitata* var. *micrantha*
Red Bluff dwarf rush	*Juncus leiospermus* var. *leiospermus*
Rose meadowfoam	*Limnanthes douglasii* ssp. *rosea*
Round woollyheads	*Psilocarphus brevissimus*
Sacramento Orcutt grass	*Orcuttia viscida*
San Joaquin Valley Orcutt grass	*Orcuttia inaequalis*
Sanford arrowhead	*Sagittaria sanfordii*
Sebastapol meadowfoam	*Limnanthes vinculans*
Semaphore grass	*Pleuropogon californicus*
Shinning navarretia	*Navarretia nigelliformis* ssp. *radians*
Showy goldfields	*Lasthenia platycarpha*
Slender Orcutt grass	*Orcuttia tenuis*
Sonoma sunshine	*Blennosperma bakeri*
Spiny-sepaled button-celery	*Eryngium spinosepalum*
Succulent owl's-clover	*Castilleja campestris* ssp. *succulenta*
Three-sepaled buttercup	*Ranunculus bonariensis*
Tin-wort	*Cicendia quadrangularis*
Tri-colored monkey flower	*Mimulus tricolor*
Vernal barley	*Hordeum intercedens*
Water pygmy	*Crassula aquatica*
Water-starwort	*Callitriche marginata*
Whiteflower navarretia	*Navarretia leucocephala*

see previous box

1B plant), Sebastapol meadowfoam (a CNPS List 1B plant), shinning navarretia (a CNPS List 1B plant), slender Orcutt grass (a CNPS List 1B plant), Sonoma sunshine (a CNPS List 1B plant), spiny-sepaled button-celery (a CNPS List 1B plant), succulent owl's-clover (a CNPS List 1B plant), vernal barley (a CNPS List 3 plant) (Skinner & Pavlik 1994).

Plot-based descriptions: Twinspan analyses of 1986 and 1987 (Holland & Dains 1990) show rather different groups of species for the same pools in different years on the Merced River fan in Merced Co. sj-CenV. They do not propose types. MacDonald (1976) describes Phoenix Park pools on American bluffs in Sacramento Co., Schlising & Sanders

(1982) define one association at Richvale pools in Butte Co. sac-CenV, Taylor *et al.* (1990) describe 15 vernal pool types in n. sj-CenV and sac-CenV.

Association:
Schlising & Sanders (1982):
　　Fremont goldfields association.
Pool types:
Taylor *et al.* (1990) in sac-CenV:
　　Coyote-thistle - perennial ryegrass pools,
　　Fremont goldfields-bladder clover pools,
　　Fremont goldfields-dwarf brodiaea pools,
　　Fremont goldfields-paintbrush pools
　　[as *Lasthenia fremontii/Castilleja campestris* type],
　　Fremont goldfields-popcorn flower pools
　　[as *Lasthenia fremontii/Allocarya stipitata micrantha* type],
　　Fremont goldfields-semaphore grass pools,
　　Fremont goldfields-whiteflower navarretia pools,
　　Popcorn flower-rose meadowfoam pools
　　[as *Allocarya stipitata micrantha/Limnanthes douglasii rosea* type],
　　Round woollyheads-Fitch tarweed pools,
　　Round woollyheads-showy goldfields pools,
　　Three-sepaled buttercup - water-starwort pools.
Taylor *et al.* (1990) in n. sj-CenV:
　　Coyote-thistle - bladder clover pools,
　　Coyote-thistle - rose meadowfoam pools,
　　Coyote-thistle - tin-wort pools,
　　Coyote-thistle - tricolored monkeyflower pools.

Northern volcanic ashflow vernal pools

Boggs Lake hedge-hyssop, Bolander horkelia, bromes, cuspidate downingia, few-flowered navarretia, legenere, medusa-head, needle-leaf navarretia, slender Orcutt grass, skunk-weed, two-crowned downingia, and/or other herbs and grasses which bloom after water evaporates from pools. Species composition varies among pools. Herbs < 0.25 m; canopy intermittent or open.

Wetlands: habitat seasonally flooded, seasonally saturated. Water chemistry: mixosaline, fresh. Pools form after winter rains in settings of impeded water over volcanic ash flow pans. Cowardin class: Palustrine persistent or nonpersistent emergent wetland.

Distribution: i-NorCo.
Elevation: 500-1000 m.

NDDB/Holland type and status: Vernal pool (44000). Northern volcanic ash flow vernal pool (44133) G1 S1.1.

Other types:
Cheatham & Haller type: Great Valley vernal pools.
Thorne type: Vernal pool ephemeral.
WHR habitat type: Annual grassland.

General references: Cheatham (1981), Holland (1976, 1978), Holland & Jain (1977a, 1981), Jain & Moyle (1981).

Comments: These pools are only known from three locations in Lake Co. They occur as relatively large single pools surrounded by Ponderosa pine series and other forests. The largest known occurrence is protected at Boggs Lake Preserve (over 50 ha) by The Nature Conservancy. Pools may remain at deepest portions through the summer where Bulrush and Spikerush series characteristic of more

permanent wetlands exist [*see* Bulrush series, Spikerush series].

The following rare taxa occur Northern volcanic ash flow vernal pools: alkali milkvetch (a CNPS List 1B plant), Baker navarretia (a CNPS List 1B plant), Boggs Lake hedge-hyssop (a CNPS List 1B plant), Bolander horkelia (a CNPS List 1B plant), Burke goldfields (a CNPS List 1B plant), Calistoga popcorn flower (a CNPS List 1B plant), Contra Costa goldfields (a CNPS List 1B plant), dwarf downingia (a CNPS List 1B plant), few-flowered navarretia (a CNPS List 1B plant), Lake County stonecrop (a CNPS List 1B plant), legenere (a CNPS List 1B plant), Loch Lomond button-celery (a CNPS List 1B plant), slender Orcutt grass (a CNPS List 1B plant) (Skinner & Pavlik 1994).

Species mentioned in text	
Alkali milkvetch	*Astragalus tener* var. *tener*
Baker navarretia	*Navarretia leucocephala* ssp. *bakeri*
Boggs Lake hedge-hyssop	*Gratiola heterosepala*
Bolander horkelia	*Horkelia bolanderi*
Bromes	*Bromus* species
Burke goldfields	*Lasthenia burkei*
Calistoga popcorn flower	*Plagiobothrys strictus*
Contra Costa goldfields	*Lasthenia conjugens*
Cuspidate downingia	*Downingia cuspidata*
Dwarf downingia	*Downingia pusilla*
Few-flowered navarretia	*Navarretia leucocephala* ssp. *plieantha*
Lake County stonecrop	*Parvisedum leiocarpum*
Legenere	*Legenere limosa*
Loch Lomond button-celery	*Eryngium constancei*
Medusa-head	*Taeniatherum caput-medusae*
Needleleaf navarretia	*Navarretia intertexta*
Skunkweed	*Navarretia squarrosa*
Slender Orcutt grass	*Orcuttia tenuis*
Two-crowned downingia	*Downingia bicornuta*

Plot-based descriptions: NDDB has nested frequency monitoring plot data focusing on rare species.

Northern volcanic mudflow vernal pools

California goldfields, dwarf blennosperma, round woollyheads, two-crowned downingia, water pygmy, and/or whiteflower navarretia may be present with other herbs and grasses. Species composition varies among pools. Herbs < 0.25 m; canopy intermittent or open.

Wetlands: habitat seasonally flooded, seasonally saturated. Water chemistry: mixosaline, fresh. Pools form after winter rains in settings of impeded water over rock-bound depressions. Cowardin class: Palustrine nonpersistent emergent wetland.

Distribution: CenV, f-CasR, n. f-SN.
Elevation: sea level-1100 m.

NDDB/Holland type and status: Vernal pool (44000).
Northern volcanic mudflow vernal pool (44132) G1 S1.1.

Other types:
Cheatham & Haller type: Great Valley vernal pools.
Thorne type: Vernal pool ephemeral.
WHR habitat type: Annual grassland.

General references: Holland (1978a,b), Holland & Jain (1977, 1981), Ikeda & Schlising (1990), Jain (1976), Jain & Moyle (1981), Taylor & Clifton (1992).

Comments: These pools occur on Tertiary volcanic mudflows called lahars. In f-CasR they are found on the Tuscan Formation; in n. f-SN primarily on the Merhten Formation. The pools are small, forming in irregular depressions in gently sloping surfaces.

The following rare taxa occur in Northern volcanic ash flow vernal pools: Ahart dwarf rush (a CNPS List 1B plant), Boggs Lake hedge-hyssop (a CNPS List 1B plant), dwarf downingia (a CNPS List 2 plant), Egg Lake monkeyflower (a CNPS List 1B plant), legenere (a CNPS List 1B plant), Modoc County knotweed (a CNPS List 1B plant), pincushion navarretia (a CNPS List 1B plant), profuse-flowered pogogyne (a CNPS List 1B plant), Red Bluff dwarf rush (a CNPS List 1B plant), Sacramento Orcutt grass (a CNPS List 1B plant), Sanford arrowhead (a CNPS List 1B plant), slender Orcutt grass (a CNPS List 1B plant) (Skinner & Pavlik 1994).

Species mentioned in text	
Ahart dwarf rush	*Juncus leiospermus* var. *ahartii*
Boggs Lake hedgehyssop	*Gratiola heterosepala*
California goldfields	*Lasthenia californica*
Dwarf blennosperma	*Blennosperma nana*
Dwarf downingia	*Downingia pusilla*
Egg Lake monkeyflower	*Mimulus pygmaeus*
Legenere	*Legenere limosa*
Modoc County knotweed	*Polygonum polygaloides* ssp. *esotericum*
Pincushion navarretia	*Navarretia myersii*
Profuse-flowered pogogyne	*Pogogyne floribunda*
Red Bluff dwarf rush	*Juncus leiospermus* var. *leiospermus*
Round woollyheads	*Psilocarphus brevissimus*
Sacramento Orcutt grass	*Orcuttia viscida*
Sanford arrowhead	*Sagittaria sanfordii*
Slender Orcutt grass	*Orcuttia tenuis*
Two-crowed downingia	*Downingia bicornuta*
Water pygmy	*Crassula aquatica*
Whiteflower navarretia	*Navarretia leucocephala*

Plot-based descriptions: Jokerst (1990) describes five vernal pool types for the Johnson Ranch pools in Placer Co., Taylor *et al.* (1990) two vernal pools in sac-CenV.

Pool types:
Jokerst (1990):
Deeper basin-type pools,
Shallow basin-type pools,
Small basin-type pools,
Swale-type pools,
Wide-margined pools.
Taylor *et al.* (1990):
California goldfields - two-crowned downingia pools,
Whiteflower navarretia-dwarf blennosperma pools.

San Diego mesa vernal pools

Annual hairgrass, downingias, low navarretia, Orcutt brodiaea, quillworts, round woollyheads, and/or San Diego mesa mint may be present with other herbs and grasses. Species composition varies among pools. Herbs < 0.25 m; canopy intermittent or open (Plate 32).

Wetlands: habitat seasonally flooded, seasonally saturated. Water chemistry: mixosaline, fresh. Pools form after winter rains in settings of impeded water over hardpans or claypans. Cowardin class: Palustrine non persistent emergent wetland.

Distribution: o-SoCo, m-PenR.
Elevation: sea level-1400 m.

NDDB/Holland type and status: **Vernal pools** (44000).
San Diego Mesa hardpan vernal pools (44321) G2 S2.1.
San Diego Mesa claypan vernal pools (44322) G2 S2.1.

Other types:
Cheatham & Haller type: San Diego mesa vernal pools.
Thorne type: Vernal pool ephemeral.
WHR type: Annual grassland.

General references: Bauder (1987), Witham (1976), Zedler (1987).

Comments: Coastal San Diego mesa pools may be divided into claypan and hardpan types. The latter are more restricted to Otay Mesa. Floristic variation between the types is not well understood. Other pools and pool-like wetlands exist in San Diego Co. m-PenR.

Pools and swales at Cuyamaca Lake in m-PenR differ from the low elevation mesa pools in floristic composition. Blue-eyed Mary, buttercups, rushes, and San Diego gumplant are common.

The following rare taxa occur in San Diego Co. pools: California Orcutt grass (a CNPS List 1B plant), Cuyamaca Lake downingia (a CNPS List 1B plant), little mousetail (a CNPS List 3 plant), Otay Mesa mint (a CNPS List 1B plant), Parish meadowfoam (a CNPS List 1B plant), spreading navarretia (a CNPS List 1B plant), San Diego button-celery (a CNPS List 1B plant), San Diego gumplant (a CNPS List 1B plant), San Diego mesa mint (a CNPS List 1B plant) (Skinner & Pavlik 1994).

Species mentioned in text	
Annual hairgrass	*Deschampsia danthonioides*
Blue-eyed Mary	*Collinsia parviflora*
Buttercups	*Ranunculus* species
California Orcutt grass	*Orcuttia californica*
Cuyamaca Lake downingia	*Downingia concolor* var. *brevior*
Downingias	*Downingia* species
Little mousetail	*Myosurus minimus* ssp. *apus*
Low navarretia	*Navarretia prostrata*
Orcutt brodiaea	*Brodiaea orcuttii*
Otay Mesa mint	*Pogogyne nudiuscula*
Parish meadowfoam	*Limnanthes gracilis* ssp. *parishii*
Quillworts	*Isoetes* species
Round woollyheads	*Psilocarphus brevissimus*
Rushes	*Juncus* species
San Diego button-celery	*Eryngium aristulatum* var. *parishii*
San Diego gumplant	*Grindelia hirsutula* var. *hallii*
San Diego mesa mint	*Pogogyne abramsii*
Spreading navarretia	*Navarretia fossalis*

Plot-based descriptions: Bauder (1987) describes plant species arrangement along gradients. The species did not arrange themselves into distinct assemblages when comparing non-pool, edge, and pool zones. Species composition varied among years.

San Jacinto Valley vernal pools

California Orcutt grass, little mousetail, round woollyheads, spikerush, spreading navarretia, threadleaf brodiaea, toad rush, and/or water pygmy may be present with other herbs and grasses. Species composition varies among pools. Herbs < 0.25 m; canopy intermittent or open.

Wetlands: habitat seasonally flooded, seasonally saturated. Water chemistry: mixosaline. Pools form after winter rains in settings of impeded water over alluvial claypans. Cowardin class: Palustrine nonpersistent emergent wetland.

Distribution: o-SoCo.
Elevation: sea level 1000 m.

NDDB/Holland type and status: Vernal pools (44000).
   Southern vernal pools (44300) G1 S1.1.

Other types:
Cheatham & Haller type: Interior cismontane vernal pools.
Thorne type: Vernal pool ephemeral.
WHR type: Annual grassland.

General references: Ferren & Fiedler (1993), White (1994).

Comments: These pools are only known from the Perris Basin area of the Lower San Jacinto River Valley. They have been variously considered as vernal pools or vernal alkali plains. These pools are remnants of a once broader alkali wetland area which includes, in addition to vernal pools, alkali plains and brackish marshes. Pools occur on alkali soils heavy in clay or silt soil (Domino-travers-willows soil association) and are underlain by an impermeable calcareous "cemented" cliche.

The following rare taxa occur in San Jacinto Valley vernal pools: California Orcutt grass (a CNPS List 1B plant), little mousetail (a CNPS List 3 plant), spreading navarretia (a CNPS List 1B plant), threadleaf brodiaea (a CNPS List 1B plant) (Skinner & Pavlik 1994).

Species mentioned in text	
California Orcutt grass	*Orcuttia californica*
Little mousetail	*Myosurus minimus* ssp. *apus*
Round woollyheads	*Psilocarphus brevissimus*
Spikerush	*Eleocharis* species
Spreading navarretia	*Navarretia fossalis*
Threadleaf brodiaea	*Brodiaea filifolia*
Toad rush	*Juncus bufonius*
Water pygmy	*Crassula aquatica*

Plot-based descriptions: not available.

Santa Rosa Plateau vernal pools

Annual hairgrass, downingias, low navarretia, quillworts, spikerushes, water pygmy, and/or water-starwort may be present with other herbs and grasses. Species composition varies among pools. Herbs < 0.25 m; canopy intermittent or open.

Wetlands: habitat seasonally flooded, seasonally saturated. Water chemistry: mixosaline, fresh. Pools form after winter rains in settings of impeded water over rock-bound depressions. Cowardin class: Palustrine nonpersistent emergent wetland.

Distribution: o-SoCo.
Elevation: 50 m.

NDDB/Holland type and status: Vernal pools (44000), **Marshes and swamps** (52000).
    Southern interior basalt flow vernal pools (44321) G1, S1.2.
    Vernal marsh (52500) G2 S2.1.

Other types:
Cheatham & Haller type: Interior cismontane vernal pools.
Thorne type: Vernal pool ephemeral.
WHR type: Annual grassland.

General references: Weitkamp (1992), Witham (1976), Zedler (1987).

Comments: These pools occur on basalt flows as a sequence of playa-like temporary lakes. Shallow soil may prevent woody plants from establishing. Lathrop & Thorne (1976), Witkamp (1992) describe soils, Collie & Lathrop (1976) chemical characteristics of standing water, Lathrop & Thorne (1983), Thorne & Lathrop (1969) the flora.

The following rare plants may occur in individual pools: California Orcutt grass (a CNPS List 1B Plant), Coulter goldfields (a CNPS List 1B Plant), little mousetail (a CNPS List 3 Plant), Orcutt bro-

diaea (a CNPS List 1B Plant), Parish meadow-foam (a CNPS List 1B Plant), San Diego button-celery (a CNPS List 1B Plant), spreading navarretia (a CNPS List 1B Plant), threadleaf brodiaea (a CNPS List 1B Plant), Wright trichocoronis (a CNPS List 2 Plant) (Skinner & Pavlik 1994).

Species mentioned in text

Annual hairgrass	*Deschampsia danthonioides*
California Orcutt grass	*Orcuttia californica*
Coulter goldfields	*Lasthenia glabrata* ssp. *coulteri*
Downingia	*Downingia* species
Little mousetail	*Myosurus minimus* ssp. *apus*
Low navarretia	*Navarretia prostrata*
Orcutt brodiaea	*Brodiaea orcuttii*
Parish brittlescale	*Atriplex parishii*
Parish meadowfoam	*Limnanthes gracilis* ssp. *parishii*
Quillworts	*Isoetes* species
San Diego button-celery	*Eryngium aristulatum* var. *parishii*
Spikerush	*Eleocharis* species
Spreading navarretia	*Navarretia fossalis*
Threadleaf brodiaea	*Brodiaea filifolia*
Water pygmy	*Crassula aquatica*
Water-starwort	*Callitriche marginata*
Wright trichocoronis	*Trichocoronis wrightii* var. *wrightii*

Plot-based descriptions: Kopecko & Lathrop (1975) define five vegetation zones; only one is not assigned to a series [*see* California grassland series, Spikerush series] for Santa Rosa Plateau in Riverside Co. o-SoCo.

Associations:
Kepecko & Lathrop (1975):
    Dry marsh bed zone.

Literature Cited

Adams, A.W. 1975. A brief history of juniper and shrub populations in southern Oregon. Wildlife Research Report No. 6. Oregon State Wildlife Commission, Corvallis, OR.

Alexander, R.R. 1980. Engelmann spruce – subalpine fir. Pages 86-87 *in* F.H. Eyre, editor. Forest cover types of the United States and Canada. Society of American Foresters, Washington, D.C.

Alexander, R.R., J.E. Lotan, M.J. Larson, and L.A. Volland. 1983. Lodgepole pine. Pages 63-66 *in* R.M. Burns, technical compiler. Silviculture systems for the major forest types of the United States. Agriculture Handbook No. 445. USDA, Forest Service, Washington, D.C.

Alexander, R.R., R.C. Shearer, and W.D. Shepperd. 1990. *Abies lasiocarpa* – subalpine fir. Pages 60-70 *in* R.M. Burns and B.H. Honkala, technical coordinators. Silvics of North America, Volume 1. Conifers. Agriculture Handbook 654. USDA, Forest Service, Washington, D.C.

Alexander, R.R. and W.D. Shepperd. 1990. *Picea engelmannii* – Engelmann spruce. Pages 187-203 *in* R.M. Burns and B.H. Honkala, technical coordinators. Silvics of North America, Volume 1. Conifers. Agriculture Handbook 654. USDA, Forest Service, Washington, D.C.

Allen, B.H. 1987. Ecological type classification for California. General Technical Report PSW-98. USDA, Forest Service, Pacific Southwest Research Station, Berkeley, CA.

Allen, B.H., R.R. Evett, B.A. Holtzman, and A.J. Martin. 1989. Report on rangeland cover type descriptions for California hardwood rangelands. California Department of Forestry and Fire Protection, Sacramento, CA.

Allen, B.H., B.A. Holtzman, and R.R. Evett. 1991. A classification system for California's hardwood rangelands. Hilgardia 59:1-45.

Allen-Diaz, B.H. and B.A. Holtzman. 1991. Blue oak communities in California. Madroño 38:80-95.

Allen-Diaz, B.H. 1991. Water table and plant species relationships in Sierra Nevada meadows. American Midland Naturalist 126:30-43.

Allen-Diaz, B.H. and J.W. Bartolome. 1992. Survival of *Quercus douglasii* (Fagaceae) seedlings under the influence of fire and grazing. Madroño 39:47-53.

Allen-Diaz, B.H. 1994. Montane meadows – SRM 216. Page 25 *in* T.N. Shiflet, editor. Rangeland cover types of the United States. Society for Range Management, Denver, CO.

Anderson, M.V. and R.L. Pasquinelli. 1984. Ecology and management of the northern oak woodland community, Sonoma County, California. Master's thesis. Sonoma State University, Rohnert Park, CA.

Antos, J.A. and G.A. Allen. 1990. Habitat relationships of the Pacific Coast shrub *Oemleria cerasiformis* (Rosaceae). Madroño 37:249-260.

Armstrong, W.P. 1978. Southern California's vanishing cypresses. Fremontia 6:24-29.

Armstrong, W.P. 1982. Duckweeds, California's smallest wildflowers. Fremontia 10:16-22.

Armstrong, W.P. and R.F. Thorne. 1984. The genus *Wolffia* (Lemnaceae) in California. Madroño 31:171-179.

Arno, S.F. 1980. Whitebark pine. Page 88 *in* F.H. Eyre, editor. Forest cover types of the United States and Canada. Society of American Foresters, Washington, D.C.

Arno, S.F. and R.J. Hoff. 1990. *Pinus albicaulis* – whitebark pine. Pages 268-279 *in* R.M. Burns and B.H. Honkala, technical coordinators. Silvics of North America, Volume 1. Conifers. Agriculture Handbook 654. USDA, Forest Service, Washington, D.C.

Atwater, B.F., S.G. Conrad, J.N. Dowden, C.W. Hedel, R.L. MacDonald, and W. Savage. 1979. History, landforms and vegetation of the estuary's tidal marshes. Pages 347-400 *in* T.J. Conomas, editor. San Francisco Bay: the urbanized estuary. Pacific Division of the American Association of Advanced Science, San Francisco, CA.

Atzet, T. and D.L. Wheeler. 1984. Preliminary plant associations of the Siskiyou Mountains Province. Unpublished report. USDA, Forest Service, Siskiyou National Forest, Portland, OR.

Atzet, T., D.L. Wheeler, B. Smith, J. Franklin, G. Riegel, and D. Thornburgh. 1992. Chapter 5, Vegetation. Pages 92-113 in S.D. Hobbs, S.D. Tesch, P.W. Owston, R.E. Stewart, J.C. Tappeiner, and G.E. Wells, editors. Reforestation practices in southwestern Oregon and northern California. Oregon State University, Forest Research Laboratory, Corvallis, OR.

Axelrod, D.I. 1976. Evolution of the Santa Lucia fir (*Abies bracteata*) ecosystem. Annals of the Missouri Botanical Garden 63:24-41.

Axelrod, D.I. 1977. Outline history of California vegetation. Pages 139-193 in M.G. Barbour and J. Major, editors. Terrestrial vegetation of California. Wiley–Interscience, reprinted by the California Native Plant Society 1988, Sacramento, CA.

Axelrod, D.I. 1978. The origin of coastal sage vegetation, Alta and Baja California. American Journal of Botany 65:1117-1131.

Axelrod, D.I. 1980a. History of the maritime closed-cone pines, Alta and Baja California. University of California Press, Berkeley, CA.

Axelrod, D.I. 1980b. Age and origin of the Monterey endemic area. American Journal of Botany 65:1117-1131.

Bagley, M. 1986. Baseline data for a sensitive plant monitoring study of the Eureka Dunes in Inyo County, California. Unpublished report. USDI, Bureau of Land Management, Sacramento, CA.

Bahre, C.J. and T.H. Whitlow. 1982. Floristic and vegetational patterns in a dredge field. Journal of Biogeography 9:7990.

Bailey, R.G. 1994. Description of the ecoregions of the United States. 2nd edition. Miscellaneous Publication Number 1391, revised. USDA, Forest Service, Washington, D.C.

Baker, H.G. 1972. A fen on the northern California coast. Madroño 21:405-416.

Baker, G.A., P.W. Rundel, and D.J. Parsons. 1981. Ecological relationships of *Quercus douglasii* (Fagaceae) in the foothill zone of Sequoia National Park, California. Madroño 28:1-12.

Ball, J.T. 1976. Ecological survey Last Chance Meadow candidate Research Natural Area, Mount Whitney Ranger District, Inyo National Forest. Unpublished report. USDA, Forest Service, Pacific Southwest Research Station, Berkeley, CA.

Barbour, M.G. 1970. The flora and plant communities of Bodega Head, California. Madroño 20:289-336.

Barbour, M.G. and J. Major, editors. 1977a. Terrestrial vegetation of California. Wiley Interscience, reprinted by the California Native Plant Society 1988. Sacramento, CA.

Barbour, M.G. and J. Major. 1977b. Introduction. Pages 3-10 in M.G. Barbour and J. Major, editors. Terrestrial vegetation of California. Wiley–Interscience, reprinted by the California Native Plant Society 1988, Sacramento, CA.

Barbour, M.G. and A.F. Johnson. 1977. Beach and dune. Pages 223-261 in M.G. Barbour and J. Major, editors. Terrestrial vegetation of California. Wiley–Interscience, reprinted by the California Native Plant Society 1988, Sacramento, CA.

Barbour, M.G., A. Schmida, A.F. Johnson, and B. Holton. 1981a. Comparison of coastal dune scrub in Israel and California: physiognomy, association patterns, species richness, phytogeography. Israel Journal of Botany 30:181-198.

Barbour, M.G., T.C. Wainwright, and C. Manansala. 1981b. Preparation of a vegetation map of Calaveras Big Trees State Park. Unpublished report. State of California, The Resources Agency, Department of Parks and Recreation, Sacramento, CA.

Barbour, M.G. 1984. Can a red fir forest be restored? Fremontia 11:18-19.

Barbour, M.G. and R.A. Woodward. 1985. The Shasta red fir forest of California. Canadian Journal of Forestry 15:570-576.

Barbour, M.G. 1988. Californian upland forests and woodlands. Pages 131-164 *in* M.G. Barbour and W.D. Billings, editors. North American terrestrial vegetation. Cambridge University Press, New York, NY.

Barbour, M.G. and W.B. Billings editors. 1988. North American terrestrial vegetation. Cambridge University Press, New York, NY.

Barbour, M.G. 1994a. North coastal scrub – SRM 204. Page 14 *in* T.N. Shiflet, editor. Rangeland cover types of the United States. Society for Range Management, Denver, CO.

Barbour, M.G. 1994b. Coastal sage shrub – SRM 205. Page 15 *in* T.N. Shiflet, editor. Rangeland cover types of the United States. Society for Range Management, Denver, CO.

Barney, C.W. 1980. Limber pine. Pages 98-99 *in* F.H. Eyre, editor. Forest cover types of the United States and Canada. Society of American Foresters, Washington, D.C.

Barnhart, R.A., M.J. Boyd, and J.E. Pequegnat. 1992. The ecology of Humboldt Bay, California: an estuarine profile. Biological Report 1. USDI, Fish and Wildlife Service, Washington, D.C.

Barrett, J.W., P.M. McDonald, F. Ronco, and R.A. Ryker. 1980. Interior ponderosa pine. Pages 114-115 *in* F.H. Eyre, editor. Forest cover types of the United States and Canada. Society of American Foresters, Washington, D.C.

Barrett, J.W., R.E. Martin, and D.C. Wood. 1983. Northwestern Ponderosa pine and associated species. Pages 16-18 *in* R.M. Burns, technical compiler. Silviculture systems for the major forest types of the United States. Agriculture Handbook No. 445. USDA, Forest Service, Washington, D.C.

Barrows, K. 1984. Old-growth Douglas-fir forests. Fremontia 11:20-23.

Barry, W.J. 1971. The ecology of *Populus tremuloides*, a monographic approach. Dissertation, University of California, Davis, CA.

Barry, W.J. 1972. The Central Valley prairie. Unpublished report. State of California, The Resources Agency, Department of Parks and Recreation, Sacramento, CA.

Barry, W.J., and E.I. Schlinger. 1977. Ingelnook Fen, a study and plan. Unpublished report. State of California, The Resources Agency, Department of Parks and Recreation, Sacramento, CA.

Barry, W.J. 1989a. A hierarchical vegetation classification system with emphasis on California plant communities. State of California, The Resources Agency, Department of Parks and Recreation, Sacramento, CA.

Barry, W.J. 1989b. The classification and analysis of natural vegetation with emphasis on California vegetation. Unpublished report. State of California, The Resources Agency, Department of Parks and Recreation, Sacramento, CA.

Bartell, J.A. 1980. A study of the distribution and ecology of piute cypress (*Cupressus nevadensis*). Master's thesis, California State University, Fresno, CA.

Bartolome, J.W. 1981. *Stipa pulchra*, survivor from pristine prairie. Fremontia 9:3-6.

Bartolome, J.W. and B. Gemmill. 1981. The ecological status of *Stipa pulchra* (Poaceae) in California. Madroño 28:172-184.

Bartolome, J.W. 1989. Local temporal and spatial structure. Pages 73-80 *in* L.F. Huenneke and H.A. Mooney, editors. Grassland structure and function: California annual grassland. Kluwer Academic Publishers, Boston, MA.

Bartolome, J.W., D.C. Erman, and C.F. Schwarz. 1990. Stability and change in minerotrophic peatlands, Sierra Nevada of California and Nevada. Research paper PSW-198. USDA, Forest Service, Pacific Southwest Research Station, Berkeley, CA.

Bartolome, J.W. 1994. Coastal prairie – SRM 214. Page 23 *in* T.N. Shiflet, editor. Rangeland cover types of the United States. Society for Range Management, Denver, CO.

Bartolome, J., and J. Brown. 1994. Valley grassland – SRM 215. Page 24 *in* T.N. Shiflet, editor. Rangeland cover types of the United States. Society for Range Management, Denver, CO.

Bates, P.C. 1984. The role and use of fire in black bush communities in California. Master's thesis, University of California, Davis, CA.

Bauder, E.T. 1987. Species assortment along a small-scale gradient in San Diego vernal pools. Dissertation. University of California, Davis, CA.

Bauer, H.L. 1930. Vegetation of the Tehachapi Mountains, California. Ecology 11:263-280.

Bauer, H.L. 1936. Moisture relations in the chaparral of the Santa Monica Mountains, California. Ecological Monographs 6:424-430.

Baxter, J.W. 1992. The role of canopy gaps in a coastal bluff community in California. Master's thesis, San Francisco State University, San Francisco, CA.

Beasley, R.S. and J.O. Klemmedson. 1980. Ecological relationships of bristlecone pine. American Midland Naturalist 104:242-252.

Beatley, J.C. 1965. Ecology of the Nevada test site. IV. Effects of the sedan detonations on desert shrub vegetation in northeastern Yucca Flat. Report UCLA 12-571. University of California, School of Medicine, Los Angeles, CA.

Beatley, J.C. 1966. Ecological status of introduced brome grasses (*Bromus* spp.) in desert vegetation of southern Nevada. Ecology 47:548-554.

Beatley, J.C. 1975. Climates and vegetation patterns across the Mojave/Great Basin desert transition of southern Nevada. American Midland Naturalist 93: 53-70.

Beauchamp, R.M. 1977. Survey of sensitive plants of the Algodones Dunes. Unpublished report. USDI, Bureau of Land Management, Riverside, CA.

Beauchamp, R.M. 1986. A flora of San Diego County, California. Sweetwater Press. National City, CA.

Beguin, C.N. and J. Major. 1975. Contribution a l'etude phytosociologique et ecologique des marais de la Sierra Nevada (California). Phytocoenologia 2:349-367.

Benedict, N.B. 1982. Mountain meadows: stability and change. Madroño 29:148-153.

Benedict, N.B. and J. Major. 1982. A physiographic classification of subalpine meadows of the Sierra Nevada, California. Madroño 29:1-12.

Benedict, N.B. 1983. Plant associations of subalpine meadows, Sequoia National Park, California, U.S.A. Arctic and Alpine Research 15:383-396.

Billings, W.D. 1949. The shadscale vegetation zone of Nevada and eastern California in relation to climate and soils. American Midland Naturalist 42:87-109.

Billings, W.D. and J.H. Thompson. 1957. Composition of a stand of old bristlecone pines in the White Mountains of California. Ecology 38:158-160.

Bingham, B.B. and J.O. Sawyer 1988. Volume and mass of decaying logs in an upland old-growth redwood forest. Canadian Journal of Forest Research 18:1649-1651.

Bingham, B.B. and J.O. Sawyer. 1991. Distinctive features and definitions of young, mature, and old-growth Douglas-fir/hardwood forests. Pages 363279 in L.F. Ruggierro, K.B. Audry, A.B. Cary, and M.H. Huff, technical coordinators. Wildlife and vegetation of unmanaged Douglas-fir forests. PNW-GTR-285. USDA, Forest Service, Pacific Northwest Research Station, Portland, OR.

Bingham, B.B. and J.O. Sawyer. 1992. Canopy status and tree condition of young, mature and old-growth Douglas-fir/hardwood forests. Pages 141-149 in R.R. Harris, D.E. Erman and H.M. Kerner, editors. Biodiversity of northwestern California, Wildland Resource Center Report No. 29. University of California, Berkeley, CA.

Bingham, P.B. 1993. Structure and dynamics of Brewer spruce enriched mixed conifer forest of the Klamath Mountains Province, northern California. Master's thesis, Humboldt State University, Arcata, CA.

Biswell, H.H., H. Buchanan, and R.P. Gibbens. 1966. Ecology of the vegetation of a second-growth sequoia forest. Ecology 47:630-633.

Bittman, R. 1985. National natural landmark evaluation, Phases I, II, and III. Unpublished report. State of California, The Resources Agency, Department of Fish and Game, Natural Heritage Section, Natural Diversity Data Base, Sacramento, CA.

Bluestone, V. 1981. Strand and dune vegetation at Salinas River State Beach, California. Madroño 28:49-60.

Bolsinger, C.L. and A.E. Jaramillo. 1990. *Taxus brevifolia* – Pacific yew. Pages 573-579 *in* R.M. Burns and B.H. Honkala, technical coordinators. Silvics of North America, Volume 1. Conifers. Agriculture Handbook 654. USDA, Forest Service, Washington, D.C.

Bolton, H.E. 1926. Historical memoirs of new California, Fray Francisco Palou. Volume 3. University of California Press, Berkeley, CA.

Bolton, H.E. 1927. Fray Juan Crespi, missionary explorer of the Pacific Coast, 1769-1774. University of California Press, Berkeley, CA.

Bonham, C.D. 1989. Measurements for terrestrial vegetation. John Wiley & Sons, New York, NY.

Bonnicksen, T.M. 1982. Reconstruction of a pre-settlement giant sequoia – mixed conifer forest community using the aggregation approach. Ecology 63:1134-1148.

Borchert, M. and M. Hibberd. 1984. Gradient analysis of a north slope montane forest in the western Transverse Ranges of southern California. Madroño 31:129-139.

Borchert, M. 1987. Establishment record for the Cone Peak Gradient Research Natural Area within Los Padres National Forest, Monterey County, California. Unpublished report. USDA, Forest Service, Pacific Southwest Research Station, Berkeley, CA.

Borchert, M., D. Segotta, and M.D. Purser. 1988. Coast redwood ecological types of southern Monterey County, California. General Technical Report PSW-107. USDA, Forest Service, Pacific Southwest Research Station, Berkeley, CA.

Borchert, M.I., F.W. Davis, J. Michaelsen, and L.D. Owyler. 1989. Interactions of factors affecting seedling establishment of blue oak (*Quercus douglasii*) in California. Ecology 70:389-404.

Borchert, M., F.W. Davis, and B. Allen-Diaz. 1991. Environmental relationships of herbs in blue oak (*Quercus douglasii*) woodlands of central coastal California. Madroño 38:249-266.

Borchert, M.I., N.D. Cunha, P.C. Krosse, and M.L. Lawrence. 1993. Blue oak plant communities of southern San Luis Obispo and northern Santa Barbara Counties, California. General Technical Report PSW-GTR-139. USDA, Forest Service, Pacific Southwest Research Station, Berkeley, CA.

Borchert, M., H. Gordon, and T. White. 1993. Preliminary definitions of Transverse and Peninsular Range series and association. Unpublished report. USDA, Forest Service, Los Padres National Forest, Goleta, CA.

Borchert, M. 1994. Blue oak woodland – SRM 201. Page 11 *in* T.N. Shiflet, editor. Rangeland cover types of the United States. Society for Range Management, Denver, CO.

Bourgeron, P.S. and L.D. Engelking. 1994. A preliminary vegetation classification of the western United States. Unpublished report. The Nature Conservancy, Western Heritage Task Force, Boulder, CO.

Bowler, P.A. 1989. Riparian woodlands: an endangered habitat in Southern California. Pages 80-97 *in* A.A. Schoenherr, editor. Endangered plant communities of southern California. Southern California Botanists, California State University, Fullerton, CA.

Boyd, D. 1985. Status reports on invasive weeds: gorse. Fremontia 12:16.

Boyd, D. 1985. Status reports on invasive weeds: eucalyptus. Fremontia 12:19.

Boyd, R.J. 1980a. Grand fir. Page 94 *in* F.H. Eyre, editor. Forest cover types of the United States and Canada. Society of American Foresters, Washington, D.C.

Boyd, R.J. 1980b. Western white pine. Pages 94-95 *in* F.H. Eyre, editor. Forest cover types of the United States and Canada. Society of American Foresters, Washington, D.C.

Boyd, S.D. 1983. A flora of the Gavilan Hills, western Riverside County, California. Master's thesis, University of California, Riverside, CA.

Boyd, R.S. 1992. Influence of *Ammophila arenaria* on foredune plant microdistributions at Point Reyes National Seashore, California. Madroño 39:67-76.

Bradley, W.G. 1970. The vegetation of Saratoga Springs, Death Valley National Monument. Southwest Naturalist 15:111-129.

Braun-Blanquet, J. 1932. Plant sociology: the study of plant communities. McGraw Hill Co., New York, NY.

Brayshaw, T.C. 1976. Catkin bearing plants (*Amentiferae*) of British Columbia. British Columbia Provincial Museum, Victoria, BC.

Breckon, G.J. 1974. Review of North America Pacific Coast beach vegetation. Madroño 22:333-360.

Brothers, T.S. 1984. Historical vegetation change in the Owens River riparian woodland. Pages 75-84 *in* R.E. Warner and K.M. Hendrix, editors. California riparian systems: ecology, conservation and productive management. University of California Press, Berkeley, CA.

Brothers, T.S. 1985. Riparian species distributions in relation to stream dynamics, San Gabriel River, California. Dissertation. University of California, Los Angeles, CA.

Brown, D.E., C.H. Lowe, and C.P. Pase. 1980. A digitized systematic classification for ecosystems with an illustrated summary of the natural vegetation of North America. General Technical Report RM-73. USDA, Forest Service, Rocky Mountain Research Station, Fort Collins, CO.

Brown, D.E. 1982a. Great Basin conifer woodland. Desert Plants 4:52-57.

Brown, D.E. 1982b. Californian evergreen forest and woodland. Desert Plants 4:66-69.

Brown, D.E. 1982c. Relict conifer forests and woodlands. Desert Plants 4:70-71.

Brown, D.E. 1982d. Californian valley grassland. Desert Plants 4:132-135.

Brown, D.E. 1982e. Semidesert grassland. Desert Plants 4:123-131.

Brown, D.E. 1982f. Biotic communities of the American Southwest – United States and Mexico. The University of Arizona, Tucson, Arizona.

Brown, D.R. 1993. Sonoma coast state beaches grassland monitoring program. Unpublished report. State of California, The Resources Agency, Department of Parks and Recreation, Duncans Mills, CA.

Brum, G.D. 1973. Ecology of the saguaro (*Carnegiea gigantea*): phenology and establishment in marginal populations. Madroño 22:195-204.

Bullock, S.H. and E. Chamela. 1991. Herbivory and the demography of the chaparral shrub *Ceanothus greggii* (Rhamnaceae). Madroño 38:63-72.

Bulman, T.L. 1988. The eucalyptus in California. Fremontia 16:9-12.

Burcham, L.T. 1957. California range land: an historico-ecological study of the range resource of California. Unpublished report. State of California, The Resources Agency, Department of Natural Resources, Sacramento, CA.

Burk, J.H. 1977. Sonoran Desert vegetation. Pages 869-889 *in* M.G. Barbour and J. Major, editors. Terrestrial vegetation of California. Wiley–Interscience, reprinted by the California Native Plant Society 1988, Sacramento, CA.

Burkart, A. 1976. A monograph of the genus *Prosopis* (*Leguminosae* subfamily *Mimosoideae*). Journal of the Arnold Arboretum 57:450-530.

Burke, M.T. 1982. The vegetation of the Rae Lakes Basin, southern Sierra Nevada. Madroño 29:164-176.

Burke, M.T. 1985. An ecological survey of Organ Valley candidate Research Natural Area, Cleveland National Forest, California. Unpublished report. USDA, Forest Service, Pacific Southwest Research Station, Berkeley, CA.

Burke, M.T. 1987. Grass Lake, El Dorado County candidate Research Natural Area, Lake Tahoe Basin Management Unit. Unpublished report. USDA, Forest Service, Pacific Southwest Research Station, Berkeley, CA.

Burke, M.T. 1992a. Fisherman's Camp, San Bernardino County, candidate Research Natural Area, Arrowhead Ranger District, San Bernardino National Forest, California. Unpublished report. USDA, Forest Service, Pacific Southwest Research Station, Berkeley, CA.

Burke, M.T. 1992b. Horse Meadow, San Bernardino County, candidate Research Natural Area, San Gorgonio Ranger District, San Bernardino National Forest, California. Unpublished report. USDA, Forest

Service, Pacific Southwest Research Station, Berkeley, CA.

Burns, R.M. and B.H. Honkala, technical coordinators. 1990. Silvics of North America. Agricultural Handbook No. 654. USDA, Forest Service, Washington, D.C.

Campbell, C.J. and W. Green. 1968. Perpetual succession of stream-channel vegetation in a semiarid region. Journal of the Arizona Academy of Science 5:86-98.

Campbell, B. 1980. Some mixed hardwood forest communities of the coastal ranges of southern California. Phytocoenologia 8:297-320.

Capelli, M.H. and S.J. Stanley. 1984. Preserving riparian vegetation along California's south central coast. Pages 673-686 *in* R.E. Warner and K.M. Hendrix, editors. California riparian systems: ecology, conservation and productive management. University of California Press, Berkeley, CA.

Chapman, V.J. 1977. Coastal vegetation. Second edition. Pergamon Press, Elmsford, NY.

Cheatham, N.H. and J.R. Haller. 1975. An annotated list of California habitat types. Unpublished report. University of California, Berkeley, CA.

Cheatham, N.H. 1976. Butterfly Valley Botanical Area. Fremontia 4:3-8.

Cheatham, N.H., W.J. Barry, and L. Hood. 1977. Research natural areas and related programs in California. Pages 75-108 *in* M.G. Barbour and J. Major, editors. Terrestrial vegetation of California. Wiley–Interscience, reprinted by the California Native Plant Society 1988, Sacramento, CA.

Clark, L.P. and V.J. Hannon. 1969. The mangrove swamp and salt marsh communities of the Sydney District 2: the holocoenetic complex with particular reference to physiography. Journal of Ecology 57: 213-234.

Claycomb, D.W. 1983. Vegetational changes in a tidal marsh restoration project at Humboldt Bay, CA. Master's thesis, Humboldt State University, Arcata, CA.

Clements, F.E. 1916. Plant succession: an analysis of the development of vegetation. Publication 242. Carnegie Institute of Washington, Washington, D.C.

Clements, F.E. 1920. Plant indicators: the relation of plant communities to process and practice. Publication 290. Carnegie Institute of Washington, Washington, D.C.

Cole, K. 1980. Geologic control of the vegetation in the Purisima Hills, California. Madroño 27:79-89.

Collie, N. and E.W. Lathrop. 1976. Chemical characteristics of the standing water of a vernal pool on the Santa Rosa Plateau, Riverside County, California. Pages 27-31 *in* S. Jain, editor. Vernal pools: their ecology and conservation. Publication No. 9. Institute of Ecology, University of California, Davis, CA.

Colwell, W.L. 1977. The status of vegetation mapping in California today. Pages 195-220 *in* M.G. Barbour and J. Major, editors. Terrestrial vegetation of California. Wiley–Interscience, reprinted by the California Native Plant Society 1988, Sacramento, CA.

Colwell, W.L. 1980. Knobcone pine. Pages 124-125 *in* F.H. Eyre, editor. Forest cover types of the United States and Canada. Society of American Foresters, Washington, D.C.

Combs, W.E. 1984. Stand structure and composition of the Little Lost Man Creek Research Natural Areas, Redwood National Park. Master's thesis, Humboldt State University, Arcata, CA.

Conard, S.G. and R.F. Robichaux. 1980. Ecological survey of the proposed Soda Ridge Research Natural Area, Lassen National Forest, California. Unpublished report. USDA, Forest Service, Pacific Southwest Research Station, Berkeley, CA.

Conard, S.G., and S.R. Radosevich. 1982. Post-fire succession in white fir (*Abies concolor*) vegetation of the northern Sierra Nevada. Madroño 29:42-56.

Conard, S.G., R.L. MacDonald, and R.F. Holland. 1980. Riparian vegetation and flora of the Sacramento Valley. Pages 47-55 *in* A. Sands, editor. Riparian forests in California. Publication No. 15. Institute of Ecology, University of California, Davis, CA.

Cooper, W.S. 1922. The broad-sclerophyll vegetation of California. Publication 319. Carnegie Institution of Washington, Washington, D.C.

Conquist, A., A.H. Holmgren, N.H. Holmgren, J.L. Reveal, and P.K. Holmgren. 1898. Intermountain flora;

vascular plants of the intermountain West, USA. Hafner Publishing Co., New York, NY.

Cornett, J.W. 1987. Cold tolerance in the desert fan palm, *Washingtonia filifera* (Arecaceae). Madroño 34:57-62.

Couch, E.B. 1914. Notes on the ecology of sand dune plants. Plant World 17:93-97.

Cowan, B.D. 1976. The menace of pampas grass. Fremontia 4:14-16.

Cowardin, L.M., V. Carter, F.C. Golet, and E.T. LaRoe. 1979. Classification of wetlands and deepwater habitats of the United States. FWS/OBS-79/31. USDI, Fish and Wildlife Service, Biological Services Program, Washington, D.C.

Crawford, P.D. and C.D. Oliver. 1990. *Abies amabilis* – Pacific silver fir. Pages 17-25 *in* R.M. Burns and B.H. Honkala, technical coordinators. Silvics of North America, Volume 1. Conifers. Agriculture Handbook 654. USDA, Forest Service, Washington, D.C.

Critchfield, W.B. 1971. Profiles of California vegetation. Research Paper PSW-76. USDA, Forest Service, Pacific Southwest Research Station, Berkeley, CA.

Critchfield, W.B. 1984. Crossability and relationships of Washoe pine. Madroño 31:144-170.

Cronemiller, F.P. 1959. The life history of deerbrush, a fire type. Journal of Range Management 12:2124.

Cylinder, P.D. 1995. The Monterey ecological staircase and subtypes of Monterey pine forest. Fremontia 23:713.

Dale, M.B. 1994. Do ecological communities exist? Journal of Vegetation Science 5:285-286.

DaSilva, P.G. and J.W. Bartolome. 1984. Interaction between a shrub, *Baccharis pilularis* ssp. *consanguinea* (Asteraceae), and an annual grass, *Bromus mollis* (Poaceae), in coastal California. Madroño 31:93-101.

Daubenmire, R. 1952. Forest vegetation of northern Idaho and adjacent Washington, and its bearing on concepts of vegetation classification. Ecological Monographs 22:301-330.

Daubenmire, R. 1968. Plant communities: a textbook of plant synecology. Harper & Row, Publishers, New York, NY.

Daubenmire, R. and J.B. Daubenmire. 1968. Forest vegetation of eastern Washington and northern Idaho. Technical Bulletin No. 60. Washington Agriculture Experiment Station, Pullman, WA.

Daubenmire, R. 1970. Steppe vegetation of Washington. Station Technical Bulletin No. 62. Washington Agriculture Experiment Station, Pullman, WA.

Davidson, E. and M. Fox. 1974. Effects of off-road motorcycle activity on Mojave Desert vegetation and soil. Madroño 22:381-412.

Davis, C.B. 1972. Comparative ecology of six members of the *Arctostaphylos andersonii* complex. Dissertation, University of California, Davis, CA.

Davis, F.W., and D.E. Hickson. 1988. Composition of maritime chaparral related to fire history and soil, Burton Mesa, Santa Barbara County, California. Madroño 35:169-195.

Davis, J.N. 1994a. Curlleaf mountain-mahogany – SRM 415. Page 54 *in* T.N. Shiflet, editor. Rangeland cover types of the United States. Society for Range Management, Denver, CO.

Davis, J.N. 1994b. Littleleaf mountain-mahogany – SRM 417. Page 56 *in* T.N. Shiflet, editor. Rangeland cover types of the United States. Society for Range Management, Denver, CO.

Davy, B.J. 1902. Stock ranges of northwestern California. Bulletin 12. USDA, Bureau of Plant Industry, Washington, D.C.

Dealy, J.E., J.M. Geist, and R.S. Driscoll. 1977. Communities of western juniper in the intermountain Northwest. *in* R.E. Martin, E.J. Dealy and R. S. Driscoll, editors. Proceedings of the western juniper ecology and management workshop. General Technical Report PNW-74. USDA, Forest Service, Pacific Northwest Research Station, Portland, OR.

Dealy, J.E. 1990. *Juniperus occidentalis* – western juniper. Pages 109-115 *in* R.M. Burns and B.H. Honkala, technical coordinators. Silvics of North America, Volume 1. Conifers. Agriculture Handbook 654. USDA, Forest Service, Washington, D.C.

DeBell, D.S. 1980. Black cottonwood – willow. Pages 100-101 *in* F.H. Eyre, editor. Forest cover types of the United States and Canada. Society of American Foresters, Washington, D.C.

DeBell, D.S. and T.C. Turpin. 1983. Red alder. Pages 26-28 *in* R.M. Burns, technical compiler. Silviculture systems for the major forest types of the United States. Agriculture Handbook No. 445. USDA, Forest Service, Washington, D.C.

DeBell, D.S. 1990. *Populus trichocarpa* – black cottonwood. Pages 570-576 *in* R.M. Burns and B.H. Honkala, technical coordinators. Silvics of North America, Volume 2. Hardwoods. Agricultural Handbook 654. USDA, Forest Service, Washington, D.C.

DeByle, N.V. and J.C. Zasada. 1980. Aspen. Pages 96-97 *in* F.H. Eyre, editor. Forest cover types of the United States and Canada. Society of American Foresters, Washington, D.C.

DeDecker, M. 1979. Can BLM protect the dunes? Fremontia 7:6-8.

DeDecker, M. 1982. The Eureka Valley Dunes recovery plan. Unpublished report. USDI, Fish and Game Service, San Francisco, CA.

DeDecker, M. 1984. Flora of the northern Mojave Desert, California. California Native Plant Society, Sacramento, CA.

Deghi, G.S., T. Huffman, J.W. Culver. 1995. California's native Monterey pine populations: potential for sustainability. Fremontia 23:14-23.

DeJager, W.R. 1991. *Abies lasiocarpa* in the Klamath Mountains of California and Oregon. Master's thesis, University of California, Los Angeles, CA.

Derby, J.A. and R.C. Wilson. 1978. Floristics of pavement plains of the San Bernardino Mountains. Aliso 9:374-378.

Derby, J.A. and R.C. Wilson. 1979. Phytosociology of pavement plains of the San Bernardino Mountains. Aliso 9:463-474.

des Lauriers, J. and M. Ikeda. 1986. An apparent case of introgression between pinyon pines of the New York Mountains, eastern Mojave Desert. Madroño 33:55-62.

DeSimone, S.A. and J.H. Burk. 1992. Local variation in floristics and distributional factors in California coastal sage scrub. Madroño 39:170-188.

Dodd, R. 1992. Noteworthy collections: *Cupressus bakeri* Jeps. Madroño 39:79.

Driscoll, R.S. 1964. Vegetation–soil units in the central Oregon juniper zone. Research Paper PNW-19. USDA, Forest Service, Pacific Northwest Research Station, Portland, OR.

Driscoll, R.S., D.L. Merkel, R.L. Radloff, D.E. Snyder, and J.S. Hagihara. 1984. An ecological land classification framework for the United States. Miscellaneous Publication Number 1439. USDA, Forest Service, Washington, D.C.

Duebendorfer, T.E. 1987. Vegetation–soil relations on ultramafic parent material, Pine Flat Mountain, Del Norte County, California. Master's thesis, Humboldt State University, Arcata, CA.

Duebendorfer, T.E. 1989. An integrated approach to enhancing rare plant populations through habitat restoration: II. Habitat characterization through classification of dune vegetation. Pages 478-487 *in* H.G. Hughes and T.M. Bonnicksen, editors. Restoration '89: The new management challenge. First Annual Meeting of the Society for Ecological Restoration, Society for Ecological Restoration, Oakland, CA.

Dunn, A.T. 1985. The tecate cypress. Fremontia 13:3-7.

Dunn, A.T. 1987. Population dynamics of the tecate cypress. Pages 367-376 *in* T.S. Elias, editor. Conservation and management of rare and endangered plants. California Native Plant Society, Sacramento, CA.

Dusek, K.H. 1985. Update on our rarest pine. American Forests 91:26-29.

ECOMAP. 1993. National hierarchical framework of ecological units. Unpublished document. USDA, Forest Service, Washington, D.C.

Eddleman, L.E. 1987. Establishment and stand development of western juniper in central Oregon. General Technical Report INT-215. USDA, Forest Service, Intermountain Research Station, Logan, UT.

Egler, F.E. 1954. Vegetation science concepts: I. Initial floristic composition, a factor in old field vegetation development. Vegetatio 4:412-417.

Eicher, A.L. 1987. Salt marsh vascular plant distribution in relation to tidal elevation, Humboldt Bay, California. Master's thesis, Humboldt State University, Arcata, CA.

Eicher, A.L. and J.O. Sawyer. 1989. Baseline monitoring of efforts to eradicate *Spartina alterniflora* in Humboldt Bay, California. Unpublished report. State of California, Resource Agency, Department of Fish and Game, Eureka, CA.

Elliott, H.W. and J.D. Wehausen. 1974. Vegetation succession on coastal rangeland of Point Reyes peninsula. Madroño 22:231-238.

Erman, D.C. and N.A. Erman. 1975. Macroinvertebrate composition and production in some Sierra Nevada minerotrophic peatlands. Ecology 56:591-603.

Evans, R.A. and J.A. Young. 1989. Characterization and analysis of abiotic factors and their influences on vegetation. Pages 13-28 in L.F. Huenneke and H.A. Mooney, editors. Grassland structure and function: California annual grassland. Kluwer Academic Publishers, Boston, MA.

Eyre, F.H. 1980. Forest cover types of the United States and Canada. Society of American Foresters, Washington, D.C.

Faber, P.M., E. Keller, A. Sands, and B.M. Massey. 1989. The ecology of riparian habitats of the southern California coastal region: a community profile. Biological Report 85(7.27). USDI, Fish and Wildlife Service, Washington, D.C.

Faber, P.M. 1990. Year of the oak. Fremontia 18.

Farquahr, F. 1966. Up and down California in 1860-1864: the journal of William H. Brewer, professor of agriculture in the Scheffield Scientific School from 1864 to 1903. University of California Press, Berkeley, CA.

Ferren, W.R. 1989. A preliminary and partial classification of wetlands in southern and central California with emphasis on the Santa Barbara Region. Unpublished report. State of California, The Resources Agency, Department of Fish and Game, Sacramento, CA.

Ferren, W.R. and F.W. Davis. 1991. Biotic inventory and ecosystem characterization of Fish Slough in Inyo and Mono Counties. Unpublished report. State of California, The Resources Agency, Department of Fish and Game, Sacramento, CA.

Ferren, W.R., P.L. Fiedler, and R.A. Leidy. 1994. Wetlands of the central and southern California coast and coastal watersheds. Unpublished report. Environmental Protection Agency, Region 9, San Francisco, CA.

Fiedler, P.L. 1986. Preliminary ecological survey of the Jawbone Ridge proposed Research Natural Area, Stanislaus National Forest, Tuolumne County, California. Unpublished report. USDA, Forest Service, Pacific Southwest Research Station, Berkeley, CA.

Fiedler, P.L., N. Carnal, and R. Leidy. 1986. Ecological survey of the Green Island Lake proposed Research Natural Area. Unpublished report. USDA, Forest Service, Pacific Southwest Research Station, Berkeley, CA.

Fiedler, P.L. 1987. Ecological survey of the Antelope Creek Lakes proposed Research Natural Area. Unpublished report. USDA, Forest Service, Pacific Southwest Research Station, Berkeley, CA.

Fiedler, P.L. and R.A. Leidy. 1987. Plant communities of Ring Mountain Preserve, Marin County, California. Madroño 34:173-192.

Fiedler, P.L. 1994. Rarity in vascular plants. Pages 2-3 in M.W. Skinner and B.M. Pavlik, editors. California Native plant Society's inventory of rare and endangered vascular plants of California. California Native Plant Society, Sacramento, CA.

Finch, S.J. and D. McCleery. 1980. California coast live oak. Pages 127-128 in F.H. Eyre, editor. Forest cover types of the United States and Canada. Society of American Foresters, Washington, D.C.

Finn, M.S. 1991. Ecological characteristics of California sycamore *Platanus racemosa*. Master's thesis. California State University, Los Angeles, CA.

Fites, J. 1993. Ecological guide to mixed conifer plant associations of the northern Sierra Nevada and southern Cascades. Technical Publication. USDA,

Forest Service, Pacific Southwest Region, San Francisco, CA.

Flora of North America editorial committee. 1993. Flora of North America. Oxford University Press. New York, NY.

Foiles, M.W., R.T. Graham, and D.F. Olson. 1990. *Abies grandis* – grand fir. Pages 52-59 *in* R.M. Burns and B.H. Honkala, technical coordinators. Silvics of North America, Volume 1. Conifers. Agriculture Handbook 654. USDA, Forest Service, Washington, D.C.

Foin, T.C. and M.M. Hektner. 1986. Secondary succession and the fate of native species in a California coastal prairie community. Madroño 33:189-206.

Franklin, J.F., C.T. Dyrness, and W.H. Moir. 1971. A reconnaissance method for forest site classification. Shinrin Richi 13:1-14.

Franklin, J.F. and C.T. Dyrness. 1973. Natural vegetation of Oregon and Washington. General Technical Report PNW-8. USDA, Forest Service, Pacific Northwest Research Station, Portland, OR.

Franklin, J.F. 1980a. Mountain hemlock. Pages 85-86 *in* F.H. Eyre, editor. Forest cover types of the United States and Canada. Society of American Foresters, Washington, D.C.

Franklin, J.F. 1980b. Coastal true fir – hemlock. Pages 103-104 *in* F.H. Eyre, editor. Forest cover types of the United States and Canada. Society of American Foresters, Washington, D.C.

Franklin, J.F., K. Cromack, W. Denison, A. McKee, C. Maser, J. Sedell, F. Swanson, and G. Juday. 1981. Ecological characteristics of old-growth Douglas-fir forests. General Technical Report PNW-118. USDA, Forest Service, Pacific Northwest Research Station, Portland, OR.

Franklin, J.F. and W. Emmingham. 1983. True fir – hemlock. Pages 13-15 *in* R.M. Burns, technical compiler. Silviculture systems for the major forest types of the United States. Agriculture Handbook No. 445. USDA, Forest Service, Washington, D.C.

Franklin, J.F. 1988. Pacific Northwest forests. Pages 104-130 *in* M.G. Barbour and W.D. Billings, editors. North American terrestrial vegetation. Cambridge University Press, New York, NY.

Franklin, J.F. 1990. *Abies procera* – noble fir. Pages 80-87 *in* R.M. Burns and B.H. Honkala, technical coordinators. Silvics of North America, Volume 1. Conifers. Agriculture Handbook 654. USDA, Forest Service, Washington, D.C.

Fritts, H.C. 1969. Bristlecone pine in the White Mountains of California. Growth and ring width characteristics, Paper No. 4. University of Arizona, Laboratory of Tree Ring Research, Tucson, AZ.

Gankin, R. and J. Major. 1964. *Arctostaphylos myrtifolia*, its biology and relationship to the problem of endemism. Ecology 45:792-808.

Gauch, H.G. 1982. Multivariate analysis in community ecology. Cambridge University Press, New York, NY.

Gautier, C. and P. Zedler. 1980. Research Natural Area proposal: Guatay Mountain, Descanso District, Cleveland National Forest. Unpublished report. USDA, Forest Service, Pacific Southwest Research Station, Berkeley, CA.

Gemmill, B. 1980. Radial growth of California black oak in the San Bernardino Mountains. Pages 128-135 *in* T.R. Plumb, editor. Ecology, management and utilization of California oaks. General Technical Report PSW-44. USDA, Forest Service, Pacific Southwest Research Station, Berkeley, CA.

George, M.R., J.R. Brown, and W.J. Clawson. 1992. Application of nonequilibrium ecology to management of Mediterranean grasslands. Journal of Range Management 45:436-440.

Gleason, H. 1926. The individualistic concept of the plant association. Bulletin of the Torrey Botanical Club 53:1-20.

Gleason, H. 1939. The individualistic concept of the plant association. American Midland Naturalist 21:92-110.

Gordon, D.T. 1980a. Red fir. Pages 87-88 *in* F.H. Eyre, editor. Forest cover types of the United States and Canada. Society of American Foresters, Washington, D.C.

Gordon, D.T. 1980b. White fir. Pages 92-93 *in* F.H. Eyre, editor. Forest cover types of the United States and Canada. Society of American Foresters, Washington, D.C.

Gordon, H.J. and T.C. White. 1994. Ecological guide to southern California chaparral plant series. Technical Publication R5-ECOL-TP-005. USDA, Forest Service, Pacific Southwest Region, San Francisco, CA.

Gore, A.J.E. 1983. Mires: swamp, bog, fen, and moor. Volume 25, Ecosystems of the world. Elsevier Science Publishing Co., New York, NY.

Graham, R.T. 1990. *Pinus monticola* – western white pine. Pages 385-394 *in* R.M. Burns and B.H. Honkala, technical coordinators. Silvics of North America, Volume 1. Conifers. Agriculture Handbook 654. USDA, Forest Service, Washington, D.C.

Grams, H.J., K.R. McPherson, V.V. King, S.A. MacLeod, and M.G. Barbour. 1977. Northern coastal scrub on Point Reyes Peninsula, California. Madroño 24:18-23.

Graves, G.W. 1932. Ecological relationship of *Pinus sabiniana*. Botanical Gazette 94:106-133.

Gray, J.T. 1978. The vegetation of two California mountain slopes. Madroño 25:177-185.

Gray, J.T. 1982. Community structure in *Ceanothus* chaparral and coastal sage scrub of southern California. Ecological Monographs 52:415-435.

Gray, J.T. 1983. Competition for light and a dynamic boundary between chaparral and coastal sage scrub. Madroño 30:43-49.

Gray, M.V. and J.M. Greaves. 1984. Riparian forest as habitat for the least Bell's vireo. Pages 605-611 *in* R.E. Warner and K.M. Hendrix, editors. California riparian systems: ecology, conservation and productive management. University of California Press, Berkeley, CA.

Green, H.A. 1929. Historical note on the Monterey cypress at Cypress Point. Madroño 1:197-198.

Grenier, K.H. 1989. Vegetation patterns in grasslands of Redwood National Park, California. Master's thesis, Humboldt State University, Arcata, CA.

Griffin, J.R. 1965. Digger pine seedling response to serpentinite and non-serpentinite soil. Ecology 46:801-807.

Griffin, J.R. and W.B. Critchfield. 1972. The distribution of forest trees in California. Research Paper PSW-82. USDA, Forest Service, Pacific Southwest Research Station, Berkeley, CA.

Griffin, J.R. 1974. A strange forest in San Benito County. Fremontia 2:11-15.

Griffin, J.R. 1975. Ecological survey of Teakettle Creek candidate Research Natural Area, Sierra National Forest. Unpublished report. USDA, Forest Service, Pacific Southwest Research Station, Berkeley, CA.

Griffin, J.R. 1976a. Ecological survey of South Fork Devil's Canyon candidate Research Natural Area (NW Cone Peak region), Monterey District, Los Padres National Forest. Unpublished report. USDA, Forest Service, Pacific Southwest Research Station, Berkeley, CA.

Griffin, J.R. 1976b. Regeneration on *Quercus lobata* savannas, Santa Lucia Mountains, California. American Midland Naturalist 95:422-435.

Griffin, J.R. 1977. Oak woodland. Pages 383-415 *in* M.G. Barbour and J. Major, editors. Terrestrial vegetation of California. Wiley–Interscience, reprinted by the California Native Plant Society 1988, Sacramento, CA.

Griffin, J.R. 1978. Maritime chaparral and endemic shrubs of the Monterey Bay region, California. Madroño 25:65-81.

Griffin, J.R. 1985. Isolated *Pinus ponderosa* forests on sandy soils near Santa Cruz, California. Ecology 45:410-412.

Griggs, F.T. 1974. Systematics and ecology of the genus *Orcuttia* (Gramineae). Master's thesis, Chico State University, Chico, CA.

Griggs, F.T. 1980a. Element preservation plan: alkali bunchgrass prairie. Unpublished report. The Nature Conservancy, San Francisco, CA.

Griggs, F.T. 1980b. Valley saltbush scrub element protection plan. Unpublished report. The Nature Conservancy, San Francisco, CA.

Griggs, F.T. and J. Zaninovich. 1984. Definitions of Tulare Basin plant associations. Unpublished report. The Nature Conservancy, San Francisco, CA.

Griggs, F.T. 1987. The ecological setting for the natural regeneration of Engelmann oak (*Quercus engelmannii*

Greene) on the Santa Rosa plateau, Riverside County, California. Pages 71-75 *in* T.R. Plumb and N.H. Pillsbury, editors. Multiple-use management of California's hardwood resources. General Technical Report, PSW-100. USDA, Forest Service, Pacific Southwest Research Station, Berkeley, CA.

Grossman, D.H., K.L. Goodin, and C.L. Reuss. 1994. Rare plant communities of the coterminous United States. The Nature Conservancy, Arlington, VA.

Gudmonds, K.N. and M.G. Barbour. 1987. Mixed evergreen forest stands in the northern Sierra Nevada. Pages 32-37 *in* T.R. Plumb and N.H. Pillsbury, editors. Multiple-use management of California's hardwood resources. General Technical Report, PSW 100. USDA, Forest Service, Pacific Southwest Research Station, Berkeley, CA.

Haddock, P.G. 1961. New data on distribution of some true firs of the Pacific Coast. Forest Service 7:349-350.

Haidinger, T.L. and J.E. Keeley. 1993. Role of high fire frequency in destruction of mixed chaparral. Madroño 40:141-147.

Hall, F.C. 1977. Western juniper in association with other tree species. *in* R.E. Martin, E.J. Dealy, and R.S. Driscoll, editors. Proceedings of the western juniper ecology and management workshop. General Technical Report PNW-74. USDA, Forest Service, Pacific Northwest Research Station, Portland, OR.

Haller, J.R. 1967. A comparison of the mainland and island populations of Torrey pine. Pages 79-88 *in* R.N. Philbrick, editors. Proceedings of the symposium on the biology of the California Islands. Santa Barbara Botanic Garden, Santa Barbara, CA.

Halpern, C.B. 1985. Hydric montane meadows of Sequoia National Park, California: a literature review and classification. Technical Report No. 20. USDI, National Park Service, Cooperative National Park Resources Studies Unit, University of California, Davis, CA.

Halpern, C.B. 1986. Montane meadow plant associations of Sequoia National Park, California. Madroño 33:1-23.

Hamrick, J.L. and W.J. Libby. 1972. Variation and selection in western U.S. montane species. I. White fir. Silvae genetica 21:29-35.

Hanes, T.L. 1965. Ecological studies on two closely related chaparral shrubs in southern California. Ecological Monographs 35:213-235.

Hanes, T.L. and H. Jones. 1967. Postfire chaparral succession in southern California. Ecology 48:259-264.

Hanes, T.L. 1976. Vegetation types of the San Gabriel Mountains. Pages 65-76 *in* J. Latting, editor. Plant communities of southern California. California Native Plant Society, Sacramento, CA.

Hanes, T.L. 1977. Chaparral. Pages 417-469 *in* M.G. Barbour and J. Major, editors. Terrestrial vegetation of California. Wiley–Interscience, reprinted by the California Native Plant Society 1988, Sacramento, CA.

Hanes, T.L. 1981. California chaparral. Pages 139-174 *in* F. di Castri, D.W. Goodall and R.L. Specht, editors. Ecosystems of the world 11: Mediterranean-type shrublands. Elsevier Scientific Publishing Company, New York, NY.

Hanes, T.L. 1984. Vegetation of the Santa Ana River and some flood control implications. Pages 882-888 *in* R.E. Warner and K.M. Hendrix, editors. California riparian systems: ecology, conservation and productive management. University of California Press, Berkeley, CA.

Hanes, T.L., R.D. Friesen, and K. Keane. 1989. Alluvial scrub vegetation in coastal southern California. General Technical Report PSW-110. USDA, Forest Service, Pacific Southwest Research Station, Berkeley, CA.

Hanes, T.H., B. Hecht, and L.P. Stromberg. 1990. Water relationships of vernal pools in the Sacramento Region, California. Pages 49-60 *in* D.H. Ikeda, R.A. Schlising, F.J. Fuller, L.P. Janeway, and P. Woods, editors. Vernal pool plants: their habitat and biology. Study No. 8 from the Herbarium. California State University, Chico, CA.

Hanes, T.L. 1994. Chamise chaparral – SRM 206. Page 14 *in* T.N. Shiflet, editor. Rangeland cover types of the United States. Society for Range Management, Denver, CO.

Harrington, C.A. 1990. *Alnus rubra* – red alder. Pages 116-123 *in* R.M. Burns and B.H. Honkala, technical coordinators. Silvics of North America, Volume 2. Hardwoods. Agriculture Handbook 654. USDA, Forest Service, Washington, D.C.

Harris, A.S. 1980a. Sitka spruce. Pages 101-102 *in* F.H. Eyre, editor. Forest cover types of the United States and Canada. Society of American Foresters, Washington, D.C.

Harris, A.S. 1980b. Western hemlock. Pages 102-103 *in* F.H. Eyre, editor. Forest cover types of the United States and Canada. Society of American Foresters, Washington, D.C.

Harris, A.S. and D.L. Johnson. 1983. Western hemlock – Sitka spruce. Pages 5-8 *in* R.M. Burns, technical compiler. Silviculture systems for the major forest types of the United States. Agriculture Handbook No. 445. USDA, Forest Service, Washington, D.C.

Harris, A.S. 1990a. *Chamaecyparis nootkatensis* – Alaska cedar. Pages 97-102 *in* R.M. Burns and B.H. Honkala, technical coordinators. Silvics of North America, Volume 1. Conifers. Agriculture Handbook 654. USDA, Forest Service, Washington, D.C.

Harris, A.S. 1990b. *Picea sitchensis* – Sitka spruce. Pages 260-267 *in* R.M. Burns and B.H. Honkala, technical coordinators. Silvics of North America, Volume 1. Conifers. Agriculture Handbook 654. USDA, Forest Service, Washington, D.C.

Hawksworth, F.G. and D.K. Bailey. 1980. Bristlecone pine. Pages 89-90 *in* F.H. Eyre, editor. Forest cover types of the United States and Canada. Society of American Foresters, Washington, D.C.

Heady, H.F., T.C. Foin, M.M. Hektner, D.W. Taylor, M.G. Barbour, and W.J. Barry. 1977. Coastal prairie and northern coastal scrub. Pages 733-760 *in* M.G. Barbour and J. Major, editors. Terrestrial vegetation of California. Wiley–Interscience, reprinted by the California Native Plant Society 1988, Sacramento, CA.

Heady, H.F. 1977. Valley grassland. Pages 491-514 *in* M.G. Barbour and J. Major, editors. Terrestrial vegetation of California. Wiley–Interscience, reprinted by the California Native Plant Society 1988, Sacramento, CA.

Heckard, L.R. and J.C. Hickman. 1984. The phytogeographical significance of Snow Mountain, North Coast Ranges, California. Madroño 31:30-47.

Hektner, M.M. and T.C. Foin. 1977. Vegetation analysis of a northern California coastal prairie: Sea Ranch, Sonoma County, California. Madroño 24:83-103.

Helms, J.A. and R.D. Ratliff. 1987. Germination and establishment of *Pinus contorta* var. *murrayana* (Pinaceae) in mountain meadows of Yosemite National Park, California. Madroño 34:77-90.

Hendrickson, J. and B. Prigge. 1975. White fir in the mountains of eastern Mojave Desert of California. Madroño 23:164-168.

Hendrickson, J. 1976. Ecology of southern California coastal salt marshes. Pages 49-64 *in* J. Latting, editor. Plant communities of southern California. California Native Plant Society, Sacramento, CA.

Henry, M.A. 1979. A rare grass on the Eureka Dunes. Fremontia 7:3-6.

Hermann, F.J. 1970. Manual of the carices of the Rocky Mountains and Colorado Basin, Agriculture Handbook No. 374. USDA, Forest Service, Washington, D.C.

Hermann, R.K. and D.P. Lavender. 1990. *Pseudotsuga menziesii* – Douglas-fir. Pages 527-540 *in* R.M. Burns and B.H. Honkala, technical coordinators. Silvics of North America, Volume 1. Conifers. Agriculture Handbook 654. USDA, Forest Service, Washington, D.C.

Hickman, J.C. 1993. The Jepson manual: higher plants of California. University of California Press, Berkeley, CA.

Hilu, K.W., S. Boyd, and P. Felker. 1982. Morphological diversity and taxonomy of California mesquites (*Prosopis*, Leguminosae). Madroño 29:237-254.

Hobbs, S.D., S.D. Tesch, P.W. Owston, R.E. Stewart, J.C. Tappeiner, and G. E. Wells, editors. 1992. Reforestation practices in Southwestern Oregon and Northern California. Forest Research Laboratory. Corvallis, Oregon.

Holland, D.C. 1987. *Prosopis* (Mimosaceae) in the San Joaquin Valley, California: vanishing relict or recent invader? Madroño 34:324-333.

Holland, D.C. 1988. Additional support for the recent-invasive advent of mesquite (Mimosaceae: *Prosopis*) in the San Joaquin Valley, California. Madroño 35:329.

Holland, R.F. and F.T. Griggs. 1976. A unique habitat: California's vernal pools. Fremontia 4:3-6.

Holland, R.F. 1976. Vernal pools: their ecology and conservation. Publication No. 9. Pages 11-15 *in* S. Jain, editor. Vernal pools: their ecology and conservation. Publication No. 9. Institute of Ecology, University of California, Davis, CA.

Holland, R.F. and S.K. Jain. 1977. Vernal pools. Pages 515-533 *in* M.G. Barbour and J. Major, editors. Terrestrial vegetation of California. Wiley–Interscience, reprinted by the California Native Plant Society 1988, Sacramento, CA.

Holland, R.F. 1978. Vernal pools: the geographic and edaphic distribution of vernal pools in the Great Central Valley. California Native Plant Society, Sacramento, CA.

Holland, R.F. and S.K. Jain. 1981. Insular biogeography of vernal pools in the Central Valley of California. American Naturalist 117:24-37.

Holland, R.F. 1986. Preliminary descriptions of the terrestrial natural communities of California. Unpublished report. State of California, The Resources Agency, Department of Fish and Game, Natural Heritage Division, Sacramento, CA.

Holland, R.F. and V.I. Dains. 1990. The edaphic factor in vernal pool vegetation. Pages 31-48 *in* D.H. Ikeda, R.A. Schlising, F.J. Fuller, L.P. Janeway, and P. Woods, editors. Vernal pool plants: their habitat and biology. Study No. 8 from the Herbarium. California State University, Chico, CA.

Holland, V.L. 1976. In defense of blue oaks. Fremontia 4:3-8.

Holstein, G. 1981. An update on conservation of vernal pools in California. Pages 273-297 *in* S.K. Jain and P. Moyle, editors. Vernal pools and intermittent streams. Publication No. 28. Institute of Ecology, University of California, Davis, CA.

Holstein, G. 1984. California riparian forests: deciduous islands in an evergreen sea. Pages 2-22 *in* R.E. Warner and K.M. Hendrix, editors. California riparian systems: ecology, conservation and productive management. University of California Press, Berkeley, CA.

Holton, B. and A.F. Johnson. 1979. Dune scrub communities and their correlation with environmental factors at Point Reyes National Seashore, California. Journal of Biogeography 6:317-328.

Holzman, B. 1994. Creosote bush scrub – SRM 211. Page 20 *in* T.N. Shiflet, editor. Rangeland cover types of the United States. Society for Range Management, Denver, CO.

Horton, J.S. 1960. Vegetation types of the San Bernardino Mountains. PSW Technical Paper 44. USDA, Forest Service, Pacific Southwest Research Station, Berkeley, CA.

Howard, A.Q. and R.A. Arnold. 1980. Antioch Dunes – safe at last? Fremontia 8:3-12.

Howard, A.Q. 1992. Understanding the pygmy forest through time. Fremontia 20:17-21.

Huenneke, L.F. and H.A. Mooney. 1989. Grassland structure and function: California annual grassland. Kluwer Academic Publishers, Boston, MA.

Hull, J.C. and C.H. Muller. 1977. The potential for dominance by *Stipa pulchra* in a California grassland. American Midland Naturalist 97:147-175.

Hunt, C.B. 1966. Plant ecology of Death Valley, California. Professional paper 509. USDI, Geological Survey, Washington, D.C.

Hunt, J.L. 1976. Environmental restriction of *Abies amabilis* in the English Peak area, Marble Mountains, California. Master's thesis, Humboldt State University, Arcata, CA.

Hunter, S.C. and T.E. Paysen. 1986. Vegetation classification system for California: user's guide. General Technical Report PSW-94. USDA, Forest Service, Pacific Southwest Research Station, Berkeley, CA.

Huntsinger, L. and J.W. Bartolome. 1992. Ecological dynamics of *Quercus*-dominated woodlands in southern California and southern Spain: a state transition model. Vegetatio 99-100:299-305.

Ikeda, D.H., R.A. Schlising, F.J. Fuller, L. P. Janeway, and P. Woods, editors. 1989. Vernal pool plants: their habitat and biology. Study No. 8 from the Herbarium. California State University, Chico, CA.

Imper, D.K., G.E. Hoovey, J.O. Sawyer, and S.A. Carlson. 1987. Operations and maintenance schedule. Report I: Inventory and description. Unpublished report. State of California, The Resources Agency, Department of Fish and Game, Eureka, CA.

Imper, D.K. 1988a. Ecological survey of the proposed Haypress Meadows Research Natural Area, SAF type 207, Klamath National Forest. Unpublished report. USDA, Forest Service, Pacific Southwest Research Station, Berkeley, CA.

Imper, D.K. 1988b. Ecological survey of the proposed Shasta Red Fir Research Natural Area, SAF type 207 (red fir), Shasta-Trinity National Forests. Unpublished report. USDA, Forest Service, Pacific Southwest Research Station, Berkeley, CA.

Imper, D.K. 1991. Ecological survey of the proposed Mayfield Research Natural Area, SAF type 248 (Knobcone pine), Lassen National Forest. Unpublished report. USDA, Forest Service, Pacific Southwest Research Station, Berkeley, CA.

Ives, J.D. and R.G. Barry. 1976. Arctic and alpine environments. Methuen & Co., London, U.K.

Jackson, L.E. and L.C. Bliss. 1982. Distribution of ephemeral herbaceous plants near treeline in the Sierra Nevada, California, U.S.A. Arctic and Alpine Research 14:33-42.

Jain, S.K. 1976. Vernal pools: their ecology and conservation. Publication No. 9. Institute of Ecology, University of California, Davis, CA.

Jain, S.K. and P. Moyle. 1981. Vernal pools and intermittent streams. Publication No. 28. Institute of Ecology, University of California, Davis, CA.

Jenkinson, J.L. 1980. Jeffrey pine. Page 123 *in* F.H. Eyre, editor. Forest cover types of the United States and Canada. Society of American Foresters, Washington, D.C.

Jenkinson, J.L. 1990. *Pinus jeffreyi* – Jeffrey pine. Pages 359-369 *in* R.M. Burns and B.H. Honkala, technical coordinators. Silvics of North America, Volume 1. Conifers. Agriculture Handbook 654. USDA, Forest Service, Washington, D.C.

Jenny, H., R.J. Arkley, and A.M. Schultz. 1969. The pygmy forest podsol ecosystem and its dune associates of the Mendocino coast. Madroño 20:60-74.

Jensen, D.B. and K.A. Schierenbeck. 1990. An ecological survey of the proposed Raider Creek Research Natural Area, Modoc National Forest,

California. Unpublished report. USDA, Forest Service, Pacific Southwest Research Station, Berkeley, CA.

Jensen, D.B. 1992. An ecological survey of the proposed Home Camp Creek Research Natural Area, Sierra National Forest, Fresno County, California. Unpublished report. USDA, Forest Service, Pacific Southwest Research Station, Berkeley, CA.

Jensen, H.A. 1947. A system for classifying vegetation in California. California Fish and Game 33:199-266.

Jimerson, T.M. and R.M. Creasy. 1991. Variation in Port Orford–cedar plant communities along primary environmental gradients in northwest California. Pages 122-133 *in* H.M. Kerner, editor. Biodiversity of Northwestern California. Report 29. University of California, Division of Agriculture and Natural Resources, Wildlife Resources Center, Berkeley, CA.

Jimerson, T.M. 1993. Preliminary plant associations of the Klamath province, Six Rivers and Klamath National Forests. Unpublished report. USDA, Forest Service, Six Rivers National Forest, Eureka, CA.

Jimerson, T.M. 1994. A field guide to Port Orford cedar plant associations in northwest California. Technical Publication R5-ECOL-TP-002. USDA, Forest Service, Pacific Southwest Region, San Francisco, CA.

Johnson, C.G. 1994. Green fescue – SRM 103. Page 4 *in* T.N. Shiflet, editor. Rangeland cover types of the United States. Society for Range Management, Denver, CO.

Johnson, H.B. 1976. Vegetation and plant communities of southern California deserts. Pages 125-162 *in* J. Latting, editor. Plant communities of southern California. California Native Plant Society, Sacramento, CA.

Johnson, J.W. 1963. Ecological study of dune flora, Humboldt Bay. Master's thesis, Humboldt State College, Arcata, CA.

Johnson, S. 1987. Can tamarisk be controlled. Fremontia 15:19-20.

Jokerst, J.A. 1983. The vascular plant flora of Table Mountain, Butte County, California. Madroño 30 [Supplement]:1-19.

Jokerst, J.D. 1987. An ecological reconnaissance survey of the Twin Rocks candidate Research Natural Area. Unpublished report. USDA, Forest Service, Pacific Southwest Research Station, Berkeley, CA.

Jokerst, J.D. 1990. Floristic analysis of volcanic mudflow vernal pools. Pages 1-29 in D.H. Ikeda, R.A. Schlising, F.J. Fuller, L.P. Janeway, and P. Woods, editors. Vernal pool plants: their habitat and biology. Study No. 8 from the Herbarium. California State University, Chico, CA.

Jones, K.G. 1984. The Nipomo Dunes. Fremontia 11: 3-10.

Jones and Stokes Associates. 1993. Methods used to survey the vegetation of Orange County parks and open space areas and The Irvine Company property. Report JSA 92-032. Sacramento, CA.

Jones and Stokes Associates. 1994a. The Monterey pine ecological staircase: the nature of vegetation and soils on different geomorphic surfaces on the Monterey Peninsula with an emphasis on Monterey pine forest. Report JSA 94083. Sacramento, CA.

Jones and Stokes Associates. 1994b. Monterey pine forest ecological assessment; historical distribution, ecology, and current status of Monterey pine. Report JSA 94083. Sacramento, CA.

Jongman, R.H., C.J. ter Braak, and O.F. van Tongeren. 1987. Data analysis in community and landscape ecology. Pudoc, Wageningen, The Netherlands.

Josselyn, M. 1983. The ecology of San Francisco Bay tidal marshes; a community profile. Biological Service Program FWS/OBS-83/23. USDI, Fish and Wildlife Service, Washington, D.C.

Keddy, P. 1993. Do ecological communities exist? A reply to Bastow Wilson. Journal of Vegetation Science 4:135-136.

Keeler-Wolf, T. and V. Keeler-Wolf. 1975. A survey of the scientific values of the proposed Hosselkus Limestone Research Natural Area, Shasta-Trinity National Forest, California. Unpublished report. USDA, Forest Service, Pacific Southwest Research Station, Berkeley, CA.

Keeler-Wolf, T. and V. Keeler-Wolf. 1976. A survey of the scientific values of the proposed Whippoorwill Flat Research Natural Area, Inyo National Forest, California. Unpublished report. USDA, Forest Service, Pacific Southwest Research Station, Berkeley, CA.

Keeler-Wolf, T. and V. Keeler-Wolf. 1977. A survey of the scientific values of the proposed Limekiln Creek Research Natural Area, Monterey Ranger District, Los Padres National Forest, California. Unpublished report. USDA, Forest Service, Pacific Southwest Research Station, Berkeley, CA.

Keeler-Wolf, T. and V. Keeler-Wolf. 1981. An ecological survey of the proposed Mount Pleasant Research Natural Area, Plumas National Forest, California. Unpublished report. USDA, Forest Service, Pacific Southwest Research Station, Berkeley, CA.

Keeler-Wolf, T. 1982. An ecological survey of the proposed Cedar Basin Research Natural Area, Shasta-Trinity National Forest, California. Unpublished report. USDA, Forest Service, Pacific Southwest Research Station, Berkeley, CA.

Keeler-Wolf, T. 1983. An ecological survey of the Frenzel Creek Research Natural Area, Mendocino National Forest, Colusa County, California. Unpublished report. USDA, Forest Service, Pacific Southwest Research Station, Berkeley, CA.

Keeler-Wolf, T. 1984a. Ecological evaluation of the Rough Gulch Drainage with comparisons to the adjacent Chinquapin and Yolla Bolly candidate Research Natural Areas, Shasta-Trinity National Forest, California. Unpublished report. USDA, Forest Service, Shasta-Trinity National Forest, Redding, CA.

Keeler-Wolf, T. 1984b. Vegetation map of the upper Sugar Creek drainage, Siskiyou County, California. Unpublished report. USDA, Forest Service, Pacific Southwest Research Station, Berkeley, CA.

Keeler-Wolf, T. 1984c. An ecological survey of the Mount Shasta Mudflow Research Natural Area, Shasta-Trinity National Forest, California. Unpublished report. USDA, Forest Service, Pacific Southwest Research Station, Berkeley, CA.

Keeler-Wolf, T. 1984d. An ecological survey of the Devil's Garden Research Natural Area, Modoc National Forest, California. Unpublished report. USDA, Forest Service, Pacific Southwest Research Station, Berkeley, CA.

Keeler-Wolf, T. 1985a. Ecological survey of the proposed Bridge Creek Research Natural Area, Klamath National Forest, Siskiyou County, California. Unpublished report. USDA, Forest Service, Pacific Southwest Research Station, Berkeley, CA.

Keeler-Wolf, T. 1985b. An ecological survey of the proposed Mud Lake-Wheeler Peak Baker Cypress Research Natural Area, Plumas National Forest, Plumas County, California. Unpublished report. USDA, Forest Service, Pacific Southwest Research Station, Berkeley, CA.

Keeler-Wolf, T. 1985c. An ecological survey of the proposed Bell Meadow Research Natural Area, Stanislaus National Forest, California. Unpublished report. USDA, Forest Service, Pacific Southwest Research Station, Berkeley, CA.

Keeler-Wolf, T. 1986a. An ecological survey of the proposed Stone Corral-Josephine Peridotite Research Natural Area (L.E. Horton-*Darlingtonia* Bog RNA) on the Six Rivers National Forest, Del Norte County, California. Unpublished report. USDA, Forest Service, Pacific Southwest Research Station, Berkeley, CA.

Keeler-Wolf, T. 1986b. An ecological survey of the proposed Indian Creek Research Natural Area, Lassen National Forest. Unpublished report. USDA, Forest Service, Pacific Southwest Research Station, Berkeley, CA.

Keeler-Wolf, T. 1986c. An ecological survey of the proposed Doll Basin Research Natural Area, Mendocino National Forest, California. Unpublished report. USDA, Forest Service, Pacific Southwest Research Station, Berkeley, CA.

Keeler-Wolf, T. 1986d. An ecological survey of the proposed Cahuilla Mountain Research Natural Area, San Bernardino National Forest, California. Unpublished report. USDA, Forest Service, Pacific Southwest Research Station, Berkeley, CA.

Keeler-Wolf, T. 1986e. An ecological survey of the proposed Hall Canyon Research Natural Area, San Bernardino National Forest, Riverside County, California. Unpublished report. USDA, Forest Service, Pacific Southwest Research Station, Berkeley, CA.

Keeler-Wolf, T. 1987a. An ecological survey of the proposed Crater Creek Research Natural Area, Klamath National Forest, Siskiyou County, California.

Unpublished report. USDA, Forest Service, Pacific Southwest Research Station, Berkeley, CA.

Keeler-Wolf, T. 1987b. An ecological survey of the proposed Pearch Creek Research Natural Area, Six Rivers National Forest, Humboldt County, California. Unpublished report. USDA, Forest Service, Pacific Southwest Research Station, Berkeley, CA.

Keeler-Wolf, T. 1987c. An ecological survey of the proposed Rock Creek Butte Research Natural Area, Siskiyou County, California. Unpublished report. USDA, Forest Service, Pacific Southwest Research Station, Berkeley, CA.

Keeler-Wolf, T. 1987d. An ecological survey of the proposed Upper Goose Creek Research Natural Area, Six Rivers National Forest, Del Norte County, California. Unpublished report. USDA, Forest Service, Pacific Southwest Research Station, Berkeley, CA.

Keeler-Wolf, T. 1987e. An ecological survey of the proposed Hale Ridge Research Natural Area, Mendocino National Forest, Lake County, California. Unpublished report. USDA, Forest Service, Pacific Southwest Research Station, Berkeley, CA.

Keeler-Wolf, T. 1987f. An ecological survey of the proposed Big Grizzly Black Oak Research Natural Area, Mariposa County, California. Unpublished report. USDA, Forest Service, Pacific Southwest Research Station, Berkeley, CA.

Keeler-Wolf, T. 1988a. The role of *Chrysolepis chrysophylla* (Fagaceae) in the *Pseudotsuga*–hardwood forest of the Klamath Province of California. Madroño 35: 285-308.

Keeler-Wolf, T. 1988b. Draft establishment record for Doll Basin Research Natural Area within Mendocino National Forest, Tehama County, California. Unpublished report. USDA, Forest Service, Pacific Southwest Research Station, Berkeley, CA.

Keeler-Wolf, T. 1988c. Establishment record for Mud Lake Research Natural Area within Plumas National Forest, Plumas County, California. Unpublished report. USDA, Forest Service, Pacific Southwest Research Station, Berkeley, CA.

Keeler-Wolf, T. 1988d. An ecological survey of the proposed Millard Canyon Research Natural Area, San Bernardino National Forest, Riverside County,

California. Unpublished report. USDA, Forest Service, Pacific Southwest Research Station, Berkeley, CA.

Keeler-Wolf, T. 1988e. Establishment report for Hall Canyon Research Natural Area within San Bernardino National Forest, Riverside, County, California. Unpublished report. USDA, Forest Service, Pacific Southwest Research Station, Berkeley, CA.

Keeler-Wolf, T. 1989a. An ecological survey of the Wagon Caves proposed Research Natural Area, Los Padres National Forest, Monterey Ranger District, Monterey County, California. Unpublished report. USDA, Forest Service, Pacific Southwest Research Station, Berkeley, CA.

Keeler-Wolf, T. 1989b. An ecological survey of the Moses Mountain candidate Research Natural Area, Sequoia National Forest, Tulare County, California. Unpublished report. USDA, Forest Service, Pacific Southwest Research Station, Berkeley, CA.

Keeler-Wolf, T. 1989c. Establishment record for Bridge Creek Research Natural Area within the Klamath National Forest, Siskiyou County, California. Unpublished report. USDA, Forest Service, Pacific Southwest Research Station, Berkeley, CA.

Keeler-Wolf, T. 1989d. Establishment record for Cedar Basin Research Natural Area within Shasta-Trinity National Forest, Siskiyou County, California. Unpublished report. USDA, Forest Service, Pacific Southwest Research Station, Berkeley, CA.

Keeler-Wolf, T. 1989e. Establishment record for Hosselkus Limestone Research Natural Area within the Shasta-Trinity National Forest, Shasta County, California. Unpublished report. USDA, Forest Service, Pacific Southwest Research Station, Berkeley, CA.

Keeler-Wolf, T. 1989f. Establishment record for Sugar Creek Research Natural Area within the Klamath National Forest, Siskiyou County, California. Unpublished report. USDA, Forest Service, Pacific Southwest Research Station, Berkeley, CA.

Keeler-Wolf, T. 1989g. Establishment record for Antelope Creek Lakes Research Natural Area within the Klamath National Forest, Siskiyou County, California. Unpublished report. USDA, Forest Service, Pacific Southwest Research Station, Berkeley, CA.

Keeler-Wolf, T. 1989h. Establishment record for the Whippoorwill Flat Research Natural Area within the Inyo National Forest, Inyo County, California. Unpublished report. USDA, Forest Service, Pacific Southwest Research Station, Berkeley, CA.

Keeler-Wolf, T. 1989i. Establishment record for the Ruth Research Natural Area within the Six Rivers National Forest, Trinity County, California. Unpublished report. USDA, Forest Service, Pacific Southwest Research Station, Berkeley, CA.

Keeler-Wolf, T. 1989j. Establishment record for Babbitt Peak Research Natural Area within the Tahoe and Toiyabe National Forests, Sierra County, California. Unpublished report. USDA, Forest Service, Pacific Southwest Research Station, Berkeley, CA.

Keeler-Wolf, T. 1989k. Establishment record for Mount Pleasant Research Natural Area, Plumas National Forest, Plumas County, California. Unpublished report. USDA, Forest Service, Pacific Southwest Research Station, Berkeley, CA.

Keeler-Wolf, T. 1989l. Establishment record for Bell Meadow Research Natural Area within the Stanislaus National Forest, Tuolumne County, California. Unpublished report. USDA, Forest Service, Pacific Southwest Research Station, Berkeley, CA.

Keeler-Wolf, T. 1989m. Establishment record for Cahuilla Mountain Research Natural Area within San Bernardino National Forest, Riverside County, California. Unpublished report. USDA, Forest Service, Pacific Southwest Research Station, Berkeley, CA.

Keeler-Wolf, T. 1990a. An ecological survey of the proposed King Creek Research Natural Area, Cleveland National Forest, San Diego County, California. Unpublished report. USDA, Forest Service, Pacific Southwest Research Station, Berkeley, CA.

Keeler-Wolf, T. 1990b. An ecological survey of the proposed Timber Crater Research Natural Area, Lassen National Forest, Siskiyou County, California. Unpublished report. USDA, Forest Service, Pacific Southwest Research Station, Berkeley, CA.

Keeler-Wolf, T. 1990c. An ecological survey of the proposed Long Canyon Research Natural Area, Sequoia National Forest, Kern County, California. Unpublished report. USDA, Forest Service, Pacific Southwest Research Station, Berkeley, CA.

Keeler-Wolf, T. 1990d. Establishment record for Indian Creek Research Natural Area within Lassen National Forest, Tehama County, California. Unpublished report. USDA, Forest Service, Pacific Southwest Research Station, Berkeley, CA.

Keeler-Wolf, T. 1990e. Ecological surveys of forest service research natural areas in California. General Technical Report PSW-125. USDA, Forest Service, Pacific Southwest Research Station, Berkeley, CA.

Keeler-Wolf, T. 1991a. Ecological survey of the proposed Big Pine Mountain Research Natural Area, Los Padres National Forest, Santa Barbara County, California. Unpublished report. USDA, Forest Service, Pacific Southwest Research Station, Berkeley, CA.

Keeler-Wolf, T. 1991b. An ecological survey of the proposed Mountaineer Creek Research Natural Area, Sequoia National Forest, Tulare County, California. Unpublished report. USDA, Forest Service, Pacific Southwest Research Station, Berkeley, CA.

Keeler-Wolf, T. 1992. Ecological survey of the Graham Pinery candidate Research Natural Area, Lassen National Forest, California. Unpublished report. USDA, Forest Service, Pacific Southwest Research Station, Berkeley, CA.

Keeler-Wolf, T. 1993. Conserving California's rare plant communities. Fremontia 22:14-22.

Keeler-Wolf, T., K. Lewis, and C. Roye. 1994. The definition and location of Sycamore alluvial woodland in California. Unpublished report. State of California, The Resources Agency, Department of Fish and Game, Sacramento, CA.

Keeler-Wolf, V. and T. Keeler-Wolf. 1974. A contribution to the natural history of the Northern Yolla Bolly Mountains. Senior thesis, University of California, Santa Cruz, CA.

Keeley, J.E. 1975. Longevity of nonsprouting *Ceanothus*. American Midland Naturalist 93:504-507.

Keeley, J.E. and S.C. Keeley. 1978. Knobcone pine southward range extension in the Sierra Nevada. Madroño 25:106.

Keeley, J.E. and P.H. Zedler. 1978. Reproduction of chaparral shrubs after fire: a comparison of sprouting and seeding strategies. American Midland Naturalist 99:142-161.

Keeley, J.E. 1987a. Ten years of change in seed banks of the chaparral shrubs, *Arctostaphylos glauca* and *A. glandulosa*. American Midland Naturalist 117:446-448.

Keeley, J.E. 1987b. Fruit production patterns in the chaparral shrub *Ceanothus crassifolius*. Madroño 34: 273-282.

Keeley, J.E. and S.C. Keeley. 1988. Chaparral. Pages 165-207 *in* M.G. Barbour and W.D. Billings, editors. North American terrestrial vegetation. Cambridge University Press, New York, NY.

Keeley, J.E. 1989. The California valley grassland. Pages 3-23 *in* A.A. Schoenherr, editor. Endangered plant communities of southern California. Southern California Botanists, California State University, Fullerton, CA.

Keeley, J.E. 1990. Demographic structure of California black walnut (*Juglans californica*; Juglandaceae) woodlands in southern California. Madroño 37:237-248.

Keeley, J.E. 1992a. Demographic structure of California chaparral in the long-term absence of fire. Journal of Vegetation Science 3:79-90.

Keeley, J.E. 1992b. Recruitment of seedlings and vegetative sprouts in unburned chaparral. Ecology 73:1194-1208.

Keeley, J.E. 1993. Native grassland restoration: the initial stage, assessing suitable sites. Pages 277-281 *in* J.E. Keeley, editor. Interface between ecology and land development in California. Southern California Academy of Sciences, Los Angeles, CA.

Kellogg, A. and E.L. Greene. 1889. Illustrations of west American oaks. Published from funds provided by James N. McDonald, San Francisco, CA.

Kellogg, E.M. and J.L. Kellogg. 1990. A study on the distribution and pattern of perennial grassland on the Camp Pendleton Marine Corps Base. Unpublished report. Tierra Data Systems, Reedley, CA.

Kellogg, E.M. and J.L. Kellogg. 1991. A re-sampling of established transects for determination of trend. Unpublished report. Tierra Data Systems, Reedley, CA.

Kent, M. and P. Coker. 1992. Vegetation description and analysis: a practical approach. CRC Press, Boca Raton, FL.

Kerbavaz, J.H. 1985. Status reports on invasive weeds: pampas grass. Fremontia 12:18.

Kinloch, B.B., and W.H. Scheuner. 1990. *Pinus lambertiana* – sugar pine. Pages 370-379 *in* R.M. Burns and B.H. Honkala, technical coordinators. Silvics of North America, Volume 1. Conifers. Agriculture Handbook 654. USDA, Forest Service, Washington, D.C.

Kirkpatrick, J.B. and C.F. Hutchinson. 1977. The community composition of Californian coastal sage scrub. Vegetatio 35:21-33.

Klikoff, L.G. 1965. Microenvironmental influence on vegetational pattern near timberline in the central Sierra Nevada. Ecological Monographs 35:187-211.

Klyver, F.D. 1931. Major plant communities in a transect of the Sierra Nevada Mountains of California. Ecology 12:1-17.

Knight, W., I. Knight, and J.T. Howell. 1970. A vegetation survey of the Butterfly Botanical Area, California. Wasmann Journal of Biology 28:1-46.

Knudsen, M.D. 1987. Life history aspects of *Quercus lobata* in riparian community, Sacramento Valley, California. Pages 38-46 *in* T.R. Plumb and N.H. Pillsbury, editors. Multiple-use management of California's hardwood resources. General Technical Report, PSW-100. USDA, Forest Service, Pacific Southwest Research Station, Berkeley, CA.

Kopecko, K.J. and E.W.Lathrop. 1975. Vegetational zonation in a vernal marsh on the Santa Rosa Plateau of Riverside County, California. Aliso 8:281-288.

Koplin, J.R., A. Franklin, and G. Newton. 1984. Elk River wildlife area monitoring project, final report. Unpublished report. City of Eureka, Public Works Department, Eureka, CA.

Krantz, T. 1983. The pebble plains of Baldwin Lake. Fremontia 10:9-13.

Krantz, T. 1988. Limestone endemics of Big Bear Valley. Fremontia 16:20-21.

Kruckeberg, A.R. 1984. California serpentines: flora, vegetation, geology, soils, and management problems. University of California Press, Berkeley, CA.

Kummerow, J., B.A. Ellis, and J.N. Mills. 1985. Post-fire seedling establishment of *Adenostoma fasciculatum* and *Ceanothus greggii* in southern California chaparral. Madroño 32:148-157.

Kurmes, E.A. and D.E. Wommack. 1980. Arizona cypress. Page 117 *in* F.H. Eyre, editor. Forest cover types of the United States and Canada. Society of American Foresters, Washington, D.C.

Küchler, A.W. 1977. The map of the natural vegetation of California. Pages 909-938 *in* M.G. Barbour and J. Major, editors. Terrestrial vegetation of California. Wiley–Interscience, reprinted by the California Native Plant Society 1988, Sacramento, CA.

Laacke, R.J. and J.N. Fiske. 1983a. Red fir and white fir. Pages 41-43 *in* R.M. Burns, technical compiler. Silviculture systems for the major forest types of the United States. Agriculture Handbook No. 445. USDA, Forest Service, Washington, D.C.

Laacke, R.J. and J.N. Fiske. 1983b. Sierra Nevada mixed conifers. Pages 44-47 *in* R.M. Burns, technical compiler. Silviculture systems for the major forest types of the United States. Agriculture Handbook No. 445. USDA, Forest Service, Washington, D.C.

Laacke, R.J. 1990a. *Abies concolor* – white fir. Pages 36-46 *in* R.M. Burns and B.H. Honkala, technical coordinators. Silvics of North America, Volume 1. Conifers. Agriculture Handbook 654. USDA, Forest Service, Washington, D.C.

Laacke, R.J. 1990b. *Abies magnifica* – California red fir. *in* R.M. Burns and B.H. Honkala, technical coordinators. Silvics of North America, Volume 1. Conifers. Agriculture Handbook 654. USDA, Forest Service, Washington, D.C.

LaBanca, T. 1993. Vegetation changes at Clam Beach coastal dunes, Humboldt County, California. Master's thesis, Humboldt State University, Arcata, CA.

Laidlaw-Holmes, J.M. 1981. Forest habitat types on metasedimentary soil of the South Fork Mountain Region of California. Master's thesis, Humboldt State University, Arcata, CA.

LaMarche, V.C. and H.A. Mooney. 1967. Altithermal timberline advance in western United States. Nature 213:980-982.

Landis, F. 1993. S.E.A. database terrestrial plant community definitions. Preliminary edition. Printed by the author, Los Angeles, CA.

Langer, S. 1988. New discoveries of Pacific silver fir in the Siskiyous of California. Four Seasons 8:35-37.

Lanner, R.M. 1984. Bristlecone pine and Clark's nut-cracker: probable interaction in the White Mountains, California. Great Basin Naturalist 44:357-360.

Larigauderie, A., T.W. Hubbard, and J. Kummerow. 1990. Growth dynamics of two chaparral shrub species with time after fire. Madroño 37:225-236.

Lathrop, E.W. and R.F. Thorne. 1976. Vernal pools of the Santa Rosa Plateau. Fremontia 4:9-11.

Lathrop, E.W. and B.D. Martin. 1982. Response of understory vegetation to prescribed burning in yellow pine forests of Cuyamaca Rancho State Park, California. Aliso 10:329-343.

Lathrop, E.W. and R.F. Thorne. 1983. A flora of the vernal pools on the Santa Rosa Plateau, Riverside County, California. Aliso 10:449-469.

Lathrop, E.W. and H.A. Zuill. 1984. Southern oak woodlands of the Santa Rosa Plateau, Riverside County, California. Aliso 10:603-611.

Lathrop, E.W. and M.J. Arct. 1987. Age structure of Engelmann oak populations on the Santa Rosa Plateau. Pages 47-52 in T.R. Plumb and N.H. Pillsbury, editors. Multiple-use management of California's hardwood resources. General Technical Report, PSW-100. USDA, Forest Service, Pacific Southwest Research Station, Berkeley, CA.

Latting, J. 1976. Plant communities of southern California. California Native Plant Society, Sacramento, CA.

Ledig, F.T. 1984. Gene conservation, endemics, and California's Torrey pine. Fremontia 12:9-13.

Leitner, B.M. and P. Leitner. 1988. An ecological survey of the proposed Soldier Research Natural Area, Six Rivers National Forest, Trinity County, California.

Unpublished report. USDA, Forest Service, Pacific Southwest Research Station, Berkeley, CA.

Lenihan, J.M. 1990. Forest associations of Little Lost Man Creek, Humboldt County, California: reference level in the hierarchical structure of old-growth coastal redwood vegetation. Madroño 37:69-87.

Leonard and Associates. 1993. Mount Tamalpais area vegetation management plan. Unpublished report. Marin Municipal Water District, Marin County Open Space District, San Rafael, CA.

Lewis, P.A. 1966. Plant communities of the Marble Mountains Wilderness, Siskiyou County, California. Master's thesis, Pacific Union College, Angwin, CA.

Lin, J. 1970. The floristic and plant succession in vernal pools vegetation. Master's thesis, San Francisco State College, San Francisco, CA.

Lippmann, M.C. 1977. More on the weedy "pampas grass" in California. Fremontia 4:25-27.

Liu, T.S. 1971. A monograph of the Genus *Abies*. Taipei.

Lloret, F. and P.H. Zedler. 1991. Recruitment pattern of *Rhus integrifolia* populations in periods between fire in chaparral. Journal of Vegetation Science 2:217-230.

Lloyd, R.M. and R.S. Mitchell. 1973. A flora of the White Mountains, California and Nevada. University of California Press, Berkeley, CA.

Lotan, J.E. and W.B. Critchfield. 1990. *Pinus contorta* – lodgepole pine. Pages 302-315 in R.M. Burns and B.H. Honkala, technical coordinators. Silvics of North America, Volume 1. Conifers. Agriculture Handbook 654. USDA, Forest Service, Washington, D.C.

MacDonald, K.B. and M.G. Barbour. 1974. Beach and salt marsh vegetation of the North American coast. Pages 175-234 in R.J. Reimold and W.H. Queen, editors. Ecology of halophytes. Academic Press, New York, NY.

MacDonald, K.B. 1977. Coastal salt marsh. Pages 263-294 in M.G. Barbour and J. Major, editors. Terrestrial vegetation of California. Wiley–Interscience, reprinted by the California Native Plant Society 1988, Sacramento, CA.

MacDonald, R. 1976. Vegetation of the Phoenix Park vernal pools on the American River Bluffs, Sacramento County, California. Pages 69-76 *in* S. Jain, editor. Vernal pools: their ecology and conservation. Publication No. 9. Institute of Ecology, University of California, Davis, CA.

Mack, R.N. 1981. Invasion of *Bromus tectorum* L. into western North America: an ecological chronicle. Agroecosystems 7:145-165.

MacMahon, J.A., and F.H. Wagner. 1985. The Mojave, Sonoran and Chihuahuan deserts of North America. Pages 139-174 *in* M. Evenari, I. Noy-Meir, and D.W. Goodall, editors. Ecosystems of the world 12A: hot deserts and arid shrublands. Elsevier Scientific Publishing Company, New York, NY.

MacMahon, J.A. 1988. Warm deserts. Pages 231-264 *in* M.G. Barbour and W.D. Billings, editors. North American terrestrial vegetation. Cambridge University Press, New York, NY.

Magney, D.L. 1992. Descriptions of three new southern California vegetation types: southern cactus scrub, southern coastal needlegrass grassland, and scalebroom scrub. Crossosoma 18:1-9.

Major, J. and D.W. Taylor. 1977. Alpine. Pages 601-675 *in* M.G. Barbour and J. Major, editors. Terrestrial vegetation of California. Wiley–Interscience, reprinted by the California Native Plant Society 1988, Sacramento, CA.

Major, J. 1977. California climate in relation to vegetation. Pages 11-74 *in* M.G. Barbour and J. Major, editors. Terrestrial vegetation of California. Wiley–Interscience, reprinted by the California Native Plant Society 1988, Sacramento, CA.

Major, J. 1994. Alpine grassland – SRM 213. Page 23 *in* T.N. Shiflet, editor. Rangeland cover types of the United States. Society for Range Management, Denver, CO.

Malanson, G.P. 1984. Fire history and patterns of Venturan subassociation of Californian coastal sage scrub. Vegetatio 57:121-128.

Mallory, J.I. 1980. Canyon live oak. Pages 125-126 *in* F.H. Eyre, editor. Forest cover types of the United States and Canada. Society of American Foresters, Washington, D.C.

Marchand, D.E. 1973. Edaphic control of plant distribution in the White Mountains, eastern California. Ecology 54:233-250.

Marks, M., B. Lapin, and J. Randall. 1994. *Phragmites australis* (*P. communis*): threats, management, and monitoring. Natural Areas Journal 14:285-294.

Martin, R.E. 1980a. Western juniper. Pages 115-116 *in* F.H. Eyre, editor. Forest cover types of the United States and Canada. Society of American Foresters, Washington, D.C.

Martin, R.E. 1994. Blackbush – SRM 212. Page 21 *in* T.N. Shiflet, editor. Rangeland cover types of the United States. Society for Range Management, Denver, CO.

Martin, S.C. 1980b. Mesquite. Page 118 *in* F.H. Eyre, editor. Forest cover types of the United States and Canada. Society of American Foresters, Washington, D.C.

Mason, H.L. 1957. A flora of the marshes of California. University of California Press, Berkeley, CA.

Matthew, S.C. 1986a. Old-growth forest associations of the Bull Creek watershed, Humboldt Redwoods State Park, California. Master's thesis, Humboldt State University, Arcata, CA.

Matthew, S.C. 1986b. Vegetation map and inventory of old-growth forest in Humboldt Redwoods State Park. Unpublished report. State of California, The Resources Agency, Department of Parks and Recreation, Sacramento, CA.

Mathews, M.A. and N. Nedeff. 1995. California's native Monterey pine forest: can it be saved? Fremontia 23:36.

Mayer, K. and W. Landenslayer. 1988. A guide to wildlife habitats of California. State of California, The Resources Agency, Department of Forestry and Fire Protection, Sacramento, CA.

McBride, J.R. 1974. Plant succession in the Berkeley Hills, California. Madroño 22:317-380.

McBride, J.R. and E.C. Stone. 1976. Plant succession of the sand dunes of the Monterey Peninsula, California. American Midland Naturalist 96:118-131.

McBride, J.R. 1994. Riparian woodland – SRM 203. Page 13 *in* T.N. Shiflet, editor. Rangeland cover types of the United States. Society for Range Management, Denver, CO.

McClaran, M.P. 1987. Blue oak age structure relation to livestock grazing history in Tulare County, California. Pages 358-360 *in* T.R. Plumb and N.H. Pillsbury, editors. Multiple-use management of California's hardwood resources. General Technical Report, PSW-100. USDA, Forest Service, Pacific Southwest Research Station, Berkeley, CA.

McClaran, M.P. and J.W. Bartolome. 1989a. Effect of *Quercus douglasii* (Fagaceae) on herbaceous understory along a rainfall gradient. Madroño 36:141-153.

McClaran, M.P. and J.W. Bartolome. 1989b. Fire related recruitment in stagnant *Quercus douglasii* populations. Canadian Journal of Forest Research 19:580-585.

McClaran, M.P. and J.W. Bartolome. 1990. Comparison of actual and predicted blue oak age structures. Journal of Range Management 1:61-63.

McClintock, E. 1978. The Washington fan palm. Fremontia 6:3-5.

McClintock, E. 1985a. Escaped exotic weeds of California. Fremontia 12:3-6.

McClintock, E. 1985b. Status reports on invasive weeds: brooms. Fremontia 12:17.

McCormick, J. 1968. Succession. Via:22-35, 131-132.

McCormick, J.F. and R.B. Blatt. 1980. Recovery of an Appalachian forest following the chestnut blight or Catherine Keever – you were right! American Midland Naturalist 104:264-273.

McDonald, P.M. and E.E. Littrell. 1976. The bigcone Douglas fir – canyon live oak community in southern California. Madroño 23:310-320.

McDonald, P.M. 1980a. California black oak. Page 122 *in* F.H. Eyre, editor. Forest cover types of the United States and Canada. Society of American Foresters, Washington, D.C.

McDonald, P.M. 1980b. Pacific ponderosa pine. Pages 120-121 *in* F.H. Eyre, editor. Forest cover types of the

United States and Canada. Society of American Foresters, Washington, D.C.

McDonald, P.M. 1980c. Pacific ponderosa pine – Douglas-fir. Page 120 *in* F.H. Eyre, editor. Forest cover types of the United States and Canada. Society of American Foresters, Washington, D.C.

McDonald, P.M., D. Minore, and T. Atzet. 1983. Southwestern Oregon – northern California hardwoods. Pages 29-32 *in* R.M. Burns, technical compiler. Silviculture systems for the major forest types of the United States. Agriculture Handbook No. 445. USDA, Forest Service, Washington, D.C.

McDonald, P.M. and R.J. Laacke. 1990. *Pinus radiata* – Monterey pine. Pages 433-441 *in* R.M. Burns and B.H. Honkala, technical coordinators. Silvics of North America, Volume 1. Conifers. Agriculture Handbook 654. USDA, Forest Service, Washington, D.C.

McDonald, P.M. and J.C. Tappeiner. 1990. *Arbutus menziesii* – Pacific madrone. Pages 124-132 *in* R.M. Burns and B.H. Honkala, technical coordinators. Silvics of North America, Volume 2. Hardwoods. Agriculture Handbook 654. USDA, Forest Service, Washington, D.C.

McDonald, P.M. 1990a. *Pseudotsuga macrocarpa* – bigcone Douglas-fir. Pages 520-526 *in* R.M. Burns and B.H. Honkala, technical coordinators. Silvics of North America, Volume 1. Conifers. Agriculture Handbook 654. USDA, Forest Service, Washington, D.C.

McDonald, P.M. 1990b. *Quercus douglasii* – blue oak. Pages 631-639 *in* R.M. Burns and B.H. Honkala, technical coordinators. Silvics of North America, Volume 2. Hardwoods. Agriculture Handbook 654. USDA, Forest Service, Washington, D.C.

McDonald, P.M. 1990c. *Quercus kelloggii* – California black oak. Pages 661-671 *in* R.M. Burns and B.H. Honkala, technical coordinators. Silvics of North America, Volume 2. Hardwoods. Agriculture Handbook 654. USDA, Forest Service, Washington, D.C.

McHargue, L.T. 1973. A vegetational analysis of the Coachella Valley, California. Dissertation, University of California, Irvine, CA.

McKee, A. 1990. *Castanopsis chrysophylla* – giant chinkapin. Pages 234-239 *in* R.M. Burns and B.H. Honkala, technical coordinators. Silvics of North

America, Volume 2. Hardwoods. Agriculture Handbook 654. USDA, Forest Service, Washington, D.C.

Means, J.E. 1990. *Tsuga mertensiana* – mountain hemlock. Pages 623-634 *in* R.M. Burns and B.H. Honkala, technical coordinators. Silvics of North America, Volume 1. Conifers. Agriculture Handbook 654. USDA, Forest Service, Washington, D.C.

Medeiros, J.L. 1979. San Luis Island: The last of the great valley. Fremontia 7:3-9.

Meeuwig, R.O. and R.L. Bassett. 1983. Pinyon–juniper. Pages 84-86 *in* R.M. Burns, technical compiler. Silviculture systems for the major forest types of the United States. Agriculture Handbook No. 445. USDA, Forest Service, Washington, D.C.

Meeuwig, R.O., J.D. Budy, and R.L. Everett. 1990. *Pinus monophylla* – singleleaf pinyon. Pages 380-384 *in* R.M. Burns and B.H. Honkala, technical coordinators. Silvics of North America, Volume 1. Conifers. Agriculture Handbook 654. USDA, Forest Service, Washington, D.C.

Meier, L. 1979. A vegetative survey of the Fern Canyon Research Natural Area, San Dimas Experimental Forest. Unpublished report. USDA, Forest Service, Pacific Southwest Research Station, Berkeley, CA.

Mendocino National Forest. 1987. Proposed land and resource management plan. Unpublished report. USDA, Forest Service, Mendocino National Forest, Colusa, CA.

Mensing, S. 1990. Blue oak regeneration in the Tehachapi Mountains. Fremontia 18:38-41.

Miller, P.C. and D.K. Poole. 1979. Patterns of water use by shrubs in southern California. Forest Science 25:84-98.

Miller, R.E. 1980. Red alder. Page 100 *in* F.H. Eyre, editor. Forest cover types of the United States and Canada. Society of American Foresters, Washington, D.C.

Miller, L. 1988. How yellow bush lupine came to Humboldt. Fremontia 16:6-7.

Minckley, W.L. and D.E. Brown. 1982. Part 6. Wetlands. Page 342 *in* F.S. Crosswhite, editor. Biotic communities of the American Southwest – United States and Mexico. The University of Arizona Press, Tucson, AZ.

Minnich, R.A. 1976. Vegetation of the San Bernardino mountains. Pages 99-125 *in* J. Latting, editor. Plant communities of southern California. California Native Plant Society, Sacramento, CA.

Minnich, R.A. 1980. Vegetation of Santa Cruz and Santa Catalina Islands. Pages 123-137 *in* D.M. Power, editor. The California Islands. Santa Barbara Museum of Natural History, Santa Barbara, CA.

Minnich, R.A. 1980b. Wildfire and the geographic relationships between canyon live oak, Coulter pine, and bigcone Douglas-fir forests. Pages 55-61 *in* T.R. Plumb, editor. Ecology, management, and utilization of California oaks. General Technical Report PSW-44. USDA, Forest Service, Pacific Southwest Research Station, Berkeley, CA.

Minnich, R.A. 1982. *Pseudotsuga macrocarpa* in Baja California? Madroño 29:22-31.

Minnich, R.A., A. Sanders, S. Wood, K. Barrows, and J. Lyman. 1993. Natural resources management plan, Marine Corps Air-Ground Combat Center, Twentynine Palms, California. Unpublished report. University of California, Riverside, CA.

Minnich, R.A., M.G. Barbour, J.H. Burk, and R.F. Fernau. In press. Sixty years of change in conifer forests of the San Bernardino Mountains. Journal of Conservation Biology.

Minore, D. 1980. Western hemlock – Sitka spruce. Page 103 *in* F.H. Eyre, editor. Forest cover types of the United States and Canada. Society of American Foresters, Washington, D.C.

Minore, D. and D. Kingsley. 1983. Mixed conifers of southwestern Oregon. Pages 23-25 *in* R.M. Burns, technical compiler. Silviculture systems for the major forest types of the United States. Agriculture Handbook No. 445. USDA, Forest Service, Washington, D.C.

Minore, D. 1990. *Thuja plicata* – western red-cedar. Pages 590-600 *in* R.M. Burns and B.H. Honkala, technical coordinators. Silvics of North America, Volume 1. Conifers. Agriculture Handbook 654. USDA, Forest Service, Washington, D.C.

Mize, C.W. 1973. Vegetation types of lower elevation forests in the Klamath Region, California. Master's thesis, Humboldt State University, Arcata, CA.

Monroe, G.W. 1973. The natural resources of Humboldt Bay. Coastal Wetland Series 6. State of California, The Resources Agency, Department of Fish and Game, Eureka, CA.

Montygierd-Loba, T.M. and J.E. Keeley. 1987. Demographic structure of *Ceanothus megacarpus* chaparral in the long absence of fire. Ecology 68:211-213.

Mooney, H.A., G. St. Andre, and R.D. Wright. 1962. Alpine and subalpine vegetation patterns in the White Mountains of California. American Midland Naturalist 68:257-273.

Mooney, H.A. 1973. Plant communities and vegetation. *in* R.M. Lloyd and R.S. Mitchell, editors. A flora of the White Mountains, California and Nevada. University of California Press, Berkeley, CA.

Mooney, H.A. 1977. Southern coastal scrub. Pages 471-489 *in* M.G. Barbour and J. Major, editors. Terrestrial vegetation of California. Wiley–Interscience, reprinted by the California Native Plant Society 1988, Sacramento, CA.

Mountjoy, J.H. 1979. Broom – threat to native plants. Fremontia 6:11-15.

Mueggler, W.F. 1988. Aspen community types of the intermountain region. General Technical Report INT-250. USDA, Forest Service, Intermountain Research Station, Ogden, UT.

Mueggler, W.F. 1994. Aspen woodland – SRM 411. Page 50 *in* T.N. Shiflet, editor. Rangeland cover types of the United States. Society for Range Management, Denver, CO.

Mueller-Dombois, D. and H. Ellenberg. 1974. Aims and methods of vegetation ecology. John Wiley & Sons, New York, NY.

Muldavin, E.H. 1982. Forest communities of Shinbone Ridge in the Yolla Bolly Mountains of northern California. Master's thesis, Humboldt State University, Arcata, CA.

Mullally, D.P. 1992. Distribution and environmental relations of California black walnut (*Juglans californica*) in the eastern Santa Susana Mountains, Los Angeles County. Crossosoma 18:1-17.

Mullally, D.P. 1993. Mixed hardwood forests and other woodlands of the Santa Susana Mountains of Los Angeles County. Unpublished paper. Submitted to Crossosoma.

Muller, C.H. 1967. Relictual origins of insular endemics in *Quercus*. *in* R.N. Philbrick, editor. Proceedings of the symposium on the biology of the California Islands. Santa Barbara Botanic Garden, Santa Barbara, CA.

Munz, P.A. and D.D. Keck. 1949. California plant communities. Aliso 2:87-105.

Munz, P.A. and D.D. Keck. 1950. California plant communities; part two. Aliso 2:199-202.

Munz, P.A. 1959. California plant communities. *in* P.A. Munz. A California flora. University of California Press, Berkeley, CA.

Munz, P.A. 1968. A California flora and supplement. University of California Press, Berkeley, CA.

Munz, P.A. 1974. A flora of southern California. University of California Press, Berkeley, CA.

Murray, M.D. 1988. Baker cypress. Fremontia 16:17-18.

Murray, M.P. 1991. Meadow vegetation change in the subalpine zone of the Marble Mountain Wilderness. Master's thesis, Humboldt State University, Arcata, CA.

Myatt, R.G. 1980. Canyon live oak vegetation in the Sierra Nevada. Pages 86-91 *in* T.R. Plumb, editor. Ecology, management and utilization of California oaks. General Technical Report PSW-44. USDA, Forest Service, Pacific Southwest Research Station, Berkeley, CA.

Nachlinger, J.L. 1988. An ecological survey of the candidate Lyon Peak/Needle Lake Research Natural Area, Tahoe National Forest, California. Unpublished report. USDA, Forest Service, Pacific Southwest Research Station, Berkeley, CA.

Neal, D.L. 1980. Blue oak – digger pine. Pages 126-127 *in* F.H. Eyre, editor. Forest cover types of the United

States and Canada. Society of American Foresters, Washington, D.C.

Neal, D.L. 1994. Bitterbrush – SRM 210. Page 19 *in* T.N. Shiflet, editor. Rangeland cover types of the United States. Society for Range Management, Denver, CO.

Neill, W.M. 1985. Status reports on invasive weeds: tamarisk. Fremontia 12:22.

Newton, G.B. 1987a. Ecological survey of the Devil's Basin proposed Research Natural Area. Unpublished report. USDA, Forest Service, Pacific Southwest Research Station, Berkeley, CA.

Newton, G.B. 1987b. The ecology and management of three salt marsh species of Humboldt Bay. Pages 263-266 *in* T.S. Elias, editor. Conservation and management of rare and endangered plants. California Native Plant Society, Sacramento, CA.

Newton, G.B. 1989. Evaluation of restoration and enhancement at Elk River wildlife area, a wetland mitigation site. Master's thesis, Humboldt State University, Arcata, CA.

Nichols, H.T. 1989. Managing fire in Sequoia and Kings Canyon National Parks. Fremontia 16:11-14.

Odion, D.C., R.M. Callaway, W.R. Ferren, and F.W. Davis. 1992. Vegetation of Fish Slough, an Owens Valley wetland ecosystem. Pages 171-196 *in* C.A. Hall and B. Widawski, editors. The history of water: eastern Sierra Nevada, Owens Valley, White-Inyo mountains. White Mountains Research Station Symposium 4. University of California, White Mountain Research Station, Los Angeles, CA.

O'Leary, J.F. and R.A. Minnich. 1981. Postfire recovery of creosote bush scrub vegetation in the western Colorado Desert. Madroño 28:61-66.

O'Leary, J.F. 1989. Californian coastal sage scrub: general characteristics and considerations for biological conservation. Pages 24-41 *in* A.A. Schoenherr, editor. Endangered plant communities of southern California. Southern California Botanists, California State University, Fullerton, CA.

Oliver, C.D. and B.C. Larson. 1990. Forest stand dynamics. McGraw–Hill, Inc., New York, NY.

Oliver, W.W.and R.F. Powers. 1983. Pacific ponderosa pine. Pages 48-52 *in* R.M. Burns, technical compiler. Silviculture systems for the major forest types of the United States. Agriculture Handbook No. 445. USDA, Forest Service, Washington, D.C.

Oliver, W.W. and R.A. Ryker. 1990. *Pinus ponderosa* – ponderosa pine. Pages 413-424 *in* R.M. Burns and B.H. Honkala, technical coordinators. Silvics of North America, Volume 1. Conifers. Agriculture Handbook 654. USDA, Forest Service, Washington, D.C.

Olson, D.F. 1983. Redwood. Pages 37-40 *in* R.M. Burns, technical compiler. Silviculture systems for the major forest types of the United States. Agriculture Handbook No. 445. USDA, Forest Service, Washington, D.C.

Olson, D.F., D.F. Roy, and G.A. Walters. 1990. *Sequoia sempervirens* – redwood. Pages 541-551 *in* R.M. Burns and B.H. Honkala, technical coordinators. Silvics of North America, Volume 1. Conifers. Agriculture Handbook 654. USDA, Forest Service, Washington, D.C.

Olson, M. 1992. Vernal pools of northern Santa Barbara Co., California. Unpublished report. Santa Barbara County Resource Management Department, Santa Barbara, CA.

Oosting, H.J., and W.D. Billings. 1943. *Abietum magnificae*: the red fir forest of the Sierra Nevada. Ecological Monographs 13:259-274.

Ornduff, R. 1974. California plant life. University of California Press, Berkeley, CA.

Packee, E.C. 1990. *Tsuga heterophylla* – western hemlock. Pages 613-622 *in* R.M. Burns and B.H. Honkala, technical coordinators. Silvics of North America, Volume 1. Conifers. Agriculture Handbook 654. USDA, Forest Service, Washington, D.C.

Palmer, J.S. 1979. Vegetation on quartz diorite in the Bear Lakes area, Trinity County, California. Master's thesis, Humboldt State University, Arcata, CA.

Palmer, R. 1981. Ecological survey of the vegetation of the Sugar Pine Point Area, Tahoe National Forest, California. Unpublished report. USDA, Forest Service, Pacific Southwest Research Station, Berkeley, CA.

Palmer, M.W. and P.S. White. 1994. On the existence of ecological communities. Journal of Vegetation Science 5:279-282.

Parikh, A. 1993a. Ecological survey report, San Emigdio Mesa candidate Research Natural Area, Los Padres National Forest, Mount Pinos Ranger District, Ventura County, California. Unpublished report. USDA, Forest Service, Pacific Southwest Research Station, Berkeley, CA.

Parikh, A. 1993b. Ecological survey report, Sawmill Mountain candidate Research Natural Area, Los Padres National Forest, Mount Pinos Ranger District, Ventura County, California. Unpublished report. USDA, Forest Service, Pacific Southwest Research Station, Berkeley, CA.

Parker, E. 1988. Those amazing Siskiyou firs and a new discovery. Four Seasons 8:5-16.

Parker, G.R. and D.G. Leopold. 1983. Replacement of *Ulmus americana* L. in a mature east-central Indiana woods. Bulletin of the Torrey Botanical Club 110: 482-488.

Parker, I. and W. Matayas. 1979. Calveg: a classification of Californian vegetation. Unpublished map and report. USDA, Forest Service, Pacific Southwest Region, Ecology Group, San Francisco, CA.

Parker, J. 1974. Coastal dune systems between Mad River and Little River, Humboldt County, California. Master's thesis, Humboldt State College, Arcata, CA.

Parker, V.T. and C.H. Muller. 1982. Vegetational and environmental changes beneath isolated live oak trees (*Quercus agrifolia*) in a California annual grassland. American Midland Naturalist 107:69-81.

Parker, V.T. 1990. The vegetation of the Mount Tamalpais watershed of the Marin Municipal Water District and those on the adjacent lands of the Marin County Open Space District. Unpublished report. Marin Municipal Water District, San Raphael, CA.

Parker, V.T. 1994. Coast live oak woodland – SRM 202. Page 12 *in* T. N. Shiflet, editor. Rangeland cover types of the United States. Society for Range Management, Denver, CO.

Parsons, D.J. 1972. The southern extensions of *Tsuga mertensiana* (mountain hemlock) in the Sierra Nevada. Madroño 21:536-539.

Parsons, D.J. 1976. The role of fire in natural communities: an example from the southern Sierra Nevada, California. Environmental Conservation 3: 41-99.

Parsons, D.J. 1980. California mixed subalpine. Pages 90-91 *in* F.H. Eyre, editor. Forest cover types of the United States and Canada. Society of American Foresters, Washington, D.C.

Parsons, D.J. 1981. The historical role of fire on the foothill communities of Sequoia National Park. Madroño 28:111-120.

Parsons, D.J. and T.J. Stohlgren. 1989. Effects of varying fire regimes on annual grasslands in the southern Sierra Nevada of California. Madroño 36: 154-168.

Pase, C.P. and D.E. Brown. 1982. Californian coastal scrub. Desert Plants 4:86-90.

Pase, C.P. 1982a. California (coastal) chaparral. Desert Plants 4:91-94.

Pase, C.P. 1982b. Sierran subalpine conifer forest. Desert Plants 4:40-41.

Pase, C.P. 1982c. Sierran montane conifer forest. Desert Plants 4:49-51.

Patric, J.H. and T.L. Hanes. 1964. Chaparral succession in a San Gabriel Mountain area of California. Ecology 45:353-360.

Pavlik, B.M. 1985. Sand dune flora of the Great Basin and Mojave Deserts of California, Nevada, and Oregon. Madroño 32:197-213.

Pavlik, B.M., P.C. Muick, S.G. Johnson, and M. Popper. 1991. Oaks of California. Cachuma Press, Inc., Los Olivos, CA.

Pavlik, B.M. and M.W. Skinner. 1994. Ecological characteristics of California's rare plants. Pages 4-6 *in* M.W. Skinner and B.M. Pavlik, editors. California Native Plant Society's inventory of rare and endangered vascular plants of California. California Native Plant Society, Sacramento, CA.

Paysen, T.E., J.A. Derby, H. Black, V.C. Bleich, and J.W. Mincks. 1980. A vegetation classification system applied to southern California. General Technical Report PSW-45. USDA, Forest Service, Pacific Southwest Research Station, Berkeley, CA.

Paysen, T.E., J.A. Derby, and C.E. Conrad. 1982. A vegetation classification system for use in California: its conceptual basis. General Technical Report PSW-63. USDA, Forest Service, Pacific Southwest Research Station, Berkeley, CA.

Paysen, T.E. 1982. Vegetation classification – California. *in* C.E. Conrad and W.C. Oechel, editors. Dynamics and management of Mediterranean-type ecosystems. General Technical Report PSW-58. USDA, Forest Service, Pacific Southwest Research Station, Berkeley, CA.

Peart, D.R. 1982. Experimental analysis of succession in a grassland at Sea Ranch, California. Dissertation, University of California, Davis, CA.

Peinado, M., F. Alcaraz, J. Delgadillo, M. De La Cruz, J. Alvarez, and J.L. Aquirre. 1994. The coastal salt marshes of California and Baja California. Vegetatio 110:55-66.

Perala, D.A. 1990. *Populus tremuloides* Michx. Pages 555-569 *in* R.M. Burns and B.H. Honkala, technical coordinators. Silvics of North America, Volume 2. Hardwoods. Agriculture Handbook 654. USDA, Forest Service, Washington, D.C.

Pfister, R.D. and P.M. McDonald. 1980. Lodgepole pine. Pages 97-98 *in* F.H. Eyre, editor. Forest cover types of the United States and Canada. Society of American Foresters, Washington, D.C.

Pfister, R.D. and S.F. Arno. 1980. Classifying forest habitat types based on potential climax vegetation. Forest Science 26:52-70.

Philbrick, R.N. and J.R. Haller. 1977. The southern California islands. Pages 893-906 *in* M.G. Barbour and J. Major, editors. Terrestrial vegetation of California. Wiley–Interscience, reprinted by the California Native Plant Society 1988, Sacramento, CA.

Phillips, E.A., K.K. Page, and S.D. Knapp. 1980. Vegetational characteristics of two stands of Joshua tree woodland. Madroño 27:43-47.

Phillips, D.L. and J.A. MacMahon. 1981. Competition and spacing patterns in desert shrubs. Ecology 69:97-115.

Pickart, A. 1987. A classification of northern foredune and its relationship to Menzies' wallflower of the north spit of Humboldt Bay, California. Unpublished report. The Nature Conservancy, San Francisco, CA.

Pickart, A., L. Miller, and T.E. Duebendorfer. 1989. An integrated approach to enhancing rare plant populations through habitat restoration: II. Restoration of altered coastal dunes. Pages 488-500 *in* H.G. Hughes and T.M. Bonnicksen, editors. Restoration '89: The new management challenge. First Annual Meeting of the Society for Ecological Restoration, Society for Ecological Restoration, Oakland, CA.

Pillers, M.D., and J.D. Stuart. 1993. Leaf-litter accretion and decomposition in interior and coastal old-growth redwood stands. Canadian Journal of Forest Research 23:552-557.

Plumb, T.R. and N.H. Pillsbury. 1987. Multiple-use management of California's hardwood resources. General Technical Report PSW-100. USDA, Forest Service, Pacific Southwest Research Station, Berkeley, CA.

Potter, D.A. 1994. Guide to forested communities of the upper montane in the central and southern Sierra Nevada. Technical Publication R5-ECOL-TP-003. USDA, Forest Service, Pacific Southwest Region, San Francisco, CA.

Powers, R.F. and W.W. Oliver. 1990. *Libocedrus decurrens* – incense-cedar. Pages 173-180 *in* R.M. Burns and B.H. Honkala, technical coordinators. Silvics of North America, Volume 1. Conifers. Agriculture Handbook 654. USDA, Forest Service, Washington, D.C.

Powers, R.F. 1990. *Pinus sabiniana* – digger pine. Pages 463-469 *in* R.M. Burns and B.H. Honkala, technical coordinators. Silvics of North America, Volume 1. Conifers. Agriculture Handbook 654. USDA, Forest Service, Washington, D.C.

Purer, E.A. 1939. Ecological study of vernal pools, San Diego County, California. Ecology 20:217-229.

Quinn, R.D. 1989. The status of walnut forests and woodlands (*Juglans californica*) in southern California. Pages 42-54 *in* A.A. Schoenherr, editor. Endangered plant communities of southern California. Southern California Botanists, California State University, Fullerton, CA.

Rae, R.W. 1970. Studies in the ecology of Mason Bog, Sagehen Creek. Master's thesis, University of California, Davis, CA.

Randall, D.C. 1972. An analysis of some desert shrub vegetation of Saline Valley, California. Dissertation, University of California, Davis, CA.

Rantz, S.E. 1972. Mean annual precipitation in the California region. U.S.G.S. map. USDI, Geological Survey, Washington, D.C.

Ratliff, R.D. 1979. Meadow sites of the Sierra Nevada, California: classification and species relationships. Dissertation, New Mexico State University, Las Cruces, NM.

Ratliff, R.D. 1982. A meadow site classification for the Sierra Nevada, California. General Technical Report PSW-60. USDA, Forest Service, Pacific Southwest Research Station, Berkeley, CA.

Ratliff, R.D. 1985. Meadows in the Sierra Nevada of California: state of knowledge. General Technical Report PSW-84. USDA, Forest Service, Pacific Southwest Research Station, Berkeley, CA.

Raven, P.H. 1977. The California flora. Pages 109-137 *in* M.G. Barbour and J. Major, editors. Terrestrial vegetation of California. Wiley–Interscience, reprinted by the California Native Plant Society 1988, Sacramento, CA.

Raven, P.H. and D.I. Axelrod. 1978. Origin and relationships of the California flora. University of California Publications in Botany 72:1-134.

Read, R.A. and J. Sprackling. 1980. Cottonwood – willow. Page 113 *in* F.H. Eyre, editor. Forest cover types of the United States and Canada. Society of American Foresters, Washington, D.C.

Read, E.A. 1994. The importance of community classification to mitigation and restoration of coastal sage scrub. Restoration Ecology 2:80-86.

Reed, P.B. 1988. National list of plant species that occur in wetlands: California (Region 0). Biological Report 88(26.10). USDI, Fish and Wildlife Service, Washington, D.C.

Reveal, J.L. 1978. A report on the autecology and status of Cuyamaca cypress (*Cupressus arizonica* var. *stephensonii*), Cleveland National Forest, California. Unpublished report. USDA, Forest Service, Pacific Southwest Research Station, Berkeley, CA.

Riegel, G.M., D.A. Thornburgh, and J.O. Sawyer. 1990. Forest habitat types of the south Warner Mountains, Modoc County, California. Madroño 37:88-112.

Riegel, G.M., B.G. Smith, and J.F. Franklin. 1991. Foothill oak woodlands of the interior valleys of southwestern Oregon. Northwest Science 66:66-76.

Rieger, J.P. and D.A. Kreager. 1989. Giant reed (*Arundo donax*): a climax community of the riparian zone. General Technical Report PSW-110. USDA, Forest Service, Pacific Southwest Research Station, Berkeley, CA.

Rigg, G.B. 1922. A bog forest. Ecology 3:207-213.

Riser, R.J. and M. Fry. 1994. Montane shrubland – SRM 209. Page 18 *in* T.N. Shiflet, editor. Rangeland cover types of the United States. Society for Range Management, Denver, CO.

Roberts, W.G., J.G. Howe, and J. Major. 1980. A survey of riparian forest flora and fauna in California. Pages 3-19 *in* A. Sands, editor. Riparian forests in California. Publication No. 15. Institute of Ecology, University of California, Davis, CA.

Roberts, R.C. 1984. The transitional nature of northwestern California riparian systems. Pages 85-91 *in* R.E. Warner and K.M. Hendrix, editors. California riparian systems: ecology, conservation and productive management. University of California Press, Berkeley, CA.

Rodgers, R.S. 1980. Hemlock stands from Wisconsin to Nova Scotia: transitions in understory composition along a floristic gradient. Ecology 61:178-193.

Rodgers, R.S. 1981. Mature mesophytic hardwood forest: community transitions, by layer, from east-central Minnesota to southeastern Michigan. Ecology 62:1634-1647.

Rowlands, P.G. 1978. The vegetation dynamics of the Joshua tree (*Yucca brevifolia* Engelm.) in the southwestern United States of America. Dissertation, University of California, Riverside, CA.

Roy, D.F. 1980. Redwood. Pages 109-110 in F.H. Eyre, editor. Forest cover types of the United States and Canada. Society of American Foresters, Washington, D.C.

Rundall, D.C. 1972. An analysis of some desert shrub vegetation of Saline Valley, California. Dissertation, University of California, Davis, CA.

Rundel, P.W. 1971. Community structure and stability in the giant sequoia groves of the Sierra Nevada, California. American Midland Naturalist 84:478-492.

Rundel, P.W. 1972a. An annotated checklist of the groves of *Sequoiadendron giganteum* in the Sierra Nevada, California. Madroño 21:319-383.

Rundel, P.W. 1972b. Habitat restriction in giant sequoia: the environmental control of grove boundaries. American Midland Naturalist 87:81-99.

Rundel, P.W. 1975. Primary succession on granite outcrops in the montane southern Sierra Nevada. Madroño 23:209-220.

Rundel, P.W., D.J. Parsons, and D.T. Gordon. 1977. Montane and subalpine vegetation of the Sierra Nevada and Cascade Ranges. Pages 559-599 in M.G. Barbour and J. Major, editors. Terrestrial vegetation of California. Wiley–Interscience, reprinted by the California Native Plant Society 1988, Sacramento, CA.

Russell, E.W. 1983. Pollen analysis of past vegetation at Point Reyes National Seashore, California. Madroño 30:1-11.

Ryerson, A.D. 1983. Population structure of *Pinus balfouriana* Grev. and Balf. along the margins of its distribution area in the Sierran and Klamath regions of California. Master's thesis, California State University, Sacramento, CA.

Saenz, L. 1983. *Quercus garryana* woodland/grassland mosaic dynamics in northern California. Master's thesis, Humboldt State University, Arcata, CA.

Saenz, L. and J.O. Sawyer. 1986. Grassland as compared to adjacent *Quercus garryana* woodland

understories exposed to different grazing regimes. Madroño 33:40-46.

Sands, A. 1980. Riparian forests in California. Publication No. 15. University of California, Institute of Ecology, Davis, CA.

Sawyer, J.O. and D.A. Thornburgh. 1969. Ecological reconnaissance of relict conifers in the Klamath Region. Unpublished report. USDA, Forest Service, Pacific Southwest Research Station, Berkeley, CA.

Sawyer, J.O. and D.A. Thornburgh. 1970. The ecology of relict conifers in the Klamath Region, California. Unpublished report. USDA, Forest Service, Pacific Southwest Research Station, Berkeley, CA.

Sawyer, J.O. and D.A. Thornburgh. 1971. Vegetation types on granodiorite in the Klamath Mountains, California. Unpublished report. USDA, Forest Service, Pacific Southwest Research Station, Berkeley, CA.

Sawyer, J.O. and K.T. Stillman. 1977. An ecological survey of the proposed Specimen Creek Research Natural Area, Siskiyou County, California. Unpublished report. USDA, Forest Service, Pacific Southwest Research Station, Berkeley, CA.

Sawyer, J.O. and D.A. Thornburgh. 1977. Montane and subalpine vegetation of the Klamath Mountains. Pages 699-732 in M.G. Barbour and J. Major, editors. Terrestrial vegetation of California. Wiley–Interscience, reprinted by the California Native Plant Society 1988, Sacramento, CA.

Sawyer, J.O., D.A. Thornburgh, and J.R. Griffin. 1977. Mixed evergreen forest. Pages 359-381 in M.G. Barbour and J. Major, editors. Terrestrial vegetation of California. Wiley–Interscience, reprinted by the California Native Plant Society 1988, Sacramento, CA.

Sawyer, J.O., J. Palmer, and E. Cope. 1978. An ecological survey of the proposed Preacher Meadows Research Natural Area, Trinity County, California. Unpublished report. USDA, Forest Service, Pacific Southwest Research Station, Berkeley, CA.

Sawyer, J.O. and K.T. Stillman. 1978. An ecological survey of the proposed William's Point Research Natural Area, Siskiyou County, California. Unpublished report. USDA, Forest Service, Pacific Southwest Research Station, Berkeley, CA.

Sawyer, J.O., K.T. Stillman, and P. Siekel. 1978. An ecological survey of the proposed Indian Creek Brewer Spruce Research Natural Area, Siskiyou County, California. Unpublished report. USDA, Forest Service, Pacific Southwest Research Station, Berkeley, CA.

Sawyer, J.O. 1980. Douglas-fir – tanoak – Pacific madrone. Pages 111-112 in F.H. Eyre, editor. Forest cover types of the United States and Canada. Society of American Foresters, Washington, D.C.

Sawyer, J.O. 1981a. Ecological survey of the Adorni candidate Research Natural Area, Humboldt County, California. Unpublished report. USDA, Forest Service, Pacific Southwest Research Station, Berkeley, CA.

Sawyer, J.O. 1981b. An ecological survey of the North Trinity Mountain candidate Research Natural Area, Humboldt County, California. Unpublished report. USDA, Forest Service, Pacific Southwest Research Station, Berkeley, CA.

Sawyer, J.O. and E.M. Cope. 1982. Noteworthy collections, California: *Abies lasiocarpa* (Hook.) Nuttall. Madroño 29:219.

Sawyer, J.O. 1986. Darlingtonia seeps. Fremontia 14:18.

Sawyer, J.O. 1987. The problem of the Salmon Mountains. Pages 155-158 in T.S. Elias, editor. Conservation and management of rare and endangered plants. California Native Plant Society, Sacramento, CA.

Schierenbeck, K.A. and D.B. Jensen. 1994. Vegetation of the Upper Raider and Hornback creek basins, South Warner Mountains: Northwestern limit of *Abies concolor* var. *lowiana*. Madroño 41:53-64.

Schlesinger, W.H., J.T. Gray, D.S. Gill, and B.E. Mahall. 1982. *Ceanothus megacarpus* chaparral: a synthesis of ecosystem processes during development and annual growth. Botanical Review 48:71-117.

Schlising, R.A. and E.L. Sanders. 1982. Quantitative analysis of vegetation at the Richvale vernal pools, California. American Journal of Botany 69:734-742.

Schmida, A. and R.H. Whittaker. 1981. Pattern and microsite effects in two shrub communities, southern California. Ecology 62:234-251.

Schorr, P.K. 1970. The effects of fire on manzanita chaparral in the San Jacinto Mountains of southern California. Master's thesis, California State University, Los Angeles, CA.

Schulman, E. 1954. Longevity under adversity in conifers. Science 119:395-399.

Scott, J.M., F. Davis, B. Csuti, R. Noss, B. Butterfield, C. Groves, H. Anderson, S. Caicco, F. D'Erchia, T.C. Edwards, J. Ulliman, and R.G. Wright. 1993. Gap analysis: a geographic approach to protecting biological diversity. Wildlife Monographs No. 123:141 (Supplement to the Journal of Wildlife Management 57(1)).

Scott, R., K. Buer, and S. James. 1980. South Fork Trinity River watershed study. Symposium on watershed management. Volume I. American Society of Civil Engineers, New York, NY.

Scott, T. 1991. The distribution of Engelmann oak (*Quercus engelmannii*) in California. General Technical Report PSW-126. USDA, Forest Service, Pacific Southwest Research Station, Berkeley, CA.

Shapiro & Associates. 1980. Humboldt Bay wetlands review and baylands, final report. U.S. Army Corps of Engineers, San Francisco, CA.

Sharf, M.R., E.T. Nilsen, and P.W. Rundel. 1982. Biomass and net primary production of *Prosopis glandulosa* (Fagaceae) in the Sonoran Desert of California. American Journal of Botany 69:760-767.

Sholars, R.E. 1979. Water relations in the pygmy forest of Mendocino County, California. Dissertation, University of California, Davis, CA.

Sholars, R.E. 1982. The pygmy forest and associated plant communities of coastal Mendocino County, California. Printed by the author, Mendocino, CA.

Sholars, R.E. 1984. The pygmy forest of Mendocino. Fremontia 12:3-8.

Shreve, F. 1927. The vegetation of a coastal mountain range. Ecology 8:27-44.

Shreve, F. and I.L. Wiggins. 1964. Vegetation and flora of the Sonoran Desert. Stanford University Press, Stanford, CA.

Simpson, L.G. 1980. Forest types on ultramafic parent materials of the southern Siskiyou Mountains in the Klamath Region of California. Master's thesis, Humboldt State University, Arcata, CA.

Sims, P.L. 1988. Grasslands. Pages 265-286 *in* M.G. Barbour and W.D. Billings, editors. North American terrestrial vegetation. Cambridge University Press, New York, NY.

Singer, M.J. 1978. The soils of Ione. Fremontia 6:11-13.

Skinner, M.W. and B.M. Pavlik. 1994. California Native Plant Society's inventory of rare and endangered vascular plants of California. Fifth edition. California Native Plant Society, Sacramento, CA.

Skolmen, R.G. and F.T. Ledig. 1990. *Eucalyptus globulus* – bluegum eucalyptus. *in* R.M. Burns and B.H. Honkala, technical coordinators. Silvics of North America, Volume 2. Hardwoods. Agriculture Handbook 654. USDA, Forest Service, Washington, D.C.

Skolmen, R.G. 1990. *Eucalyptus saligna* – saligna eucalyptus. Pages 318-324 *in* R.M. Burns and B.H. Honkala, technical coordinators. Silvics of North America, Volume 2. Hardwoods. Agriculture Handbook 654. USDA, Forest Service, Washington, D.C.

Smith, R.L. 1980. Alluvial scrub vegetation of the San Gabriel River floodplain, California. Madroño 27:126-138.

Smith, J.P. and K. Berg. 1988. Inventory of rare and endangered vascular plants of California. Fourth edition. California Native Plant Society, Sacramento, CA.

Smith, T. and M. Huston. 1989. A theory of the spatial and temporal dynamics of plant communities. Vegetatio 83:49-69.

Smith, S. 1994. Ecological guide to eastside pine plant associations: northeastern California. Technical Report R5-ECOL-TP-004. USDA, Forest Service, Pacific Southwest Region, San Francisco, CA.

Snow, G.E. 1973. Some factors controlling the establishment and distribution of *Quercus agrifolia* Nee and *Q. engelmannii* Greene in certain southern California oak woodlands. Dissertation, Oregon State University, Corvallis, OR.

Solinas, P., G. Spycher, and C. Topik. 1985. Processes of soil organic matter accretion at a mudflow chronosequence, Mt. Shasta, California. Ecology 64:1273-1282.

Solis, D.M. 1983. Summer habitat ecology of spotted owls in northwestern California. Master's thesis, Humboldt State University, Arcata, CA.

Spenger, C. 1985. The northernmost stand of tecate cypress. Fremontia 13:8-10.

Spolsky, A.M. 1979. An overview of the plant communities of Anza-Borrego Desert State Park. Unpublished report. State of California, The Resources Agency, Department of Parks and Recreation, Anza-Borrego Desert State Park, Borrego Springs, CA.

Sposito, G. 1992. How plants affect soils: Hans Jenny and the "biotic factor" of soil formation. Fremontia 20:12-16.

Sproul, F.T. 1981. Ecological survey of Falls Canyon candidate Research Natural Area, Angeles National Forest, California. Unpublished report. USDA, Forest Service, Pacific Southwest Research Station, Berkeley, CA.

Stebbins, G.L. and J. Major. 1965. Endemism and speciation in the California flora. Ecological Monographs 35:1-35.

Stebbins, G.L. 1976. Ecological islands and vernal pools. Fremontia 4:12-18.

Stein, W.I. 1980a. Port Orford–cedar. Page 108 *in* F.H. Eyre, editor. Forest cover types of the United States and Canada. Society of American Foresters, Washington, D.C.

Stein, W.I. 1980b. Oregon white oak. Pages 110-111 *in* F.H. Eyre, editor. Forest cover types of the United States and Canada. Society of American Foresters, Washington, D.C.

Stein, W.I. 1990a. *Quercus garryana* – Oregon white oak. Pages 650-660 *in* R.M. Burns and B.H. Honkala, technical coordinators. Silvics of North America, Volume 2. Hardwoods. Agriculture Handbook 654. USDA, Forest Service, Washington, D.C.

Stein, W.I. 1990b. *Umbellularia californica* – California-laurel. Pages 826-834 *in* R.M. Burns and

B.H. Honkala, technical coordinators. Silvics of North America, Volume 2. Hardwoods. Agriculture Handbook 654. USDA, Forest Service, Washington, D.C.

Stillman, K.T. 1980. Meadow vegetation on meta-sedimentary and metavolcanic parent materials in the north central Marble Mountains, California. Master's thesis, Humboldt State University, Arcata, CA.

Stoddart, L.A., A.D. Smith, and T.W. Box. 1975. Range management. McGraw–Hill, New York, NY.

Stohlgren, T.J., D.J. Parsons, and P.W. Rundel. 1984. Population structure of *Adenostoma fasciculatum* in mature stands of chamise chaparral in the southern Sierra Nevada, California. Oecologia 64:87-91.

Stone, R.D. and V.A. Sumida. 1983. The Kingston range of California: a resource survey. Publication No. 10. University of California, Environmental Field Program, Santa Cruz, CA.

Stone, R.D. 1990. California's endemic vernal pool macrophytes: relation of structure and function. Pages 89-107 *in* D.H. Ikeda, R.A. Schlising, F.J. Fuller, L.P. Janeway, and P. Woods, editors. Vernal pool plants: their habitat and biology. Study No. 8 from the Herbarium. California State University, Chico, CA.

Stromberg, J.C. and D.T. Patten. 1992. Mortality and age of black cottonwood stands along diverted and undiverted streams in the eastern Sierra Nevada, California. Madroño 39:205-223.

Stuart, J.D., S. Matthews, and A. Nilson. 1986. Resource inventory and management plan for the Stagecoach Hill Azalea Preserve. Unpublished report. California State Parks Foundation, Sacramento, CA.

Stuart, J.D., T. Worley, and A.C. Buell. 1992. Vegetation classification, disturbance history, and successional interpretations in Castle Crags State Park. Unpublished report. State of California, The Resources Agency, Department of Parks and Recreation, Sacramento, CA.

Stuart, J.D., M.C. Grifantini, and L. Fox III. 1993. Early successional pathways following wildfire and subsequent silvicultural treatment in Douglas-fir/hardwood forest, NW California. Forest Science 39: 561-572.

Sudworth, G.B. 1908. Forest trees of the Pacific slope. 1967 reprint, Dover Press, New York, NY.

Sugihara, N.G. and L.J. Reed. 1987. Vegetation of the Bald Hills oak woodlands, Redwood National Park, California. Madroño 34:193-208.

Swanson, J.C. 1967. The ecology and distribution of *Juglans californica* Wats. in southern California. Master's thesis, California State College, Los Angeles, CA.

Talley, S.N. 1974. The ecology of Santa Lucia fir (*Abies bracteata*), a narrow endemic of California. Dissertation, Duke University, Durham, NC.

Talley, S.N. 1976a. An ecological survey of the Bourland Meadow candidate Research Natural Area on the Stanislaus National Forest. Unpublished report. USDA, Forest Service, Pacific Southwest Research Station, Berkeley, CA.

Talley, S.N. 1976b. An ecological survey of the Onion Creek candidate Research Natural Area on the Tahoe National Forest, California. Unpublished report. USDA, Forest Service, Pacific Southwest Research Station, Berkeley, CA.

Talley, S.N. 1977. An ecological survey of the Babbitt Peak candidate Research Natural Area on the Tahoe National Forest. Unpublished report. USDA, Forest Service, Pacific Southwest Research Station, Berkeley, CA.

Talley, S.N. 1978. An ecological summary of the Sentinel Meadow candidate Research Natural Area on the Inyo National Forest, California. Unpublished report. USDA, Forest Service, Pacific Southwest Research Station, Berkeley, CA.

Talley, S.N. and J.R. Griffin. 1980. Fire ecology of a montane pine forest, Junipero Serra Peak, California. Madroño 27:49-60.

Talley, S.N. 1981. Vegetational growth, and structure of a low elevation mixed conifer forest in the central Sierra Nevada, Merced River candidate Research Natural Area, Sierra National Forest, California. Unpublished report. USDA, Forest Service, Pacific Southwest Research Station, Berkeley, CA.

Tappeiner, J.C. 1980. Sierra Nevada mixed conifer. Pages 118-119 *in* F.H. Eyre, editor. Forest cover types of the United States and Canada. Society of American Foresters, Washington, D.C.

Tappeiner, J.C., P.M. McDonald, and D.F. Roy. 1990. *Lithocarpus densiflora* – tanoak. Pages 417-425 *in* R.M. Burns and B.H. Honkala, technical coordinators. Silvics of North America, Volume 2. Hardwoods. Agriculture Handbook 654. USDA, Forest Service, Washington, D.C.

Taylor, D.W. 1975a. Site evaluation: Yolla Bolly Research Natural Area. Unpublished report. USDA, Forest Service, Pacific Southwest Research Station, Berkeley, CA.

Taylor, D.W. 1975b. Composition of an old-growth Douglas-fir forest in northwestern California. Unpublished report. USDA, Forest Service, Pacific Southwest Research Station, Berkeley, CA.

Taylor, D.W. 1976a. Vegetation reconnaissance of the White Mountain Scientific Area and vicinity, Inyo National Forest. Unpublished report. USDA, Forest Service, Pacific Southwest Research Station, Berkeley, CA.

Taylor, D.W. 1976b. Disjunction of Great Basin plants in the northern Sierra Nevada. Madroño 23:301-364.

Taylor, D.W. and D.C. Randall. 1977a. Ecological survey of the vegetation of the proposed Peavine Research Natural Area, El Dorado National Forest, California. Unpublished report. USDA, Forest Service, Pacific Southwest Research Station, Berkeley, CA.

Taylor, D.W. 1977b. Ecological survey of the vegetation of the proposed Bald Mountain (Station Creek) Research Natural Area, El Dorado National Forest, California. Unpublished report. USDA, Forest Service, Pacific Southwest Research Station, Berkeley, CA.

Taylor, D.W. and D.C. Randall. 1978. Ecological survey of the vegetation of the Cub Creek watershed, Lassen National Forest, California. Unpublished report. USDA, Forest Service, Pacific Southwest Research Station, Berkeley, CA.

Taylor, D.W. 1979. Ecological survey of the vegetation of White Mountain Natural Area, Inyo National Forest, California. Unpublished report. USDA, Forest Service, Pacific Southwest Research Station, Berkeley, CA.

Taylor, D.W. and K.A. Teare. 1979a. Ecological survey of the vegetation of the proposed Trelorita Research Natural Area, Shasta-Trinity National Forest, Trinity County, California. Unpublished report. USDA, Forest

Service, Pacific Southwest Research Station, Berkeley, CA.

Taylor, D.W. and K.A. Teare. 1979b. Ecological survey of the vegetation of the proposed Smokey Creek Research Natural Area, Shasta-Trinity National Forest, Trinity County, California. Unpublished report. USDA, Forest Service, Pacific Southwest Research Station, Berkeley, CA.

Taylor, D.W. 1980. Ecological survey of the vegetation of Indiana Summit Research Natural Area, Inyo National Forest, California. Unpublished report. USDA, Forest Service, Pacific Southwest Research Station, Berkeley, CA.

Taylor, D.W. 1982. Ecological survey of the vegetation of Yurok Research Natural Area, California. Unpublished report. USDA, Forest Service, Pacific Southwest Research Station, Berkeley, CA.

Taylor, D.W. 1984. Vegetation of the Harvey Monroe Hall Research Natural Area, Inyo National Forest, California. Unpublished report. USDA, Forest Service, Pacific Southwest Research Station, Berkeley, CA.

Taylor, D.W., G.L. Clifton, R.F. Holland, and C.W. Witham. 1992. Vernal pools along the PGT-PG&E pipeline expansion project, California. Unpublished report. BioSystems Analysis, Inc., Tiberon, CA.

Thomas, T.W. 1987. Population structure of the valley oak in the Santa Monica Mountains National Recreation Area. Pages 335-340 *in* T.R. Plumb and N.H. Pillsbury, editors. Multiple-use management of California's hardwood resources. General Technical Report, PSW-100. USDA, Forest Service, Pacific Southwest Research Station, Berkeley, CA.

Thompson, K. 1961. Riparian forests of the Sacramento Valley, California. Annals of the Association of American Geographers 51:294-315.

Thornburgh, D.A. 1981. An ecological survey of the proposed Ruth Research Natural Area, Trinity County, California. Unpublished report. USDA, Forest Service, Pacific Southwest Research Station, Berkeley, CA.

Thornburgh, D.A. 1982. Succession in the mixed evergreen forests of northwestern California. Pages 87-91 *in* J.E. Means, editors. Forest succession and stand development research in the Northwest. Forest

Research Laboratory, Oregon State University, Corvallis, OR.

Thornburgh, D.A. 1987. Ecological survey of the proposed Hennessy Ridge Research Natural Area, Six Rivers National Forest, California. Unpublished report. USDA, Forest Service, Pacific Southwest Research Station, Berkeley, CA.

Thornburgh, D.A. 1990a. *Picea breweriana* – Brewer spruce. Pages 181-186 *in* R.M. Burns and B.H. Honkala, technical coordinators. Silvics of North America, Volume 1. Conifers. Agriculture Handbook 654. USDA, Forest Service, Washington, D.C.

Thornburgh, D.A. 1990b. *Quercus chrysolepis* – canyon live oak. Pages 618-624 *in* R.M. Burns and B.H. Honkala, technical coordinators. Silvics of North America, Volume 2. Hardwoods. Agriculture Handbook 654. USDA, Forest Service, Washington, D.C.

Thorne, R.F. and E.W. Lathrop. 1969. A vernal marsh on the Santa Rosa Plateau of Riverside County, California. Aliso 7:85-95.

Thorne, R.F. 1969. A supplement to the floras of Santa Catalina and San Clemente Islands, Los Angeles County, California. Aliso 7:73-83.

Thorne, R.F. 1976. The vascular plant communities of California. Pages 1-31 *in* J. Latting, editor. Plant communities of southern California. California Native Plant Society, Sacramento, CA.

Thorne, R.F. 1977. Montane and subalpine forests of the Transverse and Peninsular Ranges. Pages 537-557 *in* M.G. Barbour and J. Major, editors. Terrestrial vegetation of California. Wiley Interscience, reprinted by the California Native Plant Society 1988, Sacramento, CA.

Thorne, R.F. 1982. The desert and other transmontane plant communities of southern California. Aliso 10:219-257.

Tisdale, E.W. 1994a. Bluebunch wheatgrass – SRM 101. Page 1 *in* T.N. Shiflet, editor. Rangeland cover types of the United States. Society for Range Management, Denver, CO.

Tisdale, E.W. 1994b. Idaho fescue – SRM 102. Page 2 *in* T.N. Shiflet, editor. Rangeland cover types of the United States. Society for Range Management, Denver, CO.

Tisdale, E.W. 1994c. Basin bigbrush – SRM 401. Page 40 *in* T.N. Shiflet, editor. Rangeland cover types of the United States. Society for Range Management, Denver, CO.

Tisdale, E.W. 1994d. Black sagebrush – SRM 405. Page 44 *in* T.N. Shiflet, editor. Rangeland cover types of the United States. Society for Range Management, Denver, CO.

Tisdale, E.W. 1994e. Low sagebrush – SRM 406. Page 45 *in* T.N. Shiflet, editor. Rangeland cover types of the United States. Society for Range Management, Denver, CO.

Tomback, D.F. 1986. Post-fire regeneration of krummholz whitebark pine: a consequence of nutcracker seed caching. Madroño 33:100-110.

Trombulak, S.C. and M.L. Cody. 1980. Elevational distributions of *Pinus edulis* and *P. monophylla* (Pinaceae) in the New York Mountains, eastern Mojave Desert. Madroño 27:61-67.

Trowbridge, B.A. 1975. Plants of upper Sagehen Creek drainage basin, Nevada County, California. Master's thesis, San Francisco State University, San Francisco, CA.

Tueller, P.T., C.D. Beeson, R.J. Tausch, N. E. West, and K.H. Rea. 1979. Pinyon–juniper woodlands of the Great Basin. General Technical Report. USDA, Forest Service, Intermountain Research Station, Ogden, UT.

Tueller, P.T. 1994. Salt desert scrub – SRM 414. Page 53 *in* T.N. Shiflet, editor. Rangeland cover types of the United States. Society for Range Management, Denver, CO.

Turner, R.M. and D.E. Brown. 1982. Sonoran Desert scrub. Desert Plants 4:181-221.

Turner, R.M. 1982a. Great Basin desertscrub. Desert Plants 4:145-155.

Turner, R.M. 1982b. Mohave desertscrub. Desert Plants 4:157-168.

Twisselmann, E.C. 1956. A flora of the Temblor Range and the neighboring part of the San Joaquin Valley. Wasmann Journal of Biology 14:161-195.

Twisselmann, E.C. 1967. A flora of Kern County, California. Wasmann Journal of Biology 21:1-395.

Unsicker, J.E. 1974. Synecology of the California bay tree, *Umbellularia californica* (H.& A.) Nutt. in the Santa Cruz Mountains. Dissertation, University of California, Santa Cruz, CA.

Vale, T.R. 1975. Invasion of big sagebrush (*Artemisia tridentata*) by white fir (*Abies concolor*) on the southeastern slopes of the Warner Mountains, California. Great Basin Naturalist 35:319-324.

Vale, T.R. 1979. *Pinus coulteri* and wildfire on Mount Diablo, California. Madroño 261:135-140.

Vankat, J.L. 1970. Vegetation change in Sequoia National Park, California. Dissertation, University of California, Davis, CA.

Vankat, J.L., and J. Major. 1978. Vegetation changes in Sequoia National Park, California. Journal of Biogeography 5:377-402.

Vankat, J.L. 1982. A gradient perspective on the vegetation of Sequoia National Park, California. Madroño 29:164-176.

Vasek, F.C. 1966. The distribution and taxonomy of three western junipers. Brittonia 18:350-372.

Vasek, F.C., H.B. Johnson, and D.H. Eslinger. 1975. Effects of pipeline construction on creosote bush scrub vegetation of the Mojave Desert. Madroño 23:1-13.

Vasek, F.C. and J.F. Clovis. 1976. Growth forms in *Arctostaphylos glauca*. American Journal of Botany 63:189-195.

Vasek, F.C. and R.F. Thorne. 1977. Transmontane coniferous vegetation. Pages 797-832 *in* M.G. Barbour and J. Major, editors. Terrestrial vegetation of California. Wiley–Interscience, reprinted by the California Native Plant Society 1988, Sacramento, CA.

Vasek, F.C. and M.G. Barbour. 1977. Mojave Desert scrub vegetation. Pages 835-867 *in* M.G. Barbour and J. Major, editors. Terrestrial vegetation of California.

Wiley–Interscience, reprinted by the California Native Plant Society 1988, Sacramento, CA.

Vasek, F.C. 1978. Jeffrey pine and vegetation of the southern Modoc National Forest. Madroño 25:9-30.

Vasek, F.C. and L.J. Lund. 1980. Soil characteristics associated with a primary plant succession on a Mojave Desert dry lake. Ecology 61:1013-1018.

Vasek, F.C. 1980. Creosote bush: long-lived clones in the Mojave Desert. American Journal of Botany 67:246-255.

Vasek, F.C. 1985. Southern California white fir (Pinaceae). Madroño 32:65-77.

Viers, S.D. 1982. Coast redwood forest: stand dynamics, successional status, and the role of fire. Pages 119-141 *in* J.E. Means, editors. Forest succession and stand development research in the Northwest. Forest Research Laboratory, Oregon State University, Corvallis, OR.

Vogl, R.J. and L.T. McHargue. 1966. Vegetation of California fan palm oases on the San Andreas fault. Ecology 47:532-540.

Vogl, R.J. 1966. Salt-marsh vegetation of Upper Newport Bay, California. Ecology 47:80-87.

Vogl, R.J. and P.K. Schorr. 1972. Fire and manzanita chaparral in the San Jacinto Mountains, California. Ecology 53:1179-1188.

Vogl, R.J. 1973. Ecology of knobcone pine in the Santa Ana Mountains, California. Ecological Monographs 43:125-143.

Vogl, R.J. 1976. An introduction to the plant communities of the Santa Ana and San Jacinto mountains. Pages 77-98 *in* J. Latting, editor. Plant communities of southern California. California Native Plant Society, Sacramento, CA.

Vogl, R.J., W.P. Armstrong, K.L. White, and K.L. Cole. 1977. The closed-cone pines and cypresses. Pages 295-358 *in* M.G. Barbour and J. Major, editors. Terrestrial vegetation of California. Wiley–Interscience, reprinted by the California Native Plant Society 1988, Sacramento, CA.

Volland, L.A. 1982. Plant associations of the central Oregon pumice zone. Report R6-ECOL-104-1985. USDA, Forest Service, Pacific Northwest Research Station, Portland, OR.

Vora, R.S. 1988. Species frequency in relation to timber harvest methods and elevation in the pine type of northeast California. Madroño 35:150-158.

Waddell, D.R. 1982. Montane forest vegetation–soil relationships in the Yolla Bolly Mountains, northern California. Master's thesis, Humboldt State University, Arcata, CA.

Wainwright, T.C. and M.G. Barbour. 1984. Characteristics of mixed evergreen forest in the Sonoma mountains of California. Madroño 31:219-230.

Waring, R.H. 1969. Forest plants of the eastern Siskiyous: their environmental and vegetational distribution. Northwest Science 43:1-17.

Warner, R.E. and K.M. Hendrix. 1984. California riparian systems. University of California Press, Berkeley, CA.

Weatherspoon, C.P. 1990. *Sequoiadendron giganteum* – giant sequoia. Pages 552-562 *in* R.M. Burns and B.H. Honkala, technical coordinators. Silvics of North America, Volume 1. Conifers. Agriculture Handbook 654. USDA, Forest Service, Washington, D.C.

Weaver, J.E. and F.E. Clements. 1938. Plant ecology. McGraw Hill, New York, NY.

Webster, L. 1981. Composition of native grasslands in the San Joaquin Valley, California. Madroño 28:231-241.

Weitkamp, W.A. 1992. Processes affecting soil development in a Vernal Pool catena in southern Riverside County, California. Master's thesis, University of California, Riverside, CA.

Wells, P.V. 1962. Vegetation in relation to geological substratum and fire in the San Luis Obispo quadrangle, California. Ecological Monographs 32:79-103.

Werschkull, G.D., F.T. Griggs, and J.M. Zaninovich. 1984. Tulare Basin protection plan. Report 103. The Nature Conservancy, San Francisco, CA.

West, N.E. 1988. Intermountain deserts, shrub steppes, and woodlands. Pages 209-230 *in* M.G. Barbour and W.D. Billings, editors. North American terrestrial vegetation. Cambridge University Press, New York, NY.

West, N.E. 1994. Juniper–pinyon woodland – SRM 412. Page 51 *in* T.N. Shiflet, editor. Rangeland cover types of the United States. Society for Range Management, Denver, CO.

Westman, W.E. and R.H. Whittaker. 1975. The pygmy forest region of northern California: studies on biomass and primary productivity. Ecology 63:453-520.

Westman, W.E. 1979. A potential role of coastal sage scrub understories in the recovery of chaparral after fire. Madroño 26:64-68.

Westman, W.E. 1981a. Diversity relations and succession in Californian coastal sage scrub. Ecology 62:170-184.

Westman, W.E. 1981b. Factors influencing the distribution of species of Californian coastal sage scrub. Ecology 62:439-455.

Westman, W.E. 1981c. Seasonal dimorphism of foliage in Californian coastal scrub. Oecologia 51:385-388.

Westman, W.E. 1983. Xeric Mediterranean-type shrubland associations of Alta and Baja California and the community/continuum debate. Vegetatio 52:3-19.

Westoby, M., B. Walker, and I. Noy-Meir. 1989. Opportunistic management for rangelands not at equilibrium. Journal of Range Management 42:266-274.

Whipple, J. and E. Cope. 1979. An ecological survey of the proposed Mount Eddy Research Natural Area. Unpublished report. USDA, Forest Service, Pacific Southwest Research Station, Berkeley, CA.

White, K.L. 1966a. Structure and composition of foothill woodland in central coastal California. Ecology 47:229-237.

White, K.L. 1966b. Old field succession on Hastings Reservation, California. Ecology 47:865-868.

White, K.L. 1967. Native bunchgrass (*Stipa pulchra*) on Hastings Reservation, California. Ecology 48:949-955.

White, S.D. 1991. *Quercus wislizenii* forest and shrubland in the San Bernardino Mountains, California. Master's thesis, Humboldt State University, Arcata, CA.

White, S.D. 1994a. Cleghorn Canyon candidate Research Natural Area: ecological survey and bioregional analysis. Unpublished report. USDA, Forest Service, Pacific Southwest Research Station, Berkeley, CA.

White, S.D. 1994b. Coastal sage scrub classification for western Riverside County, California. Unpublished report. Tierra Madre Consultants Inc., Riverside, CA.

White, S.D. 1994c. Vernal pools in the San Jacinto Valley. Fremontia 22:17-19.

White, S.D. 1994d. Vegetation descriptions, site characteristics, and plant ecology in Puente Hills shrublands. Natural resources in the Puente Hills-Chino Hills corridor: implications for land use and planning. Unpublished report. Whittier College, Whittier, CA.

White, S.D. and J.O. Sawyer. 1995. *Quercus wislizenii* forest and shrubland in the San Bernardino Mountains, California. Madroño 41:302-315.

White, T., H.J. Gordon, and M. Borchart. 1993. Chaparral types in south zone forests. Unpublished report. USDA, Forest Service, Cleveland National Forest, San Diego, CA.

White, T. 1994a. Scrub oak mixed chaparral – SRM 207. Page 17 *in* T.N. Shiflet, editor. Rangeland cover types of the United States. Society for Range Management, Denver, CO.

White, T. 1994b. Ceanothus mixed chaparral – SRM 208. Page 18 *in* T.N. Shiflet, editor. Rangeland cover types of the United States. Society for Range Management, Denver, CO.

Whittaker, R.H. 1960. Vegetation of the Siskiyou Mountains, Oregon and California. Ecological Monographs 30:279-338.

Wieslander, A.E. 1935. A vegetation type map of California. Madroño 3:140-144.

Wieslander, A.E. and H.A. Jensen. 1946. Forest areas, timber volumes, and vegetation types in California. Release Number 4 with map. USDA, Forest Service, Pacific Southwest Research Station, Berkeley, CA.

Williams, J.E., G.C. Kobetich, and C.T. Benz. 1984. Management aspects of relict populations inhabiting the Amargosa Canyon ecosystem. Pages 706-715 *in* R.E. Warner and K.M. Hendrix, editors. California riparian systems: ecology, conservation and productive management. University of California Press, Berkeley, CA.

Williams, W.T. and J.R. Potter. 1972. The coastal strand community at Morro Bay State Park, California. Bulletin of the Torrey Botanical Club 111:145-152.

Williams, W.T. 1985. The Morro Bay sand spit, a California treasure. Fremontia 13:11-16.

Williamson, R.L. 1980. Pacific Douglas-fir. Pages 106-107 *in* F.H. Eyre, editor. Forest cover types of the United States and Canada. Society of American Foresters, Washington, D.C.

Williamson, R.L. and A.D. Twombly. 1983. Pacific Douglas-fir. Pages 9-12 *in* R.M. Burns, technical compiler. Silviculture systems for the major forest types of the United States. Agriculture Handbook No. 445. USDA, Forest Service, Washington, D.C.

Wilson, J.B. 1991. Does vegetation science exist? Journal of Vegetation Science 2:289-290.

Wilson, J.L. 1986. A study of plant species diversity and vegetation patterns associated with the Pine Hill gabbro formation and adjacent substrata, El Dorado County, California. Master's thesis, California State University, Sacramento, CA.

Wilson, R.C. and R.J. Vogl. 1965. Manzanita chaparral in the Santa Ana Mountains, California. Madroño 18:47-62.

Winward, A.H. 1994. Snowbrush – SRM 420. Page 58 *in* T.N. Shiflet, editor. Rangeland cover types of the United States. Society for Range Management, Denver, CO.

Witham, H. 1976. Rings of bright color and puddles of sky. Fremontia 4:7-8.

Wolfram, H.W. and M.A. Martin. 1965. Big sagebrush in Fresno County, California. Journal of Range Management 18:285-286.

Wood, M.K. and V.T. Parker. 1988. Management of *Arctostaphylos myrtifolia* Parry at the Apricum Hill

Ecological Reserve. Unpublished report. State of California, The Resources Agency, Department of Fish and Game, Rancho Cordova, CA.

Wright, R.D. and H.A. Mooney. 1965. Substrate-oriented distribution of bristlecone pine in the White Mountains of California. American Midland Naturalist 73:257-284.

Yeaton, R.I., R.W. Yeaton, and J.E. Horenstein. 1980. The altitudinal replacement of digger pine by ponderosa pine on the western slopes of the Sierra Nevada. Bulletin of the Torrey Botanical Club 107:487-495.

Yoder, V., M.G. Barbour, R.S. Boyd, and R.A. Woodward. 1983. Vegetation of the Alabama Hills region, Inyo County, California. Madroño 30:118-126.

Young, J.A., R.A. Evans, and J. Major. 1977. Sagebrush steppe. Pages 763-796 *in* M.G. Barbour and J. Major, editors. Terrestrial vegetation of California. Wiley-Interscience, reprinted by the California Native Plant Society 1988, Sacramento, CA.

Young, J.A. and R.A. Evans. 1989. Seed production and germination dynamics in California annual grasslands. Pages 39-45 *in* L.F. Huenneke and H.A. Mooney, editors. Grassland structure and function: California annual grassland. Kluwer Academic Publishers, Boston, MA.

Zammit, C.A. and P.H. Zedler. 1988. The influence of dominant shrubs, fire, and time since fire on soil seed banks in mixed chaparral. Vegetatio 75:175-187.

Zammit, C.A. and P.H. Zedler. 1992. Size structure and seed production of *Ceanothus greggii* in mixed chaparral. Ecology 81:499-511.

Zedler, J.B. 1982. The ecology of southern California coastal salt marshes: a community profile. Pages 123-127 *in* C.E. Conrad and W.C. Oechel, editors. Dynamics and management of Mediterranean-type ecosystems, General technical report PSW-58. USDA, Forest Service, Pacific Southwest Research Station, Berkeley, CA.

Zedler, J.B., C.S. Nordby, and B.E. Kus. 1992. The ecology of Tijuana estuary, California: a national estuarine research reserve. Unpublished report. NOAA, Office of Coastal Resources Management, Sanctuaries and Reserves Division, Washington, D.C.

Zedler, P.H. 1977. Life history attributes of plants and the fire cycle: A case study in chaparral dominated by *Cupressus forbesii*. *in* H.A. Mooney and C.E. Conrad, editors. Environmental consequences of fire and fuel management in Mediterranean ecosystems. General Technical Report WO-3. USDA, Forest Service, Pacific Southwest Research Station, Berkeley, CA.

Zedler, P.H. 1981. Vegetation change in chaparral and desert communities in San Diego County, California. Pages 406-430 *in* D.C. West, H.H. Shugart and D.B. Botkin, editors. Forest succession concepts and application. Springer–Verlag, New York, NY.

Zedler, P.H. 1982. Demography and chaparral management in southern California. Pages 123-127 *in* C.E. Conrad and W.C. Oechel, editors. Dynamics and management of mediterranean-type ecosystems. General Technical Report PSW-58. USDA, Forest Service, Pacific Southwest Research Station, Berkeley, CA.

Zedler, P.H., C.R. Gautier, and G.S. McMaster. 1983. Vegetation change in response to extreme events: the effect of a short interval between fires in California chaparral and coastal shrub. Ecology 64:809-818.

Zedler, P.H. 1987. The ecology of southern California vernal pools: a community profile. Biological Report 85 (7.11). USDI, United States Fish and Wildlife Service, Washington, D.C.

Zedler, P.H. and G.A. Scheid. 1988. Invasion of *Carpobrotus edulis* and *Salix lasiolepis* after fire in a coastal chaparral site in Santa Barbara County, California. Madroño 35:196-201.

Zedler, P.H. 1990. Life histories of vernal pool vascular plants. Pages 123-146 *in* D.H. Ikeda, R.A. Schlising, F.J. Fuller, L.P. Janeway, and P. Woods, editors. Vernal pool plants: their habitat and biology. Study No. 8 from the Herbarium. California State University, Chico, CA.

Zembal, R. 1989. Riparian habitat and breeding birds along the Santa Margarita and Santa Ana rivers of southern California. Pages 98-113 *in* A.A. Schoenherr, editor. Endangered plant communities of southern California. Southern California Botanists, California State University, Fullerton, CA.

Zinke, P.J. 1977. The redwood forest and associated north coast forests. Pages 679-698 *in* M.G. Barbour and J. Major, editors. Terrestrial vegetation of California. Wiley–Interscience, reprinted by the California Native Plant Society 1988, Sacramento, CA.

Zobel, B. 1952. Jeffrey pine in the south coast ranges of California. Madroño 11:283-284.

Zobel, D.B., and G.M. Hawk. 1980. The environment of *Chamaecyparis lawsoniana*. American Midland Naturalist 103:280-297.

Zobel, D.B., L.F. Roth, and G.M. Hawk. 1985. Ecology, pathology, and management of Port Orford-cedar (*Chamaecyparis lawsoniana*). General Technical Report PNW-184. USDA, Forest Service, Pacific Northwest Research Station, Portland, OR.

Zobel, D.B. 1990. *Chamaecyparis lawsoniana* – Port Orford-cedar. Pages 88-96 *in* R.M. Burns and B.H. Honkala, technical coordinators. Silvics of North America, Volume 1. Conifers. Agriculture Handbook 654. USDA, Forest Service, Washington, D.C.

Zuckerman, S. 1990. Social and ecological prospects for second-growth forestry in the Mattole Valley, Humboldt County, California. Master's thesis, University of California, Berkeley, CA.

Zuill, H.A. 1967. Structure of two cover types of southern oak woodland in California. Master's thesis, Loma Linda University, Loma Linda, CA.

CALIFORNIA NATIVE PLANT SOCIETY
CALIFORNIA PLANT COMMUNITIES FIELD DATA FORM
See code list for italicized fields
Rev. 95/05/19

Temporary plot/transect number:	FOR OFFICE USE ONLY
	Permanent number:
Date:_____/_____/_____ DD MMM YY	Community name: Community number: Occ. no.:
County:	Source code:
USGS Quad 7.5'/15' *(circle one)*	Quad code: Quad name: Map index no.:
CNPS Chapter: (or other affiliation)	Update: Yes No

Landowner: _____

Contact Person: _____

Address: _____

City: _____ Zip: _____ Phone: _____

Observers: _____

Elevation (ft.):_____ Slope (°):_____ Aspect (°):_____ **Topography:** *Macro*_____ *Micro*_____

Legal: MERIDIAN_____TWP_____RNG_____SEC_____1/4_____1/16_____

LAT _____ LONG _____ UTMN _____ UTME _____ UTM Zone _____

Community type: _____ *Dominant vegetation form* _____
      Wetland or Upland

If Community Type = W (see Artificial Keys to Cowardin Systems and Names)

Cowardin System _____ Subsystem _____ Class _____

Distance to water (m): Vertical _____ Horizontal _____ Channel form (if riverine) _____
                                                    Straight, Meandering, Braided

Vegetation Description

Dominant layer _____ Plant community _____
Ground, Shrub, Tree

                    Patch size (acres): _____ Community size (acres) _____

Adjacent series: _____

_____

_____

Trend code _____	*Impact codes* ____ ____ ____ ____ ____ ____ ____
1. Increasing **2.** Stable **3.** Decreasing	(list codes in order, with most significant first)
4. Fluctuating **5.** Unknown	Intensity ____ ____ ____ ____ ____ ____ ____ 1. Light **2.** Moderate **3.** Heavy (List beneath each impact code)

Layer Height (nearest decimeter - 0.1 m): Ground _____ Shrub _____ Tree _____

Structure: Ground_____ Shrub _____ Tree _____   **Phenology:** Ground _____Shrub _____ Tree _____
1. Continuous 2. Intermittent 3. Open   Early, Peak, Late

Transect Length (m) _____ Transect Direction (°) _____

Sampling Interval (m): Ground _____ Shrub _____ Tree _____ Bare _____

414

Coarse fragments and soils information

Percent cover: Bedrock _____ Gravel _____ Cobble _____ Stone _____ Litter _____
                                       <3 in. dia.   3-10 in. dia.   >10 in. dia.   organic matter

Soil texture _____ Soil series (from soils map) _____ *Parent material* _____

Site Location and Plot Description

Photographs

Site History

Unknown Specimens *List code, identification notes (e.g. Genus, condition of specimen) of unknowns*

Additional Comments

CALIFORNIA NATIVE PLANT SOCIETY
CALIFORNIA PLANT COMMUNITIES FIELD DATA FORM (PART 2)
Page _____ of _____

G = Ground
S = Shrub
T = Tree

W = Within plot, no hits
X = Outside plots

Layer	Bare, Rock, Litter, Moss/Lichen, Sky or Plant name	Hits	Tally	%
Total Canopy (BARE)			Total =	

FIELD SAMPLING PROTOCOL
California Native Plant Society
Plant Communities of California
Rev. 95/5/19

INTRODUCTION

This document describes the procedures used for vegetation sampling by CNPS. The samples will provide information for the classification and description of selected plant communities in California. The sampling method is based on a 50 m long point-transect centered in a 50 m x 5 m *plot*. At each 0.5 m interval along the transect (beginning at the 50 cm mark and ending at 50.0 m), a point is projected vertically into the vegetation. Each species intercepted by a point is recorded, providing a tally of hits for each species in the herb, shrub, and tree canopies. In so far as it is possible, it is important to take care to stretch the tape taut, in order to maintain a consistent sampling area. Percent cover for each species according to vegetation layer (herb, shrub, and tree) can be calculated from these data. Finally, a list of all additional species within the 250 m^2 plot is made.

Often, the composition and abundance of the species within a type will vary with seasonality or in response to disturbance, such as fire. The optimal time to sample vegetation is determined by flowering dates such that as many species as possible can be identified. This becomes of greater concern in herbaceous vegetation types as opposed to those dominated by woody species.

PLOT LOCATION

Plots are located within subjectively chosen patches of homogeneous vegetation. Once such an area has been chosen and approximate boundaries defined, the transect is objectively located. The observer may walk to the center of the patch and then determine the center of the transect in an arbitrary manner (e.g. by tossing an object over the shoulder). The direction of the transect line from this center point is chosen randomly, using a wrist watch: the position of the second hand can refer to a compass direction, with noon equivalent to north.

For unusual cases such as narrow bands or small patches of vegetation which do not lend themselves to the placement of a straight 50 m long transect, the transect may be bent or curved. However, this should be avoided whenever possible in order to maintain consistency among the plots and to avoid observer bias in establishing the transects. In a narrow riparian corridor, for instance, locate the center of the patch which is long enough to accommodate a transect and flip a coin to determine the direction of the transect parallel to the axis of the patch.

REPLICATION

Determining how many plots to establish in a given patch of vegetation involves an assessment of the size and floristic variability of the patch, the time available to the field team, and the proximity of additional patches of the same vegetation type. Here the volunteers must make a decision, which will be based on these considerations after spending enough time in the field to gain a familiarity with the type. In some patches, one plot will adequately capture the composition and structure of the vegetation type; in others, additional plots will be necessary. For example, if a team establishes a plot in a patch of forest vegetation, and it is evident to the members of the team that the floristic composition and structure of the plot does not adequately represent that of the patch, additional plots should be established. If there are a number of individual patches of the same type in an area, it may be preferable to spread the sampling among them, thus capturing the variability among adjacent stands. Before embarking on a sampling campaign, contact the Department of Fish and Game/CNPS plant ecologist (916-324-6857) for assistance with developing a strategy for sampling a given vegetation type.

CNPS Field Sampling Protocol

GENERAL PLOT INFORMATION

The following items are included on each datasheet. Where indicated, refer to attached code list.

Temporary plot/transect number: Assigned in the field, using a unique number for each patch and for each replicate plot within a patch. Permanent plot numbers will be assigned by CNPS.

Date: Date of sampling.

County: County plot is located in.

USGS Map: The name of the USGS map the plot is located on; note series (15' or 7½').

CNPS Chapter: CNPS chapter, or other organization or agency if source of data is other than CNPS chapter.

Landowner: Name of landowner or agency acronym if known; else list as 'private'.

Contact Person: Name, address and phone number of individual responsible for data collection on the plot.

Observers: Names of individuals assisting on the plot.

Elevation: Recorded in feet or meters; please indicate units.

Slope: Degrees, read from clinometer or compass or estimated; averaged over plot.

Aspect: Degrees from true north, read from a compass or estimated; averaged over plot.

Macrotopography: Characterize the large-scale topographic position of the site. This is the general position of the sample along major topographic features of the area. *See attached code list.*

Microtopography: Characterize the local relief of the site. This is the general shape or lay of the ground along minor topographic features of the area. *See attached code list.*

Legal description: Township/Range/Section/Quarter section/Quarter-Quarter section/Meridian Legal map location of site; this is useful for land ownership determination. Meridian designations for California: Humboldt; Mt. Diablo; San Bernardino.

Latitude and longitude: Degrees latitude (northing) and longitude (easting).

UTMN and UTME: Northing and easting coordinates using the Universal Transverse Mercator (UTM) grid as delineated on the USGS topographic map; to the nearest 0.01 of a km. See sample map for an example of determining coordinates.

UTM zone: Universal Transverse Mercator zone. Zone 10 for western part of California (west of the 120th latitude); zone 11 for eastern part of California (east of the 120th latitude).

Community type: Indicate if the sample is in a wetland or an upland; note that a site need not be officially delineated as a wetland to qualify as such in this context.

Dominant vegetation form: *See attached code list.*

CNPS Field Sampling Protocol

WETLAND COMMUNITY TYPES

<u>Cowardin class</u>: See "Artificial Keys to Cowardin Systems and Names" (attached). If the plot is located in a wetland, record the proper Cowardin system name. Systems are described in detail in: Cowardin et al. 1979. Classification of wetlands and deepwater habitats of the United States. US Dept. of Interior, Fish and Wildlife Service, Office of Biological Services, Washington D.C.

Marine: habitats exposed to the waves and currents of the open ocean (subtidal and intertidal habitats).

Estuarine: includes deepwater tidal habitats and adjacent tidal wetlands that are usually semi-enclosed by land but have open, partly obstructed, or sporadic access to the open ocean, and in which ocean water is at least occasionally diluted by freshwater runoff from the land (i.e. estuaries and lagoons).

Riverine: includes all wetlands and deepwater habitats contained within a channel, excluding any wetland dominated by trees, shrubs, persistent emergent plants, emergent mosses, or lichens, and any channels that contain oceanic-derived salts greater than 0.5%.

Lacustrine: includes wetlands and deepwater habitats with all of the following characteristics: 1) situated in a topographic depression or a dammed river channel; 2) lacking trees, shrubs, persistent emergents, emergent mosses or lichens with greater than 30% areal coverage; and 3) total area exceeds 8 ha (20 acres). Similar areas less than 8 ha are included in the lacustrine system if an active wave-formed or bedrock shoreline feature makes up all or part of the boundary, or if the water depth in the deepest part of the basin exceeds 2 m (6.6 feet) at low tide. Oceanic derived salinity is always less than 0.5%.

Palustrine: includes all nontidal wetlands dominated by trees, shrubs, persistent emergents, emergent mosses or lichens, and all such wetlands that occur in tidal areas where salinity derived from oceanic salts is less than 0.5%. Also included are areas lacking vegetation, but with all of the following four characteristics: 1) areas less than 8 ha (20 acres); 2) active wave-formed or bedrock shoreline features lacking; 3) water depth in the deepest part of the basin less than 2 m (6.6 feet) at low water; and 4) salinity due to ocean-derived salts less than 0.5%.

<u>Vertical distance from high water mark of active stream channel</u>: If the plot is in or near a wetland community, record to the nearest meter or foot the estimated vertical distance from the middle of the plot to the average water line of the channel, basin, or other body of water.

<u>Horizontal distance from high water mark of active stream channel</u>: If the plot is in or near a wetland community, record to the nearest meter or foot the estimated horizontal distance from the middle of the plot to the average water line of the channel, basin, or other body of water.

<u>Stream channel form</u>: If the plot is located in or near a community along a stream, river, or dry wash, record the channel form of the waterway. The channel form is considered S (single channeled) if it consists of predominantly a single primary channel, M (meandering) if it is a meandering channel, and B (braided) if it consists of multiple channels interwoven or braided.

VEGETATION DESCRIPTION

<u>Dominant layer</u>: Indicate whether the community is dominated by the ground layer (G), shrubs (S), or trees (T).

<u>Plant community</u>: Name of series, stand or habitat according to the CNPS classification (Sawyer and Keeler-Wolf 1995); if the type is not known, or is not defined by the CNPS classification, leave the space blank.

<u>Patch size</u>: Estimated size (in acres) of patch being sampled.

CNPS Field Sampling Protocol

Community size: Estimated area (in acres) covered by the vegetation being sampled; include all areas within 1 km of the sample site.

Adjacent series: Adjacent vegetation series, stands or habitats according to CNPS classification; list in order of most extensive to least extensive.

Vegetation trend: Characterize the community as either increasing (expanding), stable, decreasing, fluctuating or unknown.

Impacts: Enter codes for potential or existing impacts on the stability of the plant community. Characterize each as either 1. light, 2. moderate or 3. heavy in the space provided beneath each code. *See attached code list.*

Layer height: Average height of each layer, in meters. See *Assessment of Layers*, below.

Vegetation structure: Characterize the structure of each layer.
If more than three layers are evident, e.g. sublayers are present, describe these as well.

> Continuous = continuously interlocking or touching crowns
> Intermittent = interlocking or touching crowns interrupted by openings
> Open = infrequently interlocking or touching crowns

Phenology: Characterize the phenology as either early (E), peak (P), or late (L) for each layer.

Transect length: Length of transect sampled in meters; standard length is 50 m.

Transect direction: Direction of the transect in degrees.

Sampling interval: The standard sampling interval is every 0.5 m; if a non-standard interval is used, please indicate so and explain why under the additional comments section (e.g. if data are from a source other than CNPS sampling efforts, or the vegetation ecologist has prescribed a different interval for a specific vegetation type).

COARSE FRAGMENTS AND SOIL INFORMATION (optional; contact CNPS plant ecologists for community types for which this information is critical).

Coarse fragments, litter: Estimate the percent coverage of each size class at or near the ground surface averaged over the 250 m^2 plot.

> Bedrock: continuous, exposed, non-transported rock

> Gravel: rounded and angular fragments < 3 inches in diameter

> Cobble: rounded and angular fragments 3 - 10 inches in diameter

> Stone: rounded and angular coarse fragments > 10 inches in diameter

> Litter: extent of undecomposed litter on surface of plot (this includes all organic matter, e.g. fallen logs, branches, and twigs down to needles and leaves).

Soil texture: Record the texture of the upper soil horizon, below the organic layer if one is present. *See attached code list.*

Soil series: Soil series based on the USDA system of soil classification recorded from local soil map.

Parent material: Geologic parent material of site. *See attached code list.*

Site location and description: A careful description which makes revisiting the vegetation patch and plots possible; give landmarks and directions. Indicate location on a photocopy of a USGS topographic map (preferably 7.5') and attach to field survey form; if possible, draw a boundary around the patch on the map. It is also helpful to *briefly* describe the topography, aspect and vegetation structure of the site.

Photographs: (optional). Describe view direction of color slides taken of the site. It is helpful to take a photograph from each end of the transect, and one across the center to provide a record of the site.

Site history: Briefly describe the history of the site, including type and year of disturbance (e.g. fire, landslides or avalanching, drought or pest outbreak). Also note the nature and extent of land use such as grazing, timber harvest, or mining.

Unknown plant specimens: List the numbers of any unknown plant specimens here, noting any information such as family or genus (if known), important characters, and whether or not there is adequate material for identification.

Additional comments: Feel free to note any additional observations of the site, or deviations from the standard sampling protocol. If additional data were recorded, e.g. if tree diameters were measured, please indicate so here.

VEGETATION DATA

Point-intercept transect: A 50 m long tape is laid along the center of the plot and secured at both ends. The observer uses a 1 meter length of steel roundbar to sight along a vertical line at every 0.5 m interval from the 0.5 to the 100 meter mark. Each species intercepted by the vertical line is tallied by vegetation layer. A total of 100 points along the transect are thus sampled.

Assessment of Layers. Estimates of the maximum height of the herb and shrub layers, and the minimum height of the tree layer, are recorded (see above). These estimates are made after a quick assessment of the vegetation and its structure; these need not be overly precise, and will vary among vegetation types. A caveat: if a number of plots are being established within the same community type, it is important to be consistent when assigning layers. This is not difficult after 2 or 3 transects have been established. Some types will have more than three layers (e.g. two tree layers of different maximum height); this should be indicated in the plot description. However, data are recorded for only three layers (herb, shrub and tree) whenever possible. The manner in which a species is recorded on the data sheet depends on the layer it occupies. The layer a species occupies will usually be determined by growth form, but exceptions will occur. For instance, a plot may contain a shrubby, multi-stemmed form of a tree species which occupies the shrub layer.

Because the species occupies the shrub layer, even though nominally a tree, it is treated as a shrub and recorded in the shrub layer on the data sheet. Similarly, a shrub occupying space in the tree canopy is recorded in the tree layer. Seedlings of woody plants, shorter than the maximum height of the herb layer, are recorded in the herb layer. An individual plant is recorded within only one layer, depending on the height of the tallest part of the individual. A species may, however, be represented in more than one layer on a plot depending on the height of each individual. For example, a single transect may contain seedlings of a tree species in the "herb", or lowest layer; saplings in the "shrub", or second layer; and mature trees in a third layer. For each species listed, indicate the layer it is occupying with either G (ground), S (shrub), or T (tree).

Determining Hits. It is important not to bias the location of the point to include a plant; this will result in overestimation of plant cover. This bias is most likely to be a problem with the herbaceous species. Take care to record hits along the same side of the tape within a plot; which side is unimportant, as long as one is consistent. The roundbar provides a line which can be projected into the vegetation layer. Only hits which fall within the canopy outline (delineated by visually rounding out the canopy edges) of a tree, shrub, or herb, or which directly hit an annual grass or other linear growth form, are valid (see Figure 1a). If two species within a single layer are intercepted by a point, both are recorded for that layer (see Figure 1b). If no vascular plant is hit by a point, a non-plant category (bare, rock, or litter in the herb layer; sky in the shrub or tree layers) is recorded as a hit for that layer. If the tree and shrub layers are both bare, *and* the herb layer is either bare or occupied by a non-vascular plant (rock, moss, lichen, litter) then the category BARE at the bottom of the page also receives a tally. Although this may seem redundant, recording non-hits in this manner allows for the calculation of absolute plant cover for the entire plot as well as for each separate layer. Plant names are recorded as Latin binomials (not common names) and should be consistent with the Jepson Manual (Hickman 1993).

It may be helpful to consider the above as a series of decision rules. In the herb layer: IF the point intercepts a grass, or the canopy outline of an herbaceous or woody species, record a hit for that plant. If more than one species is intercepted, record a hit for each within that layer. IF AND ONLY IF no vascular plant is intercepted in the herb layer, one and only one non-vascular plant category receives a hit: the options are bare, litter, rock or moss/lichen.

In the shrub and tree layers: IF the point intercepts the sphere of influence of an individual, that species receives a hit for the layer which the highest point of the individual occurs within.

Data Sheets: In order to accommodate different styles of recording, two types of datasheet have been prepared. Some observers may find it more convenient to use the optional long form, which provides a prompt for which point is being recorded. This form, which is available from the CNPS plant ecologist, must then be summarized on the short form by summing the hits for each species and recording them by layer. Alternatively, the short form may be used directly; please take the time to sum the tallies as indicated on the sample data sheet.

Additional Species: All vascular plants not recorded for the transect are listed by layer after searching the entire 250 m^2 plot (2.5 m on each side of the 50 m transect). A careful and exhaustive search is required to be sure that no species are missed. Place a 'W' in the tally column to indicate that the plant occured in the plot, but not along the transect. If there are species of interest (e.g. characteristic species which are low in cover or infrequent, or sensitive species) present adjacent to the plot, but not within it, these may also be listed, with an "X" indicating that they did not occur within the plot itself. Bear in mind that this may indicate that additional samples are needed to pick up these elements.

Unknown specimens: Plant specimens which cannot be determined to species in the field, or which need further verification, are collected and pressed according to standard procedure. Each specimen is assigned a field number made up of the plot number and a sequential number unique to each unknown plant on the plot. For example, unknown number CNPS4-2-6 is the sixth unknown specimen collected on the second plot established in patch number 4. This number is recorded on the datasheet in lieu of a species name. When in doubt, it is preferable to record a species as unknown rather than guessing.

DATA HANDLING AND MANAGEMENT

After the plot is completed, the data sheets must be proofed before submitting them to CNPS. Please make sure that all site information fields are filled out and complete; that the field data form is complete, with nomenclature consistent with Hickman (1993). If data are missing, make a note under the comments section. Please take a moment to make sure that the writing is legible, and that all hits are tallied. This is best done either on-site, or immediately after the fieldwork is completed. After proofing the forms, make a xerox copy in case they are accidentally misplaced. These steps take only a few moments relative to the time invested in sampling, and may make the difference between the data being useful or not. Some chapters have appointed an individual to collect and proof the data forms before submitting them to the Department of Fish and Game/CNPS plant ecologist.

422

Protocol

Figure 1. Determining hits along a transect

1a. Determining canopy outline (bird's eye view of plot).

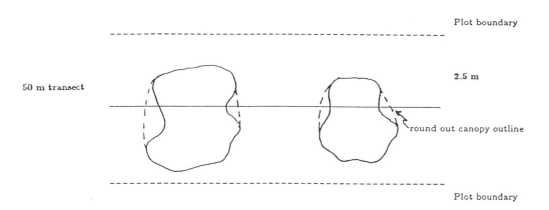

1b. Determining hits within multiple layers.

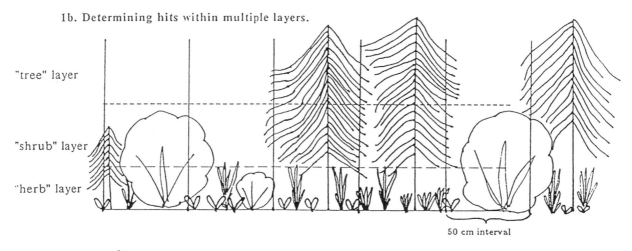

species 1 (♡): 4 hits in herb layer

species 2 (🖐): 2 hits in herb layer

species 3 (🌲): 4 hits in tree layer, 1 hit in shrub layer

species 4 (🌱): 2 hits in shrub layer, 1 hit in herb layer

CNPS Field Sampling Protocol

EQUIPMENT

50 m tape	clipboard/data sheets	*Optional:*
steel roundbar	topographic map	clinometer
compass	surveyor stakes (for marking corners)	watch with second hand

California Native Plant Society
Plant Communities Field Data Form Code List

MACRO TOPOGRAPHY

00	Bench
01	Ridge Top
02	Upper 1/3 of slope
03	Middle 1/3 of slope
04	Lower 1/3 of slope
05	Toeslope
06	Bottom
07	Edge of basin/wetland
08	Draw
09	Other
10	Terrace

MICRO TOPOGRAPHY

01	Convex or rounded
02	Linear or even
03	Concave or depression
04	Undulating pattern
05	Hummock or swale patterned
06	Mounded
07	Other

DOMINANT VEGETATION FORM

BRDF	Broadleaf deciduous forest
BREF	Broadleaf evergreen forest
COFO	Conifer forest
MBCF	Mixed broadleaf-conifer forest
MOCH	Montane chaparral
LOCH	Lowland chaparral
COSH	Coastal scrub
INSH	Interior scrub
GRAS	Grassland
CODU	Coastal dune
INDU	Interior dune
ALPI	Alpine
MIDE	Mixed desert

IMPACTS

01	Development
02	ORV activity
03	Agriculture
04	Grazing
05	Competition from exotics
06	Logging
07	Insufficient population/stand size
08	Altered flood/tidal regime
09	Mining
10	Hybridization
11	Groundwater pumping
12	Dam/inundation
13	Other
14	Surface water diversion
15	Road/trail construct/maint
16	Biocides
17	Pollution
18	Unknown
19	Vandalism/dumping
20	Foot traffic/trampling
21	Improper burning regime
22	Over-collecting/poaching
23	Erosion or runoff
24	Altered thermal regime
25	Landfill
26	Degraded water quality
27	Wood cutting
28	Military operations
29	Recreational use (non-ORV)
30	Nest parasitism
31	Non-native predators
32	Rip-rap, bank protection
33	Channelization (human caused)
34	Feral pigs

SOIL TEXTURE

COSA	Coarse, sand
COLS	Coarse, loamy sand
MCLS	Moderately coarse, sandy loam
MESA	Medium to very fine, sandy loam
MELO	Medium, loam
MESL	Medium, silt loam
MESI	Medium, silt
MFCL	Moderately fine, clay loam
MFSA	Moderately fine, sandy clay loam
MFSL	Moderately fine, silty clay loam
FISA	Fine, sandy clay
FISC	Fine, silty clay
FICL	Fine, clay
SAND	Sand (class unknown)
LOAM	Loam (class unknown)
CLAY	Clay (class unknown)
UNKN	Unknown

PARENT MATERIAL

ANDE	Andesite
ASHT	Ash (of any origin)
GRAN	Grandiorite
GREE	Greenstone
IGTU	Igneous (type unknown)
MONZ	Monzonite
PYFL	Pyroclastic flow
QUDI	Quartz diorite
RHYO	Rhyolite
TRSY	Trachyte and syenite
VOFL	Volcanic flow
VOMU	Volcanic mud
WETU	Welded tuff (tufa)
BLUE	Blueschist
CHER	Chert
DOLO	Dolomite
FRME	Franciscan melange
GNBG	Gneiss and biotite gneiss
HORN	Hornfels
MARB	Marble
METU	Metamorphic (type unknown)
PHYL	Phyllite
QUAR	Quartzite
SCHI	Schist
SESC	Semi-schist
SLAT	Slate
BREC	Breccia (non-volcanic)
CACO	Calcareous conglomerate
CASA	Calcareous sandstone
CASH	Calcareous shale
CASI	Calcareous siltstone
CONG	Conglomerate
FANG	Fanglomerate
GLTI	Glacial till, mixed origin, moraine
LALA	Large landslide (unconsolodated)
LIME	Limestone
SAND	Sandstone
SETU	Sedimentary (type unknown)
SHAL	Shale
SILT	Siltstone
DIAB	Diabase
GABB	Gabbro
PERI	Peridotite
SERP	Serpentine
ULTU	Ultramafic (type unknown)
CALU	Calcarious (origin unknown)
CIAL	Clayey alluvium
DUNE	Sand dunes
GRAL	Gravelly alluvium
LOES	Loess
MIAL	Mixed alluvium
MIIG	Mixed igneous
MIME	Mixed metamorphic
MIRT	Mix of two or more rock types
MISE	Mixed sedimentary
OTHE	Other than on the list
SAAL	Sandy alluvium
SIAL	Silty alluvium

California Natural Community Field Survey Form

Mail to:
Natural Diversity Data Base
California Dept. of Fish and Game
1416 Ninth Street
Sacramento, CA 95814
(916) 324-6857

For office use only	
Source Code_____	Quad Code_____
Community Code _____	Occ #_____
Map Index #_____	Update Y _____ N _____

Please provide as much of the following information as you can. Please attach a map (if possible, based on the USGS 7.5 minute series) showing the site's location and boundaries. Use the back if needed.

Community name:_____

Reporter:_____ Phone _____
Affiliation and Address_____
Date of field work:_____ County:_____
Location:_____
_____

Quad name: _____T_____R _____ _____ ¼ of____ ¼ sec____ Meridian_____
UTM Zone __ __ Northing __ __ __ __ __ __ __ Easting __ __ __ __ __ __
Landowner/Manager: _____ Photographs: Slide ☐ Print ☐
Elevation: _____Aspect: _____ Slope (indicate % or °)_____ Drainage:_____
Site acreage: _____ Evidence of disturbance/threats: _____

Current land use: _____
Substrate/Soils: _____
_____

General description of community: _____
_____
_____

Any Special Plants or Animals present: _____
_____
_____

Successional status/Evidence of regeneration of dominant taxa:_____
_____
_____

Overall site quality: Excellent____ Good ____Fair ____Poor ____Comments: _____

Basis for report: Remote image ____Binocular/Telescopic survey _____
Windshield survey ____ Brief walk-thru ____ Detailed survey ____ Other _____
Relevé: In the space below, indicate each species cover % within the following growth form categories:

Trees	Shrubs	Herbs/Graminoids

Continue on back if needed. Thank you for your contribution.

426

Trees	Shrubs	Herbs/Graminoids

Indices

The *Manual* has three indices.

Index to Vegetation Names and Codes

We list the *Manual's* series, stand, habitat, and vernal pool types alphabetically. Following the page number is a code of up to twelve letters of the type. This code is used in other indices.

Index to NDDB/Holland Types

We compare NDDB/Holland categories to *Manual's* series, stand, habitat, and vernal pool types. The five numbers are used to indicate relation of hierarchy in the classification (Holland 1986). For example:

21000 Coastal dunes
    21100 Active coastal dunes
    21200 Fordunes
        21210 Northern foredunes
            21211 Northern foredune grassland
    21300 Dune scrub
        21310 Northern dune scrub
        21320 Central dune scrub
        21230 Southern foredunes

We list the *Manual's* codes below the Holland classes to indicate relationships.

Index to Plant Names

We list common and scientific names alphabetically.

Index to Vegetation Names and Codes

Index to NDDB/Holland Types

175:Mojyucca
34230 Mojave yucca scrub and steppe
    33:Biggalleta, 175:Mojyucca
34240 Mojave mixed woody and succulent scrub
    114:Brittlebush, 115:Briwhibur, 129:Catacacia,
    145:Crebuswhibur, 168:Jostree, 175:Mojyucca,
    181:Nolina
34250 Mojave wash scrub
    129:Catacacia
34300 Blackbush scrub
    108:Blabush
34400 Nolina scrub
    181:Nolina
35000 Great Basin scrubs
    100:Bigsagebrush, 107:Bitterbrush, 110:Blasagebrush,
    170:Lowsagebrush, 183:Parrabbitbru, 189:Rubrabbitbru
35100 Great Basin mixed scrub
    100:Bigsagebrush, 107:Bitterbrush
35200 Sagebrush scrub
    100:Bigsagebrush, 107:Bitterbrush, 110:Blasagebrush,
    170:Lowsagebrush
35210 Big sagebrush
    100:Bigsagebrush, 107:Bitterbrush
35220 Subalpine sagebrush scrub
    110:Blasagebrush, 170:Lowsagebrush
35300 Sagebrush steppe
    100:Bigsagebrush, 107:Bitterbrush
35400 Rabbitbrush scrub
    189:Rubrabbitbru
35410 Mono pumice flat
    183:Parrabbitbru
36000 Chenopod scrubs
    98:Allscale, 118:Busseepweed, 149:Desholly,
    153:Fousaltbush, 154:Greasewood, 159:Hopsage,
    166:Iodbush, 173:Mixsaltbush, 199:Shadscale,
    201:Spinescale, 210:Winfat
36100 Desert chenopod scrub
    98:Allscale, 118:Busseepweed, 149:Desholly,
    153:Fousaltbush, 154:Greasewood, 159:Hopsage,
    166:Iodbush, 173:Mixsaltbush, 199:Shadscale,
    201:Spinescale, 210:Winfat
36110 Desert saltbush scrub
    98:Allscale, 149:Desholly, 153:Fousaltbush,
    173:Mixsaltbush, 201:Spinescale
36120 Desert sink scrub
    118:Busseepweed, 153:Fousaltbush, 154:Greasewood,
    166:Iodbush, 173:Mixsaltbush
36130 Desert greasewood scrub
    118:Busseepweed, 154:Greasewood, 166:Iodbush
36140 Shadscale scrub
    159:Hopsage, 199:Shadscale, 210:Winfat
36200 Great Valley chenopod scrub
    98:Allscale, 118:Busseepweed, 166:Iodbush,
    201:Spinescale
36210 Valley sink scrub
    118:Busseepweed, 166:Iodbush
36220 Valley saltbush scrub
    98:Allscale, 201:Spinescale
36300 Foothill chenopod scrub
    98:Allscale
36310 Sierra-Tehachapi saltbush scrub

98:Allscale
36320 Interior Coast Range saltbush scrub
    98:Allscale
37000 Chaparrals
    102:Bigmanzanita,103:Bigceanothus, 104:Bigceabirmou,
    105:Bigceaholred, 106:BirmoumahCal, 113:Breoak,
    122:Calbucwhisag, 128:Canlivoakshr, 130:Chamise,
    132:Chabigman, 133:Chablasag, 134:Chacupcea,
    135:ChaEasman, 136:Chahoacea, 137:Chamismanwoo,
    138:Chawedcea, 139:Chawhisag, 140:Chawhitethor,
    142:Coybrush, 146:Cupceafreoak, 151:Easmanzanita,
    155:Gremanzanita, 156:Haiceanothus,
    157:Hoaceanothus, 160:Hucoak, 161:Intlivoakshr,
    163:Intlivoakcan, 164:Intlivoakcha, 165:Intlivoakscr,
    167:Ionmanzanita, 169:Leaoak, 174:Mixscroak,
    178:Mouwhitethor, 185:Redshank, 186:Redshabir,
    187:Redshacha, 190:Sadoak, 191:Salblahuc,
    193:Scalebroom, 196:Scroakbirmou, 197:Scroakcha,
    202:Sumac, 205:Tobbrush, 206:Wedceanothus,
    208:Whisage, 209:Whimanzanita, 211:Woomanzanita,
    224:Birmoumah
37100 Upper Sonoran mixed chaparral
    102:Bigmanzanita, 130:Chamise, 135:ChaEastman,
    136:Chahoacea, 137:Chamismanwoo, 151:Easmanzanita,
    174:Mixscroak, 194:Scroak, 196:Scroakbirmou,
    197:Scroakcha, 198:Scroakchawhi, 202:Sumac,
    209:Whimanzanita
37110 Northern mixed chaparral
    102:Bigmanzanita, 130:Chamise, 151:Easmanzanita,
    209:Whimanzanita
37111 Gabbroic northern mixed chaparral
    130:Chamise, 209:Whimanzanita
37120 Southern mixed chaparral
    135:ChaEastman, 136:Chahoacea, 137:Chamismanwoo,
    174:Mixscroak, 194:Scroak, 196:Scroakbirmou,
    197:Scroakcha, 198:Scroakchawhi, 202:Sumac
37121 Granitic southern mixed chaparral
    136:Chahoacea, 137:Chamismanwoo, 174:Mixscroak,
    194:Scroak, 196:Scroakbirmou, 197:Scroakcha,
    198:Scroakchawhi
37122 Mafic southern mixed chaparral
    135:ChaEasman, 136:Chahoacea, 137:Chamismanwoo,
    174:Mixscroak, 194:Scroak, 196:Scroakbirmou,
    197:Scroakcha, 198:Scroakchawhi, 202:Sumac
37200 Chamise chaparral
    130:Chamise, 132:Chabigman, 133:Chablasag,
    134:Chacupcea, 135:ChaEasman, 136:Chahoacea,
    137:Chamismanwoo, 138:Chawedcea, 139:Chawhisag
37300 Red shank chaparral
    185:Redshank, 186:Redshabir, 187:Redshacha
37400 Semi-desert chaparral
    106:BirmoumahCal, 146:Cupceafreoak
37500 Montane chaparral
    113:Breoak, 117:Buschinquapi148:Deerbrush,
    155:Gremanzanita, 160:Hucoak, 178:Mouwhitethor,
    205:Tobbrush, 206:Wedceanothus, 209:Whimanzanita
37510 Mixed montane chaparral
    155:Gremanzanita, 160:Hucoak, 178:Mouwhitethor,
    205:Tobbrush, 206:Wedceanothus, 209:Whimanzanita
37520 Montane manzanita chaparral
    155:Gremanzanita, 209:Whimanzanita

37530 Montane ceanothus chaparral
148:Deerbrush, 178:Mouwhitethor, 205:Tobbrush
37531 Deer brush chaparral
148:Deebrush
37532 Whitethorn chaparral
178:Mouwhitethor
37533 Tobacco brush chaparral
205:Tobbrush
37540 Montane scrub oak chaparral
113:Breoak, 160:Hucoak
37541 Shin oak brush
113:Breoak
37542 Huckleberry oak chaparral
160:Hucoak
37540 Bush chinquapin chaparral
117:Buschinquapi
37600 Serpentine chaparral
169:Leaoak, 209:Whimanzanita
37610 Mixed serpentine chaparral
169:Leaoak
37620 Leather oak chaparral
169:Leaoak
37700 Island chaparral
194:Scroak
37800 Upper Sonoran ceanothus chaparral
103:Bigceanothus, 104:Bigceabirmou, 105:Bigceaholred,
112:Blublossom, 156:Haiceanothus, 206:Wedceanothus
37810 Buck brush chaparral
206:Wedceanothus
37820 Blue brush chaparral
112:Blublossom
37830 Ceanothus crassifolius chaparral
157:Hoaceanothus
37840 Ceanothus megacarpus chaparral
103:Bigceanothus, 104:Bigceabirmou, 105:Bigceaholred
37900 Scrub oak chaparral
174:Mixscroak, 194:Scroak, 196:Scroakbirmou,
197:Scroakcha, 198:Scroakchawhi
37A00 Interior live oak chaparral
128:Canlivoakshr, 161:Intlivoakshr, 163:Intlivoakcan,
164:Intlivoakcha, 165:Intlivoakscr
37B00 Upper Sonoran manzanita chaparral
102:Bigmanzanita, 130:Chamise, 132:Chabigman,
134:Chacupcea, 135:ChaEasman, 138:Chawedcea,
151:Easmanzanita, 155:Gremanzanita,
178:Mouwhitethor, 209:Whimanzanita
37C00 Maritime chaparral
130:Chamise, 132:Chabigman, 137:Chamiswooman,
194:Scroak, 211:Woomanzanita
37C10 Northern maritime chaparral
130:Chamise, 132:Chabigman, 194:Scroak,
211:Woomanzanita
37C20 Central maritime chaparral
211:Woomanzanita
37C30 Southern maritime chaparral
130:Chamise, 137:Chamiswooman
37D00 Ione chaparral
167:Ionmanzanita
37E00 Mesic north slope chaparral
128:Canlivoakshr, 130:Chamise, 132:Chabigman,
134:Chacupcea, 161:Intlivoakshr, 163:Intlivoakcan,

164:Intlivoakcha, 165:Intlivoakscr, 174:Mixscroak,
194:Scroak, 196:Scroakbirmou, 197:Scroakcha,
198:Scroakchawhi
37E10 Northern north slope chaparral
130:Chamise, 132:Chabigman, 134:Chacupcea,
161:Intlivoakshr, 194:Scroak
37E20 Southern north slope chaparral
128:Canlivoakshr, 161:Intlivoakshr, 163:Intlivoakcan,
164:Intlivoakcha, 165:Intlivoakscr, 174:Mixscroak,
194:Scroak, 196:Scroakbirmou, 197:Scroakcha,
198:Scroakchawhi
37F00 Poison-oak chaparral
130:Chamise, 132:Chabigman, 140:Chawhitethor,
142:Coybrush, 161:Intlivoakshr, 191:Salblahuc,
194:Scroak, 206:Wedceanothus, 211:Woomanzanita
37G00 Coastal sage-chaparral scrub
133:Chablasag, 139:Chawhisag
37H00 Alluvial fan chaparral
120:Calbuckwheat, 122:Calbucwhisag, 193:Scalebroom,
208:Whisage
37J00 Flannel bush chaparral
106:BirmoumahCal, 146:Cupceafreoak
38000 Montane dwarf scrubs
356:Subuplshrhab
39000 Upper Sonoran subshrub scrubs
111:BlaCalephnar
41000 Coastal prairies
42:Caloatgrass, 59:Idafescuc, 61:Intpregrass,
69:Pacreedgrass, 88:Tufhairgrass
41100 Coastal terrace prairie
69:Pacreedgrass, 88:Tufhairgrass
41200 Bald hills prairie
42:Caloatgrass, 59:Idafescue
42000 Valley and foothill grasslands
30:Alksacaton, 47:Creryegrass, 50:Desneedlegra,
55:Fooneedlegra, 59:Idafescue, 67:Nodneedlegra,
68:Onesidblu, 75:Purneedlegra
42100 Valley needlegrass grassland
50.Desneedlegra, 67.Nodneedlegra, 68:Onesidblu,
75:Purneedlegra
42120 Valley sacaton grassland
130:Alksacaton
42130 Serpentine bunchgrass
55:Fooneedlegra, 59:Idafescue, 68:Onesidblu
42140 Valley wildrye grassland
147:Creryegrass
42150 Pine bluegrass grassland
68:Onesidblu
43000 Great Basin grasslands
31:Ashryegrass, 34:Bluwheatgras, 42:Caloatgrass,
48:Crewheatgras, 59:Idafescue, 60:Indricegrass,
61:Intpergrassl, 62:Kenbluegrass, 66:Neeandthr,
68:Onesidblu
44000 Vernal pools
360:Norbasflover,362:Norclaverpoo, 364:Norharverpoo,
366:Norvolashver, 367:Norvolmudver,
368:SanDiemesver, 369:SanJacValver, 370:SanRosPlaver
44110 Northern hardpan vernal pool
364:Norharverpoo
44120 Northern claypan vernal pool
362:Norclaverpoo

44131 Northern basalt flow vernal pool
    360:Norbasflover
44132 Northern volcanic mudflow vernal pool
    367:Norvolmudver
44133 Northern volcanic ash flow vernal pool
    366:Norvolashver
44300 Southern vernal pool
    369:SanJacValver
44310 Southern interior basalt flow vernal pool
    370:SanRosPlaver
44321 San Diego Mesa hardpan vernal pool
    368:SanDiemesver
44322 San Diego Mesa claypan vernal pools
    368:SanDiemesver
45000 Meadows and seeps
    30:Alksacaton, 57:Grefescue, 65:Nebsedge,
    77:RocMoused, 78:Saltgrass, 82:Sedge, 84:Shoreedgrass,
    85:Shosedge, 86:Spikerush, 88:Tufhairgrass,
    177:Mouheabil, 348:Alpinehab, 353:Monmeahab,
    355:Submeahab, 356:Subuplshrhab
45100 Montane meadow
    65:Nebsedge, 77:RocMoused, 82:Sedge,
    84:Shoreedgrass, 86:Spikerush, 177:Mouheabil,
    353:Monmeahab, 355:Submeahab
45110 Wet montane meadow
45120 Dry montane meadow
    84:Shoreedgrass, 353:Monmeahab
45200 Subalpine and alpine meadow
    82:Sedge, 84:Shoreedgrass, 88:Tufhairgrass,
    177:Mouheabil, 348:Alpinehab
45210 Wet subalpine or alpine meadow
    57:Grefescue, 82:Sedge, 84:Shoreedgrass, 85:Shosedge,
    88:Tufhairgrass, 177:Mouheabil, 348:Alpinehab,
    355:Submeahab, 356:Subuplshrhab
45220 Dry subalpine or alpine meadow
    57:Grefescue, 82:Sedge, 84:Shoreedgrass, 85:Shosedge,
    348:Alpinehab, 355:Submeahab, 356:Subuplshrhab
45300 Alkali meadows and seeps
    30:Alksacaton, 52:Ditgrass, 78:Saltgrass
45310 Alkali meadow
    30:Alksacaton, 78:Saltgrass
45320 Alkali seep
    52:Ditgrass
45400 Freshwater seep
    65:Nebsedge, 77:RocMoused, 82:Sedge, 86:Spikerush
46000 Alkali playa communities
    118:Busseepweed, 154:Greasewood, 166:Iodbush
47000 Pavement plain communities
    110:Blasagebrush
51000 Bogs and fens
51100 Bog and fen
    49:Dargog, 352:Fenhab
51110 Sphagnum bog
    352:Fenhab
51120 Darlingtonia bog
    49:Darbog
51120 Fen
    352:Fenhab
52000 Marshes and swamps
    35:Bulrush, 37:Bulcattail, 39:Burreed, 43:Cattail,
    46:Cordgrass, 52:Ditgrass, 53:Duckweed, 63:Mosfern,

    71:Pickleweed, 73:Ponwitflolea, 74:Ponwitsublea,
    76:Quillwort, 78:Saltgrass, 90:Yelponlil, 262:Hoowillow,
    279:Mixwillow, 287:Pacwillow, 306:Sitspruce,
    307:Sitwillow
52100 Coastal salt marsh
    46:Cordgrass, 71:Pickleweed, 78:Saltgrass
52110 Northern coastal salt marsh
    46:Cordgrass, 71:Pickleweed, 78:Saltgrass
52120 Southern coastal salt marsh
    46:Cordgrass, 71:Pickleweed, 78:Saltgrass
52200 Coastal brackish marsh
    35:Bulrush, 37:Bulcattail, 43:Cattail, 52:Ditgrass
53300 Alkali marsh
    35:Bulrush, 37:Bulcattail, 43:Cattail, 52:Ditgrass
52310 Cismontane alkali marsh
    35:Bulrush, 37:Bulcattail, 43:Cattail, 52:Ditgrass
52320 Transmontane alkali marsh
    35:Bulrush, 37:Bulcattail, 43:Cattail, 52:Ditgrass
52400 Freshwater marsh
    335:Bulrush, 37:Bulcattail, 39:Burreed, 43:Cattail,
    53:Duckweed, 63:Mosfern, 73:Ponwitflolea,
    74:Ponwitsublea, 76:Quillwort, 90:Yelponlil
52410 Coast and valley freshwater marsh
    35:Bulrush, 37:Bulcattail, 43:Cattail, 53:Duckweed,
    63:Mosfern, 73:Ponwitflolea, 74:Ponwitsublea,
    76:Quillwort, 90:Yelponlil
52420 Transmontane freshwater marsh
    35:Bulrush, 37:Bulcattail, 43:Cattail, 53:Duckweed,
    63:Mosfern, 73:Ponwitflolea, 74:Ponwitsublea,
    90:Yelponlil
52430 Montane freshwater marsh
    335:Bulrush, 37:Bulcattail, 39:Burreed, 43:Cattail,
    53:Duckweed, 63:Mosfern, 73:Ponwitflolea,
    74:Ponwitsublea, 76:Quillwort, 90:Yelponlil
52500 Vernal marsh
    82:Sedge, 86:Spikerush, 370:SanRosPlaver
5251A Ledum swamp
    352:Fen
52600 Freshwater swamp
    262:Hoowillow, 279:Mixwillow, 287:Pacwillow,
    306:Sitspruce, 307:Sitwillow,
61000 Riparian forests
    180:Narwillow, 219:Arrwillow, 220:Aspen,
    226:Blacottonwoo, 229:Blawillow, 237:Calsycamore,
    239:Canlivoak, 241:Coalivoak, 259:Frecottonwoo,
    262:Hoowillow, 274:Mesquite, 279:Mixwillow,
    287:Pacwillow, 295:Redalder, 299:Redwillow,
    300:Redwood, 307:Sitwillow, 312:Valoak, 314:Watbirch,
    319:Whialder
61100 North Coast riparian forest
    226:Blacottonwoo, 262:Hoowillow, 279:Mixwillow,
    287:Pacwillow, 295:Redalder, 300:Redwood
61110 North Coast black cottonwood riparian forest
    226:Blacottonwoo
61120 North coast alluvial redwood forest
    300:Redwood
61130 Red alder riparian forest
    262:Hoowillow, 279:Mixwillow, 287:Pacwillow,
    295:Redalder
61200 Central coast riparian forest
    219:Arrwillow, 237:Calsycamore, 241:Coalivoak,

259:Frecottonwoo, 279:Mixwillow, 287:Pacwillow, 299:Redwillow
61210 Central Coast cottonwood-sycamore riparian forest
237:Calsycamore, 259:Frecottonwoo, 279:Mixwillow, 287:Pacwillow, 299:Redwillow
61220 Central coast live oak riparian forest
241:Coalivoak
61230 Central Coast arroyo willow riparian forest
219:Arrwillow, 279:Mixwillow
61300 Southern riparian forest
180:Narwillow, 219:Arrwillow, 229:Blawillow, 239:Canlivoak, 241:Coalivoak, 259:Frecottonwoo, 279:Mixwillow, 287:Pacwillow
61310 Southern coast live oak forest
241:Coalivoak
61320 Southern arroyo willow riparian forest
219:Arrwillow, 279:Mixwillow
61330 Southern cottonwood-willow riparian forest
180:Narwillow, 219:Arrwillow, 229:Blawillow, 259:Frecottonwoo, 279:Mixwillow, 287:Pacwillow
61350 Canyon live oak ravine forest
239:Canlivoak
61400 Great Valley riparian forest
180:Narwillow, 229:Blawillow, 259:Frecottonwoo, 279:Mixwillow, 287:Pacwillow, 299:Redwillow, 312:Valoak, 319:Whialder
61410 Great Valley cottonwood riparian forest
180:Narwillow, 229:Blawillow, 259:Frecottonwoo, 279:Mixwillow
61420 Great Valley mixed riparian forest
229:Blawillow, 259:Frecottonwoo, 279:Mixwillow, 287:Pacwillow, 299:Redwillow
61430 Great Valley valley oak riparian forest
312:Valoak
61510 White alder riparian forest
319:Whialder
61500 Montane riparian forest
220:Aspen, 226:Blacottonwoo, 319:Whialder
61510 White alder forest
319:Whialder
61520 Aspen riparian forest
220:Aspen
61530 Montane black cottonwood riparian forest
226:Blacottonwoo
61600 Modoc-Great Basin riparian forest
80:Narwillow, 259:Frecottonwoo, 279:Mixwillow, 299:Redwillow, 314:Watbirch
61610 Modoc-Great Basin cottonwood-willow riparian forest
80:Narwillow, 259:Frecottonwoo, 279:Mixwillow, 299:Redwillow, 314:Watbirch
61700 Mojave riparian forest
180:Narwillow, 229:Blawillow, 259:Frecottonwoo, 279:Mixwillow, 299:Redwillow
61800 Colorado riparian forest
180:Narwillow, 229:Blawillow, 259:Frecottonwoo, 274:Mesquite, 279:Mixwillow
61810 Sonoran cottonwood-willow riparian forest
180:Narwillow, 229:Blawillow, 259:Frecottonwoo, 279:Mixwillow
61820 Mesquite bosque
274:Mesquite

62000 Riparian woodlands
219:Arrwillow, 232:Blupalveriro, 237:Calsycamore, 256:Fanpalm, 319:Whialder
62100 Sycamore alluvial woodland
237:Calsycamore
62200 Desert dry wash woodland
232:Blupalveriro
62300 Desert fan palm oasis woodland
256:Fanpalm
62400 Southern sycamore-alder riparian woodland
237:Calsycamore, 319:Whialder
63000 Riparian scrubs
99:Arrweed, 119:Buttonbush, 129:Catacacia, 171:Mexelderberr, 176:Moualder, 177:Mouheabil, 179:Mulefat, 180:Narwillow, 192:Sanwillow, 193:Scalebroom, 200:Sitalder, 203:Tamarisk, 219:Arrwillow, 229:Blawillow, 262:Hoowillow, 274:Mesquite, 279:Mixwillow, 287:Pacwillow, 295:Redalder, 299:Redwillow, 307:Sitwillow, 314:Watbirch, 354:Monwetshrhab, 357:Subwetshrhab
63100 North Coast riparian scrub
180:Narwillow, 192:Sanwillow, 219:Arrwillow, 262:Hoowillow, 279:Mixwillow, 295:Redalder, 307:Sitwillow
63110 Woodwardia thicket
295:Redalder
63200 Central Coast riparian scrub
179:Mulefat, 180:Narwillow, 219:Arrwillow, 229:Blawillow, 279:Mixwillow, 287:Pacwillow, 299:Redwillow, 307:Sitwillow
63300 Southern riparian scrub
179:Mulefat, 180:Narwillow, 219:Arrwillow, 229:Blawillow, 279:Mixwillow, 287:Pacwillow, 299:Redwillow
63310 Mulefat scrub
179:Mulefat
63320 Southern willow scrub
180:Narwillow, 219:Arrwillow, 229:Blawillow, 279:Mixwillow, 287:Pacwillow, 299:Redwillow
63400 Great Valley riparian scrub
119:Buttonbush, 171:Mexelderberr, 180:Narwillow, 219:Arrwillow, 274:Mesquite, 279:Mixwillow, 287:Pacwillow
63410 Great Valley willow scrub
219:Arrwillow, 180:Narwillow, 279:Mixwillow, 287:Pacwillow
63420 Great Valley mesquite scrub
274:Mesquite
63430 Buttonbush scrub
119:Buttonbush
63440 Elderberry savanna
171:Mexelderberr
63500 Montane riparian scrub
176:Moualder, 177:Mouheabil, 200:Sitalder, 354:Monwetshrhab, 357:Subwetshrhab
63700 Mojave desert wash scrub
129:Catacacia, 193:Scalebroom
63800 Colorado riparian scrub
99:Arrweed, 203:Tamarisk
63810 Tamarisk scrub
203:Tamarisk

63820 Arrowweed scrub
    99:Arrweed
71000 Cismontane woodlands
    224:Birmoumah, 227:Blaoak, 230:Bluoak,
    238:Calwalnut, 241:Coalivoak, 253:Engoak,
    257:Foopine, 264:Intlivoak, 265:Isloak, 277:Mixoak,
    285:Orewhioak, 312:Valoak, 337:Hinwalsta
71100 Oak woodland
    227:Blaoak, 230:Bluoak, 230:Bluoak, 241:Coalivoak,
    264:Intlivoak, 285:Orewhioak, 312:Valoak
71110 Oregon oak woodland
    285:Orewhioak
71120 Black oak woodland
    227:Blaoak
71130 Valley oak woodland
    312:Valoak
71140 Blue oak woodland
    230:Bluoak
71150 Interior live oak woodland
    264:Intlivoak
71160 Coast live oak woodland
    241:Coalivoak
71170 Alvord oak woodland
    230:Bluoak
71180 Engelmann oak woodland
    253:Engoak
71181 Open Engelmann oak woodland
    253:Engoak
71182 Dense Engelmann oak woodland
    253:Engoak
71190 Island oak woodland
    265:Isloak
71200 Walnut woodland
    238:Calwalnut, 337:Hinwalsta
71210 California walnut woodland
    238:Calwalnut
71220 Hinds walnut woodland
    337:Hinwalsta
71300 Digger pine woodland
    230:Bluoak, 257:Foopine
71310 Open digger pine woodland
    230:Bluoak
71321 Serpentine Digger pine chaparral
    257:Foopine
71322 Non-serpentine Digger pine chaparral
    257:Foopine
71400 Mixed cismontane woodland
    230:Bluoak, 264:Intlivoak, 277:Mixoak, 285:Orewhioak
71410 Digger pine-oak woodland
    230:Bluoak, 264:Intlivoak, 277:Mixoak
71420 Mixed north cismontane woodland
    285:Orewhioak
71430 Juniper-oak cismontane woodland
    230:Bluoak
72000 Pinyon and juniper woodlands
    236:Caljuniper, 283:Moujuniper, 288:Parpinyon,
    304:Sinpinyon, 305:SinpinUtajun, 316:Wesjuniper
72110 Northern juniper woodland
    316:Wesjuniper
72120 Great Basin pinyon and jupiper woodland
    283:Moujuniper, 304:Sinpinyon, 305:SinpinUtajun,

    316:Wesjuniper
72121 Great Basin pinyon juniper woodland
    305:SinpinUtajun
72122 Great Basin pinyon woodland
    304:Sinpinyon
72123 Great Basin juniper woodland and scrub
    283:Moujuniper, 316:Wesjuniper
72200 Mojavean pinyon and juniper woodland
    236:Caljuniper, 304:Sinpinyon
72210 Mojavean pinyon woodland
    304:Sinpinyon
72220 Mojavean juniper woodland and scrub
    236:Caljuniper
72300 Peninsular pinyon and juniper woodlands
    236:Caljuniper, 288:Parpinyon
72310 Peninsular pinyon woodland
    288:Parpinyon
72320 Peninsular juniper woodland and scrub
    236:Caljuniper, 288:Parpinyon
72400 Cismontane juniper woodland and scrub
    236:Caljuniper
73000 Joshua tree woodlands
    168:Jostree
75000 Sonoran thorn woodlands
    152:Foopalversag, 327:Allthosta, 331:Cruthosta,
    333:Eletresta
75100 Elephant tree woodland
    333:Eletresta
75200 Crucifixion thorn woodland
    331:Cruthosta
75300 All-thorn woodland
    327:Allthosta
75400 Arizonan woodland
    152:Foopalversag
81000 Broadleaved upland forests
    220:Aspen, 224:Birmoumah, 227:Blaoak,
    235:Calbuckeye, 238:Calwalnut, 239:Canlivoak,
    241:Coalivoak, 245:Curmoumah, 250:Doufirtan,
    264:Intlivoak, 277:Mixoak, 295:Redalder, 330:Catirosta,
    338:Holchesta
81100 Mixed evergreen forest
    250:Doufirtan
81200 California bay forest
    234:Calbay
81300 Oak forest
    239:Canlivoak, 241:Coaoak, 264:Intlivoak
81310 Coast live oak forest
    241:Coaoak
81320 Canyon live oak forest
    239:Canlivoak
81330 Interior live oak forest
    264:Intlivoak
81430 Black oak forest
    227:Blaoak
81400 Tan-oak forest
    250:Doufirtan
81500 Mixed north slope forest
    235:Calbuckeye
81600 Walnut forest
    238:Calwalnut
81700 Island ironwood forest

330:Catirosta
81800 Cherry forest
   338:Holchesta
81810 Island cherry forest
   338:Holchesta
81820 Mainland cherry forest
   235:Calbuckeye, 338:Holchesta
81900 Silktassel forest
   234:Calbay
81A00 Red alder forest
   295:Redalder
81B00 Aspen forest
   220:Aspen
82000 North Coast coniferous forests
   246:Doufir, 250:Doufirtan, 261:Grafir, 292:PorOrfced,
   300:Redwood, 306:Sitspruce, 315:Weshemlock
82100 Sitka spruce-grand fir forest
   261:Grafir, 306:Sitspruce
82200 Western hemlock forest
   315:Weshemlock
82300 Redwood forest
   300:Redwood
82310 Alluvial redwood forest
   300:Redwood
82320 Upland redwood forest
   300:Redwood
82400 Douglas-fir forest
   246:Doufir, 250:Doufirtan, 315:Weshemlock
82410 Coastal Douglas-fir - western hemlock forest
   315:Weshemlock
82420 Upland Douglas-fir forest
   246:Doufir, 250:Doufirtan
82500 Port Orford-cedar forest
   292:PorOrfced
83000 Closed-cone coniferous forests
   221:Beapine, 225:Bispine, 269:Knopine,
   273:McNcypress, 280:Monpine, 297:Menpyncress,
   329:Bakcypsta, 332:Cuycypsta, 337:Gowcyrsta,
   339:Moncypsta, 341:Piucypsta, 343:SanCrucypsta,
   345:Torpinsta
83100 Coastal closed-cone coniferous forest
   221:Beapine, 225:Bispine, 269:Knopine, 280:Monpine,
   297:Menpyncress, 337:Gowcyrsta, 339:Moncypsta,
   345:Torpinsta
83110 Beach pine forest
   221:Beapine
83121 Northern Bishop pine forest
   225:Bispine
83122 Southern Bishop pine forest
   225:Bispine
83130 Monterey pine forest
   280:Monpine
83140 Torrey pine forest
   345:Torpinsta
83150 Monterey cypress forest
   339:Moncypsta
83160 Pygmy cypress forest
   297:Menpymcypres, 337:Gowcypsta
83161 Mendocino pymgy cypress forest
   297:Menpymcypres
83162 Monterey pygmy cypress forest

337:Gowcypsta
83200 Interior closed-cone coniferous forest
   269:Knopine, 273:McNcypress, 329:Bakcypsta,
   332:Cuycypsta, 343:SanCrucypsta, 341:Piucypsta
83210 Knobcone pine forest
   269:Knopine
83220 Northern interior cypress forest
   273:McNcypress, 329:Bakcypsta, 343:SanCrucypsta
83330 Southern interior cypress forest
   332:Cuycypsta, 341:Piucypsta, 345:Teccrysta
84000 Lower montane coniferous forests
   222:BigDoufir, 223:BigDoufircan, 227:Blaoak,
   243:Coupine, 244:Coupincanliv, 246:Doufir,
   249:Doufirponpin, 260:Giasequoia, 263:Inccedar,
   266:Jefpine, 268:Jefpinponpin, 289:Ponpine,
   302:SanLucfir, 317:Weswhipin, 321:Whifir,
   342:SanBenMousta
84100 Coast Range and Klamath coniferous forest
   222:BigDoufir, 223:BigDoufircan, 243:Coupine,
   244:Coupincanliv, 246:Doufir, 249:Doufirponpin,
   266:Jefpine, 289:Ponpine, 302:SanLucfir,
   317:Weswhipin, 342:SanBenMousta
84110 Coast range mixed coniferous forest
   246:Doufir, 249:Doufirponpin
84120 Santa Lucia fir forest
   302:SanLucfir
84130 Coast Range ponderosa pine forest
   289:Ponpine
84131 Upland Coast Range ponderosa pine forest
   289:Ponpine
84132 Maritime Coast Range ponderosa pine forest
   289:Ponpine
84140 Coulter pine forest
   243:Coupine, 244:Coupincanliv
84150 Bigcone spruce-canyon oak forest
   222:BigDoufir, 223:BigDoufircan
84160 Ultramafic white pine forest
   317:Weswhipin
84170 Ultramafic Jeffrey pine forest
   266:Jefpine, 342:SanBenMousta
84171 Northern ultramafic Jeffrey pine forest
   266:Jefpine
84172 Southern ultramafic Jeffrey pine forest
   342:SanBenMousta
84180 Ultramafic mixed coniferous forest
   266:Jefpine, 342:SanBenMousta
84182 Southern ultramafic mixed coniferous forest
   342:SanBenMousta
84200 Sierran coniferous forest
   227:Blaoak, 249:Doufirponpin, 260:Giasequoia,
   263:Inccedar, 275:Mixconifer, 263:Inccedar,
   268:Jefpinponpin, 289:Ponpine, 321:Whifir
84210 Westside ponderosa pine forest
   289:Ponpine
84220 Eastside ponderosa pine forest
   268:Jefpinponpin, 289:Ponpine
84221 Ponderosa dune forest
   289:Ponpine
84230 Sierran mixed coniferous forest
   227:Blaoak, 249:Doufirponpin, 263:Inccedar,
   275:Mixconifer, 289:Ponpine, 321:Whifir

84240 White fir forest
 263:Inccedar, 321:Whifir
84250 Big tree forest
 260:Giasequoia
85000 Upper montane coniferous forests
 254:Engspruce, 266:Jefpine, 268:Jefpinponpin,
 296:Redfir, 308:Subfir, 313:Waspine, 317:Weswhipin,
 321:Whifir, 326:Alayelcedsta, 334:KlaMouEnrsta
85100 Jeffrey pine forest
 266:Jefpine, 268:Jefpinponpin
85200 Upper montane mixed coniferous forest
 266:Jefpine, 268:Jefpinponpin
85210 Jeffrey pine-fir forest
 266:Jefpine, 268:Jefpinponpin
85220 Washoe pine-fir forest
 313:Waspine
85300 Upper montane fir forest
 296:Redfir, 321:Whifir
85310 Red fir forest
 296:Redfir
85320 Southern California white fir forest
 321:Whifir
85330 Desert mountain white fir forest
 321:Whifir
85400 Klamath enriched coniferous forest
 254:Engspruce, 308:Subfir, 326:Alayelcedsta,
 334:KlaMouEnrsta
85410 Siskiyou enriched coniferous forest
 326:Alayelcedsta, 334:KlaMouEnrsta
85420 Salmon-Scott enriched coniferous forest
 254:Engspruce, 308:Subfir, 334:KlaMouEnrsta
86000 Subalpine coniferous forests
 233:Bripine, 258:Foxpine, 270:Limpine, 271:Lodpine,
 278:Mixsubfor, 281:Mouhemlock, 324:Whipine,
 340:Pacsilfirsta
86100 Lodgepole pine forest
 271:Lodpine
86200 Sierran mixed subalpine coniferous forest

271:Lodpine, 278:Mixsubfor, 281:Mouhemlock,
 324:Whipine
86210 Whitebark pine-mountain hemlock forest
 278:Mixsubfor, 281:Mouhemlock, 324:Whipine
86220 Whitebark pine-lodgepole pine forest
 271:Lodpine, 278:Mixsubfor, 324:Whipine
86300 Foxtail pine forest
 258:Foxpine
86400 Bristlecone pine forest
 233:Bripine
86500 Southern California subalpine forest
 278:Mixsubfor
86600 Whitebark pine forest
 324:Whipine
86700 Limber pine forest
 270:Limpine
91000 Alpine boulder and rock fields
 82:Sedge, 170:Lowsagebrush, 348:Alpinehab,
 356:Subuplshrhab
91100 Alpine fell-field
 170:Lowsagebrush, 482:Sedge, 348:Alpinehab,
 356:Subuplshrhab
91110 Klamath Cascade fell-field
 482:Sedge, 348:Alpinehab
91120 Sierra Nevada fell-field
 82:Sedge, 348:Alpinehab
91130 Southern California fell-field
 82:Sedge, 348:Alpinehab
91140 White Mountains fell-field
 282:Sedge, 170:Lowsagebrush, 348:Alpinehab,
 356:Subuplshrhab
91200 Alpine glacier
 348:Alpinehab
91210 Wet alpine talus and scree slope
 348:Alpinehab
91220 Dry alpine talus and scree slope
 348:Alpinehab
93100 Alpine snowfield
 348:Alpinehab

Index to Plant Names

232:Blupalveriro
Blue wheatgrass
  289:Ponpine
Blue wildrye
  61:Intpergra,
  62:Kenbluegrass,
  67:Nodneedlegra,
  75:Purneedlegra,
  148:Deerbrush,
  353:Monmeahab
Blue-eyed Mary
  230:Bluoak,
  369:SanDiemesver
Bluebunch wheatgrass
  31:Ashryegrass,
  34:Bluwheatgras,
  66:Neeandthr,
  100:Bigsagebrush
Bluejoint reedgrass
  353:Monmeahab
Bluff cholla
  141:Coapripea
Bog bilberry
  352:Fenhab
Bog-bean
  352:Fenhab
Boggs Lake hedge-hyssop
  361:Norbasflover,
  363:Norclaverpoo,
  365:Norharverpoo,
  367:Norvolashver,
  368:Norvolmudver
Bolander bedstraw
  275:Mixconifer,
  289:Ponpine
Bolander horkelia
  361:Norbasflover,
  367:Norvolashver
Bolander pine
  225:Bispine,
  294:Pygcypress
Bonnie Doon manzanita
  211:Woomanzanita
Booth willow
  358:Subwetshrhab
Bouteloua eriopoda
  33:Biggalleta
Bowlesia
  230:Bluoak
Bowlesia incana
  230:Bluoak
Box elder
  226:Blacottonwoo
Box-thorn
  141:Coapripea,
  145:Crebuswhibur,
  159:Hopsage,
  168:Jostree,
  182:Ocotillo
Bracken
  42:Caloatgrass,

61:Intpergra,
62:Kenbluegrass,
69:Pacreedgrass,
227:Blaoak,
241:Coalivoak,
250:Doufirtan,
296:Redfir,
300:Redwood,
321:Whifir,
353:Monmeahab
Bracted manzanita
  211:Woomanzanita
Brasenia schreberi
  39:Burreed,
  90:Yelponlil
Brassica
  40:Calanngra
Brazilian pepper
  202:Sumac
Brewer bitter-cress
  88:Tufhairgrass
Brewer cinquefoil
  177:Mouheabil
Brewer heather
  84:Shoreedgrass,
  177:Mouheabil
Brewer oak
  113:Breoak,
  285:Orewhioak
Brewer sedge
  82:Sedge
Brewer spruce
  113:Breoak,
  254:Engspruce,
  296:Redfir,
  308:Subfir,
  321:Whifir,
  327:Alayelcedsta,
  335:KlaMouEnrsta,
  341:Pacsilfirsta
Bristlecone pine
  233:Bripine,
  270:Limpine
Brittlebush
  114:Brittlebush,
  115:Briwhibur,
  120:Calbuckwheat,
  124:Calsagebrush,
  129:Catacacia,
  144:Crebush,
  149:Desholly,
  152:Foopalversag,
  172:Mixsage,
  175:Mojyucca,
  181:Nolina,
  193:Scalebroom,
  204:Tedbeacho,
  207:Whibursage,
  328:Allthosta,
  332:Cruthosta,
  334:Eletresta

Brittleleaf manzanita
  211:Woomanzanita
Brittlescale
  173:Mixsaltbush
Broadleaf cattail
  35:Bulrush,
  37:Bulcattail,
  43:Cattail
Broadleaf pondweed
  73:Ponwitflolea
Brodiaea filifolia
  370:SanJacValver,
  371:SanRosPlaver
Brodiaea minor
  365:Norharverpoo
Brodiaea orcuttii
  369:SanDiemesver,
  371:SanRosPlaver
Bromes
  40:Calanngra,
  47:Creryegrass,
  367:Norvolashver
Bromus
  47:Creryegrass,
  367:Norvolashver
Bromus californica
  285:Orewhioak
Bromus carinatus
  61:Intpergra,
  62:Kenbluegrass,
  289:Ponpine
Bromus diandrus
  40:Calanngra,
  44:Cheatgrass,
  67:Nodneedlegra,
  75:Purneedlegra,
  213:Yelbuslup
Bromus hordeaceus
  40:Calanngra,
  55:Fooneedlegra,
  61:Intpergra,
  67:Nodneedlegra,
  75:Purneedlegra,
  237:Calsycamore
Bromus madritensis
  33:Biggalleta,
  40:Calanngra,
  67:Nodneedlegra,
  75:Purneedlegra
Bromus tectorum
  31:Ashryegrass,
  34:Bluwheatgras,
  44:Cheatgrass,
  48:Crewheatgras,
  60:Indricegrass,
  66:Neeandthr,
  100:Bigsagebrush
Buckwheats
  84:Shoreedgrass
Budsage
  108:Blabush,

199:Shadscale
Bulrushes
  32:Beasedge,
  39:Burreed,
  65:Nebsedge,
  71:Pickleweed,
  77:RocMoused,
  86:Spikerush,
  82:Sedge,
  90:Yelponlil,
  349:Alpinehab,
  352:Fenhab,
  353:Monmeahab
Burke goldfields
  361:Norbasflover,
  365:Norharverpoo
Bursera microphylla
  334:Eletresta
Bush buckwheat
  98:Allscale,
  181:Nolina
Bush chinquapin
  117:Buschinquapi,
  155:Gremanzanita,
  158:Holodiscus,
  160:Hucoak,
  178:Mouwhitethor,
  190:Sadoak,
  205:Tobbrush,
  266:Jefpine,
  296:Redfir,
  321:Whifir
Bush monkeyflower
  120:Calbuckwheat,
  123:Calencelia,
  124:Calsagebrush,
  127:CalsagCalbuc,
  172:Mixsage,
  184:Pursage,
  191:Salblahuc,
  202:Sumac,
  208:Whisage
Bush poppy
  146:Cupceafreoak
Bush seepweed
  30:Alksacaton,
  118:Busseepweed,
  149:Desholly,
  154:Greasewood,
  166:Iodbush
Bush tanoak
  246:Doufir,
  317:Weswhipin
Bush-penstemon
  124:Calsagebrush
Butte County meadow-
  foam
  361:Norbasflover
Buttercups
  369:SanDiemesver
Butterwort

137:Chamismanwoo,
38:Chawedcea,
146:Cupceafreoak,
174:Mixscroak,
185:Redshank,
186:Redshabir,
187:Redshacha,
196:Scroakbirmou,
197:Scroakcha,
347:Twopinsta
Ceanothus hearstiorum
211:Woomanzanita
Ceanothus incanus
112:Blublossom
Ceanothus integerrimus
112:Blublossom,
128:Canlivoakshr,
133:Chablasag,
140:Chawhitethor,
148:Deerbrush,
163:Intlivoakcan,
174:Mixscroak,
206:Wedceanothus,
227:Blaoak,
239:Canlivoak
Ceanothus jepsonii
169:Leaoak
Ceanothus megacarpus
103:Bigceanothus,
104:Bigceabirmou,
105:Bigceaholred,
106:BirmoumahCal,
132:Chabigman,
224:Birmoumah
Ceanothus oliganthus
241:Coalivoak
Ceanothus palmeri
196:Scroakbirmou
Ceanothus prostratus
275:Mixconifer,
289:Ponpine
Ceanothus thyrsiflorus
112:Blublossom,
194:Scroak
Ceanothus tomentosus
130:Chamise,
132:Chabigman,
134:Chacupcea,
135:ChaEasman,
136:Chahoacea,
137:Chamismanwoo,
67:Ionmanzanita,
174:Mixscroak
Ceanothus velutinus
112:Blublossom,
113:Breoak,
117:Buschinquapi,
148:Deerbrush,
155:Gremanzanita,
158:Holodiscus,
160:Hucoak,

178:Mouwhitethor,
190:Sadoak,
205:Tobbrush,
209:Whimanzanita
Ceanothus verrucosus
136:Chahoacea,
137:Chamismanwoo,
74:Mixscroak,
194:Scroak,
196:Scroakbirmou,
197:Scroakcha,
198:Scroakchawhi,
346:Torpinsta
Centaurea
40:Calanngra
Cephalanthus occidentalis
119:Buttonbush,
219:Arrwillow
Ceratophyllum
52:Ditgrass
Cercidium floridum
129:Catacacia,
182:Ocotillo,
232:Blupalveriro
Cercidium microphyllum
152:Foopalversag
Cercocarpus betuloides
102:Bigmanzanita,
103:Bigceanothus,
104:Bigceabirmou,
105:Bigceaholred,
106:BirmoumahCal,
134:Chacupcea,
135:ChaEasman,
146:Cupceafreoak,
157:Hoaceanothus,
161:Intlivoakshr,
164:Intlivoakcha,
165:Intlivoakscr,
174:Mixscroak,
185:Redshank,
186:Redshabir,
187:Redshacha,
193:Scalebroom,
194:Scroak,
197:Scroakcha,
211:Woomanzanita,
224:Birmoumah,
230:Bluoak,
257:Foopine
Cercocarpus ledifolius
107:Bitterbrush,
245:Curmoumah,
266:Jefpine,
270:Limpine,
283:Moujuniper,
289:Ponpine,
316:Wesjuniper
Chain fern
295:Redalder,
300:Redwood

Chamaebatia foliolosa
275:Mixconifer
Chamaecyparis lawsoniana
49:Darlingtonia,
250:Doufirtan,
292:PorOrfced,
296:Redfir,
315:Weshemlock,
319:Whialder,
327:Alayelcedsta,
335:KlaMouEnrsta
Chamaecyparis nootkatensis
327:Alayelcedsta,
335:KlaMouEnrsta
Chamaesyce hooveri
363:Norclaverpoo,
365:Norharverpoo
Chamise
80:Sanverbeabur,
102:Bigmanzanita,
103:Bigceanothus,
104:Bigceabirmou,
105:Bigceaholred,
106:BirmoumahCal,
112:Blublossom,
114:Brittlebush,
122:Calbucwhisag,
124:Calsagebrush,
126:Calsagblasag,
127:CalsagCalbuc,
133:Chablasag,
134:Chacupcea,
136:Chahoacea,
137:Chamismanwoo,
38:Chawedcea,
139:Chawhisag,
140:Chawhitethor,
146:Cupceafreoak,
151:Easmanzanita,
156:Haiceanothus,
157:Hoaceanothus,
163:Intlivoakcan,
164:Intlivoakcha,
165:Intlivoakscr,
167:Ionmanzanita,
169:Leaoak,
185:Redshank,
186:Redshabir,
187:Redshacha,
194:Scroak,
196:Scroakbirmou,
197:Scroakcha,
198:Scroakchawhi,
206:Wedceanothus,
209:Whimanzanita,
211:Woomanzanita,
224:Birmoumah,
241:Coalivoak,
288:Parpinyon,
333:Cuycypsta,

343:SanBenMousta
Chaparral mallow
109:Blasage,
120:Calbuckwheat,
208:Whisage
Chaparral pea
194:Scroak
Chaparral whitethorn
102:Bigmanzanita,
122:Calbucwhisag,
128:Canlivoakshr,
132:Chabigman,
134:Chacupcea,
135:ChaEasman,
140:Chawhitethor,
151:Easmanzanita,
157:Hoaceanothus,
161:Intlivoakshr,
163:Intlivoakcan,
164:Intlivoakcha,
174:Mixscroak,
194:Scroak,
196:Scroakbirmou,
197:Scroakcha,
198:Scroakchawhi
Chaparral yucca
102:Bigmanzanita,
106:BirmoumahCal,
109:Blasage,
122:Calbucwhisag,
123:Calencelia,
124:Calsagebrush,
126:Calsagblasag,
127:CalsagCalbuc,
130:Chamise,
132:Chabigman,
133:Chablasag,
134:Chacupcea,
135:ChaEasman,
136:Chahoacea,
137:Chamismanwoo,
38:Chawedcea,
139:Chawhisag,
140:Chawhitethor,
146:Cupceafreoak,
151:Easmanzanita,
164:Intlivoakcha,
186:Redshabir,
187:Redshacha,
193:Scalebroom,
208:Whisage,
236:Caljuniper
Cheatgrass
31:Ashryegrass,
34:Bluwheatgras,
44:Cheatgrass,
48:Crewheatgras,
60:Indricegrass,
66:Neeandthr,
100:Bigsagebrush
Cheesebush

50:Desneedlegra,
98:Allscale,
129:Catacacia,
153:Fousaltbush,
168:Jostree,
193:Scalebroom,
201:Spinescale,
203:Tamarisk
Chile lotus
230:Bluoak
Chilopsis linearis
129:Catacacia,
232:Blupalveriro
Chimaphila menziesii
321:Whifir
Chimaphila umbellata
250:Doufirtan,
296:Redfir
Chinquapin
246:Doufir,
250:Doufirtan,
321:Whifir
Choke cherry
245:Curmoumah,
283:Moujuniper,
289:Ponpine
Cholla
114:Brittlebush,
115:Briwhibur,
144:Crebush,
145:Crebuswhibur,
175:Mojyucca,
182:Ocotillo,
207:Whibursage
Chrysolepis chrysophylla
246:Doufir,
250:Doufirtan,
321:Whifir
Chrysolepis sempervirens
117:Buschinquapi,
155:Gremanzanita,
158:Holodiscus,
160:Hucoak,
178:Mouwhitethor,
190:Sadoak,
205:Tobbrush,
266:Jefpine,
296:Redfir,
321:Whifir
Chrysothamnus
168:Jostree,
305:SinpinUtajun,
311:Utajuniper
Chrysothamnus albidus
78:Saltgrass
Chrysothamnus nauseosus
30:Alksacaton,
100:Bigsagebrush,
107:Bitterbrush,
154:Greasewood,
158:Holodiscus,

170:Lowsagebrush,
189:Rubrabbitbru,
316:Wesjuniper
Chrysothamnus parryi
183:Parrabbitbru
Chrysothamnus viscidiflorus
100:Bigsagebrush,
107:Bitterbrush,
153:Fousaltbush,
154:Greasewood,
170:Lowsagebrush,
183:Parrabbitbru,
189:Rubrabbitbru,
199:Shadscale,
210:Winfat
Cinquefoils
85:Shosedge
Claytonia sibirica
295:Redalder
Cleome sparsifolia
51:Dessanver
Cleveland monkeyflower
136:Chahoacea,
137:Chamismanwoo,
74:Mixscroak,
194:Scroak,
196:Scroakbirmou,
197:Scroakcha,
198:Scroakchawhi
Cliff spurge
141:Coapripea
Cliffrose
146:Cupceafreoak,
236:Caljuniper,
305:SinpinUtajun
Cneoridium dumosum
141:Coapripea
Coast ceanothus
211:Woomanzanita
Coast goldenbush
120:Calbuckwheat,
123:Calencelia,
124:Calsagebrush,
150:Dunlupgol,
184:Pursage,
208:Whisage
Coast live oak
122:Calbucwhisag,
184:Pursage,
211:Woomanzanita,
27:Blaoak,
230:Bluoak,
234:Calbay,
237:Calsycamore,
238:Calwalnut,
241:Coalivoak,
243:Coupine,
244:Coupincanliv,
253:Engoak,
257:Foopine,

277:Mixoak,
280:Monpine,
309:Tanoak,
312:Valoak,
329:Antdunsta
Coast mugwort
69:Pacreedgrass,
191:Salblahuc
Coast prickly-pear
109:Blasage,
141:Coapripea,
150:Dunlupgol
Coast silktassel
191:Salblahuc,
234:Calbay
Coast whitethorn
112:Blublossom
Coleogyne ramosissima
100:Bigsagebrush,
108:Blabush,
159:Hopsage,
168:Jostree,
175:Mojyucca,
181:Nolina,
210:Winfat,
236:Caljuniper,
305:SinpinUtajun,
311:Utajuniper
Collinsia parviflora
369:SanDiemesver
Collinsia sparsiflora
230:Bluoak
Colonial bent
88:Tufhairgrass
Columbia needlegrass
268:Jefpinponpin,
289:Ponpine
Colusa grass
363:Norclaverpoo,
365:Norharverpoo
Common fiddleneck
230:Bluoak
Common manzanita
112:Blublossom,
148:Deerbrush,
206:Wedceanothus,
209:Whimanzanita,
264:Intlivoak
Common pickleweed
46:Cordgrass,
78:Saltgrass
Common reed
45:Comreed
Common snowberry
241:Coalivoak,
285:Orewhioak
Common three-square
35:Bulrush,
37:Bulcattail,
43:Cattail
Common tule

35:Bulrush,
37:Bulcattail,
43:Cattail
Common vetch
300:Redwood
Common water-nymph
52:Ditgrass
Compact phlox
357:Subuplshrhab
Cone flower
49:Darlingtonia
Congdon sedge
82:Sedge,
357:Subuplshrhab
Congdon silktassel
343:SanBenMousta
Contra Costa goldfields
361:Norbasflover,
363:Norclaverpoo,
367:Norvolashver
Contra Costa wallflower
329:Antdunsta
Cooper rush
78:Saltgrass
Cordgrass
46:Cordgrass,
71:Pickleweed
Cordilleran arnica
349:Alpinehab
Coreopsis maritima
346:Torpinsta
Cornus nuttallii
246:Doufir
Cornus sericea
119:Buttonbush,
176:Moualder,
192:Sanwillow,
200:Sitalder,
219:Arrwillow,
287:Pacwillow,
307:Sitwillow,
319:Whialder,
355:Monwetshrhab
Cornus sessilis
319:Whialder
Cortaderia
70:Pamgrass
Cortaderia jubata
70:Pamgrass
Cortaderia selloana
70:Pamgrass
Corylus cornuta
241:Coalivoak,
246:Doufir,
250:Doufirtan,
275:Mixconifer,
306:Sitspruce,
321:Whifir
Coulter goldfields
363:Norclaverpoo,
365:Norharverpoo,

227:Blaoak,
239:Canlivoak
Del Norte iris
266:Jefpine
Delphinium parryi
230:Bluoak
Delphinium trolliifolium
285:Orewhioak
Dendromecon rigida
146:Cupceafreoak
Dense draba
349:Alpinehab
Dense reedgrass
194:Scroak,
197:Scroakcha
Deschampsia cespitosa
42:Caloatgrass,
61:Intpergra,
69:Pacreedgrass,
84:Shoreedgrass,
85:Shosedge,
88:Tufhairgrass,
142:Coybrush,
191:Salblahuc,
353:Monmeahab
Deschampsia
danthonioides
369:SanDiemesver,
371:SanRosPlaver
Desert agave
115:Briwhibur,
175:Mojyucca,
181:Nolina,
182:Ocotillo
Desert barberry
129:Catacacia
Desert bitterbrush
107:Bitterbrush,
146:Cupceafreoak
Desert ceanothus
174:Mixscroak
Desert dicoria
51:Dessanver
Desert gold
51:Dessanver
Desert needlegrass
50:Desneedlegra
Desert peach
107:Bitterbrush
Desert sand-verbena
51:Dessanver
Desert scrub oak
146:Cupceafreoak,
168:Jostree,
174:Mixscroak,
194:Scroak,
196:Scroakbirmou,
197:Scroakcha,
198:Scroakchawhi,
236:Caljuniper,
347:Twopinsta

Desert snowberry
100:Bigsagebrush,
266:Jefpine,
289:Ponpine,
313:Waspine,
347:Twopinsta
Desert-almond
146:Cupceafreoak
Desert-apricot
146:Cupceafreoak
Desert-holly
144:Crebush,
145:Crebuswhibur,
149:Desholly
Desert-lavender
129:Catacacia,
152:Foopalversag,
181:Nolina
Desert-olive
129:Catacacia
Desert-willow
129:Catacacia,
232:Blupalveriro
Devil's-lettuce
329:Antdunsta
Dewey sedge
82:Sedge
Dichelostemma ida-maia
277:Mixoak
Dicoria canescens
51:Dessanver
Diego bentgrass
82:Sedge,
353:Monmeahab
Disporum hookeri
321:Whifir
Distichlis spicata
30:Alksacaton,
31:Ashryegrass,
35:Bulrush,
37:Bulcattail,
43:Cattail,
46:Cordgrass,
47:Creryegrass,
71:Pickleweed,
78:Saltgrass,
80:Sanverbeabur,
98:Allscale,
166:Iodbush,
201:Spinescale
Ditch-grasses
52:Ditgrass,
71:Pickleweed
Diverseleaf pondweed
73:Ponwitflolea
Dodder
46:Cordgrass
Dodecatheon alpinum
82:Sedge
Dodecatheon jeffreyi
177:Mouheabil,

356:Submeahab
Dogtail
40:Calanngra,
61:Intpergra,
285:Orewhioak
Douglas iris
300:Redwood
Douglas pogogyne
361:Norbasflover
Douglas-fir
42:Caloatgrass,
70:Pamgrass,
148:Deerbrush,
221:Beapine,
225:Bispine,
227:Blaoak,
239:Canlivoak,
246:Doufir,
249:Doufirponpin,
250:Doufirtan,
254:Engspruce,
260:Giasequoia,
261:Grafir,
263:Inccedar,
266:Jefpine,
275:Mixconifer,
277:Mixoak,
280:Monpine,
285:Orewhioak,
289:Ponpine,
292:PorOrfced,
295:Redalder,
300:Redwood,
309:Tanoak,
315:Weshemlock,
317:Weswhipin,
319:Whialder,
321:Whifir,
327:Alayelcedsta,
330:Bakcypsta,
335:KlaMouEnrsta
Downingia
361:Norbasflover,
369:SanDiemesver,
371:SanRosPlaver
Downingia bacigalupii
361:Norbasflover
Downingia bicornuta
367:Norvolashver,
368:Norvolmudver
Downingia cuspidata
367:Norvolashver
Downingia pusilla
363:Norclaverpoo,
365:Norharverpoo,
367:Norvolashver,
368:Norvolmudver
Draba densiflora
349:Alpinehab
Draba oligosperma
349:Alpinehab

Drosera rotundifolia
49:Darlingtonia,
352:Fenhab
Drummond cinquefoil
177:Mouheabil
Drummond windflower
258:Foxpine
Duckmeats
53:Duckweed,
63:Mosfern
Duckweeds
53:Duckweed
Dudleya
141:Coapripea
Dune buckwheat
58:Iceplant,
80:Sanverbeabur
Dune lupine
80:Sanverbeabur,
150:Dunlupgol
Dune sagebrush
80:Sanverbeabur,
150:Dunlupgol
Dunn mariposa lily
136:Chahoacea,
137:Chamismanwoo,
74:Mixscroak,
194:Scroak,
196:Scroakbirmou,
198:Scroakchawhi
Dusky willow
355:Monwetshrhab
Dwarf bilberry
281:Mouhemlock
Dwarf blennosperma
363:Norclaverpoo,
368:Norvolmudver
Dwarf brodiaea
365:Norharverpoo
Dwarf downingia
361:Norbasflover,
363:Norclaverpoo,
365:Norharverpoo,
367:Norvolashver,
368:Norvolmudver
Dwarf silktassel
317:Weswhipin

Eastwood manzanita
102:Bigmanzanita,
130:Chamise,
132:Chabigman,
134:Chacupcea,
135:ChaEasman,
137:Chamismanwoo,
38:Chawedcea,
151:Easmanzanita,
161:Intlivoakshr,
164:Intlivoakcha,
186:Redshabir,
187:Redshacha,

363:Norclaverpoo,
365:Norharverpoo
Eryngium constancei
361:Norbasflover,
367:Norvolashver
Eschscholzia californica
40:Calanngra
Eucalyptus camaldulensis
255:Eucalyptus
Eucalyptus citriodora
255:Eucalyptus
Eucalyptus cladocalyx
255:Eucalyptus
Eucalyptus globulus
255:Eucalyptus
Eucalyptus polyanthemos
255:Eucalyptus
Eucalyptus pulverulenta
255:Eucalyptus
Eucalyptus sideroxylon
255:Eucalyptus
Eucalyptus tereticornis
255:Eucalyptus
Eucalyptus viminalis
255:Eucalyptus
Euphorbia misera
141:Coapripea
European beachgrass
54:Eurbeachgras,
64:Natdunegrass,
80:Sanverbeabur,
142:Coybrush
European hairgrass
40:Calanngra
Evening-primrose
51:Dessanver

Fairy duster
129:Catacacia
Fallugia paradoxa
146:Cupceafreoak
False lily-of-the-valley
306:Sitspruce
Fan palm
256:Fanpalm
Fat-hen
71:Pickleweed
Fendler meadow-rue
271:Lodpine
Fennelleaf pondweed
74:Ponwitsublea
Ferocactus cylindraceus
115:Briwhibur,
175:Mojyucca,
181:Nolina,
182:Ocotillo
Festuca arundinacea
69:Pacreedgrass,
88:Tufhairgrass
Festuca californica
55:Fooneedlegra,

67:Nodneedlegra,
75:Purneedlegra,
113:Breoak,
246:Doufir,
250:Doufirtan,
266:Jefpine,
285:Orewhioak
Festuca idahoensis
42:Caloatgrass,
69:Pacreedgrass,
170:Lowsagebrush,
266:Jefpine,
268:Jefpinponpin,
289:Ponpine
Festuca rubra
42:Caloatgrass,
61:Intpergra,
62:Kenbluegrass,
69:Pacreedgrass,
88:Tufhairgrass
Festuca viridula
57:Grefescue,
61:Intpergra,
62:Kenbluegrass
Few-flowered navarretia
361:Norbasflover,
367:Norvolashver
Few-flowered spikerush
86:Spikerush
Fig-marigold
58:Iceplant
Filaree
40:Calanngra
Firecracker flower
277:Mixoak
Fitch tarweed
365:Norharverpoo
Flat sagebrush
347:Twopinsta
Floatingleaf pondweed
73:Ponwitflolea
Fluff grass
33:Biggalleta
Foothill ash
196:Scroakbirmou,
235:Calbuckeye,
238:Calwalnut
Foothill needlegrass
50:Desneedlegra,
55:Fooneedlegra,
67:Nodneedlegra,
75:Purneedlegra
Foothill palo verde
152:Foopalversag
Foothill pine
161:Intlivoakshr,
167:Ionmanzanita,
169:Leaoak,
194:Scroak,
223:BigDoufircan,
224:Birmoumah,

230:Bluoak,
235:Calbuckeye,
243:Coupine,
244:Coupincanliv,
257:Foopine,
264:Intlivoak,
269:Knopine,
273:McNcypress,
277:Mixoak,
303:Sarcypress,
342:Piucypsta,
343:SanBenMousta
Foothill sedge
42:Caloatgrass,
61:Intpergra,
62:Kenbluegrass
Forest red gum
255:Eucalyptus
Forestiera pubescens
129:Catacacia
Fouquieria splendens
115:Briwhibur,
145:Crebuswhibur,
181:Nolina,
182:Ocotillo,
204:Tedbeacho
Fourwing saltbush
114:Brittlebush,
153:Fousaltbush,
154:Greasewood,
159:Hopsage,
173:Mixsaltbush,
201:Spinescale
Foxtail
230:Bluoak
Foxtail chess
33:Biggalleta,
67:Nodneedlegra,
75:Purneedlegra
Foxtail pine
188:Rotsagebrush,
245:Curmoumah,
258:Foxpine,
266:Jefpine,
270:Limpine,
271:Lodpine,
278:Mixsubfor,
317:Weswhipin,
324:Whipine,
335:KlaMouEnrsta
Fragile fern
349:Alpinehab
Fragrant bedstraw
319:Whialder
Fragrant fritillary
363:Norclaverpoo
Frankenia salina
46:Cordgrass,
71:Pickleweed,
118:Busseepweed,
154:Greasewood,

166:Iodbush,
363:Norclaverpoo
Fraxinus dipetala
196:Scroakbirmou,
235:Calbuckeye,
238:Calwalnut
Fraxinus latifolia
171:Mexelderberr,
226:Blacottonwoo,
268:Jefpinponpin,
312:Valoak,
319:Whialder
Fraxinus velutina
256:Fanpalm
Fremont box-thorn
129:Catacacia
Fremont cottonwood
171:Mexelderberr,
180:Narwillow,
192:Sanwillow,
193:Scalebroom,
219:Arrwillow,
226:Blacottonwoo,
229:Blawillow,
237:Calsycamore,
256:Fanpalm,
262:Hoowillow,
279:Mixwillow,
287:Pacwillow,
299:Redwillow,
307:Sitwillow,
314:Watbirch
Fremont goldfields
363:Norclaverpoo,
365:Norharverpoo
Fremont silktassel
117:Buschinquapi,
155:Gremanzanita,
158:Holodiscus,
160:Hucoak,
206:Wedceanothus,
209:Whimanzanita
Fremont tidytips
363:Norclaverpoo
Fremontia
106:BirmoumahCal,
146:Cupceafreoak
*Fremontodendron
    californicum*
106:BirmoumahCal,
146:Cupceafreoak
French broom
116:Broom
Fritillaria falcata
343:SanBenMousta
Fritillaria liliacea
363:Norclaverpoo
Fritillaria viridea
343:SanBenMousta

Galium andrewsii

230:Bluoak
Galium bolanderi
   275:Mixconifer
Galium clementis
   302:SanLucfir
Galium trifolium
   319:Whialder
Galleta
   33:Biggalleta
Garrya
   169:Leaoak
Garrya buxifolia
   317:Weswhipin
Garrya congdonii
   343:SanBenMousta
Garrya elliptica
   191:Salblahuc,
   234:Calbay
Garrya flavescens
   106:BirmoumahCal,
   347:Twopinsta
Garrya fremontii
   117:Buschinquapi,
   155:Gremanzanita,
   158:Holodiscus,
   160:Hucoak,
   206:Wedceanothus,
   209:Whimanzanita
Garrya veatchii
   146:Cupceafreoak,
   185:Redshank
Gaultheria shallon
   69:Pacreedgrass,
   112:Blublossom,
   142:Coybrush,
   191:Salblahuc,
   250:Doufirtan,
   292:PorOrfced,
   295:Redalder,
   300:Redwood,
   352:Fenhab
Gayophytum diffusum
   183:Parrabbitbru
Genista monspessulana
   116:Broom
Gentian
   353:Monmeahab
Gentiana newberryi
   353:Monmeahab
Geraea canescens
   51:Dessanver
Giant reed
   56:Giareed
Giant sequoia
   260:Giasequoia
Glyceria elata
   353:Monmeahab
Goldenback fern
   230:Bluoak,
   239:Canlivoak
Goldfields

40:Calanngra
Goodyera oblongifolia
   275:Mixconifer
Gordon ivesia
   349:Alpinehab
Gorse
   116:Broom
Gowen cypress
   294:Pygcypress,
   337:Gowcypsta
Grand fir
   221:Beapine,
   261:Grafir,
   295:Redalder,
   300:Redwood,
   306:Sitspruce,
   321:Whifir
Granite-gilia
   60:Indricegrass,
   357:Subuplshrhab
Grass nut
   227:Blaoak
Grass-of-parnassus
   49:Darlingtonia
Grassleaf pondweed
   73:Ponwitflolea
Gratiola heterosepala
   363:Norclaverpoo,
   365:Norharverpoo,
   368:Norvolmudver
Gray lovage
   271:Lodpine
Grayia spinosa
   108:Blabush,
   144:Crebush,
   153:Fousaltbush,
   159:Hopsage,
   199:Shadscale,
   305:SinpinUtajun,
   311:Utajuniper
Grayleaf willow
   355:Monwetshrhab,
   358:Subwetshrhab
Greasewood
   78:Saltgrass,
   118:Busseepweed,
   154:Greasewood,
   166:Iodbush,
   199:Shadscale
Green ephedra
   100:Bigsagebrush,
   107:Bitterbrush,
   110:Blasagebrush,
   153:Fousaltbush,
   304:Sinpinyon,
   347:Twopinsta
Green fescue
   57:Grefescue,
   61:Intpergra,
   62:Kenbluegrass
Greene tuctoria

361:Norbasflover,
   363:Norclaverpoo,
   365:Norharverpoo
Greenleaf manzanita
   113:Breoak,
   117:Buschinquapi,
   148:Deerbrush,
   155:Gremanzanita,
   158:Holodiscus,
   160:Hucoak,
   178:Mouwhitethor,
   190:Sadoak,
   205:Tobbrush,
   206:Wedceanothus,
   209:Whimanzanita,
   227:Blaoak,
   245:Curmoumah,
   246:Doufir,
   269:Knopine,
   289:Ponpine,
   317:Weswhipin
Grindelia hirsutula
   369:SanDiemesver
Grindelia stricta
   71:Pickleweed
Ground juniper
   327:Alayelcedsta,
   335:KlaMouEnrsta
Gumplant
   71:Pickleweed
Gutierrezia californica
   329:Antdunsta

Hairy Orcutt grass
   361:Norbasflover,
   363:Norclaverpoo,
   365:Norharverpoo
Hairy draba
   349:Alpinehab
Hairy honeysuckle
   250:Doufirtan
Hairy manzanita
   269:Knopine
Hairy oatgrass
   61:Intpergra
Hairy woodrush
   62:Kenbluegrass
Hairy yerba santa
   193:Scalebroom
Hairyleaf ceanothus
   156:Haiceanothus,
   174:Mixscroak,
   241:Coalivoak
Haplopappus apargioides
   349:Alpinehab
Hareleaf
   61:Intpergra
Hazel
   241:Coalivoak,
   246:Doufir,
   250:Doufirtan,

275:Mixconifer,
   306:Sitspruce,
   321:Whifir
Hearst ceanothus
   211:Woomanzanita
Hearst manzanita
   211:Woomanzanita
Heartleaf arnica
   281:Mouhemlock,
   289:Ponpine,
   296:Redfir,
   321:Whifir
Heather goldenbush
   80:Sanverbeabur,
   150:Dunlupgol,
   213:Yelbuslup
Hedgehog cactus
   115:Briwhibur,
   175:Mojyucca
Helianthemum
   suffrutescens
   167:Ionmanzanita
Heller sedge
   82:Sedge
Hemizonia fitchii
   365:Norharverpoo
Henderson bentgrass
   363:Norclaverpoo,
   365:Norharverpoo
Heracleum lanatum
   69:Pacreedgrass,
   191:Salblahuc,
   353:Monmeahab
Heretic penstemon
   188:Rotsagebrush,
   349:Alpinehab,
   356:Submeahab
Heteromeles arbutifolia
   103:Bigceanothus,
   104:Bigceabirmou,
   105:Bigceaholred,
   112:Blublossom,
   130:Chamise,
   132:Chabigman,
   135:ChaEasman,
   136:Chahoacea,
   137:Chamismanwoo,
   39:Chawhisag,
   140:Chawhitethor,
   151:Easmanzanita,
   156:Haiceanothus,
   157:Hoaceanothus,
   163:Intlivoakcan,
   167:Ionmanzanita,
   169:Leaoak,
   174:Mixscroak,
   194:Scroak,
   196:Scroakbirmou,
   197:Scroakcha,
   198:Scroakchawhi,
   202:Sumac,

206:Wedceanothus,
235:Calbuckeye,
241:Coalivoak,
257:Foopine,
277:Mixoak
Heterotheca grandiflora
329:Antdunsta
Heuchera rubescens
349:Alpinehab
Hillside gooseberry
230:Bluoak
Himalaya berry
319:Whialder
Hinds walnut
338:Hinwalsta
Hoary coffeeberry
167:Ionmanzanita
Hoaryleaf ceanothus
130:Chamise,
132:Chabigman,
133:Chablasag,
134:Chacupcea,
135:ChaEasman,
136:Chahoacea,
137:Chamismanwoo,
57:Hoaceanothus,
197:Scroakcha
Holcus lanatus
42:Caloatgrass,
61:Intpergra,
62:Kenbluegrass,
69:Pacreedgrass,
88:Tufhairgrass
Hollyleaf cherry
106:BirmoumahCal,
161:Intlivoakshr,
165:Intlivoakscr,
202:Sumac,
235:Calbuckeye,
339:Holchesta
Hollyleaf redberry
102:Bigmanzanita,
103:Bigceanothus,
104:Bigceabirmou,
105:Bigceaholred,
106:BirmoumahCal,
137:Chamismanwoo,
39:Chawhisag,
157:Hoaceanothus,
161:Intlivoakshr,
163:Intlivoakcan,
165:Intlivoakscr,
174:Mixscroak,
185:Redshank,
187:Redshacha,
194:Scroak,
196:Scroakbirmou,
197:Scroakcha,
198:Scroakchawhi,
206:Wedceanothus,
224:Birmoumah

Holodiscus discolor
148:Deerbrush,
155:Gremanzanita,
158:Holodiscus,
160:Hucoak,
178:Mouwhitethor,
190:Sadoak,
205:Tobbrush,
209:Whimanzanita,
227:Blaoak,
241:Coalivoak,
317:Weswhipin,
324:Whipine
Holodiscus microphyllus
158:Holodiscus,
357:Subuplshrhab
Honey mesquite
98:Allscale,
153:Fousaltbush,
232:Blupalveriro,
256:Fanpalm,
274:Mesquite
Hook three-awn
67:Nodneedlegra
Hooker fairybells
321:Whifir
Hooker manzanita
211:Woomanzanita
Hooker willow
69:Pacreedgrass,
80:Sanverbeabur,
180:Narwillow,
192:Sanwillow,
226:Blacottonwoo,
229:Blawillow,
262:Hoowillow,
279:Mixwillow,
287:Pacwillow,
295:Redalder,
307:Sitwillow
Hoover button-celery
363:Norclaverpoo
Hoover manzanita
211:Woomanzanita
Hoover spurge
363:Norclaverpoo
Hop-sage
108:Blabush,
153:Fousaltbush,
159:Hopsage,
199:Shadscale
Hordeum leporinum
230:Bluoak
Horkelia bolanderi
361:Norbasflover,
367:Norvolashver
Horkelia parryi
167:Ionmanzanita
Hornworts
52:Ditgrass
Horsebrush

100:Bigsagebrush,
107:Bitterbrush,
170:Lowsagebrush,
199:Shadscale
Hottentot-fig
58:Iceplant,
64:Natdunegrass
Huckleberry oak
113:Breoak,
117:Buschinquapi,
155:Gremanzanita,
158:Holodiscus,
160:Hucoak,
178:Mouwhitethor,
190:Sadoak,
205:Tobbrush,
246:Doufir,
266:Jefpine,
268:Jefpinponpin,
269:Knopine,
275:Mixconifer,
281:Mouhemlock,
292:PorOrfced,
296:Redfir,
321:Whifir
Humboldt Bay wallflower
54:Eurbeachgras
Humboldt wallflower
80:Sanverbeabur
Hymenoclea salsola
50:Desneedlegra,
98:Allscale,
129:Catacacia,
153:Fousaltbush,
168:Jostree,
193:Scalebroom,
201:Spinescale,
203:Tamarisk
Hypericum anagalloides
352:Fenhab
Hyptis emoryi
129:Catacacia,
152:Foopalversag,
181:Nolina

Iceplants
80:Sanverbeabur
Idaho fescue
42:Caloatgrass,
69:Pacreedgrass,
170:Lowsagebrush,
266:Jefpine,
268:Jefpinponpin,
289:Ponpine
Incense-cedar
206:Wedceanothus,
222:BigDoufir,
227:Blaoak,
239:Canlivoak,
246:Doufir,
249:Doufirponpin,

254:Engspruce,
260:Giasequoia,
263:Inccedar,
266:Jefpine,
268:Jefpinponpin,
275:Mixconifer,
285:Orewhioak,
289:Ponpine,
296:Redfir,
321:Whifir,
330:Bakcypsta,
335:KlaMouEnrsta,
343:SanBenMousta
Indian manzanita
239:Canlivoak
Indian paintbrush
353:Monmeahab
Indian rhubarb
319:Whialder
Indian ricegrass
31:Ashryegrass,
33:Biggalleta,
34:Bluwheatgras,
44:Cheatgrass,
48:Crewheatgras,
50:Desneedlegra,
51:Dessanver,
60:Indricegrass,
66:Neeandthr
Indigo bush
144:Crebush
Inflated duckweed
63:Mosfern
Inflated sedge
82:Sedge
Inside-out flower
246:Doufir
Interior live oak
102:Bigmanzanita,
128:Canlivoakshr,
130:Chamise,
132:Chabigman,
140:Chawhitethor,
146:Cupceafreoak,
148:Deerbrush,
151:Easmanzanita,
161:Intlivoakshr,
163:Intlivoakcan,
164:Intlivoakcha,
165:Intlivoakscr,
167:Ionmanzanita,
174:Mixscroak,
186:Redshabir,
194:Scroak,
196:Scroakbirmou,
197:Scroakcha,
198:Scroakchawhi,
230:Bluoak,
234:Calbay,
235:Calbuckeye,
243:Coupine,

257:Foopine,
264:Intlivoak,
266:Jefpine,
269:Knopine,
273:McNcypress,
277:Mixoak,
303:Sarcypress
Iodine bush
30:Alksacaton,
78:Saltgrass,
118:Busseepweed,
154:Greasewood,
166:Iodbush
Ione buckwheat
167:Ionmanzanita
Ione manzanita
167:Ionmanzanita
Ipomopsis congesta
349:Alpinehab,
357:Subuplshrhab
Irises
275:Mixconifer
Iris douglasiana
300:Redwood
Iris innominata
266:Jefpine
Irish Hill buckwheat
167:Ionmanzanita
Ironwood
129:Catacacia,
182:Ocotillo,
232:Blupalveriro
Island oak
265:Isloak
Isocoma acradenia
98:Allscale,
111:BlaCalephnar
Isocoma menziesii
120:Calbuckwheat,
123:Calencelia,
124:Calsagebrush,
150:Dunlupgol,
184:Pursage,
208:Whisage
Isoetes
76:Quillwort,
369:SanDiemesver,
371:SanRosPlaver
Isomeris arborea
98:Allscale,
111:BlaCalephnar,
123:Calencelia,
141:Coapripea,
168:Jostree,
193:Scalebroom
Italian ryegrass
67:Nodneedlegra,
75:Purneedlegra
Ivesia gordonii
349:Alpinehab
Ivesia lycopodioides

82:Sedge

Jaumea
71:Pickleweed,
78:Saltgrass
Jaumea carnosa
71:Pickleweed,
78:Saltgrass
Jeffrey pine
100:Bigsagebrush,
107:Bitterbrush,
110:Blasagebrush,
226:Blacottonwoo,
227:Blaoak,
245:Curmoumah,
246:Doufir,
249:Doufirponpin,
260:Giasequoia,
266:Jefpine,
268:Jefpinponpin,
270:Limpine,
283:Moujuniper,
285:Orewhioak,
288:Parpinyon,
289:Ponpine,
292:PorOrfced,
296:Redfir,
304:Sinpinyon,
313:Waspine,
316:Wesjuniper,
317:Weswhipin,
321:Whifir,
330:Bakcypsta,
335:KlaMouEnrsta,
343:SanBenMousta
Jeffrey shooting star
177:Mouheabil,
356:Submeahab
Jepson barberry
343:SanBenMousta
Jepson willow
176:Moualder,
200:Sitalder,
355:Monwetshrhab,
358:Subwetshrhab
Johnny-jump-up
230:Bluoak
Jojoba
115:Briwhibur,
129:Catacacia,
175:Mojyucca,
181:Nolina
Joshua tree
33:Biggalleta,
100:Bigsagebrush,
107:Bitterbrush,
108:Blabush,
144:Crebush,
145:Crebuswhibur,
159:Hopsage
Juglans californica

184:Pursage,
193:Scalebroom,
202:Sumac,
223:BigDoufircan,
238:Calwalnut,
338:Hinwalsta
Juncus
32:Beasedge,
65:Nebsedge,
369:SanDiemesver
Juncus balticus
30:Alksacaton,
66:Neeandthr,
78:Saltgrass,
349:Alpinehab
Juncus bufonius
370:SanJacValver
Juncus cooperi
78:Saltgrass
Juncus leiospermus
361:Norbasflover,
363:Norclaverpoo,
365:Norharverpoo,
368:Norvolmudver
Juncus mertensianus
84:Shoreedgrass,
85:Shosedge,
177:Mouheabil,
356:Submeahab
Juncus nevadensis
86:Spikerush
Juncus parryi
281:Mouhemlock
Juncus patens
62:Kenbluegrass
Junegrass
31:Ashryegrass,
34:Bluwheatgras,
60:Indricegrass,
66:Neeandthr,
67:Nodneedlegra,
75:Purneedlegra,
82:Sedge
Junipers
100:Bigsagebrush,
107:Bitterbrush,
183:Parrabbitbru,
189:Rubrabbitbru
Juniperus
100:Bigsagebrush,
107:Bitterbrush,
183:Parrabbitbru,
189:Rubrabbitbru
Juniperus californica
108:Blabush,
146:Cupceafreoak,
161:Intlivoakshr,
169:Leaoak,
181:Nolina,
193:Scalebroom,
224:Birmoumah,

236:Caljuniper,
304:Sinpinyon
Juniperus communis
327:Alayelcedsta,
335:KlaMouEnrsta
Juniperus occidentalis
100:Bigsagebrush,
170:Lowsagebrush,
226:Blacottonwoo,
257:Foopine,
283:Moujuniper
Juniperus osteosperma
100:Bigsagebrush,
107:Bitterbrush,
110:Blasagebrush,
304:Sinpinyon,
305:SinpinUtajun,
311:Utajuniper

Kalmia polifolia
84:Shoreedgrass,
177:Mouheabil
Keckiella antirrhinoides
202:Sumac
Keckiella cordifolia
124:Calsagebrush
Keckiella corymbosa
158:Holodiscus
Kelloggia
321:Whifir
Kelloggia galioides
321:Whifir
Kentucky bluegrass
61:Intpergra,
62:Kenbluegrass,
353:Monmeahab
King ricegrass
357:Subuplshrhab
Klamath gooseberry
285:Orewhioak
Knobcone pine
161:Intlivoakshr,
167:Ionmanzanita,
169:Leaoak,
197:Scroakcha,
223:BigDoufircan,
227:Blaoak,
266:Jefpine,
269:Knopine,
273:McNcypress,
280:Monpine,
303:Sarcypress,
317:Weswhipin,
330:Bakcypsta,
335:KlaMouEnrsta,
344:Sancrucypsta
Kochia californica
118:Busseepweed,
154:Greasewood,
166:Iodbush
Koeberlinia spinosa

260:Giasequoia,
263:Inccedar,
266:Jefpine,
269:Knopine,
270:Limpine,
271:Lodpine,
278:Mixsubfor,
281:Mouhemlock,
283:Moujuniper,
289:Ponpine,
296:Redfir,
308:Subfir,
313:Waspine,
317:Weswhipin,
321:Whifir,
324:Whipine,
327:Alayelcedsta,
335:KlaMouEnrsta,
341:Pacsilfirsta
Lolium
40:Calanngra
Lolium multiflorum
40:Calanngra,
67:Nodneedlegra,
75:Purneedlegra
Longbeak sedge
82:Sedge
Longleaf pondweed
73:Ponwitflolea
Longstalk clover
353:Monmeahab
Lonicera hispidula
250:Doufirtan
Lost Hills crownscale
365:Norharverpoo
Lotus longifolius
353:Monmeahab
Lotus scoparius
120:Calbuckwheat,
122:Calbucwhisag,
123:Calencelia,
124:Calsagebrush,
126:Calsagblasag,
127:CalsagCalbuc,
167:Ionmanzanita,
193:Scalebroom
Lotus subpinnatus
230:Bluoak
Low navarretia
369:SanDiemesver,
371:SanRosPlaver
Low sagebrush
110:Blasagebrush,
170:Lowsagebrush,
304:Sinpinyon,
316:Wesjuniper,
357:Subuplshrhab
Lupines
40:Calanngra,
57:Grefescue
Lupinus

40:Calanngra,
57:Grefescue
Lupinus albifrons
296:Redfir
Lupinus arboreus
54:Eurbeachgras,
69:Pacreedgrass,
80:Sanverbeabur,
112:Blublossom,
142:Coybrush,
191:Salblahuc,
213:Yelbuslup
Lupinus caudatus
313:Waspine
Lupinus chamissonis
80:Sanverbeabur,
150:Dunlupgol
Lupinus concinnus
230:Bluoak
Luzula comosa
62:Kenbluegrass
Luzulaleaf sedge
82:Sedge
Lycium
141:Coapripea,
145:Crebuswhibur,
159:Hopsage,
168:Jostree,
182:Ocotillo
Lycium andersonii
152:Foopalversag
Lycium fremontii
129:Catacacia
Lyon cherry
265:Isloak,
339:Holchesta
Lyonothamnus floribundus
331:Catirosta

MacDonald oak
265:Isloak
Mackenzie willow
176:Moualder,
200:Sitalder,
355:Monwetshrhab
Madrone
148:Deerbrush,
221:Beapine,
225:Bispine,
227:Blaoak,
234:Calbay,
239:Canlivoak,
241:Coalivoak,
246:Doufir,
250:Doufirtan,
264:Intlivoak,
277:Mixoak,
280:Monpine,
285:Orewhioak,
300:Redwood,
302:SanLucfir,

309:Tanoak,
315:Weshemlock,
321:Whifir
Mahala carpet
275:Mixconifer,
289:Ponpine
Maianthemum dilatatum
306:Sitspruce
Malacothamnus
fasciculatum
109:Blasage,
120:Calbuckwheat,
208:Whisage
Malephora crocea
58:Iceplant
Malosma laurina
109:Blasage,
127:CalsagCalbuc,
133:Chablasag,
136:Chahoacea,
137:Chamismanwoo,
72:Mixsage,
184:Pursage,
193:Scalebroom,
202:Sumac,
208:Whisage,
241:Coalivoak
Malus fusca
306:Sitspruce
Man root
300:Redwood
Manna gum
255:Eucalyptus
Many-flowered navarretia
365:Norharverpoo
Many-nerved sedge
82:Sedge,
356:Submeahab
Manzanita
130:Chamise,
132:Chabigman,
135:ChaEasman,
136:Chahoacea,
138:Chawedcea,
146:Cupceafreoak,
161:Intlivoakshr,
163:Intlivoakcan,
194:Scroak,
196:Scroakbirmou,
197:Scroakcha,
198:Scroakchawhi,
257:Foopine
Marah fabaceus
300:Redwood
Marsh arrow-grass
86:Spikerush
McNab cypress
169:Leaoak,
273:McNcypress,
303:Sarcypress
Meadow onion

358:Subwetshrhab
Medusa-head
34:Bluwheatgras,
66:Neeandthr,
367:Norvolashver
Melica californica
42:Caloatgrass,
55:Fooneedlegra,
67:Nodneedlegra,
75:Purneedlegra
Melica imperfecta
75:Purneedlegra
Melica subulata
321:Whifir
Menodora spinescens
108:Blabush
Menyanthes trifoliata
352:Fenhab
Merten rush
84:Shoreedgrass,
85:Shosedge,
177:Mouheabil,
356:Submeahab
Mesquite
144:Crebush
Mexican elderberry
123:Calencelia,
124:Calsagebrush,
141:Coapripea,
171:Mexclderberr,
172:Mixsage,
184:Pursage,
193:Scalebroom,
202:Sumac,
219:Arrwillow,
229:Blawillow,
238:Calwalnut,
287:Pacwillow,
299:Redwillow,
307:Sitwillow
Mexican mosquito fern
63:Mosfern
Milfoils
52:Ditgrass
Milkwort
275:Mixconifer
Mimulus aurantiacus
120:Calbuckwheat,
123:Calencelia,
124:Calsagebrush,
127:CalsagCalbuc,
172:Mixsage,
184:Pursage,
191:Salblahuc,
208:Whisage
Mimulus clevelandii
136:Chahoacea,
137:Chamismanwoo,
74:Mixscroak,
194:Scroak,
196:Scroakbirmou,

197:Scroakcha,
198:Scroakchawhi
Mimulus primuloides
86:Spikerush,
352:Fenhab
Mimulus pygmaeus
361:Norbasflover,
368:Norvolmudver
Mimulus suksdorfii
158:Holodiscus
Miner dogwood
319:Whialder
Mirabilis
123:Calencelia,
141:Coapripea
Mirabilis californica
114:Brittlebush
Mission star
230:Bluoak
Mission-manzanita
134:Chacupcea,
137:Chamismanwoo,
39:Chawhisag
Mock-orange
285:Orewhioak
Modoc County knotweed
368:Norvolmudver
Modoc coffeeberry
268:Jefpinponpin
Mojave yucca
168:Jostree,
175:Mojyucca,
236:Caljuniper
Monardella odoratissima
188:Rotsagebrush
Mono clover
170:Lowsagebrush
Monterey cypress
340:Moncypsta
Monterey pine
225:Bispine,
269:Knopine,
280:Monpine,
337:Gowcypsta
Mosquito fern
53:Duckweed,
63:Mosfern
Mount Dana sedge
82:Sedge,
88:Tufhairgrass
Mountain alder
176:Moualder,
200:Sitalder,
355:Monwetshrhab
Mountain big sagebrush
266:Jefpine,
289:Ponpine
Mountain heather
84:Shoreedgrass,
358:Subwetshrhab
Mountain heathers

177:Mouheabil
Mountain hemlock
254:Engspruce,
258:Foxpine,
271:Lodpine,
278:Mixsubfor,
281:Mouhemlock,
296:Redfir,
308:Subfir,
324:Whipine,
327:Alayelcedsta,
335:KlaMouEnrsta,
341:Pacsilfirsta
Mountain juniper
170:Lowsagebrush,
245:Curmoumah,
283:Moujuniper,
304:Sinpinyon,
316:Wesjuniper
Mountain maple
176:Moualder,
200:Sitalder
Mountain misery
275:Mixconifer
Mountain spikerush
86:Spikerush
Mountain timothy
86:Spikerush
Mountain whitethorn
113:Breoak,
117:Buschinquapi,
148:Deerbrush,
155:Gremanzanita,
160:Hucoak,
178:Mouwhitethor,
190:Sadoak,
205:Tobbrush,
209:Whimanzanita,
266:Jefpine
Mud-midgets
53:Duckweed,
63:Mosfern
Mugwort
219:Arrwillow
Muhlenbergia asperifolia
78:Saltgrass,
86:Spikerush
Muhlenbergia filiformis
42:Caloatgrass
Mule's ears
266:Jefpine,
268:Jefpinponpin
Mulefat
179:Mulefat,
193:Scalebroom,
219:Arrwillow,
229:Blawillow,
237:Calsycamore,
299:Redwillow,
319:Whialder
Muller oak

174:Mixscroak,
185:Redshank,
194:Scroak,
197:Scroakcha,
288:Parpinyon
Musk bush
169:Leaoak
Myosurus minimus
363:Norclaverpoo,
365:Norharverpoo,
369:SanDiemesver,
370:SanJacValver,
371:SanRosPlaver
Myrica californica
142:Coybrush,
213:Yelbuslup,
219:Arrwillow,
226:Blacottonwoo,
262:Hoowillow
Myriophyllum
52:Ditgrass
Najas guadalupensis
52:Ditgrass

Naked cleome
51:Dessanver
Naked-stemmed buck-
wheat
275:Mixconifer
Narrower sedge
62:Kenbluegrass
Narrowleaf cattail
35:Bulrush,
37:Bulcattail,
43:Cattail,
99:Arrweed
Narrowleaf goldenbush
111:BlaCalephnar,
181:Nolina
Narrowleaf sword fern
239:Canlivoak
Narrowleaf willow
99:Arrweed,
119:Buttonbush,
171:Mexelderberr,
179:Mulefat,
180:Narwillow,
226:Blacottonwoo,
256:Fanpalm,
279:Mixwillow,
314:Watbirch
Native dunegrass
54:Eurbeachgras,
64:Natdunegrass
Navarretia fossalis
369:SanDiemesver,
370:SanJacValver,
371:SanRosPlaver
Navarretia intertexta
367:Norvolashver
Navarretia leucocephala

368:Norvolmudver
Navarretia myersii
368:Norvolmudver
Navarretia prostrata
369:SanDiemesver,
371:SanRosPlaver
Navarretia squarrosa
367:Norvolashver
Nebraska sedge
32:Beasedge,
65:Nebsedge,
77:RocMoused,
88:Tufhairgrass
Needle-and-thread
31:Ashryegrass,
34:Bluwheatgras,
44:Cheatgrass,
47:Creryegrass,
60:Indricegrass,
66:Neeandthr,
68:Onesidblu
Needleleaf navarretia
367:Norvolashver
Nelson pepperwort
365:Norharverpoo
Neostapfia colusana
363:Norclaverpoo,
365:Norharverpoo
Netted willow
358:Subwetshrhab
Nevada bulrush
35:Bulrush,
37:Bulcattail,
43:Cattail
Nevada ephedra
199:Shadscale
Nevada pondweed
74:Ponwitsublea
Nevada rush
86:Spikerush
Noble fir
296:Redfir,
327:Alayelcedsta,
335:KlaMouEnrsta,
341:Pacsilfirsta
Nodding needlegrass
50:Desneedlegra,
55:Fooneedlegra,
67:Nodneedlegra,
75:Purneedlegra
Nolina bigelovii
181:Nolina
Nolina parryi
181:Nolina
Northern goldenrod
88:Tufhairgrass
Nude buckwheat
85:Shosedge,
158:Holodiscus,
329:Antdunsta
Nuphar luteum

84:Shoreedgrass,
188:Rotsagebrush
Pentagramma triangularis
230:Bluoak,
239:Canlivoak
Peruvian pepper
202:Sumac
Peucephyllum schottii
115:Briwhibur,
175:Mojyucca
Phacelia
120:Calbuckwheat
Phacelia imbricata
230:Bluoak
Phacelia ramosissima
120:Calbuckwheat
Philadelphus lewisii
285:Orewhioak
Phleum alpinum
86:Spikerush,
349:Alpinehab
Phlox covillei
349:Alpinehab
Phlox pulvinata
349:Alpinehab,
357:Subuplshrhab
Phloxleaf bedstraw
230:Bluoak
Phragmites australis
45:Comreed
Phyllodoce
84:Shoreedgrass,
177:Mouheabil,
358:Subwetshrhab
Phyllodoce breweri
84:Shoreedgrass,
177:Mouheabil
Phyllodoce empetriformis
177:Mouheabil,
281:Mouhemlock
Picea breweriana
113:Breoak,
254:Engspruce,
296:Redfir,
308:Subfir,
321:Whifir,
335:KlaMouEnrsta,
327:Alayelcedsta,
341:Pacsilfirsta
Picea engelmannii
254:Engspruce,
308:Subfir,
335:KlaMouEnrsta
Picea sitchensis
69:Pacreedgrass,
70:Pamgrass,
191:Salblahuc,
221:Beapine,
261:Grafir,
292:PorOrfced,
295:Redalder,

306:Sitspruce,
315:Weshemlock
Pickeringia montana
194:Scroak
Pickleweeds
71:Pickleweed
Pincushion navarretia
368:Norvolmudver
Pinemat manzanita
113:Breoak,
117:Buschinquapi,
155:Gremanzanita,
160:Hucoak,
246:Doufir,
266:Jefpine,
296:Redfir,
313:Waspine
Pines
100:Bigsagebrush,
107:Bitterbrush,
183:Parrabbitbru,
189:Rubrabbitbru
Pinguicula macroceras
49:Darlingtonia
Pink sand-verbena
64:Natdunegrass,
80:Sanverbeabur
Pinus
100:Bigsagebrush,
107:Bitterbrush,
183:Parrabbitbru,
189:Rubrabbitbru
Pinus albicaulis
188:Rotsagebrush,
245:Curmoumah,
258:Foxpine,
270:Limpine,
271:Lodpine,
278:Mixsubfor,
281:Mouhemlock,
324:Whipine,
335:KlaMouEnrsta
Pinus attenuata
161:Intlivoakshr,
167:Ionmanzanita,
169:Leaoak,
197:Scroakcha,
223:BigDoufircan,
227:Blaoak,
266:Jefpine,
269:Knopine,
273:McNcypress,
280:Monpine,
303:Sarcypress,
317:Weswhipin,
330:Bakcypsta,
335:KlaMouEnrsta,
344:Sancrucypsta
Pinus balfouriana
188:Rotsagebrush,
245:Curmoumah,

258:Foxpine,
266:Jefpine,
270:Limpine,
271:Lodpine,
278:Mixsubfor,
317:Weswhipin,
324:Whipine,
335:KlaMouEnrsta
Pinus contorta
188:Rotsagebrush,
221:Beapine,
225:Bispine,
226:Blacottonwoo,
245:Curmoumah,
254:Engspruce,
258:Foxpine,
260:Giasequoia,
263:Inccedar,
266:Jefpine,
269:Knopine,
270:Limpine,
271:Lodpine,
278:Mixsubfor,
281:Mouhemlock,
283:Moujuniper,
289:Ponpine,
294:Pygcypress,
296:Redfir,
308:Subfir,
313:Waspine,
317:Weswhipin,
321:Whifir,
324:Whipine,
327:Alayelcedsta,
335:KlaMouEnrsta,
341:Pacsilfirsta
Pinus coulteri
151:Easmanzanita,
161:Intlivoakshr,
164:Intlivoakcha,
222:BigDoufir,
223:BigDoufircan,
239:Canlivoak,
243:Coupine,
244:Coupincanliv,
257:Foopine,
269:Knopine,
289:Ponpine,
302:SanLucfir,
309:Tanoak,
333:Cuycypsta,
343:SanBenMousta
Pinus edulis
347:Twopinsta
Pinus flexilis
233:Bripine,
258:Foxpine,
270:Limpine,
271:Lodpine
Pinus jeffreyi
100:Bigsagebrush,

107:Bitterbrush,
110:Blasagebrush,
226:Blacottonwoo,
227:Blaoak,
245:Curmoumah,
246:Doufir,
249:Doufirponpin,
260:Giasequoia,
266:Jefpine,
268:Jefpinponpin,
270:Limpine,
275:Mixconifer,
283:Moujuniper,
285:Orewhioak,
288:Parpinyon,
289:Ponpine,
292:PorOrfced,
296:Redfir,
304:Sinpinyon,
313:Waspine,
316:Wesjuniper,
317:Weswhipin,
321:Whifir,
330:Bakcypsta,
335:KlaMouEnrsta,
343:SanBenMousta
Pinus lambertiana
222:BigDoufir,
239:Canlivoak,
249:Doufirponpin,
250:Doufirtan,
254:Engspruce,
260:Giasequoia,
263:Inccedar,
275:Mixconifer,
292:PorOrfced,
296:Redfir,
302:SanLucfir,
309:Tanoak,
330:Bakcypsta,
335:KlaMouEnrsta
Pinus longaeva
233:Bripine,
270:Limpine
Pinus monophylla
100:Bigsagebrush,
107:Bitterbrush,
108:Blabush,
110:Blasagebrush,
146:Cupceafreoak,
168:Jostree,
222:BigDoufir,
223:BigDoufircan,
224:Birmoumah,
236:Caljuniper,
245:Curmoumah,
283:Moujuniper,
288:Parpinyon,
304:Sinpinyon,
305:SinpinUtajun,
311:Utajuniper,

342:Piucypsta
Pinus monticola
  254:Engspruce,
  258:Foxpine,
  269:Knopine,
  271:Lodpine,
  278:Mixsubfor,
  281:Mouhemlock,
  292:PorOrfced,
  308:Subfir,
  313:Waspine,
  317:Weswhipin,
  324:Whipine,
  327:Alayelcedsta,
  335:KlaMouEnrsta,
  341:Pacsilfirsta
Pinus muricata
  221:Beapine,
  225:Bispine,
  261:Grafir,
  280:Monpine,
  294:Pygcypress,
  337:Gowcypsta
Pinus ponderosa
  222:BigDoufir,
  227:Blaoak,
  239:Canlivoak,
  243:Coupine,
  244:Coupincanliv,
  246:Doufir,
  249:Doufirponpin,
  250:Doufirtan,
  254:Engspruce,
  260:Giasequoia,
  263:Inccedar,
  268:Jefpinponpin,
  275:Mixconifer,
  277:Mixoak,
  280:Monpine,
  285:Orewhioak,
  292:PorOrfced,
  302:SanLucfir,
  304:Sinpinyon,
  313:Waspine,
  316:Wesjuniper,
  330:Bakcypsta,
  335:KlaMouEnrsta,
  344:Sancrucypsta
Pinus quadrifolia
  236:Caljuniper,
  288:Parpinyon
Pinus radiata
  225:Bispine,
  269:Knopine,
  280:Monpine,
  337:Gowcypsta
Pinus sabiniana
  161:Intlivoakshr,
  167:Ionmanzanita,
  169:Leaoak,
  194:Scroak,

223:BigDoufircan,
224:Birmoumah,
230:Bluoak,
235:Calbuckeye,
243:Coupine,
244:Coupincanliv,
257:Foopine,
264:Intlivoak,
269:Knopine,
273:McNcypress,
277:Mixoak,
303:Sarcypress,
342:Piucypsta,
343:SanBenMousta
Pinus torreyana
  346:Torpinsta
Pinus washoensis
  100:Bigsagebrush,
  107:Bitterbrush,
  313:Waspine,
  316:Wesjuniper
Piute cypress
  342:Piucypsta
Plagiobothrys hystriculus
  363:Norclaverpoo
Plagiobothrys strictus
  367:Norvolashver
Plantago erecta
  230:Bluoak
Platanus racemosa
  193:Scalebroom,
  219:Arrwillow,
  229:Blawillow,
  237:Calsycamore,
  256:Fanpalm,
  279:Mixwillow,
  287:Pacwillow,
  299:Redwillow,
  307:Sitwillow,
  312:Valoak,
  319:Whialder
Pleuraphis jamesii
  33:Biggalleta
Pleuraphis rigida
  33:Biggalleta,
  51:Dessanver,
  144:Crebush,
  175:Mojyucca,
  204:Tedbeacho
Pluchea sericea
  99:Arrweed
Poa cusickii
  356:Submeahab
Poa douglasii
  58:Iceplant,
  64:Natdunegrass,
  80:Sanverbeabur
Poa pratensis
  61:Intpergra,
  62:Kenbluegrass,
  353:Monmeahab

Poa secunda
  30:Alksacaton,
  31:Ashryegrass,
  34:Bluwheatgrass,
  42:Caloatgrass,
  47:Creryegrass,
  50:Desneedlegra,
  55:Fooneedlegra,
  60:Indricegrass,
  61:Intpergra,
  62:Kenbluegrass,
  66:Neeandthr,
  67:Nodneedlegra,
  68:Onesidblu,
  75:Purneedlegra,
  78:Saltgrass,
  100:Bigsagebrush,
  249:Doufirponpin,
  266:Jefpine,
  268:Jefpinponpin,
  353:Monmeahab
Poa wheeleri
  324:Whipine
Podistera
  357:Subuplshrhab
Podistera nevadensis
  357:Subuplshrhab
Pogogyne abramsii
  369:SanDiemesver
Pogogyne douglasii
  361:Norbasflover
Pogogyne floribunda
  368:Norvolmudver
Pogogyne nudiuscula
  369:SanDiemesver
Pointleaf manzanita
  187:Redshacha,
  343:SanBenMousta,
  347:Twopinsta
Poison larkspur
  285:Orewhioak
Poison-oak
  112:Blublossom,
  124:Calsagebrush,
  128:Canlivoakshr,
  130:Chamise,
  132:Chabigman,
  142:Coybrush,
  156:Haiceanothus,
  165:Intlivoakscr,
  171:Mexelderberr,
  191:Salblahuc,
  193:Scalebroom,
  202:Sumac,
  227:Blaoak,
  241:Coalivoak,
  246:Doufir,
  250:Doufirtan,
  264:Intlivoak,
  277:Mixoak,
  285:Orewhioak,

309:Tanoak,
312:Valoak
Polygala cornuta
  275:Mixconifer
Polygonum davisiae
  356:Submeahab
Polygonum minimum
  349:Alpinehab
Polygonum polygaloides
  368:Norvolmudver
Polypodium californicum
  300:Redwood,
  319:Whialder
Polystichum imbricans
  239:Canlivoak
Polystichum munitum
  142:Coybrush,
  250:Doufirtan,
  275:Mixconifer,
  300:Redwood,
  302:SanLucfir,
  306:Sitspruce
Ponderosa pine
  222:BigDoufir,
  227:Blaoak,
  239:Canlivoak,
  243:Coupine,
  244:Coupincanliv,
  246:Doufir,
  249:Doufirponpin,
  250:Doufirtan,
  254:Engspruce,
  260:Giasequoia,
  263:Inccedar,
  268:Jefpinponpin,
  275:Mixconifer,
  277:Mixoak,
  280:Monpine,
  285:Orewhioak,
  292:PorOrfced,
  302:SanLucfir,
  304:Sinpinyon,
  313:Waspine,
  316:Wesjuniper,
  330:Bakcypsta,
  335:KlaMouEnrsta,
  344:Sancrucypsta
Pondweeds
  39:Burreed,
  90:Yelponlil
Populus balsamifera
  180:Narwillow,
  192:Sanwillow,
  219:Arrwillow,
  226:Blacottonwoo,
  229:Blawillow,
  262:Hoowillow,
  279:Mixwillow,
  287:Pacwillow,
  295:Redalder,
  307:Sitwillow,

314:Watbirch
Populus fremontii
171:Mexelderberr,
180:Narwillow,
192:Sanwillow,
193:Scalebroom,
219:Arrwillow,
226:Blacottonwoo,
229:Blawillow,
237:Calsycamore,
256:Fanpalm,
262:Hoowillow,
279:Mixwillow,
287:Pacwillow,
299:Redwillow,
307:Sitwillow,
314:Watbirch
Populus tremuloides
220:Aspen,
226:Blacottonwoo
Port Orford-cedar
250:Doufirtan,
292:PorOrfced,
296:Redfir,
315:Weshemlock,
319:Whialder,
327:Alayelcedsta,
335:KlaMouEnrsta
Potamogeton
39:Burreed,
90:Yelponlil
Potamogeton alpinus
73:Ponwitflolea
Potamogeton amplifolius
73:Ponwitflolea
Potamogeton crispus
74:Ponwitsublea
Potamogeton diversifolius
73:Ponwitflolea
Potamogeton epihydrus
73:Ponwitflolea
Potamogeton filiformis
74:Ponwitsublea
Potamogeton foliosus
52:Ditgrass,
74:Ponwitsublea
Potamogeton gramineus
73:Ponwitflolea
Potamogeton illinoensis
73:Ponwitflolea
Potamogeton latifolius
74:Ponwitsublea
Potamogeton natans
73:Ponwitflolea
Potamogeton nodosus
73:Ponwitflolea
Potamogeton pectinatus
74:Ponwitsublea
Potamogeton praelongus
74:Ponwitsublea
Potamogeton pusillus

74:Ponwitsublea
Potamogeton richardsonii
74:Ponwitsublea
Potamogeton robbinsii
74:Ponwitsublea
Potamogeton zosteriformis
74:Ponwitsublea
Potentilla
84:Shoreedgrass,
85:Shosedge
Potentilla breweri
177:Mouheabil
Potentilla drummondii
177:Mouheabil
Potentilla fruticosa
357:Subuplshrhab
Potentilla gracilis
62:Kenbluegrass
Prickly-pear
114:Brittlebush,
115:Briwhibur,
145:Crebuswhibur,
172:Mixsage,
175:Mojyucca,
193:Scalebroom
Primrose monkeyflower
86:Spikerush,
352:Fenhab
Primula suffrutescens
177:Mouheabil
Prince's-pine
250:Doufirtan,
296:Redfir
Profuse-flowered
pogogyne
368:Norvolmudver
Prosopis
144:Crebush,
232:Blupalveriro
Prosopis glandulosa
98:Allscale,
153:Fousaltbush,
232:Blupalveriro,
256:Fanpalm,
274:Mesquite
Prosopis pubescens
274:Mesquite
Prunus andersonii
107:Bitterbrush
Prunus emarginata
117:Buschinquapi,
148:Deerbrush,
155:Gremanzanita,
158:Holodiscus,
160:Hucoak,
178:Mouwhitethor,
190:Sadoak,
205:Tobbrush,
275:Mixconifer,
321:Whifir
Prunus fasciculatum

146:Cupceafreoak
Prunus fremontii
146:Cupccafreoak
Prunus ilicifolia
106:BirmoumahCal,
161:Intlivoakshr,
165:Intlivoakscr,
202:Sumac,
235:Calbuckeye,
265:Isloak,
339:Holchesta
Prunus virginiana
245:Curmoumah,
283:Moujuniper,
289:Ponpine
Pseudoroegneria spicata
289:Ponpine
Pseudostellaria jamesiana
313:Waspine
Pseudotsuga macrocarpa
222:BigDoufir,
223:BigDoufircan,
239:Canlivoak,
243:Coupine,
244:Coupincanliv,
275:Mixconifer,
309:Tanoak
Pseudotsuga menziesii
42:Caloatgrass,
70:Pamgrass,
148:Deerbrush,
221:Beapine,
225:Bispine,
227:Blaoak,
239:Canlivoak,
246:Doufir,
249:Doufirponpin,
250:Doufirtan,
254:Engspruce,
260:Giasequoia,
261:Grafir,
263:Inccedar,
266:Jefpine,
275:Mixconifer,
277:Mixoak,
280:Monpine,
285:Orewhioak,
289:Ponpine,
292:PorOrfced,
295:Redalder,
300:Redwood,
309:Tanoak,
315:Weshemlock,
317:Weswhipin,
319:Whialder,
321:Whifir,
327:Alayelcedsta,
330:Bakcypsta,
335:KlaMouEnrsta
Psilocarphus brevissimus
368:Norvolmudver,

369:SanDiemesver,
370:SanJacValver
Psorothamnus schottii
144:Crebush,
145:Crebuswhibur
Psorothamnus spinosa
129:Catacacia,
182:Ocotillo,
232:Blupalveriro
Pteridium aquilinum
42:Caloatgrass,
61:Intpergra,
62:Kenbluegrass,
69:Pacreedgrass,
227:Blaoak,
241:Coalivoak,
250:Doufirtan,
296:Redfir,
300:Redwood,
321:Whifir,
353:Monmeahab
Pull-up muhly
42:Caloatgrass
Purple needlegrass
42:Caloatgrass,
55:Fooneedlegra,
67:Nodneedlegra,
75:Purneedlegra,
230:Bluoak
Purple reedgrass
357:Subuplshrhab
Purple sage
123:Calencelia,
124:Calsagebrush,
130:Chamise,
133:Chablasag,
139:Chawhisag,
184:Pursage,
202:Sumac
Purple sanicle
75:Purneedlegra
Purshia mexicana
146:Cupceafreoak,
236:Caljuniper,
305:SinpinUtajun
Purshia tridentata
100:Bigsagebrush,
107:Bitterbrush,
170:Lowsagebrush,
183:Parrabbitbru,
189:Rubrabbitbru,
245:Curmoumah,
266:Jefpine,
268:Jefpinponpin,
283:Moujuniper,
289:Ponpine,
304:Sinpinyon,
305:SinpinUtajun,
311:Utajuniper,
316:Wesjuniper
Pussypaws

151:Easmanzanita,
161:Intlivoakshr,
163:Intlivoakcan,
164:Intlivoakcha,
165:Intlivoakscr,
167:Ionmanzanita,
174:Mixscroak,
194:Scroak,
197:Scroakcha,
198:Scroakchawhi,
230:Bluoak,
234:Calbay,
235:Calbuckeye,
243:Coupine,
257:Foopine,
264:Intlivoak,
266:Jefpine,
269:Knopine,
273:McNcypress,
277:Mixoak,
303:Sarcypress
Quillworts
76:Quillwort,
369:SanDiemesver,
371:SanRosPlaver

Rabbitbrush
168:Jostree,
305:SinpinUtajun,
311:Utajuniper,
316:Wesjuniper
Raiche manzanita
169:Leaoak
Ranunculus
369:SanDiemesver
Rattail fescue
40:Calanngra
Rattlesnake-plantain
275:Mixconifer
Rayless arnica
225:Bispine
Rayless layia
343:SanBenMousta
Red Bluff dwarf rush
368:Norvolmudver
Red alder
69:Pacreedgrass,
180:Narwillow,
191:Salblahuc,
192:Sanwillow,
213:Yelbuslup,
226:Blacottonwoo,
261:Grafir,
262:Hoowillow,
279:Mixwillow,
287:Pacwillow,
295:Redalder,
306:Sitspruce,
307:Sitwillow
Red elderberry
357:Subuplshrhab

Red fescue
42:Caloatgrass,
61:Intpergra,
62:Kenbluegrass,
69:Pacreedgrass,
88:Tufhairgrass
Red fir
220:Aspen,
260:Giasequoia,
271:Lodpine,
278:Mixsubfor,
281:Mouhemlock,
296:Redfir,
313:Waspine,
317:Weswhipin,
324:Whipine,
330:Bakcypsta,
335:KlaMouEnrsta
Red flowering current
191:Salblahuc
Red gum
255:Eucalyptus
Red huckleberry
250:Doufirtan
Red iron bark
255:Eucalyptus
Red osier
119:Buttonbush,
176:Moualder,
192:Sanwillow,
200:Sitalder,
219:Arrwillow,
287:Pacwillow,
307:Sitwillow,
319:Whialder,
355:Monwetshrhab
Red shank
106:BirmoumahCal,
130:Chamise,
134:Chacupcea,
185:Redshank,
186:Redshabir,
187:Redshacha,
194:Scroak,
196:Scroakbirmou,
288:Parpinyon
Red three-awn
47:Creryegrass,
68:Onesidblu
Red willow
180:Narwillow,
192:Sanwillow,
229:Blawillow,
237:Calsycamore,
262:Hoowillow,
279:Mixwillow,
287:Pacwillow,
299:Redwillow,
307:Sitwillow,
314:Watbirch
Redwood

42:Caloatgrass,
225:Bispine,
234:Calbay,
261:Grafir,
280:Monpine,
294:Pygcypress,
295:Redalder,
300:Redwood,
306:Sitspruce,
315:Weshemlock
Redwood oxalis
300:Redwood
Rhamnus
257:Foopine
Rhamnus californica
142:Coybrush,
161:Intlivoakshr,
163:Intlivoakcan,
164:Intlivoakcha,
165:Intlivoakscr,
194:Scroak,
211:Woomanzanita,
41:Coalivoak,
277:Mixoak,
309:Tanoak,
312:Valoak
Rhamnus ilicifolia
102:Bigmanzanita,
103:Bigceanothus,
104:Bigceabirmou,
105:Bigceaholred,
106:BirmoumahCal,
137:Chamismanwoo,
39:Chawhisag,
157:Hoaceanothus,
161:Intlivoakshr,
163:Intlivoakcan,
165:Intlivoakscr,
174:Mixscroak,
185:Redshank,
187:Redshacha,
194:Scroak,
196:Scroakbirmou,
197:Scroakcha,
198:Scroakchawhi,
206:Wedceanothus,
224:Birmoumah
Rhamnus purshiana
306:Sitspruce
Rhamnus rubra
268:Jefpinponpin
Rhamnus tomentella
167:Ionmanzanita
Rhododendron
250:Doufirtan,
292:PorOrfced,
296:Redfir
*Rhododendron
macrophyllum*
250:Doufirtan,
292:PorOrfced,

296:Redfir
Rhododendron occidentale
49:Darlingtonia,
292:PorOrfced,
306:Sitspruce
Rhus integrifolia
120:Calbuckwheat,
123:Calencelia,
124:Calsagebrush,
126:Calsagblasag,
141:Coapripea,
150:Dunlupgol,
172:Mixsage,
184:Pursage,
193:Scalebroom,
202:Sumac
Rhus ovata
126:Calsagblasag,
127:CalsagCalbuc,
134:Chacupcea,
136:Chahoacea,
137:Chamismanwoo,
39:Chawhisag,
151:Easmanzanita,
156:Haiceanothus
Rhus trilobata
129:Catacacia,
140:Chawhitethor,
164:Intlivoakcha,
193:Scalebroom,
266:Jefpine
Ribbed sedge
353:Monmeahab
Ribes californica
230:Bluoak
Ribes cereum
357:Subuplshrhab
Ribes roezlii
285:Orewhioak
Ribes sanguineum
191:Salblahuc
Ribes velutinum
100:Bigsagebrush
Richardson pondweed
74:Ponwitsublea
Rigiopappus
230:Bluoak
Rigiopappus leptocladus
230:Bluoak
Ripgut
40:Calanngra,
44:Cheatgrass,
67:Nodneedlegra,
75:Purneedlegra,
213:Yelbuslup
River bulrush
35:Bulrush,
37:Bulcattail,
43:Cattail
Robbin pondweed
74:Ponwitsublea

Rock-spiraea
158:Holodiscus,
357:Subuplshrhab
Rocky Mountain sedge
32:Beasedge,
77:RocMoused
Ross sedge
271:Lodpine
Rosy everlasting
275:Mixconifer
Rothrock sagebrush
188:Rotsagebrush,
357:Subuplshrhab
Rough bentgrass
353:Monmeahab
Round woollyheads
368:Norvolmudver,
370:SanJacValver
Round-fruited sedge
300:Redwood
Rubber rabbitbrush
30:Alksacaton,
100:Bigsagebrush,
154:Greasewood,
158:Holodiscus,
170:Lowsagebrush
Rubus
239:Canlivoak,
241:Coalivoak
Rubus discolor
319:Whialder
Rubus spectabilis
306:Sitspruce
Rubus ursinus
69:Pacreedgrass,
142:Coybrush,
191:Salblahuc,
262:Hoowillow
Rudbeckia californica
49:Darlingtonia
Rumex hymenosepalus
51:Dessanver
Ruppia
52:Ditgrass,
71:Pickleweed
Rushes
32:Beasedge,
65:Nebsedge,
369:SanDiemesver
Rusty molly
118:Busseepweed,
154:Greasewood,
166:Iodbush
Ryegrass
40:Calanngra
Rynchosposa
352:Fenhab
Rynchosposa moss
352:Fenhab

Sacramento Orcutt grass

368:Norvolmudver
Sadler oak
158:Holodiscus,
160:Hucoak,
190:Sadoak,
205:Tobbrush,
281:Mouhemlock,
292:PorOrfced
296:Redfir
Sagittaria sanfordii
368:Norvolmudver
Saguaro
152:Foopalversag
Salal
69:Pacreedgrass,
112:Blublossom,
142:Coybrush,
191:Salblahuc,
250:Doufirtan,
292:PorOrfced,
295:Redalder,
300:Redwood
Salazaria mexicana
108:Blabush
Salicornia
71:Pickleweed
Salicornia subterminalis
118:Busseepweed
Salicornia virginica
78:Saltgrass
Salix
180:Narwillow,
191:Salblahuc,
192:Sanwillow,
203:Tamarisk,
219:Arrwillow,
229:Blawillow,
262:Hoowillow,
287:Pacwillow,
299:Redwillow,
307:Sitwillow,
358:Subwetshrhab
Salix arctica
349:Alpinehab,
358:Subwetshrhab
Salix boothii
358:Subwetshrhab
Salix brachycarpa
358:Subwetshrhab
Salix eastwoodiae
358:Subwetshrhab
Salix exigua
99:Arrweed,
119:Buttonbush,
171:Mexelderberr,
179:Mulefat,
180:Narwillow,
226:Blacottonwoo,
256:Fanpalm,
279:Mixwillow,
314:Watbirch

Salix gooddingii
180:Narwillow,
192:Sanwillow,
229:Blawillow,
232:Blupalveriro,
237:Calsycamore,
256:Fanpalm,
262:Hoowillow,
279:Mixwillow,
287:Pacwillow,
307:Sitwillow
Salix hookeriana
69:Pacreedgrass,
80:Sanverbeabur,
180:Narwillow,
192:Sanwillow,
226:Blacottonwoo,
229:Blawillow,
262:Hoowillow,
279:Mixwillow,
287:Pacwillow,
295:Redalder,
307:Sitwillow
Salix jepsonii
176:Moualder,
200:Sitalder,
355:Monwetshrhab,
358:Subwetshrhab
Salix laevigata
180:Narwillow,
192:Sanwillow,
229:Blawillow,
237:Calsycamore,
262:Hoowillow,
279:Mixwillow,
287:Pacwillow,
299:Redwillow,
307:Sitwillow,
314:Watbirch
Salix lasiolepis
179:Mulefat,
219:Arrwillow,
229:Blawillow,
237:Calsycamore,
256:Fanpalm,
279:Mixwillow,
295:Redalder,
314:Watbirch,
355:Monwetshrhab
Salix lemmonii
176:Moualder,
200:Sitalder,
355:Monwetshrhab,
358:Subwetshrhab
Salix ligulifolia
355:Monwetshrhab
Salix lucida
119:Buttonbush,
180:Narwillow,
192:Sanwillow,
229:Blawillow,

262:Hoowillow,
279:Mixwillow,
287:Pacwillow,
307:Sitwillow,
314:Watbirch,
355:Monwetshrhab
Salix lutea
226:Blacottonwoo,
237:Calsycamore,
358:Subwetshrhab
Salix melanopsis
355:Monwetshrhab
Salix nivalis
358:Subwetshrhab
Salix orestera
355:Monwetshrhab
Salix planifolia
358:Subwetshrhab
Salix prolixa
176:Moualder,
200:Sitalder,
355:Monwetshrhab
Salix scouleriana
176:Moualder,
200:Sitalder,
226:Blacottonwoo,
355:Monwetshrhab
Salix sessilifolia
192:Sanwillow
Salix sitchensis
180:Narwillow,
192:Sanwillow,
229:Blawillow,
262:Hoowillow,
279:Mixwillow,
287:Pacwillow,
307:Sitwillow
Salmonberry
306:Sitspruce
Salt-meadow cordgrass
46:Cordgrass
Salt-water cordgrass
46:Cordgrass
Saltbush
98:Allscale,
203:Tamarisk,
207:Whibursage,
274:Mesquite
Saltgrass
30:Alksacaton,
31:Ashryegrass,
35:Bulrush,
37:Bulcattail,
43:Cattail,
46:Cordgrass,
47:Creryegrass,
78:Saltgrass,
166:Iodbush,
201:Spinescale
Saltmarsh bulrush
35:Bulrush,

349:Alpinehab
Senecio scorzonella
  88:Tufhairgrass,
  349:Alpinehab,
  358:Subwetshrhab
Senecio triangularis
  82:Sedge,
  349:Alpinehab,
  358:Subwetshrhab
Sequoia sempervirens
  42:Caloatgrass,
  70:Pamgrass,
  225:Bispine,
  234:Calbay,
  261:Grafir,
  280:Monpine,
  294:Pygcypress,
  295:Redalder,
  300:Redwood,
  306:Sitspruce,
  315:Weshemlock
Sequoiadendron giganteum
  260:Giasequoia
Service berry
  117:Buschinquapi,
  155:Gremanzanita,
  160:Hucoak,
  209:Whimanzanita,
  245:Curmoumah,
  275:Mixconifer
Shadscale
  108:Blabush,
  110:Blasagebrush,
  154:Greasewood,
  159:Hopsage,
  166:Iodbush,
  173:Mixsaltbush,
  199:Shadscale,
  201:Spinescale,
  210:Winfat
Shasta fir
  254:Engspruce,
  258:Foxpine,
  271:Lodpine,
  278:Mixsubfor,
  281:Mouhemlock,
  292:PorOrfced,
  296:Redfir,
  308:Subfir,
  317:Weswhipin,
  324:Whipine,
  335:KlaMouEnrsta,
  341:Pacsilfirsta
Shinning pondweed
  73:Ponwitflolea
Shinning willow
  314:Watbirch,
  355:Monwetshrhab
Shorthair reedgrass
  85:Shosedge,
  358:Subwetshrhab

Shorthair sedge
  84:Shoreedgrass,
  85:Shosedge,
  177:Mouheabil,
  353:Monmeahab
Showy sedge
  358:Subwetshrhab
Shrub cinquefoil
  357:Subuplshrhab
Sibbaldia
  177:Mouheabil
Sibbaldia procumbens
  177:Mouheabil
Sierra primrose
  177:Mouheabil
Sierra ricegrass
  84:Shoreedgrass,
  85:Shosedge
Sierra willow
  358:Subwetshrhab
Silene lemmonii
  239:Canlivoak
Silktassels
  169:Leaoak
Silver cinquefoil
  62:Kenbluegrass
Silver dollar gum
  255:Eucalyptus
Silver lupine
  296:Redfir
Silverleaf gum
  255:Eucalyptus
Simmondsia chinensis
  115:Briwhibur,
  129:Catacacia,
  175:Mojyucca,
  181:Nolina
Singleleaf pinyon
  100:Bigsagebrush,
  107:Bitterbrush,
  108:Blabush,
  110:Blasagebrush,
  146:Cupceafreoak,
  222:BigDoufir,
  223:BigDoufircan,
  224:Birmoumah,
  236:Caljuniper,
  245:Curmoumah,
  283:Moujuniper,
  288:Parpinyon,
  304:Sinpinyon,
  342:Piucypsta
Sitka alder
  200:Sitalder,
  355:Monwetshrhab
Sitka spruce
  69:Pacreedgrass,
  70:Pamgrass,
  191:Salblahuc,
  221:Beapine,
  261:Grafir,

  292:PorOrfced,
  295:Redalder,
  306:Sitspruce,
  315:Weshemlock
Sitka willow
  180:Narwillow,
  192:Sanwillow,
  229:Blawillow,
  262:Hoowillow,
  279:Mixwillow,
  287:Pacwillow,
  307:Sitwillow
Skunkweed
  367:Norvolashver
Slender Orcutt grass
  367:Norvolashver,
  368:Norvolmudver
Slender arrow-grass
  78:Saltgrass
Slender buckwheat
  111:BlaCalephnar
Slender oats
  67:Nodneedlegra,
  75:Purneedlegra
Slender penstemon
  296:Redfir,
  324:Whipine
Slender wildoats
  237:Calsycamore
Slender-beaked sedge
  35:Bulrush,
  37:Bulcattail,
  43:Cattail
Slenderleaf iceplant
  58:Iceplant
Slenderleaf pondweed
  74:Ponwitsublea
Small bur-reed
  39:Burreed
Small haplopapappus
  188:Rotsagebrush
Small pondweed
  74:Ponwitsublea
Small-fruited bulrush
  353:Monmeahab
Small-fruited willow
  358:Subwetshrhab
Smoke tree
  129:Catacacia,
  182:Ocotillo,
  232:Blupalveriro
Smooth-beaked sedge
  356:Submeahab
Soft chess
  40:Calanngra,
  55:Fooneedlegra,
  61:Intpergra,
  67:Nodneedlegra,
  75:Purneedlegra,
  237:Calsycamore
Solidago multiradiata

  88:Tufhairgrass
Southern cattail
  35:Bulrush,
  37:Bulcattail,
  43:Cattail
Spanish bayonet
  168:Jostree
Spanish broom
  116:Broom
Sparganium
  39:Burreed,
  90:Yelponlil
Sparganium angustifolium
  39:Burreed
Sparganium emersum
  39:Burreed
Sparganium natans
  39:Burreed
Spartina
  46:Cordgrass,
  71:Pickleweed
Spartina alterniflora
  46:Cordgrass
Spartina densiflora
  46:Cordgrass,
  71:Pickleweed
Spartina foliosa
  46:Cordgrass
Spartina gracilis
  46:Cordgrass,
  78:Saltgrass
Spartina patens
  46:Cordgrass
Spartium junceum
  116:Broom
Sphagnum
  352:Fenhab
Spike trisetum
  84:Shoreedgrass,
  85:Shosedge
Spikenard
  292:PorOrfced,
  319:Whialder
Spikerushes
  47:Creryegrass,
  65:Nebsedge,
  353:Monmeahab,
  356:Submeahab,
  370:SanJacValver,
  371:SanRosPlaver
Spinescale
  154:Greasewood,
  159:Hopsage,
  199:Shadscale,
  201:Spinescale
Spiraea densiflora
  358:Subwetshrhab
Spiraea douglasii
  352:Fenhab
Spirodela
  53:Duckweed,

191:Salblahuc,
192:Sanwillow,
219:Arrwillow,
229:Blawillow,
262:Hoowillow,
287:Pacwillow,
299:Redwillow,
307:Sitwillow
Winter fat
108:Blabush,
110:Blasagebrush,
159:Hopsage,
199:Shadscale,
210:Winfat
Wishbone bush
114:Brittlebush,
123:Calencelia,
141:Coapripea
Wolffia
53:Duckweed,
63:Mosfern
Wolffia borealis
63:Mosfern
Wolffiella
53:Duckweed,
63:Mosfern
Woodwardia fimbriata
295:Redalder,
300:Redwood
Woody sandwort
324:Whipine
Woolly mountain-parsley
356:Submeahab
Woollyleaf ceanothus
130:Chamise,
132:Chabigman,
134:Chacupcea,
135:ChaEasman,
136:Chahoacea,
137:Chamismanwo,
167:Ionmanzanita,
174:Mixscroak
Woollyleaf manzanita
206:Wcdccanothus,
344:Sancrucypsta
Wright trichocoronis
371:SanRosPlaver
Wyethia mollis
266:Jefpine,
268:Jefpinponpin

Xerophyllum tenax
225:Bispine,
246:Doufir,
250:Doufirtan,
266:Jefpine,
317:Weswhipin,
321:Whifir
Xylococcus bicolor
130:Chamise,
132:Chabigman,

134:Chacupcea,
135:ChaEasman,
136:Chahoacea,
137:Chamismanwoo,
39:Chawhisag,
197:Scroakcha

Yarrow
57:Grefescue,
84:Shoreedgrass,
85:Shosedge,
88:Tufhairgrass,
356:Submeahab
Yellow bush lupine
54:Eurbeachgras,
69:Pacreedgrass,
80:Sanverbeabur,
112:Blublossom,
142:Coybrush,
213:Yelbuslup
Yellow bush penstemon
202:Sumac
Yellow mock aster
111:BlaCalephnar
Yellow pine
268:Jefpinponpin
Yellow pond-lily
39:Burrced,
90:Yelponlil
Yellow rabbitbrush
100:Bigsagebrush,
107:Bitterbrush,
110:Blasagebrush,
153:Fousaltbush,
154:Greasewood,
170:Lowsagebrush,
189:Rubrabbitbru,
199:Shadscale,
210:Winfat
Yellow sand-verbena
54:Eurbeachgras,
58:Iceplant,
64:Natdunegrass,
80:Sanverbeabur,
142:Coybrush,
213:Yelbuslup
Yellow willow
226:Blacottonwoo,
237:Calsycamore,
358:Subwetshrhab
Yerba mansa
35:Bulrush,
37:Bulcattail,
43:Cattail,
78:Saltgrass
Yerba santa
139:Chawhisag,
140:Chawhitethor,
164:Intlivoakcha,
167:Ionmanzanita,
169:Leaoak,

264:Intlivoak
Yucca baccata
168:Jostree
Yucca brevifolia
32:Beasedge,
33:Biggalleta,
100:Bigsagebrush,
107:Bitterbrush,
108:Blabush,
144:Crebush,
145:Crebuswhibur,
159:Hopsage,
168:Jostree,
236:Caljuniper
Yucca schidigera
168:Jostree,
175:Mojyucca,
236:Caljuniper
Yucca whipplei
102:Bigmanzanita,
106:BirmoumahCal,
109:Blasage,
122:Calbucwhisag,
123:Calencelia,
124:Calsagebrush,
126:Calsagblasag,
127:CalsagCalbuc,
130:Chamise,
132:Chabigman,
133:Chablasag,
134:Chacupcea,
135:ChaEasman,
136:Chahoacea,
137:Chamismanwo
138:Chawedcea,
139:Chawhisag,
140:Chawhitcthor,
146:Cupceafreoak,
151:Easmanzanita,
164:Intlivoakcha,
186:Redshabir,
187:Redshacha,
193:Scalebroom,
208:Whisage,
236:Caljuniper

Zannichellia palustris
52:Ditgrass